Transport Phenomena in Dispersed Media

Transport Phenomena in Dispersed Media

G. I. Kelbaliyev
D. B. Tagiyev
S. R. Rasulov

CRC Press
Taylor & Francis Group
Boca Raton London New York

CRC Press is an imprint of the
Taylor & Francis Group, an **informa** business

CRC Press
Taylor & Francis Group
6000 Broken Sound Parkway NW, Suite 300
Boca Raton, FL 33487-2742

Printed on acid-free paper

International Standard Book Number-13: 978-0-367-20230-9 (Hardback)

Visit the Taylor & Francis Web site at
http://www.taylorandfrancis.com

and the CRC Press Web site at
http://www.crcpress.com

Contents

Preface

The world around us consists of a multitude of particles of different types, natures, properties, and sizes, whose size spectrum extends from elementary particles, atoms, such as molecules, nanoparticles, colloidal, micron particles, etc., up to cosmic bodies. Multiphase disperse systems are a world of instabilities and fluctuations, ultimately responsible for the amazing diversity of the richness of the forms and structures that arise in such systems. Of great interest is that, depending on the particle size, the behavior of such systems in external fields (gravitational, electromagnetic, etc.) is described by different laws: elementary particles by the laws of quantum mechanics and quantum gravity, and rather large cosmic bodies by the general theory of relativity and other laws. The concentration of particles per unit volume and their dimensions determine the behavior, properties, and structure of the dispersed system as a whole. Depending on the concentrations and particle sizes that determine the distance between them, such heterophase systems are inherent in all substance exchange processes, physical phenomena of collision, coagulation, crushing, etc. When studying for many decades of experimental and theoretical material, it seems that a complete understanding of the behavior of structured disperse systems does not always yield to a theoretical, logical, and experimental description. Models describing the behavior of structured disperse systems do not always satisfy physical laws, with the exception of the behavior of single particles, but represent empirical or semi-empirical approximations to the true picture, which is explained by random interactions between particles and the stochastic nature of their behavior. In this study, we consider micron and millimeter particles and more (solid particles, droplets, and bubbles) whose behavior in a certain region of their existence is described in the approximation of hydrodynamics and mechanics of a continuous medium and accompanied by heat and mass exchange between them with the presence of phase transformations and a chemical reaction. The description of the behavior of disperse systems in a limited area, analysis, in particular mass and heat-transfer phenomena, requires a detailed knowledge of microphysical phenomena. Such penetration into the "microcosm of structured disperse systems," in particular, the processes of chemical technology, is currently taking place. In recent years, a very large number of studies have appeared on hydrodynamics, mass transfer, and heat transfer, taking into account the complex behavior of solid particles, droplets, and bubbles and their interaction, coagulation, agglomeration, and crushing. Mass exchange and heat exchange between particles and the environment is the basis for describing many processes of chemical technology involving the dispersed phase, the important parameters of which are the determination of the coefficients of phase transfer.

In conclusion, it should be emphasized that many of the problems outlined in the book can be subjects of broad discussions. Some of the questions are briefly presented, taking into account the reader's opportunity to become acquainted with these problems in well-known literary sources.

The authors will be very grateful to all readers, having read this book, for their priceless comments and feedback on all the issues discussed, sent to the address: AZ1143, Azerbaijan, Baku, G. Javid Ave., 29, Institute of Catalysis and Inorganic Chemistry of National Academy of Science or E-mail: kkelbaliev@yahoo.com.

Nomenclature

In this book there are many different notations and symbols of physical quantities characterizing hydrodynamics, heat, and mass transfer, in connection with which it is inevitable to use the same letters and symbols to designate various parameters. Thus, if no special reservations are made in the text, the following correspondence of symbols and quantities is implied:

a	is particle diameter;
B	is stochastic diffusion coefficient;
C_D	is drag coefficient of particle;
C	is concentration;
c_p	is specific heat capacity at constant pressure;
D_T	is coefficient of turbulent diffusion of a liquid;
D_T	is coefficient of turbulent diffusion of particles;
D_{Kn}	is coefficient of Knudsen diffusion;
D	is coefficient of molecular diffusion;
d_T	is diameter of the pipe;
D_A	is diameter of the apparatus;
D_M	is diameter of the stirrer;
F	is surface area;
F_s	is drag force;
q	is heat flow;
q_V	is source of heat;
I	is intensity of turbulence;
J	is total mass flow per unit time;
K	is total coefficients of mass transfer and heat transfer;
m	is mass of substance;
N	is total number of particles per unit volume;
n_m	is agitator velocity;
n	is an indicator of the degree of non-Newtonian fluids;
p (a)	is particle distribution function;
p_i	is the partial pressure of the component in the mixture;
p	is total pressure;
R_{mp}	is radius of the pipe;
R	is radius of the particle;
t	is time;
T	is temperature;
U	is the average flow velocity;
U_*	is dynamic flow velocity;
V'	is the pulsating velocity of the turbulent flow;
V_P	is total particle deposition rate;
V_s	is the rate of gravitational sedimentation of particles;
V_L	is velocity of lift migration of the particle;
v	is volume;
x, y	are component fractions in liquid and gas phases;
Δy	is driving force; and
x^*, y^*	are equilibrium concentrations of the substance in the liquid and gas phases.

GREEK LETTERS

α is heat transfer coefficient;
β is dimensionless thickness of particle deposits;
β_L is coefficient of mass transfer;
γ is the ratio of the dynamic viscosity of the particles to the viscosity of the medium;
δ is thickness of sediments;
δ_D is thickness of the boundary layer;
Δ is strength;
ε_R is specific energy dissipation per unit mass;
ε is porosity;
χ_T is coefficient of thermal diffusivity;
χ is the degree of deformation of a drop and bubble;
η_φ is effective viscosity of suspensions;
η_c, η_d are dynamic viscosities of the medium and particles;
η is efficiency factor;
η_S is volumetric viscosity of the porous medium;
λ is scale of turbulent pulsations;
λ_0 is Kolmogorov scale of turbulence;
λ_B is thermal conductivity of a layer of particles;
ϑ is degree of capture of particles;
θ, ϕ are polar angles;
μ_P is the degree of entrainment of particles by the pulsating medium;
μ_g is the degree of flow around the particles by the stream;
ν_c, ν_d are kinematic viscosities of the medium and particles;
ν_T is turbulent viscosity;
ψ is stream function;
ξ_T is coefficient of resistance in pipes;
ξ_s is shear viscosity of a porous medium;
ζ is parameter of tortuosity;
ρ_c, ρ_d are density of the medium and particles;
ρ_i is concentration of particles;
σ_D is coefficient of surface tension;
σ_D is deforming stress;
τ_p is relaxation time;
τ is shear stress;
τ_0 is yield strength;
φ is the volume fraction of particles in;
φ_0 is module Thiele;
ω is frequency of turbulent pulsations and crushing;
Ac is the Acrivos number;
Ar is the Archimedes number;
Bi is the Bio number;
Bn is the Bingham number;
FO is the Fourier number;
Ga is the Galilean number;
Gr is the Grachof number;
BO is the Bond number;
Kn is the Knudsen number;
Ma is the Marangoni number;
MO is the Morton number;

Nu	is the Nusselt number;
Pe	is the Peclet number;
Ra	is the Rayleigh number;
Pr	is the Prandtl number;
Re	is the Reynolds number;
Stk	is the Stokes number;
St	is the Stanton number;
Sc	is the Schmidt number;
Sh	is the Sherwood number; and
We	is the Weber number.

Indices and functions: d is a solid particle; g is gas bubble; k is a drop; T is turbulence; c is medium; 0 is the initial value; $\Gamma(x)$ is gamma function; erf (x) is error integral; $\mathrm{sh}(x)$, $\mathrm{ch}(x)$, $\mathrm{th}(x)$, $\mathrm{cth}(x)$ is hyperbolic sine, cosine, tangent, and cotangent; $\mathrm{I}(x)$, $\mathrm{K}(x)$ are the Bessel functions of an imaginary argument of the first and second kind; $\mathrm{J}_0(r)$, $\mathrm{J}_1(r)$ are the Bessel functions of zero and first order; and P_k^n is the Legendre polynomial.

Authors

Gudret Isfandiyar Kelbaliyev was born in 1946 in the city of Nakhchivan of the Azerbaijan Republic. He graduated from the Azerbaijan Oil and Chemistry Institute with a degree in mechanical engineering. He defended his thesis for the degree of Doctor of Technical Sciences with a degree in processes and apparatuses of chemical technology. He has the academic title of professor. Currently, he is the head of the Laboratory for Heat and Mass Transfer and Hydrodynamics of the Institute of Catalysis and Inorganic Chemistry of the National Academy of Sciences of Azerbaijan. In 2001, he was elected as a corresponding member of the National Academy of Sciences of Azerbaijan.

G. I. Kelbaliyev is a specialist in mathematical modeling and optimization of chemical technology processes.

He is the author of more than 250 articles, inventions, and books. His work is published in the journals *Chemical Engineering Science, Journal of Displacement Science and Technology, Powder Technology, Colloid and Surface, Chemical Engineering Communications, Journal of Aerosol Science, Petroleum Science and Technology, Thermal Engineering, Theoretical Foundations of Chemical Engineering*, and *Journal of Engineering Physics and Thermophysics.*

Dilgam Babir Tagiyev was born in 1950 in the Kelbajar region of the Azerbaijan Republic. He graduated from the Baku State University with a specialization in petrochemistry. He is a specialist in the field of petrochemistry, a doctor of chemical sciences, professor, and full member of the National Academy of Sciences of Azerbaijan. Currently he works as a director of the Institute of Catalysis and Inorganic Chemistry of the National Academy of Sciences of Azerbaijan. He is the chairman of the Dissertation Council.

D.B. Tagiyev is the author of more than 400 articles, inventions, and books. His works are published in the magazines *Studies in Surface Science and Catalysis* (Elsevier), *Petrochemistry, Chemistry, Theoretical Foundations of Chemical Engineering, Chemistry and Technology of Fuels and Oils*, etc.

Sakit Rauf Rasulov was born in 1956 in the Agdash region of the Azerbaijan Republic. He graduated from the Azerbaijan Institute of Oil and Chemistry (now the Azerbaijan State Oil and Industry University [ASOIU]) and received the qualification of a chemical technologist. He is a specialist in the field of oil refining and transportation of hydrocarbon raw materials. He holds a doctorate in technical sciences and is a professor.

Currently he works as the head of the department of the ASOIU. He is the author of more than 250 articles, inventions, and books. The works belonging to him are published in the journals *Theoretical Foundations of Chemical Engineering, Journal of Engineering Physics and Thermophysics, Chemistry and Technology of Fuels and Oils, Oil and Industry*, etc.

Introduction

Mass transfer and flow of disperse systems are important problems in the processes of chemical, petrochemical, and oil-refining technologies. Mass transfer and irreversible mass transfer of matter within one or more phases are realized as a result of chaotic motion of molecules or random motion of vortices in turbulent flows. Mass exchange is the basis for a variety of separation processes (distillation, extraction, absorption, crystallization, etc.), combined into a class of mass-exchange processes. At the same time, mass exchange accompanies many thermal processes (crystallization, condensation, and evaporation), chemical processes with interphase exchange (heterogeneous and homogeneous reactions), and physical phenomena (coagulation, fragmentation, agglomeration, and sedimentation). Disperse flows are inhomogeneous systems with a phase interface, called a free surface for gas–liquid, solid–liquid, and liquid–liquid systems. When dispersed systems move, those forces that are considered in the analysis of hydrodynamic phenomena occurring in single-phase flows are manifested. However, the presence of two or more phases changes not only the forms of motion of such systems but also their nature, since interaction and substance exchange between phases have a decisive influence, and in some cases contribute to the formation of coagulation structures. In these cases, it is impossible to describe the regimes with such concepts as the laminar or turbulent flow that is usual for single-phase flows. Unlike single-phase flows, essentially new forces appear at the interface of dispersed flows—the interfacial surface tension forces. These forces produce the work of forming the surface of a liquid at the interface of its separation. In most cases, two or more phases are involved in mass exchange, forming systems: liquid–liquid, liquid–gas, liquid–solid, gas–solid, liquid–gas–solid, etc. In practice, there are processes involving three liquid–liquid–solid phases (oil emulsions and sledges) or liquid–gas–solid phase (flotation), in which the conditions for contacting the phases are extremely diverse.

In contrast to the existing monographs, where this problem is considered within the framework of the laminar flow, this book focuses on solving this problem in a turbulent flow, in connection with important tasks such as mass transfer, coalescence, deformation, fragmentation of drops and bubbles, separation phases, and separation and precipitation in an isotropic turbulent flow. The material presented in this book is built on the principle of "from simple to complex," which greatly facilitates its perception.

Chapter 1 is devoted to the main aspects of the hydrodynamics of the flow of liquids and dispersed systems. The basic equations and hydrodynamic models for viscous fluid flow are given, as well as the equations for the flow of disperse systems with boundary conditions. Models of the laminar flow of a viscoplastic fluid in pipes are considered, and formulas for calculating the velocity profile in pipes with the formation of a dense layer on the inner surface of pipes are proposed. The conditions of turbulent fluid flow in pipes are analyzed, and a single formula for calculating the velocity profile in the boundary layer is proposed. Much attention is paid to isotropic turbulence, the evaluation of its parameters (turbulence scale, energy dissipation), lifting, and turbulent migration of particles in a dispersed flow. The hydrodynamic equations for the flow of a multitude of particles in a flow, the motion of a single particle in a turbulent flow, and the flow of droplets and bubbles in a laminar flow are given, and the nature of the flow of fine particles in a turbulent flow is investigated. Various formulas are proposed for determining the effective viscosity of dispersed systems and, in particular, the change in viscosity in structured dispersed systems. The problems of hydrodynamics of flow through porous media and models of fluid filtration in an isotropic porous medium are considered. The hydrodynamics of non-stationary filtration of oil-dispersed systems is considered, the equations of hydraulic diffusion in a porous oil reservoir are solved, and appropriate expressions are obtained for estimating the coefficients of hydraulic diffusion and permeability.

Chapter 2 is devoted to the problems of heat transfer in dispersed media, the determination of heat-transfer coefficients and energy dissipation, and many other problems associated with heat transfer. Equations of heat transfer in various coordinates and boundary conditions for their solution are proposed. The stationary one-dimensional analytical solutions of heat-transfer equations with different boundary conditions for flat, cylindrical, and spherical heat-transfer surfaces are considered. Analytical solutions are proposed for one-dimensional and multidimensional unsteady heat-transfer equations for different boundary conditions using the method of separation of variables for flat, cylindrical, and spherical heat exchange surfaces. Practical examples of using these solutions for chemical technology problems are given. The solution of the convective transfer problem and the set of formulas for determining the coefficients and the Nusselt number for various characters of the flow are considered. The problem of convective heat transfer in engineering applications is proposed in order to determine the heat transfer coefficients for forced and free convection with practical examples. A solution to the problem of heat transfer in rough pipes is considered in order to estimate the effective utilization rate of rough surfaces. Many problems associated with the assessment of energy dissipation in the flow have been solved: with laminar flow on flat and cylindrical surfaces, in a liquid–liquid dispersed system (emulsion), and in a turbulent flow of a dispersed medium. The possibilities of connecting energy dissipation with the coefficient of resistance of dispersed particles are given. The solutions given in this chapter are of value for studying heat-transfer processes in order to identify the mechanism of phenomena and determining transfer coefficients and criterial equations for their calculation under various conditions of the process. A theoretical solution to the transfer problem allows us to obtain an analytical formula suitable for analyzing and establishing a more true mechanism of the process.

Chapter 3 is devoted to the analysis of mass-transfer processes of chemical technology and the peculiarities of mass transfer between such phases as liquid–gas, liquid, solid particle, gas, etc. A number of expressions are proposed for calculating the concentrations of a substance in various units of measurement and the relationship between them. The main similarity criteria used in the mass transfer processes are given; the mass transfer during phase transitions and the rate of phase transfer are analyzed; and various examples of the calculation are given. The convective and diffusive mass transfer and the equilibrium conditions between the phases are analyzed, and expressions for determining the driving force of mass transfer are proposed. Expressions are given for determining the mass-transfer coefficients between the liquid and gas phases, taking into account the first-order chemical reaction. Various examples of calculating mass-transfer coefficients are proposed.

Chapter 4 is devoted to the analysis of the laws of mass transfer and the possibilities of their use in the calculation of mass-transfer phenomena. The first Fick law is proposed in various units for measuring concentrations and taking into account the effects of thermal diffusion and pressure diffusion. Many formulas have been proposed for determining molecular and turbulent mass diffusion and Knudsen diffusion. Formulas for calculating turbulent viscosity and molecular diffusion of a multicomponent medium are proposed. Various examples of calculating the coefficients of turbulent and molecular mass diffusion are given. Formulas for calculating the turbulent diffusion of dispersed particles depending on the degree of their transport by the turbulent flow are given, which is confirmed by experimental data. Formulas for calculating the diffusion coefficients of thermal conductivity in a fixed bed of particles depending on the porosity of the layer and compaction of the layer of particles are proposed. The solution of the problem of one-dimensional diffusion transport complicated by the Stefanov flow is considered. The problem of equimolar and non-equimolar mass transfer is solved with practical application. The problem of mass transfer during evaporation of a single spherical liquid drop is considered, and the time of its complete evaporation is determined. The equation of the second Fick law is considered for different concentration units and in different coordinates with the corresponding boundary conditions for their solution.

Chapter 5 is devoted to convective mass transfer with an analysis of the mechanism and transfer model, in particular, the "fixed film" models, such as the known-film model or the Lewis and Whitman model; the turbulent boundary layer; the permeability or penetration model; the Higby model; and the infinite surface renewal model. As a result of solving differential and integral equations of convective diffusion within the boundary layer, formulas for estimating the thickness of the diffusion boundary layer and the mass-transfer coefficient are proposed. Solved problems of convective mass transfer in a turbulent flow, as well as in an isotropic turbulent flow with a first-order chemical reaction, are provided, as a result of which concentration profiles and mass-transfer coefficient are determined. Methods are proposed for estimating the Sherwood number using the method of model equations and analogies of Reynolds, Karman, Chilton–Colburn, and empirical modeling methods. Many examples and experimental data are given. Approximate methods for solving equations of convective diffusion are considered. The analogy and similarity of mass, heat, and momentum transfer processes and dimensional analysis methods are analyzed. The effect of the Marangoni effect on the convective mass transfer and capillary flow, as well as mass transfer in the capillary flow, is considered. The equations given in this section are very complex, and it is necessary to use numerical methods to solve them. However, when solving practical problems of mass transfer, it is sufficient to use the equations in the form of criteria that allow one to estimate, within the limits of accuracy, the mass-transfer coefficient.

Chapter 6 is devoted to analytical solutions to the problem of mass transfer for different Peclet numbers. Stationary solutions of mass-transfer equations are proposed for flat, cylindrical, and spherical surfaces for small Peclet numbers. Moreover, in some cases, a first-order chemical reaction is taken into account, which makes it possible to determine the Thiele modulus and the efficiency factor. An analytical solution to the problem of non-stationary mass transfer on a flat, cylindrical, and spherical surface by the method of separation of variables is considered, as a result of which the variable mass flow to the selected surface, the Sherwood number, and the mass-transfer coefficients are determined. The calculation of mass transfer coefficients in the processes of gas–liquid absorption and extraction in a system of liquid–solid particles, as well as dissolution of the solid phase, evaporation of droplets and drying of porous materials with constant and variable diffusion coefficients, is proposed. In all applications of the problem, the Sherwood number and mass-transfer coefficients are determined, and a solution to the inverse problem of mass transfer is proposed. Much attention is paid to the problem of experimental and theoretical studies of mass transfer between a drop or a gas bubble and an isotropic turbulent flow, both for small and large values of the Peclet number. Energy dissipation in a turbulent flow and its effect on mass-transfer parameters are considered.

Chapter 7 is devoted to hydrodynamics and mass transfer in non-Newtonian fluids, general characteristics, and classification of rheological models of such fluids. The problems of the steady flow of non-Newtonian fluids on a flat, cylindrical, and spherical surface are solved, and the velocity profiles for each case are determined. The problems of the formation of various coagulation structures in dispersed media, including the formation of aggregates and clusters of aggregates in dispersed oil media are considered. The main stages of the formation of structures in dispersed systems, models of deformation, and destruction of structures are proposed. The rheology of structured systems is considered, and the non-stationary rheological equation of structure formation is proposed and solved. The proposed models are used to estimate the rheological parameters of oil-dispersed systems. In this section, special attention is paid to the rheology of oil, as a result of which rheological models of oil filtration in a porous medium are obtained. The flow of non-Newtonian oil from a vertical flat surface is considered, and models are proposed for calculating the flow velocity and layer thickness. Heat and mass transfer in a non-Newtonian fluid when flowing around a flat plate are considered, as a result of which, the Sherwood and Nusselt numbers and the mass transfer and heat transfer coefficients are determined. Semi-empirical formulas are proposed for calculating the drag coefficients of particles in a non-Newtonian fluid for sufficiently large Reynolds number values. The rheology of separation processes of oil emulsions on the effect of water droplets and asphalt-resinous substances on the effective viscosity is considered.

Chapter 8 is devoted to the aggregative and sedimentation unstable dispersed systems characterized by the variability of the state of the medium, accompanied by a continuous change in the physical properties of the medium and particles, that is, changes in the volume and size of particles as a result of their interaction, collision, coalescence, crushing, and deposition at a certain concentration of particles in a closed volume. The fundamentals of the theory of coalescence and crushing of droplets and bubbles in an isotropic turbulent flow are proposed, the essence of which is to determine the minimum and maximum particle sizes, to determine the frequency of collisions, coalescence, deformation, and crushing of drops and bubbles based on solving the mass-transfer equations depending on the value of Weber, Morton, and Reynolds numbers. Many different expressions are proposed to determine the frequency of coalescence and crushing of droplets and bubbles in an isotropic turbulent flow depending on the scale of turbulence, physical properties of the medium and particles, and energy dissipation. The solution of problems associated with gravitational and gradient coagulation of particles is considered. Much attention is paid to the determination of the drag coefficients of solid particles, droplets, and bubbles for large areas of variation of the Reynolds number. This section deals with the problems of deposition and ascent of particles in a gravitational field in horizontal and vertical pipes from an isotropic turbulent flow. It is noted that the mechanism of sedimentation of particles is carried out as a result of diffusion, gravitational diffusion, and turbulent transfer of particles to a streamlined surface.

Chapter 9 is devoted to stochastic methods in mass-transfer processes in dispersed media. The change in the size and shape of particles in processes involving dispersed media caused by phase and chemical transformations, mechanical phenomena of coagulation, and fragmentation significantly deforms the density distribution function of particles in size and time, thereby having a significant impact on the phenomena of mass, heat, and momentum systems. In this regard, this section discusses the classification of the distribution functions, analysis and solution of the Fokker–Planck stochastic equations, and the kinetic coagulation and fragmentation equation, which allow us to construct the evolution of the distribution function in time and particle size. A number of analytical solutions of the Fokker–Planck and coagulation and crushing differential equations are presented, depending on the nature of the change in the average particle size in time and the type of the coagulation frequency and crushing frequency function. There are many experimental studies on the average particle size and evolution of the distribution function. It has been shown that there is an analogy between stochastic equations and mass-transfer phenomena. Practical applications of stochastic equations in the processes of separation of oil emulsions and granulation of powder materials are given. The stochastic equations proposed in this section allow us to estimate the state of a dispersed system and supplement the basic equations of mass, heat, and momentum transfer.

The authors hope that the studies in the field of mass transfer described in this book will be useful for the calculation, modeling, and design of various mass-exchange processes of chemical technology.

1 Hydrodynamics of the Flow of Liquids and Dispersion Media

Hydrodynamics of the flow of gases and liquids are an important problem in the design of technological devices, since they affect the transfer of mass and heat, the flow of chemical reactions, and physical processes. More complex is the hydrodynamics of disperse systems, which are mixtures of solid particles, liquid droplets, and gas bubbles distributed in the carrier phase. Even more complex is the flow of disperse systems with the formation of various coagulation structures and aggregates of particles contained in the liquid and conducting to change the properties of the medium and its flow.

In this section, we consider the basic equations of the laminar flow of liquid and disperse systems, criteria for similarity of flows, and flow through porous media. In addition, the features of turbulent flow are presented and, in particular, the concept of isotropic turbulence is defined; a number of formulas are given for determining the effective viscosity of disperse systems; and many other problems are considered.

The overwhelming majority of technological processes in virtually any branch of modern production are associated, to some extent, with the use of liquids, gases, or vapors, their movement, and physical phenomena of mass and heat transfer under the conditions of chemical reactions and phase transformations. It especially concerns such branches of industry as chemical, petrochemical, and oil-refining branches, including extraction, transportation, and preparation of oil and gas. The importance of this issue presentation is determined by the fact that the components of the fluid flow velocity in the equations of mass transfer and heat exchange should be determined from the solution of the corresponding hydrodynamic problem under certain flow conditions. Since the flow phenomena considered in this section have a macroscopic character, in dynamics, the liquid is represented as a continuous incompressible medium characterized by its physicochemical and thermodynamic properties. The flow of dispersed systems, with their inherent phenomena of coagulation, crushing, deformation, and deposition of particles in concentrated media, leading to phase separation and inversion, are important problems in the processes of mass and heat transfer.

It should be noted that during the flow of liquids and dispersed phases, the determination of the distribution of the field component velocities, the drag coefficients, and many other hydrodynamic parameters, depends on the nature of the flow or around the particle and the variable physical and chemical properties of the medium, which ultimately determine the magnitude of the moving forces and directions of processes. It is important to note that the analysis of the flow near the surface, that is, boundary layer, and its structure under turbulent flow are necessary for calculating the mass and heat transfer coefficients. The following types of liquids should be distinguished:

- *Ideal liquid*—possesses absolute fluidity, is absolutely incompressible, and there are completely no adhesion forces between the particles;
- *A real liquid*—has all the above properties and is characterized by fluidity and viscosity;
- *Homogeneous liquid*—a liquid consisting of one or several components that do not have an interface between themselves—the true solutions; and
- *Heterogeneous liquid*—a liquid consisting of one or several components and inclusions that have an interface between themselves, form an interphase surface, and create two- or three-phase systems (suspensions, emulsions, aerosols). Such systems are dispersed on multiphase and widely used in chemical technology, and their descriptions, which are accompanied by interparticle, interphase interaction and aggregative changes, differ from the flow of ordinary liquids. At a high concentration of the dispersed phase, such systems exhibit the properties of non-Newtonian fluids.

In generally solving this problem, hydrodynamics is distinguished by:

1. The internal problem of hydrodynamics—the main task is to study the regularities of the flow of liquids inside closed channels;
2. The external problem of hydrodynamics—the main task is to study the regularities of the external flow of bodies of different configurations around the liquid. Such a flow is typical for the flow of particles, droplets, and bubbles in disperse systems; and
3. A mixed problem—the flow of liquids is considered simultaneously from the point of view of both internal and external problems. Such flow is characteristic for the granular materials layer, and in particular for adsorption, desorption, drying, and chemical reactions in catalytic reactors, where the flow occurs simultaneously, both outside the porous particles and in the pores.

This chapter contains the basic equations of momentum transfer in liquids and disperse media in various coordinates, believing that a more detailed description can be found in the works [5,6,8–11]. The solution of many problems related to mass transfer in the case of flow around surfaces of different configurations can be found in the papers [10,11].

1.1 EQUATION OF CONTINUITY AND MOMENTUM TRANSFER IN A HOMOGENEOUS LAMINAR FLOW

The state of a moving incompressible fluid is completely characterized by specifying at each point of space (x, y, z) and at each instant of time t three velocity components V_x, V_y, V_z and some of its two thermodynamic quantities, for example, pressure $P(x, y, z, t)$ and density $\rho(x, y, z, t)$. In an incompressible fluid, the velocity of motion satisfies that flow continuity V equation expressing the law of matter conservation.

$$\frac{\partial \rho}{\partial t} + \frac{\partial}{\partial x}(\rho V_x) + \frac{\partial}{\partial y}(\rho V_y) + \frac{\partial}{\partial z}(\rho V_z) = 0 \tag{1.1}$$

This equation can also be represented in cylindrical coordinates (r, θ, z) in the form

$$\frac{\partial \rho}{\partial t} + \frac{1}{r}\frac{\partial}{\partial r}(\rho r V_r) + \frac{1}{r}\frac{\partial}{\partial \theta}(\rho V_\theta) + \frac{\partial}{\partial z}(\rho V_z) = 0 \tag{1.2}$$

and in spherical coordinates (r, θ, φ).

$$\frac{\partial \rho}{\partial t} + \frac{1}{r^2}\frac{\partial}{\partial r}(\rho r^2 V_r) + \frac{1}{r\sin\theta}\frac{\partial}{\partial \theta}(\rho V_\theta \sin\theta) + \frac{1}{r\sin\theta}\frac{\partial}{\partial \phi}(r V_\phi) = 0 \tag{1.3}$$

The differential equation of momentum transfer (ρV) in the conditions of external forces of the sources action in a vector form is represented as

$$\rho\frac{\partial \vec{V}}{\partial t} + \rho(\vec{V}\nabla)\vec{V} = -\Delta P - div\sigma + \sum_k \rho_k \vec{F}_k \tag{1.4}$$

where σ is a tensor of viscous stress.

The first member in the left part of equation (1.4) is the local change of number of the movement in a unit of time; the second member is the convective transfer of number of the movement. The first member in the right part of this equation is the pressure force calculated per unit of volume; the second member—change of number of the movement in unit of time at the expense of forces of internal friction (diffusive transfer of an impulse) and the last member—is the total action of all external forces F_k.

If in the equation of transfer of an impulse (1.4) to substitute the corresponding expression for a tensor of viscous tension, then we will receive Navier–Stokes's equation

$$\rho\frac{dV}{dt} = -\nabla P + \eta\nabla^2 V + \rho g \tag{1.5}$$

where ∇ is a Laplace's operator, and η, ρ are dynamic viscosity and density of liquid pressure. Equation (1.5) presented in various coordinates is given below:

1. In the Cartesian coordinates (x, y, z)

$$\begin{aligned}
&\rho\left(\frac{\partial V_x}{\partial t}+V_x\frac{\partial V_x}{\partial x}+V_y\frac{\partial V_x}{\partial y}+V_z\frac{\partial V_x}{\partial z}\right)=-\frac{\partial P}{\partial x}+\eta\left(\frac{\partial^2 V_x}{\partial x^2}+\frac{\partial^2 V_x}{\partial y^2}+\frac{\partial^2 V_x}{\partial z^2}\right)\\
&\rho\left(\frac{\partial V_y}{\partial t}+V_x\frac{\partial V_y}{\partial x}+V_y\frac{\partial V_y}{\partial y}+V_z\frac{\partial V_z}{\partial z}\right)=-\frac{\partial P}{\partial y}+\eta\left(\frac{\partial^2 V_y}{\partial x^2}+\frac{\partial^2 V_y}{\partial y^2}+\frac{\partial^2 V_y}{\partial z^2}\right)\\
&\rho\left(\frac{\partial V_z}{\partial t}+V_x\frac{\partial V_z}{\partial x}+V_y\frac{\partial V_z}{\partial y}+V_z\frac{\partial V_z}{\partial z}\right)=-\frac{\partial P}{\partial z}+\eta\left(\frac{\partial^2 V_z}{\partial x^2}+\frac{\partial^2 V_z}{\partial y^2}+\frac{\partial^2 V_z}{\partial z^2}\right)
\end{aligned}\tag{1.6}$$

2. In cylindrical coordinates (r,θ,z)

$$\begin{aligned}
&\rho\left(\frac{\partial V_r}{\partial t}+V_r\frac{\partial V_r}{\partial r}+\frac{V_\theta}{r}\frac{\partial V_\theta}{\partial \theta}-\frac{V_\theta^2}{r}+V_z\frac{\partial V_r}{\partial z}\right)=-\frac{\partial P}{\partial r}+\eta\left[\begin{array}{l}\frac{\partial}{\partial r}\left(\frac{1}{r}\frac{\partial}{\partial r}(rV_r)\right)+\frac{1}{r^2}\frac{\partial^2 V_r}{\partial \theta^2}-\\ -\frac{2}{r^2}\frac{\partial V_\theta}{\partial \theta}+\frac{\partial^2 V_r}{\partial z^2}\end{array}\right]+\rho g_r\\
&\rho\left(\frac{\partial V_\theta}{\partial t}+V_r\frac{\partial V_\theta}{\partial r}+\frac{V_\theta}{r}\frac{\partial V_\theta}{\partial \theta}+\frac{V_r V_\theta}{r}+V_z\frac{\partial V_\theta}{\partial z}\right)=-\frac{\partial P}{\partial \theta}+\eta\left[\begin{array}{l}\frac{\partial}{\partial r}\left(\frac{1}{r}\frac{\partial}{\partial r}(rV_\theta)\right)+\frac{1}{r^2}\frac{\partial^2 V_\theta}{\partial \theta^2}\\ +\frac{2}{r^2}\frac{\partial V_r}{\partial \theta}+\frac{\partial^2 V_\theta}{\partial z^2}\end{array}\right]+\rho g_\theta\\
&\rho\left(\frac{\partial V_z}{\partial t}+V_z\frac{\partial V_z}{\partial r}+\frac{V_\theta}{r}\frac{\partial V_z}{\partial \theta}+V_z\frac{\partial V_z}{\partial z}\right)=-\frac{\partial P}{\partial z}+\eta\left[\frac{1}{r}\frac{\partial}{\partial r}\left(r\frac{\partial V_z}{\partial r}\right)+\frac{1}{r^2}\frac{\partial^2 V_z}{\partial \theta^2}+\frac{\partial^2 V_z}{\partial z^2}\right]+\rho g_z
\end{aligned}\tag{1.7}$$

3. In spherical coordinates (r,θ,φ)

$$\rho\left(\begin{array}{l}\frac{\partial V_r}{\partial t}+V_r\frac{\partial V_r}{\partial r}+\frac{V_\theta}{r}\frac{\partial V_r}{\partial \theta}+\\ \frac{V_\phi}{r\sin\theta}\frac{\partial V_r}{\partial \phi}-\frac{V_\theta^2+V_\phi^2}{r}\end{array}\right)=-\frac{\partial P}{\partial r}+\eta\left(\begin{array}{l}\nabla^2 V_r-\frac{2V_r}{r^2}-\frac{2}{r^2}\frac{\partial V_\theta}{\partial \theta}\\ -\frac{2V_\theta}{r^2}ctg\theta-\frac{2}{r^2\sin\theta}\frac{\partial V_\phi}{\partial \phi}\end{array}\right)+\rho g_r$$

$$\rho\begin{pmatrix}\dfrac{\partial V_\theta}{\partial t}+V_r\dfrac{\partial V_\theta}{\partial r}+\dfrac{V_\theta}{r}\dfrac{\partial V_\theta}{\partial \theta}+\\ \dfrac{V_\phi}{r\sin\theta}\dfrac{\partial V_\theta}{\partial \phi}+\dfrac{V_rV_\theta}{r}-\dfrac{V_\phi^2 ctg\theta}{r}\end{pmatrix}=-\frac{1}{r}\frac{\partial P}{\partial \theta}+\eta\begin{pmatrix}\nabla^2V_\theta+\dfrac{2}{r^2}\dfrac{\partial V_r}{\partial \theta}-\\ \dfrac{V_\theta}{r^2\sin^2\theta}-\dfrac{2\cos\theta}{r^2\sin^2\theta}\dfrac{\partial V_\phi}{\partial \phi}\end{pmatrix}+\rho g_\theta$$

$$\rho\begin{pmatrix}\dfrac{\partial V_\phi}{\partial t}+V_r\dfrac{\partial V_\phi}{\partial r}+\dfrac{V_\theta}{r}\dfrac{\partial V_\phi}{\partial \theta}+\\ \dfrac{V_\phi}{r\sin\theta}\dfrac{\partial V_\phi}{\partial \phi}+\dfrac{V_\phi V_r}{r}+\dfrac{V_\theta V_\phi}{r}ctg\theta\end{pmatrix}=-\frac{1}{r\sin\theta}\frac{\partial P}{\partial \phi}+\eta\begin{pmatrix}\nabla^2V_\phi-\dfrac{V_\phi}{r^2\sin^2\theta}+\\ \dfrac{2}{r^2\sin\theta}\dfrac{\partial V_r}{\partial \phi}+\dfrac{2\cos\theta}{r^2\sin^2\theta}\dfrac{\partial V_\theta}{\partial \phi}\end{pmatrix}+\rho g_\phi \tag{1.8}$$

$\nabla^2=\frac{1}{r^2}\frac{\partial}{\partial r}\left(r^2\frac{\partial}{\partial r}\right)+\frac{1}{r^2\sin\theta}\frac{\partial}{\partial\theta}\left(\sin\theta\frac{\partial}{\partial\theta}\right)+\frac{1}{r^2\sin^2\theta}\left(\frac{\partial^2}{\partial\phi^2}\right)$, θ,φ are polar corners. For the solution of a specific hydrodynamic objective in applications, these systems of the equations can be much more simplified according to current conditions as a result of assessment of a contribution of each member and a task of regional conditions. It is important to note that a current of disperse systems, that is, liquid or gas with firm or liquid inclusions, differ from a current of uniform liquid a little.

In particular for the flat stationary movement of incompressible liquid, the Navier–Stokes (1.6) equation and the equations of a current have an appearance

$$V_x\frac{\partial V_x}{\partial x}+V_y\frac{\partial V_x}{\partial y}=-\frac{1}{\rho}\frac{\partial P}{\partial x}+\frac{\eta}{\rho}\left(\frac{\partial^2V_x}{\partial x^2}+\frac{\partial^2V_x}{\partial y^2}\right)+\frac{1}{\rho}F_x$$

$$V_x\frac{\partial V_y}{\partial x}+V_y\frac{\partial V_y}{\partial y}=-\frac{1}{\rho}\frac{\partial P}{\partial y}+\frac{\eta}{\rho}\left(\frac{\partial^2V_y}{\partial x^2}+\frac{\partial^2V_y}{\partial y^2}\right)+\frac{1}{\rho}F_y \tag{1.9}$$

$$\frac{\partial V_x}{\partial x}+\frac{\partial V_y}{\partial y}=0$$

In the equations system (1.9) to determine the three unknowns V_x,V_y and P, we have three equations. In particular, in order to enter functions of current, then from the last equation (1.9) we have

$$V_x=\frac{\partial\psi}{\partial y},\ V_y=-\frac{\partial\psi}{\partial x} \tag{1.10}$$

Consequently, the equations of transfer of an impulse $\psi(x,y)$ (1.9) will be transformed to a look

$$\frac{\partial}{\partial t}\nabla^2\psi=\frac{\partial\psi}{\partial y}\frac{\partial\nabla^2\psi}{\partial x}-\frac{\partial\psi}{\partial x}\frac{\partial\nabla^2\psi}{\partial y^2}=\nu\nabla^4\psi \tag{1.11}$$

This expression is a partial differential equation of the fourth with respect to the current function, in connection with which finding a common solution presents certain difficulties. However, for the solution of private problems of various surfaces, stationary flow, perhaps to receive only approximate decisions.

As it appears from the previously stated equations, many problems of convective transfer are nonlinear. So, the nonlinear member obviously appears in the equations of transfer of an impulse $(V\nabla)V$. Besides, in many problems of a mass transfer, nonlinearity is also defined by dependence of many unknown parameters on temperature. Additional nonlinearity arises due to the interface between various parameters and convective velocity presented in the equations of balance of weight and internal energy, respectively, by members and $\nabla(\rho V)$ and $(V\nabla)T$.

The plurality of solutions of the nonlinear equations can be quite excluded at the expense of an accurate and correct formulation of a regional task. Regional conditions for the solution of these equations can be various and will be defined by hydrodynamic character and geometry of a current of the considered task, though it should be noted that on all firm surfaces on which moving liquid borders the boundary condition it is satisfied (conditions of sticking of liquid to a firm surface) $\vec{V}=0$.

At a current of disperse systems, this condition for disperse particles on a surface of a solid body doesn't come true. The presence of a shift in the averaged flow velocity is related to the results of longitudinal movement of particles suspended in it, if the sizes of the particles are sufficiently huge. It should be noted that in the parietal area, the velocity of particles is higher; in a stream kernel, velocities of the bearing phase are lower. It means that particles slide, rather, bearing the environment in the direction of a stream (parietal area) or toward to it (a stream kernel). In general, the average velocity of particles on a section lag behind the average velocity of a stream. Pilot studies show that the effect of sliding of particles of subjects [8,12] is higher than the size of particles, and the velocity of a current of liquid is more, but it is less than the diameter of the pipe. Therefore, at a current of disperse systems, the boundary condition of the sticking of liquid to a firm surface is replaced with a sliding condition on a firm surface.

$$V_d = V_{ck} \tag{1.12}$$

where V_{ck} is the velocity of sliding of particles on a solid surface. When dispersed systems flow in horizontal channels, the rate of their deposition also affects the particle slip velocity.

At the interface of mobile phases—two immiscible liquids or a liquid–gas system—the velocity should not vanish. On the boundary of the systems under consideration, the following boundary conditions are usually satisfied:

1. The tangential component of velocity is continuous on border of both phases

$$V_t^{(1)} = V_t^{(2)} \tag{1.13}$$

This condition for a spherical surface of a drop (bubble) registers as follows

$$\eta_1\left(\frac{\partial V_{\theta 1}}{\partial r}+\frac{1}{r}\frac{\partial V_{r1}}{\partial r}-\frac{V_{\theta 1}}{r}\right)=\eta_2\left(\frac{\partial V_{\theta 2}}{\partial r}+\frac{1}{r}\frac{\partial V_{r2}}{\partial r}-\frac{V_{\theta 2}}{r}\right) \tag{1.14}$$

2. Normal components of velocity are equal to zero

$$V_n^{(1)} = V_n^{(2)} = 0 \tag{1.15}$$

3. Forces with which liquids work at each other are equal and opposite in the direction

$$F_n^{(1)} = F_n^{(2)},\ F_t^{(1)} = F_t^{(2)} \tag{1.16}$$

On the free surface of liquid, it is possible to write $F_t = 0$. The given boundary conditions can be the general for many tasks, though they can be complemented with other conditions following from a condition of the solved problem. Estimating the expressions entering in the Navier–Stokes's equation, namely the relation of forces of inertia to viscosity forces, we will receive the size of criterion of Reynolds

$$\frac{\rho(\vec{V}\nabla)\vec{V}}{\eta\Delta^2\vec{V}} \approx \frac{\rho V\dfrac{V}{L}}{\eta\dfrac{V}{L^2}} = \frac{\rho VL}{\eta} = \frac{VL}{\nu} = \text{Re} \tag{1.17}$$

where L is the characteristic size of the channel. Thus, at small numbers, the viscous laminar mode prevails, and at big is the inertial turbulent mode. The Reynolds number size, as it is subsequently shown, entirely defines the nature of transfer of a substance and the value of coefficients of transfer.

Follows from the previously stated equations, the distribution of velocity of a stream is defined by the physical parameters of a stream: dynamic viscosity η and density of a stream ρ. The relation $\nu = \eta/\rho$ is called kinematic viscosity with the dimension m^2/s The dynamic viscosity of gases at the set temperature doesn't depend on pressure; the kinematic viscosity is inversely proportional to pressure. The viscosity of liquids significantly depends on temperature.

At a large number $\text{Re} >> 1$ viscous member can be neglected, or viscous forces are small and play a supporting role. The liquid that doesn't have viscosity is usually called and ideal liquid. Thus, at neglect processes of dissipation of energy, which can take place in the current liquid owing to internal friction (viscosity), an ideal liquid Navier–Stokes's equation takes a form

$$\frac{\partial V}{\partial t} + (V\nabla)V = -\frac{\nabla P}{\rho} + g \tag{1.18}$$

Neglecting the member containing viscosity and transitioning from the Navier–Stokes's equation to the Euler's equation is represented as a very essential simplification. For a stationary potential current, equation (1.13) can be written down in the form of Bernoulli's equation

$$\frac{V^2}{2} + \frac{P}{\rho} + gz = \text{const} \tag{1.19}$$

It follows from this equation that at the stationary movement of incompressible liquid (in the absence of the field of weight), the greatest value of pressure is reached in points where velocity addresses in zero. It is important to note that during the flow of a viscous fluid, the main phenomena of mass and heat transfer are carried out in a thin boundary layer, where the flow can be both laminar and turbulent. The existence of an interface significantly changes the approach to the solution of many tasks connected with substance transfer. It, first of all, is connected with a problem of the assessment of an amount of transferable substance for a unit of time, assessment of the thickness of diffusive and thermal interfaces, and an assessment of coefficients of transfer and resistance. At flow of various surfaces, stream velocity at some distance from a wall changes from major importance to zero on a wall. In this case, in order to fulfill the boundary conditions for a liquid, the disappearance of a normal component of velocity is important. The component of velocity remains ultimate; on the other hand, in solid walls, it has to address in zero level. The falling of velocity happens almost completely in a thin boundary (parietal) layer and is caused finally by viscosity of liquid, which can't be neglected even at large Reynolds numbers here. Despite the insignificant thickness of an interface, characterized by an essential gradient of velocity, temperature and concentration play the main role in substance transfer processes.

The movement in an interface can be both laminar and turbulent. At relatively small Reynolds numbers, the boundary layer can be considered as laminar and for its study one can use the solution of equation (1.9) with the corresponding boundary conditions [6].

The thickness of a laminar interface at flow of a flat surface is estimated as

$$\delta_D \approx 5,2\sqrt{\frac{\nu x}{U_\infty}} = 5,2\frac{x}{\sqrt{\text{Re}_x}}, \ \text{Re}_x = \frac{U_\infty x}{\nu} \tag{1.20}$$

We will determine the being velocities of a stream in a look

$$V_x \approx \frac{U_\infty y}{\delta_D}, \quad V_y \approx \frac{\nu . y^2}{\delta_D^3} \tag{1.21}$$

The being stream velocities determined in equation (1.21) can be used in the equations of transfer of heat and weight for calculation of distribution of temperature, concentration of components, and coefficients of transfer in a laminar interface.

1.1.1 Laminar Flow of Disperse Systems in Pipes with Deposition Solid Phase on the Surface

The flow of disperse systems in pipes is followed by sedimentation of a firm phase on the surface of pipes and the formation of a dense bed of particles whose thickness significantly influences the phenomena of transfer of weight, heat, and an impulse. Questions of sedimentation of particles from disperse laminar and turbulent streams on a surface will be considered in Chapter 4. Here we will consider the nature of distribution of velocity of a stream of liquid on a pipe section in a laminar stream that is important. For an axisymmetric current in pipes of round section of the equation of Navier–Stokes, in cylindrical coordinates (1.7) under a condition, it will be presented as $\partial P/\partial x = \partial P/\partial y = 0$

$$\frac{1}{r}\frac{d}{dr}\left(r\frac{dV}{dr}\right) = -\frac{1}{\eta}\frac{dP}{dx} \tag{1.22}$$

The solution of this equation under the set boundary conditions will be presented in the form $r = R, V = 0$

$$V = \frac{\Delta P R^2}{4\eta l}\left(1 - \frac{r^2}{R^2}\right) \tag{1.23}$$

where ΔP is a pressure difference on length l. The liquid consumption through the section of a pipe is equal

$$Q = 2\pi\rho\int_0^R rV dr = \frac{\pi\,\Delta P}{8\nu l}R^4 \tag{1.24}$$

Distribution of velocity in pipes will decide on an elliptic section as

$$V = \frac{\Delta P}{2\eta l}\frac{a^2 b^2}{a^2 + b^2}\left(1 - \frac{y^2}{a^2} - \frac{z^2}{b^2}\right) \tag{1.25}$$

where a and b are ellipse half shafts. For the amount of the proceeding liquid we will receive for a unit of time

$$Q = \frac{\pi \cdot \Delta P}{4\nu \cdot l}\frac{a^3 b^3}{a^3 + b^3} \tag{1.26}$$

An important role in all cases of hydrodynamic flow and sedimentation of particles in dispersed environment is the presence of a boundary layer in a stream. Despite the insignificant thickness of an interface, it plays the main role in the processes of transfer of heat, weight, and an impulse. Formation of a dense bed with characteristic roughness of this surface can intensify turbulence against the wall layer and lead to total disappearance of a viscous turbulent under the layer. If particles on a surface form just a layer or some congestion of the relevant structure, then at high velocities they can collapse and be carried away by a stream, and the velocity of a separation and ablation of particles from a surface of a layer is proportional to a square of dynamic velocity of a stream. The property of the postponed layer depending generally on physical properties of particles has a significant effect on transfer of weight and heat and on distribution of velocity of a stream on a section.

In most cases, the layer postponed for wall surfaces possesses a low coefficient of heat conductivity in comparison with a wall that is reflected on transfer coefficients of weight and heat with an external stream [7]. For non-Newtonian liquids, the structure of distribution of velocity differs from (1.23) and consists of a surface of a parabolic of the rotation formed from a pipe surface to a cylindrical surface of radius r_0 and part of a flat area perpendicular to the axis of the pipe in its central part. In the central part of the tube, the non-Newtonian fluid moves like a rigid rod of radius r_0, experiencing elastic deformations (Figure 1.1).

Distribution of velocity of a current of non-Newtonian liquid in a pipe is described by expression

$$V(r) = \frac{\Delta P R^2}{4\pi l}\left(1 - \frac{2l}{R}\frac{\tau_0}{\Delta P}\right), \quad r < r_0$$
$$V(r) = \frac{\Delta P R^2}{4\pi l}\left[1 - \frac{r^2}{R^2} - \frac{4l}{R}\frac{\tau_0}{\Delta P}\left(1 - \frac{r}{r_0}\right)\right], \quad r_0 \le r < R \tag{1.27}$$

where l is pipe lengths, and τ_0 is tension corresponding to an elasticity limit. In the case of formation of a dense bed of particles on an internal surface of a pipe thickness δ, believing that $R = R_0\beta,\ \beta = 1 - \delta/R_0$, expressions (1.27) for a quasi-stationary case will be presented as

$$V(r) = \frac{\Delta P R_0^2 \beta^2}{4\pi l}\left(1 - \frac{2l}{R_0\beta}\frac{\tau_0}{\Delta P}\right), \quad r < r_0$$
$$V(r) = \frac{\Delta P R_0^2 \beta^2}{4\pi l}\left[1 - \frac{r^2}{R_0^2\beta^2} - \frac{4l}{R_0\beta}\frac{\tau_0}{\Delta P}\left(1 - \frac{r}{r_0}\right)\right], \quad r_0 \le r < R_0\beta \tag{1.28}$$

It should be noted that if $\beta = 1$, then the thickness of deposits is absent; at $\beta = 0$ or $\delta = R_0$, full obstruction of a pipe and velocity of a current is observed $V(r) \to 0$, that is, capacity almost decreases to zero.

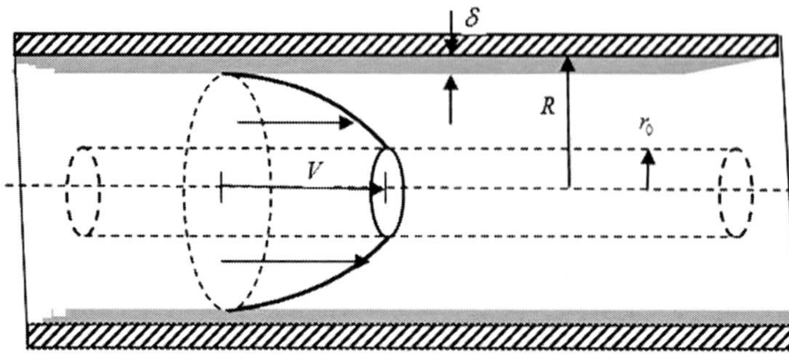

FIGURE 1.1 Distribution of viscoplastic liquid on pipe section.

Such distribution of velocities in hydrodynamics has received the name of "the structural mode of the movement." The volume consumption of viscoplastic liquid will be determined by the section of a pipe as

$$Q = \frac{\pi R^4 \Delta P}{8l\eta_c}\left(1 - \frac{4}{3}\frac{r_0}{R} + \frac{1}{3}\frac{r_0^4}{R^4}\right) \tag{1.29}$$

Thus, at full filling of section of a pipe, the liquid consumption also tends to zero $R = R_0\beta$.

$$Q = \frac{\pi R_0^3 \beta^3 \Delta P}{8l\eta_c}\left(1 - \frac{4}{3}\frac{r_0}{R_0\beta} + \frac{1}{3}\frac{r_0^4}{R_0^4\beta^4}\right) \tag{1.30}$$

For Bingams liquid of a condition of transition from a structural current to the turbulent mode, it is defined as

$$\mathrm{Re}_{cr} = \frac{1 - 4\alpha + \alpha^4}{24\alpha}\mathrm{He}\ ,\ \mathrm{He} = 16800\frac{\alpha}{(1-\alpha)^2} \tag{1.31}$$

or we can write

$$\mathrm{Re}_{cr} = 700\frac{2 + (1+\alpha)^2}{1-\alpha} \tag{1.31a}$$

$\mathrm{Re}_{cr} = V_{cr}d/\nu_c$ critical Reynolds number, $\mathrm{He} = \tau_0 d^2/\nu_c$ is Hedstrem's number, and $\alpha = \tau_0/\tau_R$, τ_R is the tangent tension on a pipe wall diameter equal d. In the presence of a dense bed on an internal surface of a pipe, it is possible to write $\mathrm{He} = \mathrm{He}_0\beta^2$, that is, with growth of thickness of deposits, attenuation of intensity of turbulence is observed. For some oils, the rheological law of flow is observed as $\tau = k\dot{\gamma}^n$, for which the Reynolds number can be represented as

$$\mathrm{Re} = \frac{8}{k}\left(\frac{n}{6n+2}\right)^n \rho_c d^n V^{2-n} = \mathrm{Re}_0\,\beta^{3n-4} \tag{1.32}$$

Thus, the transition from a structural regime to a turbulent flow occurs at certain stress values τ and thickness of sediment β. Moreover, if $n > 1$ then the intensity of turbulence decreases with increasing thickness and as $n \leq 1$ then the intensity of turbulence increases.

1.1.2 Flow of a Layer of Liquid from an Inclined Surface

Running off of various liquids on a firm inclined surface is an important task for calculation heat and a mass transfer in devices of chemical technology of various designs. The characteristic picture of a current of liquid from an inclined surface is given in Figure 1.2.

Using the hydrodynamic equation (1.6), we write the equations of flow of a thin layer along an inclined surface in the form

$$\begin{gathered}\frac{\partial V_x}{\partial t} + V_x\frac{\partial V_x}{\partial x} + V_y\frac{\partial V_x}{\partial y} = -\frac{1}{\rho}\frac{\partial P}{\partial x} + \nu\frac{\partial^2 V_x}{\partial y^2} + g\sin\theta \\ \frac{1}{\rho}\frac{\partial P}{\partial y} + g\cos\theta = 0,\quad \frac{\partial V_x}{\partial x} + \frac{\partial V_y}{\partial y} = 0\end{gathered} \tag{1.33}$$

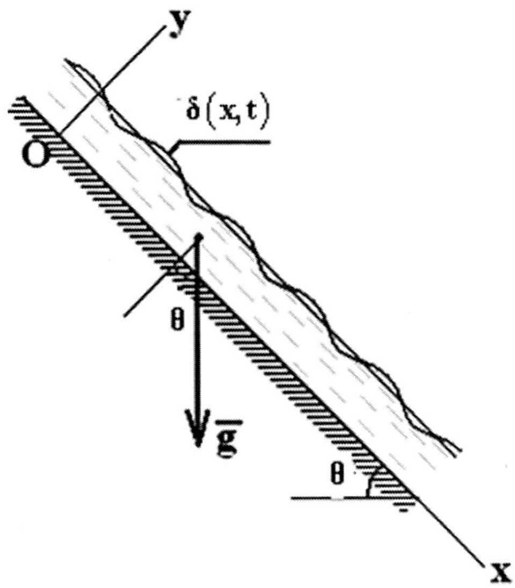

FIGURE 1.2 The scheme of running off of liquid from the inclined plane.

With boundary conditions

$$y=\delta(x,t),\quad V_y=\frac{\partial\delta}{\partial t}+V_x\frac{\partial\delta}{\partial x},\quad P+\sigma\frac{\partial^2\delta}{\partial x^2}=P_0,\ \frac{\partial V_x}{\partial y}=0 \tag{1.34}$$

$$y=0,\qquad V_x=V_y=0$$

From the last equation (1.33), we will receive

$$V_y=-\int_0^{\delta}\frac{\partial V_x}{\partial x}dy \tag{1.35}$$

From the second equation (1.33), we will receive

$$P(x,y,t)=-\rho gy\cos\theta+f(x) \tag{1.36}$$

Using a regional condition (1.28), we have

$$P(x,y,t)=-\sigma\frac{\partial^2\delta}{\partial x^2}+P_0=\rho g\delta\cos\theta+f(x) \tag{1.37}$$

From this equation we will define

$$P(x,y,t)=-\sigma\frac{\partial^2\delta}{\partial x^2}+\rho g(\delta-y)\cos\theta+P_0 \tag{1.38}$$

Thus, distribution of pressure will be determined by the thickness of a film by expression differentiating, which on x, we will receive

$$\frac{\partial P}{\partial x}=-\sigma\frac{\partial^3\delta}{\partial x^3}+\rho g\frac{\partial\delta}{\partial x}\cos\theta \tag{1.39}$$

Having substituted (1.39) in equation (1.33), taking into account expression (1.38), we will receive

$$\frac{\partial V_x}{\partial t} + V_x \frac{\partial V_x}{\partial x} - \int_0^{\delta} \frac{\partial V_x}{\partial x} dy \frac{\partial V_x}{\partial y} = \frac{\sigma}{\rho} \frac{\partial^3 \delta}{\partial x^3} + \nu \frac{\partial^2 V_x}{\partial y^2} - g \frac{\partial \delta}{\partial x} \cos\theta + g\sin\theta \tag{1.40}$$

Expression (1.40) represents the equation of a wave film current of liquid on an inclined surface. Having $q = \int_0^{\delta} V_x dy$, [$м^2/c$] accepted as a specific consumption of liquid per unit length and parabolic distribution of the being current velocity, the equation $V_x = \frac{3q}{\delta}\left(\frac{y}{\delta} - \frac{1}{2}\frac{y^2}{\delta^2}\right)$ (1.40) is representable as

$$\delta^2 \frac{\partial q}{\partial t} + 2.4 q\delta \frac{\partial q}{\partial x} - 1.2 q^2 \frac{\partial \delta}{\partial x} = \frac{\sigma \delta^3}{\rho} \frac{\partial^3 \delta}{\partial x^3} - 3\nu q - g\delta^3 \left(\frac{\partial \delta}{\partial x} \cos\theta - \sin\theta \right) \tag{1.41}$$

From equation (1.41), using conditions for a film surface (all derivatives are equal to zero), we will receive expression for the average thickness of a film

$$\delta_0 = \left(\frac{3\nu q_0}{g \sin\theta} \right)^{1/3} \tag{1.42}$$

It is necessary to mark that equations (1.40) and (1.41) do not allow the analytical decision in a general view though in case of certain assumptions it is possible to evaluate the frequency and amplitude of the refluxing wave. If $\theta = 90°$, equations (1.40) through (1.42) will be transformed to formulas for calculation of a flow of liquid on a vertical surface.

1.2 TURBULENCE AND TURBULENT FLOW

One of the main parameters characterizing the phenomena of heat, the mass, and an impulse is the regime of flow of liquid or gas. On value of criterion of Reynolds $\mathrm{Re} = \frac{U_s L}{\nu}$ in which U_s represents the average velocity of a flow of gas and liquid, where ν is the kinematic viscosity of the medium and L is the characteristic size of the flow, it is necessary to distinguish laminar and turbulent flow, characterizing the molecular and turbulent transport of the substance, respectively. The following is characteristic of a turbulent flow [5,13]:

- The presence of random chaotic velocity pulsations in the longitudinal and all other directions and the corresponding pressure pulsations at all points of the flow;
- Random mixing of pulsating volumes (moles, vortices) of gas with each other and, as a consequence, the appearance of turbulent diffusion of gas;
- The presence of a special, so-called vortex or turbulent viscosity; and
- The distribution of the average velocity over the cross section of the flow in its central part is more uniform than in laminar flow; on the contrary, its sharp drop in the near-wall region is very sharp.

The causes of the occurrence of turbulent pulsations are periodically repeating local emissions of a mass of gas and liquid from the hydrodynamics in the unstable slowed-down site of the near-wall region of a stream. In this region, the stream experiences strong braking and has a very significant gradient of the flow velocity of stream. At the mathematical description of a turbulent flow, the instantaneous velocity of a stream in the considered point for each of three of its components is longitudinal (x), transverse (y), and tangential (z) represented in the form of average velocity $\left(\bar{u}, \bar{V}, \bar{w}\right)$ and the of pulsations of velocity: $\left(u', V', w'\right)$

$$u_t = \bar{u} + u', \; V_t = \bar{V} + V', \; w_t = \bar{w} + w' \tag{1.43}$$

The order of size of velocity of turbulent pulsations is characterized by the dynamic velocity of a stream representing the average value of the longitudinal and cross-making velocity pulsations, determined by a ratio

$$U_D = \sqrt{\frac{\tau_w}{\rho}} \tag{1.44}$$

where τ_w is the tangent tension on a wall determined experimentally. All variety of profiles of average velocity of a turbulent flow in pipes and channels can be described by the universal equations, if as function and an argument to choose the dimensionless velocity $u_+ = {}^{V}\!/_{U_D}$ and dimensionless distance to a wall: $y_+ = {}^{yU_D}\!/_{\nu}$ where the value of dynamic velocity is defined generally by a formula:

$$U_D = U_S \sqrt{\frac{\xi_T}{8}} \tag{1.45}$$

Here ξ_T is a friction resistance coefficient, and U_S is average velocity. For smooth pipes, λ_T is the function of a number and defined in the range of a change of number $\mathrm{Re} = 4.10^3 - 10^5$ by a semi-empirical formula Blaziusa

$$\xi_T = {}^{0.3164}\!/_{\mathrm{Re}^{1/4}} \tag{1.46}$$

Taking into account this formula, the dynamic velocity of a stream has an appearance

$$U_D = {}^{0.2U_S}\!/_{\mathrm{Re}^{1/8}} \tag{1.47}$$

1.2.1 Turbulent Flow of Liquid in a Smooth Round Pipe

Usually at the developed turbulent flow of liquid in a pipe for restoration of a profile of velocity in an interface, the theory of a way of hashing of Prandtl is used. Classical solutions for round smooth pipes, obtained by Prandtl and Taylor and supplemented by Karman, are included by three equations corresponding to three areas of turbulent flow: a viscous under layer $y_+ \le 5$, a buffer layer $5 < y_+ < 30$, and core of stream. In the first of them the linear dependence takes place, and in the second and third, logarithmic dependences take place.

The general equation describing all curves has to meet entry conditions:

$$\begin{aligned} u_+ &= 5.0 \ln y_+ - 3.05 \\ u_+ &= 21.5 \ln y_+ + 5.5 \end{aligned} \tag{1.48}$$

A single equation describing all curves of determining velocity in the entire boundary layer has to satisfy with the initial conditions:

$$y_+ \le 5, \; U^+ = y_+ \tag{1.49}$$

and to final conditions

$$30 \le y_+ \le 1000,\ U^+ = 2.5\ln y_+ + 5.5 \quad (1.50)$$

Considering these conditions, the following expression for the description of a profile of velocity in a turbulent interface in limits is offered $0 \le y^+ \le 1000$

$$U^+ = \psi(y_+)(2.5\ln y_+ + 5.5) + \frac{y_+}{1 + 0.4y_+ \ln y_+}$$
$$\psi(y_+) = \left(1 - \frac{1}{1 + 0.2\ln y_+ + 0.07(\ln y_+)^4}\right) \quad (1.51)$$

The offered model describes, with a single equation, all area of stream for the boundary layer and core of stream $0 \le y_+ \le 1000$ (Figure 1.3).

The analysis of experimental data of cross-pulsation velocity of a turbulent flow in limits $4\times10^4 \le \text{Re} \le 5\times10^5$ has allowed the following conclusions to be drawn: (a) the maximum value of cross velocity of a stream is reached at $r_+ \approx 0.15$ and is equal $V' \approx 0.85U_*$; (b) at value $r_+ \to 1$, $\lim_{r_+\to 1}(\partial V'/\partial r_+) \to 0$ of cross velocity is established at the level $V' \approx 0.72U_*$; and (c) at $r_+ = 0$, $V' = 0$ (conditions of sticking of liquid to a wall).

Using these boundary conditions, the following empirical model for cross pulsation velocity in a pipe is offered (Figure 1.4)

$$\frac{V'}{U_*} = 0.72\left[1 - \exp(-6.25r_+) + 4.46r_+^{1/2}\exp(-7.49r_+)\right] \quad (1.52)$$

A turbulent flow in pipes at the solution of practical tasks often uses a semi-empirical formula of a look

$$V/V_0 = (y/R)^{1/n} \quad (1.53)$$

where V_0 is a current velocity in the center of a pipe, and for n the following values undertake: $n = 6$, at $\text{Re} = 4\times10^3$; $n = 7$, at $\text{Re} = 1.1\times10^5$; $n = 8.8$, at $\text{Re} = 1.1\times10^6$; $n = 10$, at $\text{Re} = 3.2\times10^6$. A lack of this formula is the big error of calculation at small distances from a wall.

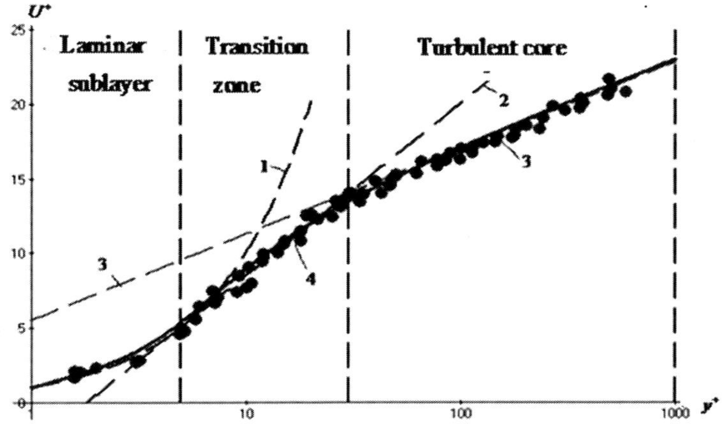

FIGURE 1.3 A universal profile of velocity for a flow of liquid in a smooth pipe: (1) $U^+ = y_+$; (2) $U^+ = 5\ln y_+ - 3.05$; (3) $U^+ = 2.5\ln y_+ + 5.5$; and (4) general distribution of velocity (1.51).

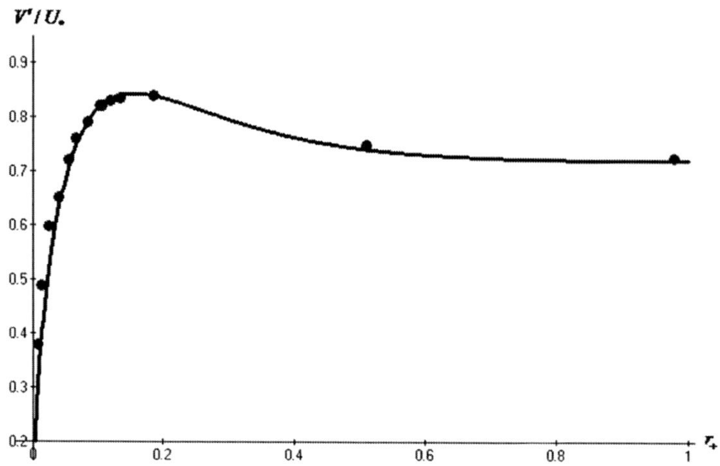

FIGURE 1.4 Distribution of transverse pulsation velocity of turbulent flow.

The emergence of the turbulent mode is connected with the unstable character of a laminar current arising at the numbers exceeding some critical value Re_{kp}. The small indignations leading to violation of stability of a laminar current can always be imposed on the main systematic movement in practical cases. At the indignations numbers $\mathrm{Re} < \mathrm{Re}_{kp}$, arising in the liquid quickly decay and at $\mathrm{Re} > \mathrm{Re}_{kp}$ they create a chaotic structure of a flow of liquid. At last, $\mathrm{Re} >> \mathrm{Re}_{kp}$, at such current, carries the name of the developed turbulence. Turbulent pulsations should be characterized by not only the size of velocity l but also those distances throughout which the velocity of pulsations undergoes noticeable change. These distances carry the name of scale of turbulent pulsations. So, for example, at a turbulent flow in a pipe, the largest scale of turbulent pulsations will coincide with the diameter of a pipe and velocity—with change of average velocity on this extent λ, that is, with the maximum value of velocity in the center of a pipe V_λ. In such large-scale pulsations, the main part of energy of a turbulent stream is concluded. Along with these large-scale pulsations in a turbulent stream, the smaller pulsations having smaller velocities are presented. Though the number of such small pulsations is very big, they to contain only a small part of kinetic energy of the current stream. Small pulsations play very important roles in the phenomena of transfer of a substance in a turbulent stream. At some value $\lambda = \lambda_0$, Reynolds number, the relevant movement $\mathrm{Re}_\lambda = \frac{V_\lambda \lambda_0}{\nu}$ is close to the unit. It means that in the area, viscous forces begin to influence the current of liquid significantly.

Large-scale currents are unstable, continuously breaking up to small-scale currents. At the same time, the energy of small-scale pulsations is continuously scooped from large-scale, that is, continuous transition of energy from large-scale pulsations to smaller pulsations is observed. At the flow liquid transition of energy has a stationary character. Pulsations of small scale turbulence receives the same energy from larger vortices and similar energy is transmitted to smaller pulsations. The turbulent flow arises only at rather large numbers and is characterized by big energy of dissipation. Loss of energy of large-scale movements is defined on creation of the smaller movement by the energy dissipation size in a unit of time.

$$\varepsilon_R = -\frac{dE}{dt} = \nu_T \left(\frac{\Delta U}{l} \right)^2 \tag{1.54}$$

where E is a dissipation of energy in unit of mass in a turbulent flow, ν_T is turbulent viscosity, ΔU is change of average rate throughout the scale of pulsations l, and ε_R is specific dissipation of energy.

The key parameters characterizing a sinuous flow in addition to number, the dynamic velocity and energy of dissipation, are the intensity or a level of turbulence, scale of turbulence, and frequency of turbulent pulsations, their distribution, and characteristic values. The scale of turbulence characterizes the average size of the pulsation volumes (moles, curls) having identical velocity and behaving as a single whole. The turbulence scale is often determined in a look

$$l = \int_0^\infty R(y)dy \tag{1.55}$$

where $R(y)$ is a coefficient of space correlation or the Euler coefficient correlation, which for two points are also defined in a look [12]

$$R(y) = \overline{u}_1'\, \overline{u}_1' \Big/ \left(\overline{u}_1'^2\right)^{1/2} \left(\overline{u}_2'^2\right)^{1/2} \tag{1.56}$$

In the statistical theory, instead of the Euler correlation coefficient, the Lagrangian temporal correlation coefficient is used

$$R_L(\vartheta) = u'(t)u'(t+\vartheta)/u'^2(t) \tag{1.57}$$

The coefficient of the Lagrangian correlation is responded by the Lagrangian time scale

$$\Theta_L = \int_0^\infty R_L(\vartheta)d\vartheta \tag{1.58}$$

This parameter characterizes an interval of time after which the pulsation movement of particles of the environment becomes independent of initial movement. Frequency of pulsations characterizes the number of changes of amplitude values of the pulsation velocity for a unit of time. Its numerical value depends on the scale of pulsations, and as are present at a turbulent flow turbulence of all scales, there is not one but a whole frequency spectrum of pulsations.

The intensity of turbulence is a measure of value of the pulsation component velocity of a current and is defined for each component by the attitude of the average square rate of pulsations toward the average rate of a current

$$I_x = \left(\overline{u}'^2\right)^{1/2} \Big/ \overline{u},\ I_y = \left(\overline{V}'^2\right)^{1/2} \Big/ \overline{u}, I_z = \left(\overline{w}'^2\right)^{1/2} \Big/ \overline{u} \tag{1.59}$$

$$I = \left[\frac{1}{3}\left(\overline{u}'^2 + \overline{V}'^2 + \overline{w}'^2\right)\right]^{1/2} \Big/ \overline{u} \tag{1.60}$$

where u', V', and w' are the pulsation component velocities of a turbulent flow. If $u'^2 = V'^2 = w'^2$, then such turbulence is called isotropic.

1.2.2 Theories of Turbulence

Now there are various approaches to the use of a turbulent flow at the quantitative assessment of transfers of a substance [14]. Among these approaches it is possible to note theories of Boussinesg, Prandtl, Taylor, Reichardt, and the theory of isotropic turbulence of Kolmogorov.

Boussinesg's theory. In this theory, it is supposed that turbulent streams of weight, heat, and an impulse are structurally similar to molecular streams, that is, are proportional to gradients of concentration, temperatures, and velocities if $\Pr = {}^{\nu}\!/_{a_T} = 1$ and $\mathrm{Sc} = {}^{\nu}\!/_{D} = 1$:

$$\tau_T = -\nu_T \frac{\partial V}{\partial y}, \ q = -c_P a_T \frac{\partial T}{\partial y}, \ j = -D_T \frac{\partial C}{\partial y} \tag{1.61}$$

Here the index of "T" belongs to a turbulent flow, and ν_T, a_T, and D_T are coefficients of turbulent viscosity, thermal diffusivity, and diffusion, respectively. Generally, the equations of transfer of heat, weight, and an impulse are described by the following equations:

$$\tau = -(\nu + \nu_T)\frac{\partial V}{\partial y}, \quad q = -c_P(a + a_T)\frac{\partial T}{\partial y}, \quad j = -(D + D_t)\frac{\partial C}{\partial y} \tag{1.62}$$

Thus, turbulent transfer is an analog of molecular transfer, though analogy is very approximate about what will be stated slightly below.

Prandtl's theory. This theory does attempt to connect sizes ν_T, χ_T, and D_T with characteristics of turbulence. According to the kinetic theory of gases, the coefficient of kinematic viscosity is proportional to the work of average square velocity of the thermal movement at the average length of a free run. The size, which is the parameter l_ν of length and proportional to the mean square value of distance on which turbulent moths of liquid keep the identity, is entered. Usually l_V calls the length of a way of shift at transfer of an impulse. At the same time, expressions of coefficients of turbulent viscosity, thermal diffusivity, and diffusion have an appearance

$$\nu_T = l_V^2\left|\frac{dV_x}{dy}\right|, \ a_T = l_m^2\left|\frac{dV_x}{dy}\right|, \ D_T = l_m^2\left|\frac{dV_x}{dy}\right| \tag{1.63}$$

where l_m is a length parameter characterizing distance at which the turbulent stream of scalar substance keeps the identity. It should be noted l_V and l_m don't make a certain physical sense and are used for smoothing and comparison of theoretical and experimental results.

Taylor's theory. This theory is based on the assumption that in a turbulent stream the turbulence has properties of transportable substance. At the same time, in full accordance with the theory of transfer of an impulse, the stream of a turbulence is represented proportional to the turbulence of a gradient with use as coefficients of proportionality of so-called coefficient of a turbulence

$$\zeta_\omega = l_\omega^2\left|\frac{dV_x}{dy}\right| \tag{1.64}$$

where l_ω is a parameter of turbulence length.

In the considered theories of turbulence, a certain hypothesis of the pulsation streams of an impulse, scalar substance, or a turbulence, on the basis of which by means of the average equations of transfer the profile of average velocity and scalar substance is removed, is accepted.

Reichard's theory. This theory has been developed for turbulent-free jets and its essence comes down to the distribution of full longitudinal velocity in the cross section of a zone of mixture follows Gauss's curve. According to this theory, turbulent transfer is statistical, and in accuracy is similar to the process of molecular transfer. It should be noted that from all described semi-empirical theories of turbulence, it is impossible to gain an impression about interrelation of average and pulsation characteristics of transfer. In this regard, it is expedient to address statistical-phenomenological methods of the description of turbulence. A feature of the statistical-phenomenological description of turbulent transfer is that turbulent fields of the considered substances are treated as stochastic functions of spatial coordinates and time. The description of processes of turbulent transfer at the same time is made through statistical characteristics of the field—distribution of probabilities for values of this field. At such approach to a turbulence problem, the task of turbulent transfer is set as follows: to express characteristics of transfer of any substance completely through statistical functions of the field of velocities and also entry and boundary conditions with attraction of phenomenological hypotheses for some characteristics of structure of turbulence. The foundation of a statistical-phenomenological approach to a problem of non-uniform turbulence is laid in Kolmogorov's work [6] in which the turbulence is characterized by two parameters: intensity and scale of turbulence. Such turbulence in literature has received the name of isotropic turbulence. Homogeneous isotropic turbulence is a peculiar case of turbulent flow, at which in all volume of liquid its average velocity remains constant.

1.3 ISOTROPIC TURBULENCE FLOW

The whirl of liquid in the case of rather great values of a Reynolds number is characteristic of the extremely irregular, chaotic changes of velocity in each point of a flow over time. In the case of developed turbulence, velocity pulsates about some value all the time. At very large Reynolds numbers at a turbulent flow, there are pulsations with scales from the biggest to very small values. The main role in a turbulent flow is played by large-scale pulsations, which scale the magnitude order of characteristics of lengths, determining the area size in which there is a whirl. Small-scale pulsations appropriate to big frequencies participate in a turbulent flow with considerably smaller amplitudes. The small-scale turbulence far from bodies has properties of homogeneity and isotropy. It means that sections where sizes are small in comparison l with the property of whirl are identical in all directions or they do not depend on the direction of velocity of average movement. In this case, substantial results may be obtained that determine local properties of turbulence directly from similarity considerations.

We will consider, first of all, the small-scale movement $(\lambda << l)$ in a volume of liquid for a case of a no viscous current $\lambda >> \lambda_0$. The size of velocity of turbulent pulsations of scale λ can depend from λ, ρ, and ξ_K. The only combination from these sizes' dimensional velocities is [6,7,12]

$$V_\lambda = (\varepsilon_R \lambda)^{1/3} \tag{1.65}$$

Therefore, the change of velocity throughout a small distance is in proportion to a cubic root from this distance. Expressing ξ_R through ΔUk in a look $\varepsilon_R = \Delta U^3/l$, we will receive

$$V_\lambda = \Delta U \left(\frac{\lambda}{l} \right)^{1/3} \tag{1.66}$$

Therefore, the value of the pulsation velocity V_λ is less than the velocity of the main stream at a size and is $\left(\lambda/l\right)^{1/3}$. A decrease in velocity and scale corresponds to a decrease in the number Re_λ

$$\mathrm{Re}_\lambda = \frac{V_\lambda \lambda}{\nu} = \frac{\Delta U \lambda^{4/3}}{\nu \cdot l^{1/3}} = \mathrm{Re}\left(\frac{\lambda}{l} \right)^{4/3} \tag{1.67}$$

where $\mathrm{Re} = \frac{\Delta U l}{\nu}$. Obviously at some internal scale λ_0 number $\mathrm{Re}_\lambda = 1$. In this connection, using the expression for specific energy of dissipation, we will receive value of the Kolmogorov scale of turbulence in a look

$$\lambda_0 \approx \frac{l}{\mathrm{Re}^{3/4}} = \left(\frac{\nu^3}{\varepsilon_R} \right)^{1/4} \tag{1.68}$$

Since this value of scale is $y < \delta_0$, carrying the name of a viscous layer, the current of liquid has viscous character, and turbulent pulsations don't disappear suddenly and fade gradually because of viscosity. We will determine thickness of a viscous layer, proceeding from a condition $\mathrm{Re}_\lambda = \frac{V_0 \delta_0}{\nu} = 1$, $V_0 \approx V_\lambda$ or

$$\delta_0 \approx \alpha \frac{\nu}{V_0} \tag{1.69}$$

The intensity of turbulence of an isotropic turbulent stream will be defined in a look

$$I = \frac{1}{4} \mathrm{Re}^{-1/4} \tag{1.70}$$

When calculating transfer of weight and heat in a turbulent stream qualitatively, important parameters are coefficients of turbulent viscosity ν_T, turbulent diffusion D_T, and turbulent heat conductivity χ_T. At transfer of weight, an important condition is the linking of the coefficient of turbulent diffusion of liquid and the disperse environment with the sizes characterizing a turbulent stream.

Turbulent boundary layer. Unlike a laminar interface, the turbulent interface has various structure. At a turbulent flow, all streams of liquid flowing around a solid body are divided into four areas: (1) *a viscous underlayer* where change of velocity is defined by value of dynamic viscosity of liquid η; (2) *transitional area* where viscous and turbulent tension becomes comparable, and at the same time dynamic and turbulent viscosity approximately become equal; (3) *fully turbulent* area where the wall still exerts an impact on the character of a current of liquid; however, turbulent pulsations reach such considerable sizes that the influence of cross shift tension significantly increases and in this area the size of turbulent viscosity becomes higher than dynamic. Usually in the theory of hydraulics, this area carries the name *logarithmic* since change of average velocity of a stream submits to the logarithmic law; and (4) *the turbulent kernel*—this area is observed as the developed turbulent stream and on turbulence scale, mainly, only the diameter of the channel exerts impact. In Figure 1.5, the structure of a turbulent interface is presented.

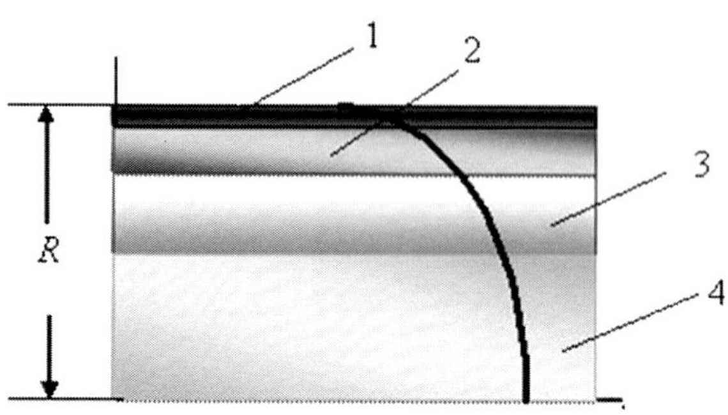

FIGURE 1.5 Structure of a turbulent interface: (1) area of a viscous sublayer, (2) transitional layer, (3) fully turbulent layer, and (4) turbulent kernel of a stream.

It is obvious that the presented model of the structure of boundary regions at the developed turbulent mode of a current is idealized, and actually sharp borders between areas don't exist. In a viscous underlayer, the turbulence gradually fades, passing on a wall into the viscous current. Such attenuation qualitatively is determined by gradual transition of turbulent viscosity in molecular, by the law [6,12]

$$\nu_T \approx \alpha_0 \nu_c \left(\frac{y}{\delta_0} \right)^3 \tag{1.71}$$

where α_0 is a proportionality coefficient. In literature, there are various assumptions of attenuation of turbulence in a viscous layer, which will be subsequently told about. The thickness of a laminar viscous underlayer is defined by the following expression

$$\delta_{Pl} / \delta \approx 191.\mathrm{Re}^{-0.7} \tag{1.72}$$

In the laminar sublayer, the transfer of mass and heat occurs in a molecular way. In a turbulent kernel, the viscosity and heat conductivity are more than the molecular viscosity and heat conductivity, and the transfer of heat and weight is carried out in generally turbulent way. In a transitional layer, the stream gradually changes from laminar to turbulent; therefore, the contribution of molecular and turbulent transfers is counterbalanced.

In pipes at critical values of number $\mathrm{Re}_{cr} > 2000$ $\left(\text{or}\,\mathrm{Re}_{cr} = 2.10^5 - 3.10^6\right)$, there comes the area of the developed turbulence. The thickness of a turbulent interface grows quicker than laminar (it is proportional x, but not for a $\sqrt{x}$ laminar current); therefore, it rather quickly fills all sections of a pipe, forming a logarithmic profile of velocity. An exception is a very thin viscous underlayer directly adjoining a wall where the current is strongly slowed down by forces of viscosity, and the linear profile of average velocity peculiar to a laminar current remains. The thickness of a viscous underlayer in pipes d_{mp} decides on the diameter as

$$\delta_{pl} = {25 d_{mp}} / {\mathrm{Re}^{7/8}} \tag{1.73}$$

In the range of change of the number $\mathrm{Re} = 4.10^3 - 10^4$ the thickness of the viscous sublayer makes up from 1.8% to 0.1% of the pipe diameter. In other parts of a layer where viscosity forces gradually give way to inertia forces, velocity increases up to stream velocity in the central part. The distance at which transition to a turbulent flow comes to the end, called by length of the initial site, depends not only on number but also on a form of an entrance and relative roughness of walls. The essential value has the existence of turbulence in the arriving stream; the interface in this case is *turbulization* rather that reduces the length of the initial site.

1.4 LIFTING AND TURBULENT MIGRATION OF PARTICLES

The existence of gradients average and the pulsation-making velocities of the longitudinal movement has the result of the emergence of a special form of the cross movement of particles. A pressure drop from the outside where the sum of tangential components of velocity of flow and rotation of a body reaching a maximum (Figure 1.6) is the cause of cross force. Cross force is always directed toward this maximum. This phenomenon of lifting the migration of particles in literature carries the name of effect of Magnus [7,12].

Experimentally, the phenomenon of lifting migration of the weighed particles is studied only for a case of the ascending laminar stream of liquid where their subsidence by gravity is the reason of flow of particles and the rotation reason—difference in velocities of flow of particles on the right and at the left, caused by existence of a gradient of velocity of a current of liquid.

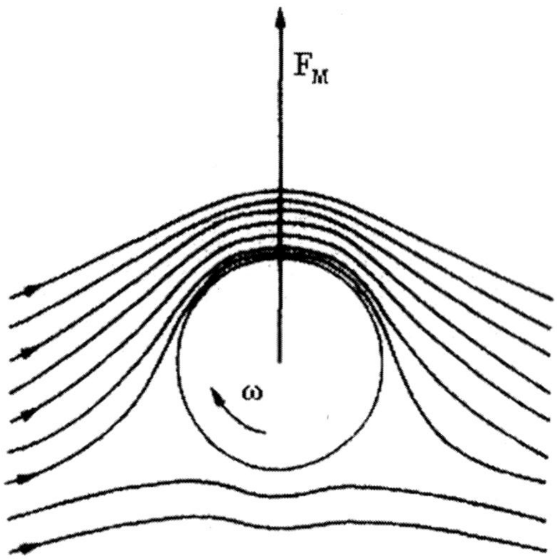

FIGURE 1.6 The schematic image of rotation of a particle in a stream.

Velocity of lifting migration of Stokes particles is defined in a look [7,12]

$$V_L = 0.17a\Delta V\left(dV/dy\right)^{1/2}\nu^{-1/2} \tag{1.74}$$

where $\Delta V = |V - V_d|$ are the velocities of flow of particles, and V_d is the velocity of particles. This equation was specified to a look subsequently

$$V_L = 0.03a^{1/2}\left(V\frac{dV}{dy}\right)^{1/2} \tag{1.75}$$

Passing to dimensionless parameters, we have

$$V_L = R_m U_+\left(\frac{dU_+}{dy_+}\right)^{1/2} \tag{1.76}$$

where $R_m = 0.03\left(\frac{aU_D^3}{\nu}\right)^{1/2}$, $y_+ = yU_D/\nu$, $U_+ = V/U_D$

Believing distribution of velocity of the bearing phase in a look [12]

$$U_+ = Ay_+^{3/2}e^{-0.05y_+} \tag{1.77}$$

$$y_+ \geq 21.5,\quad U_+ = y_+\left[\left(0.53/R_+\right)y_+^2 + 0.85y_+ + 14\right]^{-1} \tag{1.77a}$$

and having defined the corresponding derivatives, we will receive dependence of velocity of lifting migration on the coordinate

$$V_L = A^{3/2}y_+^{7/4}e^{-0.075y_+}\left(1.5 - 0.05y_+\right)^{1/2}R_m \tag{1.78}$$

$$V_L = \frac{y_+\left[14-\left(0.53\big/R_+\right)y_+^{1/2}\right]R_m}{\left[\left(\frac{0.53}{R_+}\right)y_+^2+0.85y_+ +14\right]} \tag{1.79}$$

where $A \approx 0.02$. Proceeding from a condition $\partial V_L/\partial y_+ = 0$, we will define the maximum value of velocity of lifting migration $V_{L\max}/R_m \approx 0.67$ for $y_+ = 18.68$.

Thus, the velocity of lifting migration of particles, increasing in a viscous layer, reaches the greatest value in a transitional zone, and in a zone of the developed turbulence decreases to zero.

Along with it, it is important to note one more type of transfer of particles in a turbulent stream—turbulent migration where velocity will be defined in a look [12]

$$V_T = -\frac{1}{2}\mu_p^2\tau_P V'\frac{dV'}{dy} \tag{1.80}$$

where V' is the pulsation velocities of a stream, τ_P is relaxation time, and μ_p is the extent of the degree of entrainment of particles by the pulsating medium. This equation will be presented in the dimensionless form as

$$V_T = \frac{1}{2}\frac{\mu^2\tau_P}{\nu}U_D^3U_+^2\frac{dU_+}{dy_+} \tag{1.81}$$

peak value, which is reached at $y_+ = 12.7$. Comparing maximum values of turbulent and lifting migrations for a Stokes particle we will receive

$$\frac{V_{T\max}}{V_{L\max}} \approx 0.75\mu^2\frac{\Delta\rho}{\rho_c}\left(\frac{a}{\nu_c}\right)^{3/2}U_D^{3/2} = 0.75\mu^2\frac{\Delta\rho}{\rho_c}\mathrm{Re}_+^{3/2},\ \mathrm{Re}_+ = \frac{aU_D}{\nu_c} \tag{1.82}$$

Thus, the relation of the maximum values of turbulent and lifting migrations is defined by properties of a stream, the size of particles and dynamic velocity.

Example 1.1

For assessment of this ratio, we will consider a current of particles $a = 10-50\,\mu\text{m}$ with a density $\rho_d = 1500\ \text{kg/m}^3$ in an air stream ($\rho_d = 1.29\ \text{kg/m}^3$, $\nu_c = 0.15\ \text{m}^2/\text{s}$) and in oil ($\rho_H = 850\text{kg/m}^3$, $\nu_H = 0.95\times10^{-4}\text{m}^2/\text{s}$). We will define value Re_+ at a dynamic velocity of a stream $U_D = 1\text{m/s}$ for each of the offered cases: for air $a = 10\ \mu\text{m}$, $\mathrm{Re}_+ = 0.67$, $\frac{V_{T\max}}{V_{L\max}} = 300$ and $a = 50\ \mu$r $\mathrm{Re}_+ = 3.35$, $\frac{V_{T\max}}{V_{L\max}} = 3500$; for oil $a = 10\ \mu\text{m}$, $\mathrm{Re}_+ = 0.10$, $\frac{V_{T\max}}{V_{L\max}} = 0.02$ and $a = 50\ \mu\text{m}$, $\mathrm{Re}_+ = 0.52$, $\frac{V_{T\max}}{V_{L\max}} = 0.20$.

It follows from this analysis that in gases $V_{T\max} >> V_{L\max}$, and in liquids, on the contrary, the velocity of the lifting migration is much more than a velocity of turbulent migration that is explained by great value of time of a relaxation of particles.

As appears from the previously stated ratio, the commensurability of velocities of turbulent and lifting migrations is carried out in the case

$$\mathrm{Re}_+ \approx 1.2\left(\frac{\rho_c}{\mu^2\Delta\rho}\right)^{2/3} \tag{1.83}$$

Existence of a large number of factors causes a difficult trajectory of migration of particles in a turbulent stream, though at the solution of certain problems of a mass transfer, as a result of quantitative assessment of a contribution of each component of migration, reveal the main effects defining a condition of a particle.

1.5 CLASSIFICATION AND FLOW OF DISPERSE SYSTEMS

Disperse systems are heterogeneous environments consisting at least of two and more phases, one of which is a continuous phase called the dispersive environment; the other phase is shattered and distributed in the first and is called a disperse phase. The most general classification of disperse systems is based on the difference in the aggregate state of a disperse phase and the dispersive environment. Three aggregate states (firm, liquid, and gaseous) allow us to distinguish eight types of disperse systems (Table 1.1).

Aerosols are disperse systems with a gaseous environment. On the aggregate state and the sizes of particles of a disperse phase, aerosols are divided into fogs (systems with a liquid disperse phase [the size of particles of 0.1–10 microns]), dust (systems with firm particles no more than 10 microns in size), and smokes with sizes of firm particles in limits of 0.001–10 microns. Fogs have particles of liquid of the correct spherical shape whereas dust and smokes contain firm particles of the most various form.

Suspensions are systems with a firm disperse phase and a liquid disperse environment. Depending on dispersion of a firm phase of suspension, it is subdivided into rough (diameter of particles more than 100 μm) and thin (from 0.5 μm to 100 μm). The chemical industry deals with suspensions in fluidized layers, mechanical division and purification of gases and liquids (sedimentation, filtration, etc.), dissolution of firm particles, etc.

Emulsions are the disperse systems formed by two immiscible liquids. Emulsions are divided into two types: straight lines, "oil in water," and the return "water in oil." The majority of emulsions belong to microheterogeneous systems (the size of particles more than 100 nanometers). In emulsions, if the size of the particles $a \geq a_{min}$, then a coalescence of droplets is observed and if $a > a_{max}$, then their fragmentation is observed. Depending on concentration of a disperse phase, an emulsion is subdivided into diluted (concentration of a disperse phase up to 0.1%), koncentriro-bathing, and high-concentrated. The maximum concentration of a disperse phase in the concentrated emulsions doesn't exceed 74%.

Gas emulsions are disperse systems consisting of a gaseous disperse phase and the liquid dispersive environment. Such systems often meet in processes of rectification and absorption.

In industrial practice of chemical technology for the majority of processes, the existence of a discrete phase is needed, such as fluidized layers, processes of drying and adsorption (gas–solid particles), gas–liquid reactors, processes of absorption and liquid

TABLE 1.1
Classification of Disperse Systems

Environment	Gas	Liquid	Solid
Gas	Absence systems	Aerosols Fog Clouds	Aerosols Dusts Smoke
Liquid	Gas emulsions and foam	Emulsion Petroleum and oil emulsions	Suspension Pastes
Solid	Porous bodies Absorbents Catalysts	Capillarity Soils	Composites Minerals Alloys

extraction (liquid–liquid), a flow in porous environments, and filtration (liquid–solid phase). Existence of firm, liquid, and gaseous particles in a stream, their form and the sizes, carry similar currents to the multiphase systems representing mixes of the bearing phase (gas or liquid) with the discrete educations (solid particles, drops, and bubbles) distributed in them. In the generalized sense, it is possible to call such systems suspensions, covering these all environments including suspensions in the classical sense—mixes of liquids with firm particles.

1.5.1 Hydrodynamics of the Flow of a Set of Particles in a Flow

The flow state of mechanics of disperse systems is characterized by intensive development of the theory of their current and also pilot studies in various directions of their research. If at the initial stage of the development of the theory there were many statistical descriptions, then attempts at reaching conclusions for the corresponding equations in the latest stages are made only at the macro level by averaging the results of the known movement equations. In the known works there is discussion of a current of two-phase system in approach of the interpenetrating continuums characterized by some set of final number of macrocontinual fields, and the movement of the multiphase environment is considered as the movement in the porous environment. The theoretical solution of the problem of a current of multiphase systems is connected with a certain simplification of a real picture of a current as the research of this task requires appropriate association of hydromechanics, statistical mechanics, mechanics of continuous environments, thermodynamics of irreversible processes, etc. At the same time, the systems of the differential equations for the description of the general case of the movement have to consider basic discontinuity of the environment and processes of transfer of heat and weight happening in it. The analysis of a set studies devoted to the flow of multiphase systems has shown that the most promising is such a schematization, at which the discrete environment is considered as some fictitious inseparable environment to which methods of continuum mechanics can be applied.

Among the researches devoted to a flow of multiphase systems especially, it should be noted works of Soo [8] in which important theoretical aspects of hydrodynamics of the multiphase systems, which are followed by phase and chemical transformations, are analyzed. Along with it, there should be noted works [9], the suspensions devoted to mechanics [9,13–17] devoted to a turbulent flow of particles and other works; devoted various theoretical and experimental to researches of a current of the disperse medium; *Nigmatullin* [11] devoted to mechanics of heterogeneous systems with phase to transformations; *Kutateladze* [18] devoted to the hydrodynamic resistance and heat transfer; *Mednikov* [12] devoted to mechanics of aerosols; *Bennet and Myers* [18] devoted to hydrodynamics and transfer of mass and heat; and many other works devoted to various aspects of a current of disperse systems. The general equations of the movement of multiphase systems, the equation of continuity and energy, with a certain simplification of the scheme of a current, for k – that phase *Soo* [8] is offered:

$$\frac{\partial \rho^{(k)}}{\partial t}+\frac{\partial}{\partial x_j}\left(\rho^{(k)}U_j^{(k)}\right)=W^{(k)} \quad \rho^{(k)}\frac{dU_i^{(k)}}{dt}=\frac{\partial}{\partial x_j}\left[-P^{(k)}\delta_{ji}+\eta_m^{(k)}\Delta_{ji}^{(k)}+\eta_{m2}^{(k)}\Theta^{(k)}\delta\, ji\right]+\rho^{(k)}F_i^{(k)}-$$
$$-\left(U_i^{(k)}-U_{mi}\right)W^{(k)}+\rho^{(k)}\sum_{(p)}f^{(kp)}\left(U_i^{(p)}-U_i^{(k)}\right) \tag{1.84}$$

$$\rho^{(k)}\frac{dE^{(k)}}{dt}=\frac{\partial}{\partial x_j}U_j^{(k)}\left[-P^{(k)}\delta_{ji}+\eta_m^{(k)}\Delta_{ji}^{(k)}+\eta_{m2}^{(k)}\delta_{ji}^{(k)}\right]-W^{(k)}E^{(k)}+$$
$$+\frac{\partial}{\partial x_j}\left[\frac{\lambda_m^{(k)}\partial T_{(k)}}{\partial x_j}\right]+F_E^{(k)}+c^{(k)}\rho^{(k)}\sum_{(p)}G^{(kp)}\left(T^{(p)}-T^{(k)}\right) \tag{1.85}$$

where $F_i^{(k)}$ is a i thawing a component of external force, acting per unit mass liquids and defined in a look

$$F_i^{(k)}=\frac{1}{2}\frac{\rho}{\rho^{(k)}}\frac{d}{dt}\left(U_i-U_i^{(k)}\right)+\frac{9}{2\sqrt{\pi}a_r}\frac{\rho}{\rho^{(k)}}\sqrt{\frac{\eta}{\rho}}\int_0^t\left[\frac{d}{d\tau}\left(U_i-U_i^{(k)}\right)\right]\left(t-\tau\right)^{-1/2}d\tau \tag{1.86}$$

$P^{(k)}$ is partial static pressure of a component, viscosity of a liquid phase in mix (k), η_m is viscosity of a liquid phase in mix, $\eta_{m2}=\zeta_m-\frac{2}{3}\eta_m$, ζ_m is the second viscosity, $\Delta_{ji}^{(k)}=\frac{\partial U_i}{\partial x_j}+\frac{\partial U_j}{\partial x_i}$ is shift deformation velocity, $\Theta^{(k)}=\frac{\partial U_k}{\partial x_k}$ is lengthening deformation velocity, $W^{(k)}$ is velocity of formation of a component in unit of (k) volume, x_j-j- an axis of system of coordinates, $f^{(kp)}$ a constant of time of process of exchange of the number of the movement between a set of particles (k) and (p), $F_E^{(k)}=\rho^{(k)}f^{(k)}(U-U^{(k)})^2+\rho^{(k)}\sum_{(p)}f^{(kp)}(U^{(p)}-U^{(k)})^2$ is the member considering heating of particles owing to viscous dissipation, E is the general energy of system, $G^{(kp)}$ is a power exchange time constant, both heat conductivity λ, and c are the thermal capacity of particles, $\rho^{(k)}$ is density of the component (k) of mix occupying the volume of the mix consisting n of components $(k=1,2,...n)$, and $U_j^{(k)}$ is the being velocities of that k is particle in the direction j.

The equation of energy of a discrete phase includes the ratios characterizing exchange of energy between firm particles and the bearing environment, between particles of various grade, change of thermal capacity, thermal diffusion, and convective transfer of energy. Despite essential assumptions and simplifications of equation (1.84), they have a very formal and difficult character. For simplification of these equations, their use for two-phase systems where it is possible to neglect many phenomena is possible. In a disperse stream, moving particles can receive fast rotation, especially considerable for particles of irregular shape that intensifies external exchange due to continuous change of hydrodynamic, thermal, and diffusive interfaces. Processes of collision, coagulation, and crushing at high concentrations of a disperse phase result in additional not stationary and restructuring of the specified interfaces on an external surface of particles. Therefore, real disperse system coefficients of interphase exchange can be a bit different. In a laminar stream, at great values of concentration of particles, their rotations merge and crushing can be a source of additional turbulence, though in a turbulent stream an increase in concentration of particles in a stream leads to attenuation of turbulence and changes of properties of a stream. It is important to note that above a certain concentration of particles the disperse system refers to viscoplastic fluids and obeys the laws of flow of non-Newtonian fluids. First of all, it is defined by value of the second viscosity. Thus, in a disperse stream there is a difficult multiple-factor interaction and mutual influence of both phases that can significantly change intensity of interphase exchange.

It should be noted that the behavior of a set of discrete particles in a turbulent stream of liquid substantially depends on the concentration of these particles and on their size in comparison with turbulence scales. At a small concentration of particles in a stream, it is possible to neglect direct interaction between particles in the absence of their collision and to neglect influence of them on fields of concentration, temperatures, and velocities.

1.5.2 Flow of a Single Particle in a Turbulent Flow

The movement of particles of one grade and size in a turbulent stream represents the simplest case of the movement of turbulent mix, formed by the bearing and disperse phase. If the size of particles is big in comparison with the turbulence scale, then the particle lags behind the movement of the environment. As the velocity of the movement of such particles and liquid are various, the presence of particles at a stream increases dissipation. If the sizes of particles are small in comparison with the smallest scale of turbulence, then it completely follows turbulent pulsations and reacts to all processes inherent in whirl.

Considering the above equation (1.84), the differential equation of the movement of the spherical particle, which is i – separately taken that in the turbulizirovanny environment, considered Bass, Bussinesky and Oseeny, has an appearance [8,13]

$$\frac{\pi}{6}a^3\rho_d\frac{du_{di}}{dt}=3\pi\eta a\left(V_i-u_{di}\right)-\frac{\pi}{6}a^3\left(\rho_d-\rho\right)\frac{dV_i}{dt}-\frac{1}{2}\frac{\pi}{6}a^3\rho\left(\frac{dV_i}{dt}-\frac{du_{di}}{dt}\right)$$
$$+\frac{3}{2}a^2\sqrt{\pi\rho\eta}\int_{t_0}^{t}\frac{dV_i/dt'-du_{di}/dt'}{\sqrt{t-t'}}dt'+F_e \tag{1.87}$$

where V_i and u_{di} are the velocity of the environment and a particle, respectively, considering i is that direction, t is the considered time point ρ and ρ_d are the density of the environment and a particle, a is the diameter of a particle, and F_e is the external force. The first member in the right member of equation characterizes force of resistance to the movement of a particle. The second member is caused by pressure gradient in the liquid surrounding a firm particle. The third member expresses force accelerating particles concerning liquid. The fourth member considers a current deviation from the established state (force Bass). The members containing pressure gradient, the attached weight and force Bass are essential in case density of liquid of the same order, as density of a firm particle or surpass.

Thus, if $\rho_c \ll \rho_d$, then in this case equation (1.87) significantly becomes simpler. This condition belongs generally to a current of gas and liquid suspensions (for example, aerosols) and doesn't belong to a current of emulsions (liquid–liquid systems), etc. At a current of a single drop or a single bubble in a stream at the big sizes of the last deformation and a deviation from spherical shape it's possible that it creates heterogeneity of the movement or heterogeneity of distribution of superficial (capillary) forces. Loss of spherical shape for such particles changes a picture of their hydrodynamic flow and coefficient resistance that significantly affects coefficients of transfer of weight and heat. The analytical decision (1.87) generally isn't possible, though by numerical methods it is possible to solve as the integro-differential equation.

Example 1.2

To determine resistance force for a spherical particle if the law of change of its velocity on time is set $V=V(t)$ [7]. Expression for force of resistance of a firm spherical particle can be defined from expression

$$F_T=2\pi\rho_c a^3\left[\frac{1}{3}\frac{dV}{dt}+\frac{3\nu_c}{a^2}V+\frac{3}{a}\sqrt{\frac{\nu_c}{\pi}}\int_{-\infty}^{t}\frac{dV}{dt}\frac{d\tau}{\sqrt{t-\tau}}\right] \tag{1.88}$$

If the particle moves under the law $V=\alpha t$, where α – acceleration of a particle, then from the previously stated equation we will receive

$$F_T=2\pi\rho_c a^3\alpha\left[\frac{1}{3}+\frac{3\nu_c}{a^2}t+\frac{6}{a}\sqrt{\frac{\nu_c t}{\pi}}\right] \tag{1.89}$$

At uniform motion of a particle with a velocity V_0 we have expression

$$F_T = 6\pi\rho_c v_c\, aV_0\left(1+\frac{a}{\sqrt{\pi v_c\, t}}\right) \tag{1.90}$$

which at $t \to \infty$ approaches the value given by Stokes's law.

1.5.3 Movement of Finely Dispersed Suspensions in the Turbulent Stream

In case of presentation of questions of a current of fine particles in a turbulent flow of gas, we will make the following assumptions: (a) the size of particles is small in comparison with the scale of turbulent pulsations ($1 < a < 20\ \mu\text{m}$) and therefore each particle makes movement, remaining within the initial pulsation mole; (b) particles are spherical in the form and monodispersna; (c) hydrodynamic resistance of particles to movement of the gaseous environment is described as a first approximation by the Stokes law; and (d) concentration of particles in a flow is small and they do not constrain movement of each other, do not face, do not coagulate. The pulsation movement of particles made by them within one period of pulsations of gas can be provided as change of the pulsation velocity of gas in time. We will consider movement of the particle made under the influence of environment velocity only in the longitudinal direction in the form of a monoharmonic function [12]

$$u = \bar{u} + U'\sin\omega_E t \tag{1.91}$$

where ω_E is Lagrangian frequency of pulsations, $\bar{u}$ is mean value of the pulsation velocity, U' is maximum amplitude of pulsations.

Then equation (1.87), $\text{Re}_d < 1$ for and $m_d = \frac{1}{6}\pi a^3 \rho_d$, will be presented in the form

$$\frac{du_d}{dt} + \alpha . u_d = \alpha\left(\bar{u} + U'\sin\omega . t\right) \tag{1.92}$$

The solution of this equation, in case of initial conditions $t = 0, u_d = 0$, will be presented in the form

$$u_d = \bar{u}\left(1-\exp\left(-\,{}^{t}\!/_{\tau_p}\right)\right) + \mu_p^2\omega_E\tau_p\bar{u}\exp\left(-\,{}^{t}\!/_{\tau_p}\right) + \mu_p U'\sin\left(\omega_E t - \phi\right) \tag{1.93}$$

where $\tau_p = \frac{1}{\alpha}$ is a relaxation time of particles, for Stokes particle $\tau_p = \frac{1}{18}\frac{\rho_d}{\eta}a^2$, ϕ is displacement angle of a phase of movement of a particle in the environment, determined by inertness of particles owing to which it is involved in movement of the environment with this or that lateness, $\mu_p = \frac{1}{\left(1+\omega_E^2\tau_p^2\right)^{1/2}}$ is the environment pulsating a level of hobby of particles, $\mu_g = \frac{\omega\tau_p}{\left(1+\omega^2\tau_p^2\right)^{1/2}}$ a level of flow with the μ_g pulsating environment, which is subject to the condition: $\mu_p^2 + \mu_g^2 = 1$, and ω is frequency of pulsations, relaxation time. For small-sized particles $\mu_p \to 1$, and for particles of the big size is $\mu_p \to 0$ (Figure 1.7).

The parameter of hobby of particles for the pulsating environment is the important characteristic in case of determination of coefficients of transfer in a turbulent flow. In these equations it is not considered important properties of surface phenomena in case of deformation of drops and bubbles, which are surface tension, surface energy, etc.

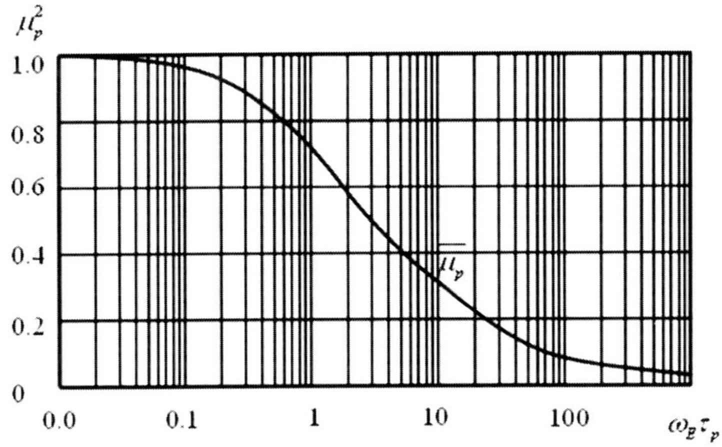

FIGURE 1.7 A level of hobby of particles for the pulsating turbulent environment.

1.6 VISCOSITY OF DISPERSE SYSTEMS

Liquid or gas in which a large number of particles is weighed can be considered as a homogeneous disperse environment if distances between particles are much more than sizes of particles, that is, are absent interaction between particles and changes of the sizes of particles as a result of physical the phenomena of coagulation, crushing, and agglomeration. The analysis of viscosity of disperse systems and a set of formulas for its calculation are provided in work [19,20], and as researches show, a considerable concentration of a disperse phase these liquids can be considered as non-Newtonian. So far, there is no uniform point of view about the mechanism of current of disperse systems about what the set of the rheological models connected with determination of viscosity and structure of system testify to. It reflects fundamental differences of views of the current, which are observed in disperse systems. However, results of numerous comparisons with experimental data draw a conclusion that approximately with an identical accuracy it is possible to describe the same equation systems, various by the physical and chemical nature, and the same disperse system—essentially different rheological models. This fact is explained by the fact that for an exception of very small concentration of particles in liquid, all rheological models are under construction on a semi-empirical or empirical basis with use of various functions, which are well-describing experimental data, not reflecting the mechanism of structuration of system. Thus, it is possible to confirm that there is no qualitative and quantitative theory connecting rheological properties of disperse system with parameters of their structure. Possibly, it is explained by existence of a set of such factors as the physical phenomena of hydrodynamic interaction and mutual collision, agglomeration, coagulations, deformation, and crushing of particles, rotation the anizometrichnykh of particles of a cylindrical, and the ellipsoidal form in volume of a stream that complicates the rheological behavior of the system. These facts can't still be considered when developing rheological models for the concentrated disperse systems, though even in attempt to consider any phenomenon creates very difficult rheological models.

With the size of effective viscosity for extremely diluted environments with the maintenance of spherical particles $\varphi \leq 0.05$ in the absence of their interaction, it is possible to determine by Einstein's formula

$$\eta_\phi = \eta_c \left(1 + 2.5\varphi\right) \tag{1.94}$$

At high concentrations of particles in volume, taking into account hydrodynamic interaction of particles, some authors use a modification of this expression

$$\eta_\phi = \eta_C\left(1 + 2.5\varphi + a_0\varphi^2 + a_1\varphi^3 + ...\right) \tag{1.95}$$

where the coefficients a_i in many works have a different meaning.

As a semi-empirical expression for calculation of effective viscosity of suspensions, which rather well-describes experimental data in the wide range of change of concentration of particles, it is possible to note a formula Moony [19,21]

$$\eta_\varphi / \eta_C = \exp\left[\kappa_1\phi / \left(1 - \kappa_2\phi\right)\right] \tag{1.96}$$

where κ_1 and κ_2 are coefficients that equal $\kappa_1 = 2.5$ and $0.75 \le \kappa_2 \le 1.5$. This formula at provides limit transition to Einstein's formula. Taylor [17] has generalized this equation for effective viscosity of emulsions

$$\eta_\varphi = \eta_C\left(1 + 2.5\phi\frac{\eta_d + 0.4\eta_C}{\eta_d + \eta_C}\right) \tag{1.97}$$

where k_1 and k_2 are viscosity of a disperse phase and medium.

For effective viscosity of the disperse [17] system, offer the following empirical formula

$$\eta_\phi / \eta_C = \left[1 - \left(\phi / \phi_P\right)\right]^{-m} \tag{1.98}$$

where φ_P is a the volume fraction of particles corresponding to their maximum packing, $\phi_P = 0.5 - 0.74$, $m = {}^{2.5\phi_P(\eta_d + 0.4\eta_c)}/_{(\eta_d + \eta_c)}$. In this formula value $\phi_P = 0.62$ is the most suitable for the majority of practical cases. Kumar et al. [17] for wide limits of change φ from 0.01 to 0.75 has offered the following formula

$$\frac{\eta_\varphi}{\eta_C} = \exp\left[2.5\frac{0.4\eta_C + \eta_d}{\eta_C + \eta_d}\left(\phi + \phi^{5/3} + \phi^{11/3}\right)\right] \tag{1.99}$$

This formula is checked for various systems of liquid–liquid and has yielded the most effective result with a relative mistake to 20%. A set of models express dependence of viscosity of the disperse system on extreme concentration of particles at which φ_P; the current stops also from the extreme tension of shift. Except the given models, in literature there is a set of empirical and semi-empirical expressions for calculation of viscosity of the concentrated systems, though the choice of any model in all cases is based not on the structurization mechanism, and on the principle of the adequate description of experimental data.

In the literature, many other rheological models can be found where various dependences are given for determining the viscosity of a dispersed medium [19].

$$\eta = \eta_\infty + \frac{\eta - \eta_\infty}{1 + \left(\alpha_0\tau\right)^m}, \quad \eta = \eta_\infty + \frac{\eta - \eta_0}{1 + \alpha_0\dot{\gamma}^m + \alpha_1\dot{\gamma}} \tag{1.100}$$

Here, $\gamma = \partial V / \partial y$ is shift velocity, τ is shift tension, and η_∞ is viscosity for suspension in the absence of interaction between particles.

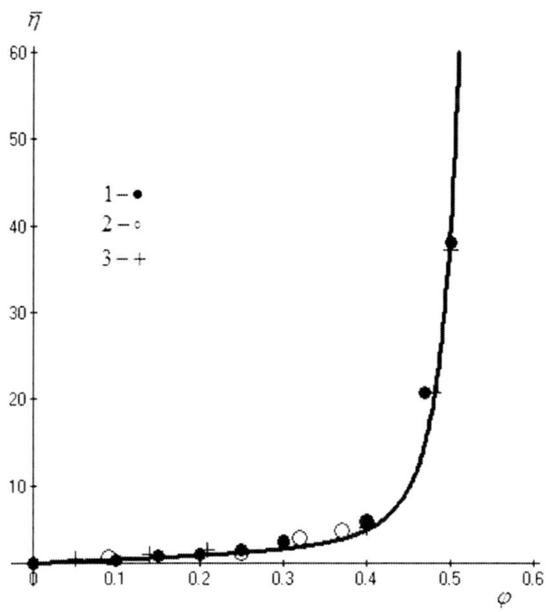

FIGURE 1.8 Dependence of effective viscosity of suspensions on a share of particles and their sizes: (1) 1,90–105 μm; (2) 2,45–80 μm; (3) 23,0–40 μm.

Figure 1.8 shows the experimental data corresponding to the dependence of the effective viscosity from the volume fraction of solid particles of the suspension for their various sizes. The effective viscosity of the dispersed system is calculated by the following formula:

$$\bar{\eta} = 1 + 2.5\varphi + 1.5\varphi \exp\left(\frac{0.45\varphi}{(\varphi - \varphi_\infty)^2}\right) \tag{1.101}$$

For the small sizes of particles, the dependence of effective viscosity on the sizes of particles becomes more noticeable where the dependence of viscosity on the sizes of particles and is described by the expression

$$\bar{\eta} = 1 + 2.5\varphi + \frac{3}{4}\varphi \exp\left(\frac{m\varphi}{(\varphi - \varphi_\infty)^2}\right), \quad m = 2.2 + 0.03a \tag{1.102}$$

Compliance of this dependence to experimental data is given in Figure 1.9.

As appears from experimental data and from this formula, the effective viscosity of disperse system significantly depends on a volume fraction and the size of particles. And, with an increase in the sizes of particles, the effective viscosity also grows. Most likely, in this case, coagulated structures and units aren't formed, and simple dense packing of particles is formed.

The effective viscosity of disperse system grows up to critical value that influences the velocity and character of a current (Figure 1.10). The viscosity is a key indicator of the rheological liquids defining their mobility, and she in disperse systems depends on tension of shift $\eta = \tau/\dot{\gamma}$, concentration, the sizes, and a form of particles. Many of these formulas don't consider structuration, which

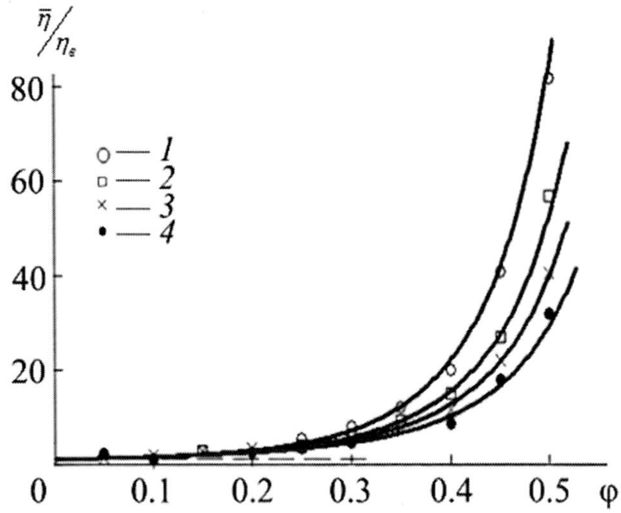

FIGURE 1.9 Settlement (continuous curves) experimental (points) of relative viscosity of disperse system from a volume fraction of firm spherical particles and their sizes: (1) 0.1 μm; (2) 0.5 μm; (3) 1.0 μm; (4) 1.5 μm.

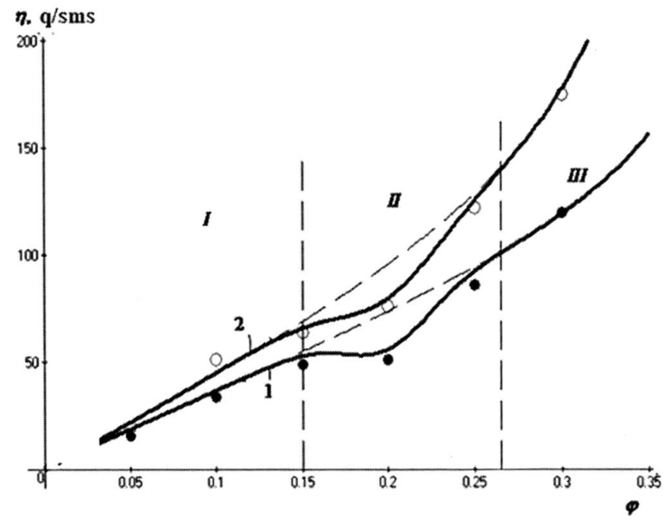

FIGURE 1.10 Dependence of viscosity of composition of asphalt on the content of polyethylene. (From Colak, Y. et al., *Petrol. Sci. Technol.*, 21, 9–10, 1427–1438, 2003.)

significantly deforms a viscosity curve. It follows from the analysis of a set of pilot studies that the viscosity in the field of the beginning of structuration submits to the semi-empirical equation of a look

$$\frac{d\eta}{d\varphi} = -k\eta\,(\varphi - \varphi_s) \tag{1.103}$$

$$\varphi \to \varphi_s, \quad \eta = \eta_0 - \eta_s$$

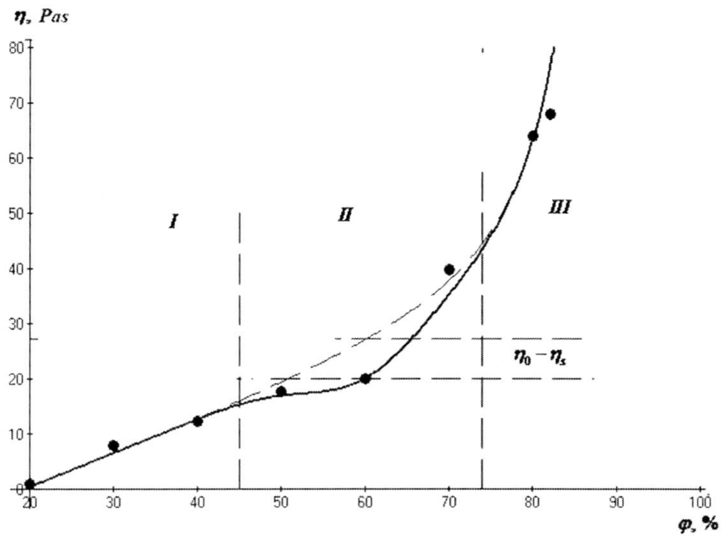

FIGURE 1.11 Dependence of viscosity of oil on the content of water.

Here, η_0 is viscosity of system without structuration, η_s is the maximum deviation of viscosity of system in the field of the beginning of structuration, and k is the coefficient. The solution of this equation will be submitted as

$$\ln \eta = -\frac{k}{2}(\varphi - \varphi_s)^2 + C_1 \tag{1.104}$$

where C_1 is the integration constant defined from entry conditions in a look $C_1 = \ln(\eta_0 - \eta_s)$. Finally, the solution of this equation will be submitted as

$$\frac{\eta_M}{\eta_c} = \frac{\eta_0 - \eta_s}{\eta_c} \exp\left[-\frac{k}{2}(\varphi - \varphi_s)^2\right] \tag{1.105}$$

This expression defines the nature of curve dependence of viscosity on volume concentration in the field of its deformation and further structuration. Here φ_∞ is the volume fraction of particles corresponding to their dense packing φ_s and is the share of particles corresponding to the beginning of formation of units or an inflection point of a curve. The area of the beginning of structuration is characterized by thixotropic properties as if the concentration of particles doesn't increase, then there is a destruction of units. Figure 1.11 gives change of viscosity of asphalt compositions with availability in them of the pure or fulfilled polyethylene below.

Existence of a certain concentration of polyethylene in asphalt compositions improves their physical properties and gives them heat stability and resistance to crack formation [22].

In Figure 1.10, dashed lines correspond to values of viscosity without structuration. Dependence of viscosity on the volume content of polyethylene in asphalt compositions for two temperatures are described by the equations

$$\begin{aligned} &1.\ T = 220^\circ\text{C},\quad \eta = 350\varphi \exp\left(0.015\varphi/(\varphi-0.5)^2\right) - A\exp\left(-850(\varphi-0.2)^2\right),\quad A = 18 \\ &2.\ T = 215^\circ\text{C},\quad \eta = 440\varphi \exp\left(0.044\varphi/(\varphi-0.5)^2\right) - A\exp\left(-640(\varphi-0.2)^2\right),\quad A = 16 \end{aligned} \tag{1.106}$$

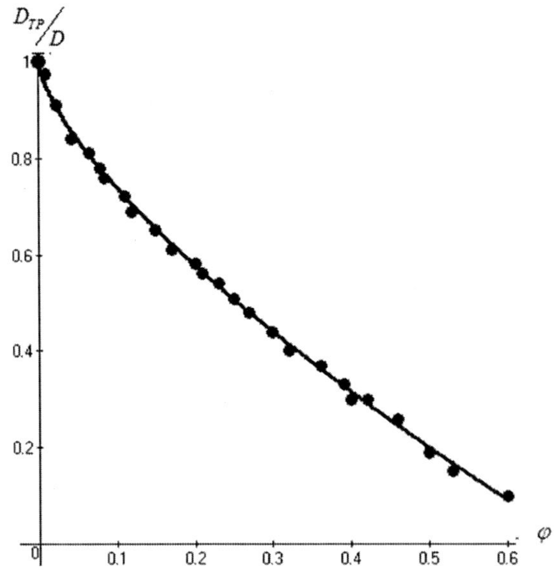

FIGURE 1.12 Dependence of coefficient of diffusion of particles on their volume fraction.

On these curves, the area II belongs to the beginning of structuration of particles of polyethylene in volume. In Figure 1.11, the dependence of viscosity on the maintenance of drops of water in an oil emulsion is given below, and it is described by the expression

$$\eta = (-11.6 + 0.6\varphi)\exp\left[\frac{2.8\varphi}{(\varphi - 100)^2}\right] - 7\exp\left[-0.01(\varphi - 60)^2\right] \tag{1.107}$$

From this formula η/η_c it is possible to calculate relative viscosity of the disperse environment in a look where the dynamic viscosity of oil of the muradkhansky field is equal $\eta_c \approx 14\Pi ac$. On these curves, areas of structuration and intensive growth of viscosity are accurately visible.

Communication of coefficient of diffusion and viscosity of the environment at the molecular level is expressed by the uniform number of Schmidt

$$\mathrm{Sc} = \frac{\nu}{D} = \frac{6\pi\eta^2 r_h}{\rho k_B T} \tag{1.108}$$

In disperse systems the coefficient of diffusion of particles, as well as effective viscosity depends on concentration of particles in unit of volume. Using literary experimental data for colloidal particles [23], it is possible to write

$$\frac{D_{TP}}{D} = 1 - 1.3\varphi^{0.7} \tag{1.109}$$

which gives the satisfactory description of an experiment (Figure 1.12).

If you consider the attitude of effective diffusion toward viscosity, then it will be presented in the form of the schedule provided in Figure 1.13.

Expression for Schmidt's number of the disperse system can be approximated in the look equation

$$\mathrm{Sc} = \mathrm{Sc}_0\left[1 + 0.45\varphi + \exp\left(\frac{1.85\varphi}{(\varphi - \varphi_\infty)^2}\right)\right] \tag{1.110}$$

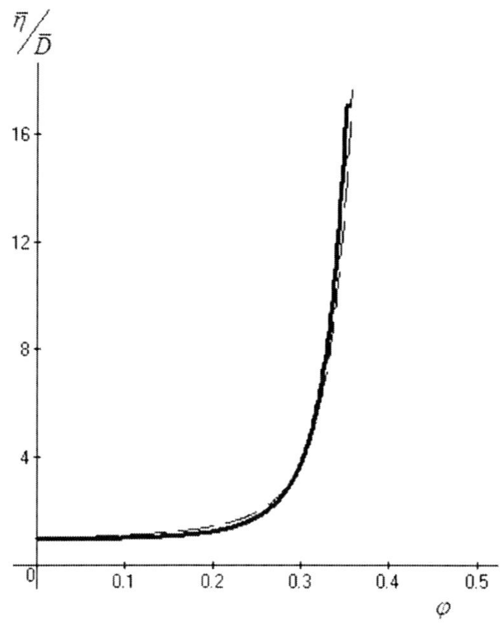

FIGURE 1.13 Communication of effective diffusion of particles and effective viscosity of the disperse environment (the dotted curve corresponds). (From Verberg, R. et al., 1996, http://citeseerx.ist.psu.edu/viewdoc/download?doi=10.1.1.285.6617&rep=rep1&type=pdf.)

It should be noted that $\varphi < 0.18$ at change of number of Schmidt depending on a volume fraction of particles will be expressed by an approximate ratio of a look $Sc = Sc_0(1 + 0.45\varphi)$.

It follows from these formulas that the condition of the structured disperse system is characterized by very great value of number of Schmidt. Destruction of the structured disperse systems leads to reduction of their effective viscosity. In Figure 1.14, the change of viscosity of the destroyed structure of the oil disperse system is given, though formation of the same structure happens also only in the opposite direction, that is, the system is characterized by thixotropy of structure.

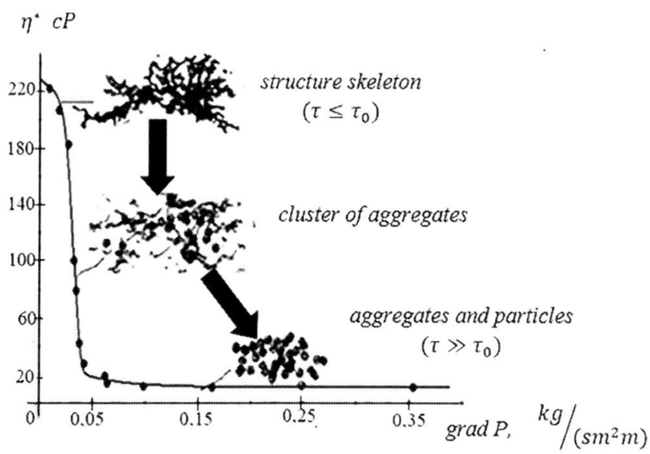

FIGURE 1.14 Dependence of viscosity of the structured system on the nature of its destruction at various values of a gradient of pressure.

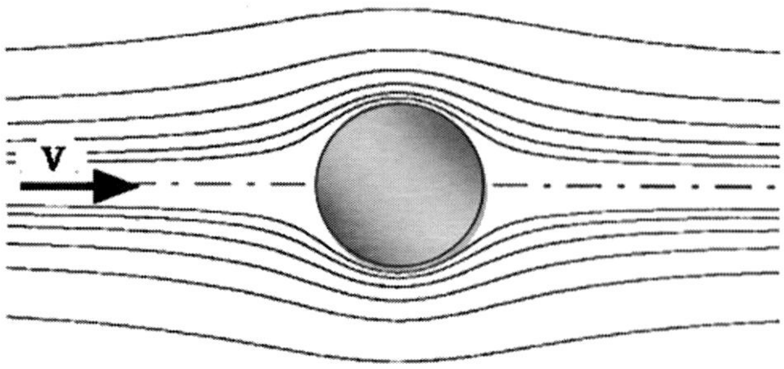

FIGURE 1.15 Flow of a firm spherical particle.

The given empirical models are used for concrete applications and represent formulas for adequate approximation of experimental data, though we will note what attempts to find the general rheological equation for different systems is considered an impossible task in advance.

1.7 HYDRODYNAMICS OF THE FLOW AROUND A SINGLE PARTICLE

For a mass transfer research in disperse environments, it is necessary to consider hydrodynamics of the movement of single particles in a stream of viscous liquid and gas in the beginning. The solution of problems of hydrodynamics assumes receiving calculated values for fields of velocities and pressure in a stream. Knowledge of these sizes allows us to have detailed information on local and average characteristics of a current. Previously, the equation of a current of a single particle in the unsteady stream without convective transfer has been considered. Subsequently, we will consider the steady flow of single spherical particles in a laminar flow, taking into account the convection terms.

1.7.1 Flow of a Firm Spherical Particle

The movement of a firm spherical particle in liquid is of the essential interest connected with its sedimentation, collision and coagulation, crushing, resistance, etc. It has been noted previously that at external flow of particles, an analogy between the phenomena of transfer of weight, heat, and an impulse is broken. Now the qualitative analysis of influence of the external hydrodynamic field (laminar and turbulent) on the solution of the previously stated tasks is carried out. The nature of the external laminar flow of a spherical particle is given in Figure 1.15.

We will consider a stationary problem of external flow of firm spherical particles of radius a_r at small numbers $\mathrm{Re}_d << 1$. In the case of an axisymmetric current and constant $(V_\theta = 0, V_\varphi = 0)$, pressure, equation (1.8) will be presented in the form [2,6,17]

$$v_c \left[\frac{1}{r^2} \frac{\partial}{\partial r} \left(r^2 \frac{\partial V_r}{\partial r} \right) + \frac{1}{r^2 \sin\theta} \frac{\partial}{\partial \theta} \left(\sin\theta \frac{\partial V_r}{\partial \theta} \right) \right] = 0 \tag{1.111}$$

For a spherical surface, having defined functions of current in a look

$$V_r = -\frac{1}{r^2 \sin\theta} \frac{\partial \psi}{\partial \theta}, \quad V_\theta = \frac{1}{r \sin\theta} \frac{\partial \psi}{\partial r} \tag{1.111a}$$

and having substituted in (1.111), in an operator form (1.11), we will receive

$$\nabla^2 \nabla^2 \psi = \nabla^4 \psi = 0 \tag{1.112}$$

where $\nabla^2\psi = \frac{\partial^2\psi}{\partial r^2} + \frac{\sin\theta}{r^2}\frac{\partial}{\partial\theta}\left(\frac{1}{\sin\theta}\frac{\partial\psi}{\partial\theta}\right)$. Boundary conditions for the decision (1.112) following

$$r = R, V_r = V_\theta = 0 \tag{1.112a}$$

$$r \to \infty, V_r = U\cos\theta \tag{1.112b}$$

Or for stream functions

$$r = a_r, \quad \frac{\partial\psi}{\partial\theta} = \frac{\partial\psi}{\partial r} = 0, \ \psi = \text{const} = 0 \tag{1.112c}$$

Under a condition $r \to \infty$, using equations (1.112) and (1.112b), we will receive

$$\frac{\partial\psi}{\partial\theta} = V_r r^2 \sin\theta = Ur^2\cos\theta\sin\theta \tag{1.112d}$$

Integrating this equation, we will receive

$$\psi = Ur^2\int\cos\theta\sin\theta d\theta = \frac{U}{2}r^2\sin^2\theta \tag{1.112e}$$

We will define the solution of equation (1.112) in a look

$$\psi = f(r)\sin^2\theta \tag{1.113}$$

Taking into account (1.113), we will define the values of derivatives entering equation (1.112)

$$\nabla^2\psi = \left[\frac{\partial^2\psi}{\partial r^2} + \frac{\sin\theta}{r^2}\frac{\partial}{\partial\theta}\left(\frac{1}{\sin\theta}\frac{\partial\psi}{\partial\theta}\right)\right] = \left(\frac{\partial^2 f}{\partial r^2} - \frac{2f}{r^2}\right)\sin^2\theta \tag{1.114}$$

We will rewrite equation (1.112) in a look

$$\nabla^4\psi = \frac{\partial^2}{\partial r^2}\nabla^2\psi + \frac{\sin\theta}{r^2}\frac{\partial}{\partial\theta}\left(\frac{1}{\sin\theta}\frac{\partial\nabla^2\psi}{\partial\theta}\right) = 0 \tag{1.115}$$

Having defined the corresponding derivatives entering this equation

$$\begin{aligned}
&\frac{\partial}{\partial r}\nabla^2\psi = \left(\frac{\partial^3 f}{\partial r^3} - \frac{2}{r^2}\frac{\partial f}{\partial r} + \frac{4f}{r^3}\right)\sin^2\theta \\
&\frac{\partial^2}{\partial r^2}\nabla^2\psi = \left(\frac{\partial^4 f}{\partial r^4} + \frac{8}{r^3}\frac{\partial f}{\partial r} - \frac{2}{r^2}\frac{\partial^2 f}{\partial r^2} - \frac{12f}{r^4}\right)\sin^2\theta \\
&\frac{\partial}{\partial\theta}\nabla^2\psi = \left(\frac{\partial^2 f}{\partial r^2} - \frac{2f}{r^2}\right)2\sin\theta\cos\theta \\
&\frac{\sin\theta}{r^2}\frac{\partial}{\partial\theta}\left(\frac{1}{\sin\theta}\frac{\partial\nabla^2\psi}{\partial\theta}\right) = -\frac{2}{r^2}\left(\frac{\partial^2 f}{\partial r^2} - \frac{2f}{r^2}\right)\sin^2\theta
\end{aligned} \tag{1.116}$$

we will write down equation (1.115) in a look

$$\frac{d^4 f}{dr^4} - \frac{4}{r^2}\frac{d^2 f}{dr^2} + \frac{8}{r^3}\frac{\partial f}{\partial r} - \frac{8f}{r^4} = 0 \tag{1.117}$$

We will look for the decision (1.117) in a look

$$f = C_n r^n \tag{1.118}$$

Having defined the derivatives entering in (1.117)

$$\frac{\partial f}{\partial r} = C_n n r^{n-1}, \quad \frac{\partial^2 f}{\partial r^2} = C_n n(n-1) r^{n-2}$$
$$\frac{\partial^3 f}{\partial r^3} = C_n n(n-1)(n-2) r^{n-3}, \quad \frac{\partial^4 f}{\partial r^4} = C_n n(n-1)(n-2)(n-3) r^{n-4} \tag{1.119}$$

We will receive the equation

$$n(n-1)(n-2)(n-3) - 4n(n-1) + 8n - 8 = 0 \tag{1.120}$$

transforming which, we have

$$(n-1)\left[n(n-2)(n-3) - 4n + 8\right] = 0 \tag{1.121}$$

The solution of this equation is

$$n - 1 = 0, \ n_1 = 1$$
$$(n-2)\left[n(n-3) - 4\right] = 0 \tag{1.122}$$
$$n_2 = 2, \ n_3 = 4, \ n_4 = -1$$

The common decision (1.112) is defined in a look

$$f(r) = C_4 r^4 + C_2 r^2 + C_1 r + C_0 r^{-1} \tag{1.123}$$

Here, C_1, C_2, C_0, and C_4 are the integration coefficients defined from boundary conditions. Using the decision (1.123), we will define the function of current in a look

$$\psi = f(r)\sin^2\theta = \left(C_4 r^4 + C_2 r^2 + C_1 r + \frac{C_0}{r}\right)\sin^2\theta \tag{1.124}$$

Using a condition (1.112e), for equation (1.124) we will receive

$$\frac{U}{2} r^2 \sin^2\theta = \left(C_4 r^4 + C_2 r^2 + C_1 r + \frac{C_0}{r}\right)\sin^2\theta \tag{1.125}$$

From this equation it is possible to write

$$\frac{U}{2} = C_4 r^2 + C_2 + {}^{C_1}\!/_{r} + {}^{C_0}\!/_{r^3} \tag{1.126}$$

At $r \to \infty$, we have expression ${}^{U}\!/_{2} = C_4 r^2 + C_2$ from which follows, as $C_4 = 0$ and $C_2 = {}^{U}\!/_{2}$. In this case, the function of current will be defined in a look

$$\psi = \left(\frac{U}{2} r^2 + C_1 r + \frac{C_0}{r} \right) \sin^2\theta \tag{1.127}$$

Using (1.112b) to this equation, we have $r = R$,

$$\begin{aligned} \frac{\partial \psi}{\partial r} &= UR + C_1 - \frac{C_0}{R^2} = 0 \\ \frac{\partial \psi}{\partial \theta} &= \left(\frac{U}{2} R^2 + C_1 R + \frac{C_0}{R} \right) 2 \sin\theta \cos\theta = 0 \end{aligned} \tag{1.128}$$

from where we will receive

$$C_1 = -{}^{3UR}\!/_{4}, \; C_0 = {}^{UR^3}\!/_{4} \tag{1.129}$$

Using values of coefficients, we will define the function of current

$$\psi = \frac{UR^2}{4} \left[2\left(\frac{r}{R} \right)^2 - 3\left(\frac{r}{R} \right) + \frac{R}{r} \right] \sin^2\theta \tag{1.130}$$

and the being stream velocities

$$\begin{aligned} V_r &= \frac{1}{r^2 \sin\theta} \frac{\partial \psi}{\partial \theta} = \frac{U}{2} \left[2 - 3\left(\frac{R}{r} \right) + \left(\frac{R}{r} \right)^3 \right] \cos\theta \\ V_\theta &= -\frac{1}{r \sin\theta} \frac{\partial \psi}{\partial r} = \frac{U}{4} \left[-4 + 3\left(\frac{R}{r} \right) + \left(\frac{R}{r} \right)^3 \right] \sin\theta \end{aligned} \tag{1.131}$$

where R is a radius of a firm spherical particle.

We will determine the being pressure upon particle surfaces

$$\frac{\partial P}{\partial r} = \frac{\eta_c}{r^2 \sin\theta} \frac{\partial}{\partial \theta} \nabla^2 \psi \tag{1.132}$$

$$\frac{\partial P}{\partial \theta} = -\frac{\eta_c}{\sin\theta} \frac{\partial}{\partial r} \nabla^2 \psi \tag{1.133}$$

$$\nabla^2 \psi = \left(10 C_4 - \frac{2C_1}{r} \right) \sin^2\theta = \frac{3}{2} U \frac{R}{r} \sin^2\theta \tag{1.134}$$

Using these components of pressure, we will determine distribution of pressure by a particle surface, under a condition, in a look $r \to \infty$, $P = P_\infty$

$$P = P_\infty - \frac{3}{2}\frac{U\eta_c R}{r^2}\cos\theta \tag{1.135}$$

Affect surfaces of a particle of normal tension

$$\tau_{rr} = -P + 2\eta_c\left(\partial V_r / \partial r\right) = -P_\infty + \eta_c \cos\theta\left(-12C_4 r - \frac{6C_1}{r^2} - \frac{12C_0}{r^4}\right) \tag{1.136}$$

and tangent (tangential) tension

$$\tau_{r\theta} = \eta_c\left(\frac{1}{r}\frac{\partial V_r}{\partial\theta} + \frac{\partial V_\theta}{\partial r} - \frac{V_\theta}{r}\right) = \eta_c \sin\theta\left(-6C_4 r - \frac{C_0}{r^4}\right) \tag{1.137}$$

We will define the distribution of normal force in a look

$$F_n = 2\pi.R^2\int_0^\pi \tau_{rr}\cos\theta\sin\theta d\theta \tag{1.138}$$

and tangent (tangential) force as

$$F_\tau = 2\pi.R^2\int_0^\pi \tau_{r\theta}\sin^2\theta d\theta \tag{1.139}$$

We will determine the total force operating on a particle as

$$F_S = F_n + F_\tau = 2\pi.R^2\int_0^\pi\left(\tau_{rr}\cos\theta - \tau_{r\theta}\sin\theta\right)\sin\theta.d\theta \tag{1.140}$$

On a drop surface at the size of this integral is equal $\left(C_1 = -\frac{3}{4}UR\right)$:

$$F_S = -8\pi\eta_c C_1 = 6\pi\eta_c RU \tag{1.141}$$

where F_S is a force of resistance of a Stokes particle $\mathrm{Re}_d << 1$. This expression represents the equation for determination of force of resistance of Stokes for page particles. The coefficient of resistance of a spherical particle is defined in a look

$$C_D = \frac{F_S}{\frac{1}{2}\rho_c U^2\pi R^2} = \frac{24}{\mathrm{Re}_d} \tag{1.142}$$

where $\mathrm{Re}_d = \frac{2RU\rho_c}{\eta_c}$ is a Reynolds number for a particle. It is important to note that this equation received in the theoretical way is suitable for small values of number $<< 1$. In the subsequent heads, some results on the determination of coefficient of resistance for moderate numbers are given, though for large numbers only empirical expressions are possible.

1.7.2 Flow of Drops and Bubbles

The movement of drops and bubbles in liquids differs from the movement of firm particles in existence of two major factors: mobility of an interface of phases and ability of drops and bubbles to change a form. At intermediate and great values of Reynolds number, these effects are shown most.

The liquid, which is flowing around a drop or gas, initiates circulation of internal liquid owing to friction—the liquid which is flowing around a drop or gas initiates, owing to friction circulation of internal liquid or substances of which the drop consists. This effect will depend, naturally, on the dimensionless parameter γ representing the relation of dynamic viscosity of external liquid to dynamic liquid of internal liquid $\gamma = {}^{\eta_c}\!/_{\eta_d}$.

Thus, unlike firm particles, at the movement of drops and bubbles in them the internal current, which is characterized by function of current ψ_d and the external current which is characterized by function of current ψ_c, is observed. In both cases, for a spherical drop the current is described by the equation of type (1.112) with boundary conditions

$$r \to \infty,\ \psi_c = \frac{1}{2}Ur^2\sin^2\theta \tag{1.143}$$

On a surface of a drop are joined an internal and external current; therefore, it is possible to write down

$$r = R, \qquad \psi_c = \psi_d \tag{1.143a}$$

$$\begin{aligned} &\frac{d\psi_d}{dr} = \frac{d\psi_c}{dr} \\ &\eta_d\frac{\partial}{\partial r}\left(\frac{1}{r^2}\frac{\partial\psi_d}{\partial r}\right) = \eta_c\frac{\partial}{\partial r}\left(\frac{1}{r^2}\frac{\partial\psi_c}{\partial r}\right) \end{aligned} \tag{1.143b}$$

For an internal and external current of the solution of equation (1.112) we will look for in a look

$$\begin{aligned} \psi_d &= \left(C_{4d}r^4 + C_{2d}r^2 + C_{1d}r + C_{0d}r^{-1}\right)\sin^2\theta \\ \psi_c &= \left(C_{4c}r^4 + C_{2c}r^2 + C_{1c}r + C_{0c}r^{-1}\right)\sin^2\theta \end{aligned} \tag{1.144}$$

Using boundary conditions, we will receive

$$\begin{aligned} &C_{4c} = 0,\ C_{2c} = {}^{U}\!/_{2},\ C_{1d} = 0,\ C_{0d} = 0, \\ &C_{4d} = \frac{\eta_c U}{4R^2(\eta_d + \eta_c)},\ C_{2d} = -\frac{\eta_c U}{4(\eta_d + \eta_c)} \\ &C_{1c} = \frac{3UR}{4}\frac{\left(\eta_d + {}^{2}\!/_{3}\eta_c\right)}{\eta_d + \eta_c},\ C_{0c} = \frac{UR^3\eta_d}{4(\eta_d + \eta_c)} \end{aligned} \tag{1.145}$$

Stream functions will be defined so

$$\psi_d = \frac{U\eta_c R^2}{4(\eta_c + \eta)}\left[\left(\frac{r}{R}\right)^4 - \left(\frac{r}{R}\right)^2\right]\sin^2\theta$$
$$\psi_c = \frac{UR^2}{4}\left[2\left(\frac{r}{R}\right)^2 - 3\frac{\eta_d + \frac{2}{3}\eta_c}{\eta_d + \eta_c}\left(\frac{r}{R}\right) + \frac{\eta_d}{\eta_d + \eta_c}\left(\frac{R}{r}\right)\right]\sin^2\theta \tag{1.146}$$

Drag force on a drop surface is equal to

$$F_S = -8\pi\eta_c C_{1c} = 6\pi\eta_c RU\frac{\eta_d + \frac{2}{3}\eta_c}{\eta_d + \eta_c} \tag{1.147}$$

Comparing to equation (1.141), the drag force at flow of drops and bubbles differs from the flow of a firm particle with function $\gamma = \eta_d/\eta_c$, the characterizing mobility of a surface of a drop. Using conditions of balance of forces of Archimedes, weight and resistance (1.147), we will determine the velocity of sedimentation of drops at $\mathrm{Re}_d < 1$

$$U_S = \frac{1}{3}\frac{R^2 g}{\eta_c}\frac{\Delta\rho}{\rho_c}\frac{\eta_d + \eta_c}{\eta_c + \frac{3}{2}\eta_d} \tag{1.148}$$

where a_r is a drop radius. Expression (1.148) is Hadamard–Rybczynski's equation and determines the velocity of sedimentation $(\Delta\rho > 0)$ or emersion $(\Delta\rho < 0)$ of drops. In special cases:

1. For firm spherical particles $\eta_d/\eta_c \to \infty$, equation (1.148) turns into Stokes's equation (1.143). We will define coefficient of resistance of a spherical drop in a look

$$C_D = \frac{F_S}{\frac{1}{2}\rho_c U^2 \pi R^2} = \frac{12\eta_c}{\rho_c UR}\frac{\eta_d + \frac{2}{3}\eta_c}{\eta_c + \eta_d} \tag{1.149}$$

If $\mathrm{Re}_d = \dfrac{2\rho_c RU}{\eta_c}$, then this expression can be written down in a look

$$C_D = \frac{24}{\mathrm{Re}_d}\frac{\gamma + \frac{2}{3}}{\gamma + 1} \tag{1.150}$$

where $\gamma = \eta_d/\eta_c$. If $\gamma \to \infty$ (for firm particles), then $C_D = 24/\mathrm{Re}_d$ $(\mathrm{Re}_d < 1)$

2. If $\gamma \to 0$, for spherical bubbles in liquids, the coefficient of resistance is equal. For the deformed drops and bubbles, the resistance coefficient sharply increases with an increase in the Reynolds number $C_D = 16/\mathrm{Re}_d$. Besides, the reasonings on the resistance coefficient, given previously at large Reynolds numbers, appear unsatisfactory as these models don't consider an interface separation, emergence of a turbulent interface, the crisis of resistance, and many other factors arising with an increase in number Re_d [24–26].

1.8 HYDRODYNAMICS OF FLUID FLOW THROUGH POROUS MEDIA

Development and designing of the determined models of the non-stationary phenomena in porous environments with more difficult internal geometry of a time and channels is limited to the essential mathematical difficulties connected with complex structure and a structure of the porous environment, which is characterized by anisotropy of structure. The very small size of the pore channels, their irregular shape, their random coordination and scatter in the volume of the reservoir, and the large surface of the rough walls mean that the properties of a pore space are not among the measured characteristics of the porous environment. Therefore, the research of streams in the porous environments considered in the form of a continuum is based on introduction of fields of the special parameters characterizing the process in makrofizichesk small elements of volume with average characteristics.

We will define the average volume value of porosity in a look

$$\varepsilon = \frac{\int \varepsilon'(\upsilon)d\upsilon}{\upsilon_0} \tag{1.151}$$

where $\varepsilon'(\upsilon)$ is the distribution of porosity on all volumes of layers, and υ_0 is the volume of the porous environment. The definition of distribution of porosity in the volume of Wednesday is connected with structure and distribution of various times, their quantity and an arrangement in the volume of the porous environment $\varepsilon'(\upsilon)$, and has absolutely casual and discrete character. The hydrodynamic theory of transfer in solid porous bodies represents quasicontinual theories, which are subject to continuous environments. For each considered real environment by means of averaging of characteristics of this process of transfer on a set of discrete elements, the average sizes characterizing the local continuous environment are entered. Thus, continuous environments are not real bodies, but their mathematical models. In the study of porous environment, it is expedient to consider a homogeneous isotropic porous environment that does not have characteristic changes in different directions.

1.8.1 Hydrodynamics of Flow and Model of Filtration and Liquid in Isotropic Porous Medium

Considering the isotropic porous environment, in works [5,6] the equation of filtration of Darcy–Forchheimer is given in a look

$$(\nabla P - \rho_c f) = \eta_c k^{-1} V + \beta \rho_c |V| V \tag{1.152}$$

In works [4,27,28] filtration is given in the isotropic porous environment, the hydrodynamics equation described by Navier–Stokes–Bricman's equation, uniting with the filtration equation

$$-\nabla(\eta_{eff} \nabla V) + (\rho_c V \nabla) V + \eta_c k^{-1} V = f - \Delta P = \rho_c g - \Delta P \tag{1.153}$$

In this equation, the first and second member define a viscous and convective current of liquid, and the subsequent members characterize filtration through the porous environment taking into account mass forces. The simplest model following from special cases is the model of linear filtration of Darci, which is written down for anisotropic systems so

$$V_x = -\frac{k_x}{\eta_c}\frac{\partial P}{\partial x}, \quad V_y = -\frac{k_y}{\eta_c}\frac{\partial P}{\partial y}$$
$$V_z = -\frac{k_z}{\eta_c}\left(\frac{\partial P}{\partial z} + \rho_c g\right) \tag{1.154}$$

In a vector form, this equation will register in a look

$$V_i = -\frac{\vec{k}}{\eta_c}\frac{\partial P}{\partial x_i} \tag{1.155}$$

Here $\vec{k} = \begin{bmatrix} k_x & 0 & 0 \\ 0 & k_y & 0 \\ 0 & 0 & k_z \end{bmatrix}$ is a tensor of coefficients of a permeability, $\left(x_i = [x\ y\ z]\right)$.

Substituting (1.155) in expression for the equation of continuity (1.1), we will receive

$$\frac{\partial \varepsilon \rho}{\partial t} - \frac{\partial}{\partial x_i}\left(\rho \frac{\vec{k}}{\eta_c}\frac{\partial P}{\partial x_i}\right) = w \tag{1.156}$$

We will assume that the permeability is a function of coordinates, and the density of liquid $\rho = \rho(P)$ and $\varepsilon = \varepsilon(P)$ porosity functions of pressure and, then it is possible to write down

$$\frac{1}{\rho}\frac{\partial \rho}{\partial P} = \beta_c, \quad \frac{\partial \varepsilon}{\partial P} = \beta_s \tag{1.157}$$

where β_c, β_s are coefficients of isothermal compressibility of a time and liquid. Then equation (1.156) will register in a look

$$\eta_c\left(\beta_c + \varepsilon\beta_s\right)\frac{\partial P}{\partial t} - \frac{\partial}{\partial x_i}\left(\vec{k}\frac{\partial P}{\partial x_i}\right) = w \tag{1.158}$$

Expression (1.158) is called the diffusivity equation, describing isothermal filtration in the elastic mode. Having entered some transformations, equation (1.156) can be written down in a look

$$\frac{\partial P}{\partial t} - \frac{\partial}{\partial x_i}\left(\vec{\chi}\frac{\partial P}{\partial x_i}\right) = w \tag{1.159}$$

Here, $\vec{\chi} = \begin{bmatrix} \chi_x & 0 & 0 \\ 0 & \chi_y & 0 \\ 0 & 0 & \chi_z \end{bmatrix}$ is a tensor of coefficients of a hydraulic diffusivity $\vec{\chi} = \frac{\vec{k}}{\eta_c(\beta_c + \varepsilon\beta_s)}$. Expression (1.159) is called the equation of a diffusivity and describes isothermal filtration in the elastic mode. Characteristic forms of equation (1.159) for flat, cylindrical, and spherical filtration are given below:

1. For flat filtration (1.159)

$$\frac{\partial P}{\partial t} + V_x\frac{\partial P}{\partial x} + V_y\frac{\partial P}{\partial y} + V_z\frac{\partial P}{\partial z} = \left(\chi_x\frac{\partial^2 P}{\partial x^2} + \chi_y\frac{\partial^2 P}{\partial y^2} + \chi_z\frac{\partial^2 P}{\partial z^2}\right) + w \tag{1.160}$$

2. For cylindrical filtration

$$\frac{\partial P}{\partial t} + V_r\frac{\partial P}{\partial r} + \frac{V_\theta}{r}\frac{\partial P}{\partial \theta} + V_z\frac{\partial P}{\partial z} = \left(\frac{\chi_r}{r}\frac{\partial}{\partial r}\left(r\frac{\partial P}{\partial r}\right) + \frac{\chi_\theta}{r^2}\frac{\partial^2 P}{\partial \theta^2} + \chi_z\frac{\partial^2 P}{\partial z^2}\right) + w \tag{1.161}$$

3. For spherical filtration

$$\frac{\partial P}{\partial t}+V_r\frac{\partial P}{\partial r}+\frac{V_\theta}{r}\frac{\partial P}{\partial \theta}+\frac{V_\varphi}{r\sin\varphi}\frac{\partial P}{\partial \varphi}=\left[\begin{array}{l}\dfrac{\chi_r}{r^2}\dfrac{\partial}{\partial r}\left(r^2\dfrac{\partial P}{\partial r}\right)+\dfrac{\chi_\theta}{r^2\sin\theta}\dfrac{\partial}{\partial \theta}\left(\sin\theta\dfrac{\partial P}{\partial \theta}\right)+\\ +\dfrac{\chi_\varphi}{r^2\sin^2\theta}\dfrac{\partial^2 P}{\partial \varphi^2}\end{array}\right]+w \quad (1.162)$$

Filtration velocity projections to the corresponding directions will be presented as

$$V_r=-\frac{k}{\eta_c}\frac{\partial P}{\partial r},\; V_\varphi=-\frac{k}{\eta_c}\left(\frac{1}{r}\frac{\partial P}{\partial \varphi}\right),\; V_\theta=-\frac{k}{\eta_c}\left(\frac{1}{r\sin\theta}\frac{\partial P}{\partial \theta}\right) \quad (1.163)$$

Usually, solutions of the equation of a diffusivity are carried out in two ways: (1) for unlimited layer, and (2) for limited layer with certain contours. Depending on these conditions for the solution of the equations of filtration, it is necessary to formulate regional (initial and boundary) conditions as for the oil layer with a limited contour and the semi-limited layer.

The entry condition is defined by a task of the law of distribution of pressure in a porous body

$$t=0,\; P=P(x,y,z) \quad (1.164)$$

Boundary conditions can be set in various ways, in particular distribution of pressure on border of a porous body and in the center, such as

$$\begin{array}{l}x=X,\; P=P_R(y,z,t);\; y=Y,\; P=P_R(x,z,t);\; z=Z,\;\; P=P_R(x,y,t)\\ x=x_0,\; P=P_0(y,z,t);\; y=0,\; P=P_0(x,z,t);\; z=0,\;\; P=P_0(x,y,t)\end{array} \quad (1.165)$$

where X,Y,Z are layer contour coordinates, P_R is pressure, and P_0,P_R is pressure at the input and output of the layer. Depending on the solution of practical tasks, these regional conditions can be changed according to features of the put problem. Depending on a task of regional conditions, the solution of a problem of filtration can significantly change.

1.8.2 Cylindrical, Non-Stationary Single-Phase Filtration Liquid in a Layer of a Porous Medium

We will consider a number of private solutions of the equations of cylindrical filtration of incompressible liquid in the elastic deformable porous environment for a current (the elastic mode of filtration) with the set various regional conditions below. Usually in calculations, layer contours in the form of a strict cylindrical form (Figure 1.16) are accepted, though in practice any forms deviating cylindrical are possible.

We accept the following assumptions: (1) the current of liquid in layer is considered only flat radial from a layer contour to a well face, and a layer arrangement strictly horizontal; otherwise, it is necessary to consider gravitation forces; (2) there are no drains and inflows and aren't considered an existence of delivery and other wells, that is, there is no interference of wells.

These factors generally influence the size and distribution of plan metric reservoir pressure; (3) filtration of liquid with the constant compressibility and constant properties which aren't depending on pressure, temperature, and other parameters is considered; and (4) there is no physical and chemical interaction between the porous environment and liquid.

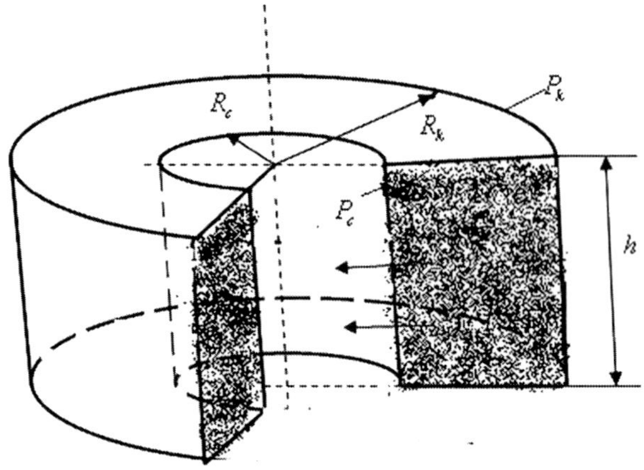

FIGURE 1.16 Schematic diagram of an oil layer: R_c, R_k – radiuses of the well and a contour of layer, respectively; P_c, P_k – pressure in a face of the well and on a layer contour, respectively.

We will consider a number of private solutions of the equations of cylindrical filtration of incompressible liquid in elastic deformable layer for an uninertial current (the elastic mode of filtration) with the set various regional conditions below.

Having taken the cylindrical form of oil layer (Figure 1.16), the general equation of filtration of liquid for the final sizes of layer taking into account closing of the well will be presented in the form [4]

$$\varepsilon\frac{\partial P}{\partial t}+V\frac{\partial P}{\partial r}=\frac{\chi}{r}\frac{\partial}{\partial r}\left(r\frac{\partial P}{\partial r}\right)-b(t)(P-P_k)^m$$
$$r=R_c,\ \ P=P_c;\ \ r=R_k,\ \ P=P_k;\ \ t=0,\ \ P=P_0(r) \tag{1.166}$$

where χ is coefficient of a hydraulic diffusivity.

This linear differential equation is removed on the basis of synthesis of a continuity equation, the dynamic equation of filtering—Darci's law, an equation of state of the porous environment and the sating liquid.

The initial condition characterizing distribution of pressure on the radius of layer of the finite sizes is very important. Obviously, the dependence of bottom hole pressure from contour is expressed as $P_c = f(P_k)$; therefore, the first condition in equation (1.50) is somewhat formal. In particular, using the Pocket equation—Kozeni for differential pressure on thickness of a porous layer [8]—the value can be calculated by the following formula

$$P_c = P_k - P_{kp} - 150(R_k - R_c)\frac{(1-\varepsilon)^2}{\varepsilon^3}\frac{\eta_s U_s}{4gr_p^2} \tag{1.167}$$

This equation calculates not only the bottom hole pressure but also the distribution of pressure on layer radius $P_0(r)$ depending on the plan metric pressure and porosity. It is necessary to mark that this expression can be used for the decision of the reverse task of assessment of porosity of a layer in case of the known values of plan metric and bottom hole pressure. The second member in the left member of equation (1.167) defines changes of pressure in case of convective transfer. Having entered the dimensionless variables

$$\upsilon=\frac{V}{V_0},\ \ \rho=\frac{r}{R_k},\ \ \tau=\frac{V_0 t}{R_k} \tag{1.168}$$

we will transform equation (1.167) to the dimensionless look

$$\frac{\partial P}{\partial \tau}+\upsilon\frac{\partial P}{\partial \rho}=\frac{1}{\text{Ke}}\frac{1}{\rho}\frac{\partial}{\partial \rho}\left(\rho\frac{\partial P}{\partial \rho}\right)-\gamma(t)(P-P_k)^m \tag{1.169}$$

Here, $\text{Ke}=\frac{V_0R_k}{\chi}=\text{Re}\,\text{Q}_\text{K}$ is the criterion characterizing the relation of convective transfer to pulse transfer by a diffusivity and also represents an analogy of number in hydrodynamics of a flow of liquid and number in processes of transfer of mass and heat, χ is coefficient of hydraulic diffusivity, $\text{Re}=\frac{V_0R_k}{v_c}$ – a Reynolds number, and $\gamma(t)=b(t)V_0/R$, $\text{Q}_\text{K}={}^{v_s}/_{\chi}$ is the criterion characterizing physical properties of liquid and layer, similar to Schmidt's number for a mass transfer and to Prandtl's number for heat transfer. The value of criterion defines the character and area of transfer and filtering liquid in layer and depends on the velocity of a current and coefficient of a hydraulic diffusivity.

If $\text{Ke}\to\infty$ that corresponds to very small values of coefficient of a hydraulic diffusivity or permeability, then equation (1.169), upon transition from a substantive derivative to complete, will be provided as

$$\varepsilon\frac{dP}{dt}=-b(t)(P-P_k)^m \tag{1.170}$$

$$t=0,\ P=P_0$$

If $\text{Kr}<<1$, that is, appropriate to great values of coefficient of a hydraulic diffusivity, then equation (1.169) represents the equation of filtering Darci in case of constant porosity

$$\frac{dP}{dt}=\frac{\chi}{r}\frac{\partial}{\partial r}\left(r\frac{\partial P}{\partial r}\right) \tag{1.171}$$

We will mark that Darci's equation in the form of (1.171) is called in question on the grounds that it proceeds from experiments with the homogeneous isotropic porous environments though in a real situation the porous environment is the non-uniform, which is inherent in each layer of characteristic distribution of pores in volume and their geometrical structure. It is obvious that such simplification often resorts, in the case of simulation of these systems, to empirical expressions.

For the solution of equation (1.171), in a general view we will enter a new variable $\theta=P-P_k$ that will allow us to write

$$\varepsilon\frac{\partial \theta}{\partial \tau}+\upsilon\frac{\partial \theta}{\partial \rho}=\frac{1}{\text{Ke}}\frac{1}{\rho}\frac{\partial}{\partial \rho}\left(\rho\frac{\partial \theta}{\partial \rho}\right)-\gamma(t)\theta^m \tag{1.172}$$

with new edge conditions

$$\begin{aligned}&r=R_c,\ \rho=\frac{R_c}{R_k},\quad \theta=P_c-P_k\\&r=R_k,\ \rho=1,\quad \theta=0\\&\tau=0,\quad \theta=P_0(r)-P_k\end{aligned} \tag{1.173}$$

For the solution of equation (1.172) in the case of smallness of convective transfer, it is feasible by a variables separation method, having supposed $m=1$ and having entered the following conversion

$$\theta(\rho,\tau)=\psi(\tau)\varphi(\rho) \tag{1.174}$$

Adding expression (1.174) in (1.172) and dividing variables into separate items, we will receive the following two equations

$$\frac{\partial\psi}{\partial t}=-\frac{1}{\varepsilon}\left(\gamma(t)+\frac{\mu^2}{\mathrm{Kr}}\right)$$
$$\frac{\partial^2\varphi}{\partial\rho^2}+\frac{1}{\rho}\frac{\partial\varphi}{\partial\rho}+\mu^2\varphi=0 \tag{1.175}$$

where μ^2 is eigenvalues. The solution of the first equation (1.175) will be provided as

$$\psi(\tau)=C_1\exp\left[-\frac{\mu^2\tau}{\varepsilon\mathrm{Kr}}-\int\frac{1}{\varepsilon}\gamma(\tau)d\tau\right] \tag{1.176}$$

The second expression (1.175) represents Bessel's equation of a zero order of a material argument. Therefore, the limited solution of the second equation (1.175) can be presented in the form

$$\varphi(\rho)=C_2\mathrm{J}_0(\mu\rho) \tag{1.177}$$

Finally, the decision (1.172) taking into account (1.80) will be presented in the form

$$\theta(\rho,\tau)=\sum_{n=0}^{\infty}A_n\mathrm{J}_0(\mu_n\rho)\exp\left[-\frac{\mu_n^2\tau}{\varepsilon\mathrm{Kr}}-\int\frac{1}{\varepsilon}\gamma(\tau)d\tau\right] \tag{1.178}$$

where μ_n is root of the equation $\mathrm{J}_0(\mu_n)=0$ received according to the second condition (1.173): $\mu_1=2.4048;\ \mu_2=5.5201;\ \mu_3=8.6537;\ldots$ Constant coefficients of a row will be defined according to the third condition (1.173) and conditions of orthogonality of Bessel functions in case of $R_c \ll R_k$ in a look

$$A_n=\frac{2\int_0^1\rho\left(P_0(r)-P_k\right)\mathrm{J}_0(\mu_n\rho)d\rho}{\left[\mathrm{J}_1(\mu_n)\right]^2} \tag{1.179}$$

Finally, the common decision (1.8) will be presented in the form

$$P(r,t)=P_k+\exp\left(-\int\frac{1}{\varepsilon}\gamma(t)dt\right)\sum_{n=1}^{\infty}A_n\mathrm{J}_0\left(\mu_n\, r/R_k\right)\exp\left(-\frac{\mu_n^2\tau}{\varepsilon\mathrm{Kr}}\right) \tag{1.180}$$

In that specific case if $P_0(r)=P_c$, then the decision (1.179) will be presented in the form

$$A_n=-\frac{2(P_k-P_c)}{\mu_n\mathrm{J}_1(\mu_n)}$$
$$P(r,\tau)=P_k-(P_k-P_c)\exp\left(-\int\frac{1}{\varepsilon}m(\tau)d\tau\right)\sum_{n=1}^{\infty}\frac{\mathrm{J}_0(\mu_n r)}{\mu_n\mathrm{J}_1(\mu_n R_k)}\exp\left(-\frac{\mu_n^2\tau}{\varepsilon\mathrm{Kr}}\right) \tag{1.181}$$

Thanks to great values μ_n^2, a row (1.181) quickly meets and therefore in case of practical calculations it is possible to use only the first members of a row. It allows us to evaluate simpler expressions of the main hydrodynamic characteristics of layer.

1.8.3 Special Cases of the Solution of the Equation of Filtering

In the case of the decision of practical tasks, equation (1.2) of certain assumptions becomes simpler. We will consider special cases of the solution of equation (1.172):

1. In case of Ke << 1, equation (1.172) will be transformed to the equation of no stationary cylindrical filtering Darcy

$$\varepsilon \frac{dP}{dt} = \frac{\chi}{r} \frac{\partial}{\partial r}\left(r \frac{\partial P}{\partial r} \right) \tag{1.182}$$

with edge conditions

$$t = 0,\ P = P_0(r);\ r = R_k,\ P = P_k;\ r = R_c,\ P = P_c \tag{1.183}$$

The decision (1.182) turns out, similar to the above, in a look

$$P(r,t) = P_k + \sum_{n=1}^{\infty} A_n \mathrm{J}_0\left(\mu_n \, {}^{r}\!/_{R_k} \right) \exp\left(-\mu_n^2 \frac{\chi}{R_k^2 \varepsilon} t \right) \tag{1.184}$$

Here, coefficient A_n is defined from equation (1.179). We will determine the velocity of refluxing of liquid so

$$V = -\frac{k(\varepsilon)}{\eta_c} \frac{\partial P}{\partial r} \tag{1.185}$$

Having defined a derivative in (1.184) taking into account (1.185) in a look

$$\frac{\partial P}{\partial r} = -\sum_{n=1}^{\infty} \frac{A_n}{R_k} \mathrm{J}_1\left(\mu_n \frac{r}{R_k} \right) \exp\left(-\mu_n^2 \frac{\chi}{R_k^2 \varepsilon} t \right) \tag{1.186}$$

we will find a slit output depending on time in case of $r = R_c$

$$q = \frac{2\pi h k(\varepsilon)}{\eta_c} \sum_{n=1}^{\infty} A_n \mathrm{J}_1\left(\mu_n \frac{R_c}{R_k} \right) \exp\left(-\mu_n^2 \frac{\chi}{R_k^2 \varepsilon} t \right) \tag{1.187}$$

2. We will consider the stationary solution of the equation of filtering, having assumed ${}^{dP}\!/_{dt} \approx 0$. Then we have the following equation

$$\frac{\chi}{r} \frac{\partial}{\partial r}\left(r \frac{\partial P}{\partial r} \right) = 0 \tag{1.188}$$

$$r = R_k,\ P = P_k; \qquad r = R_c,\ P = P_c$$

This equation can be presented in the form

$$\frac{\chi}{r} \frac{\partial}{\partial r}\left(r \frac{\partial P}{\partial r} \right) = 0 \tag{1.189}$$

Integrating expression (1.189) twice

$$P(r) = A_0 \ln r + A_1 \tag{1.190}$$

where coefficients A_0 and A_1 are also defined according to edge conditions in a look

$$A_0 = \frac{P_k - P_c}{\ln\left(R_k/R_c\right)}, \quad A_1 = P_k - \frac{P_k - P_c}{\ln\left(R_k/R_c\right)} \ln R_k \tag{1.191}$$

Adding in (1.190), after simple conversions, we will finally receive

$$P(r) = P_k - \frac{P_k - P_c}{\ln\left(R_k/R_c\right)} \ln r \tag{1.192}$$

We will determine the velocity of a current of liquid in layer in a look

$$V = -\frac{k(\varepsilon)}{\eta_c} \frac{\partial P}{\partial r} \tag{1.193}$$

Using expression (1.193), we will define

$$\frac{\partial P}{\partial r} = -\frac{P_k - P_c}{R_k \ln\left(R_k/R_c\right)} \tag{1.194}$$

Taking into account (1.193) and (1.194), we will define a liquid consumption through the porous environment in a look

$$q = 2\pi\, h R_k V = \frac{2\pi\, h k(\varepsilon)}{\eta_c} \frac{P_k - P_c}{\ln\left(R_k/R_c\right)} \tag{1.195}$$

We will give some other solutions of the equations of filtration connected with a task of various boundary conditions, in particular, of pressure gradient in the mouth of a face of the well or on a layer contour:

$$P(r)\Big|_{r=R_c} = P_c, \; \frac{\partial P}{\partial r}\Big|_{r=R_k} = \text{grad}P_K, \; P(r) = P_c + (\text{grad}P_k)\ln\left(\frac{r}{R_c}\right) \tag{1.196}$$

$$\frac{\partial P}{\partial r}\Big|_{r=R_c} = \text{grad}\, P_c, \; P(r)\Big|_{r=R_k} = P_k, \; P(r) = (\text{grad}\, P_c) + R_0 P_k \ln\left(\frac{r}{R_k}\right) \tag{1.197}$$

3. At great values of number, the Ke >>1, equation (1.172) will be transformed to a look

$$\varepsilon \frac{dP}{dt} = -b(t)(P - P_k)^m \tag{1.198}$$

$$t = 0,\; P = P_0$$

Such types of the equation also turn out at the closed well, considering a condition $\partial P/\partial r = 0$ in the non-stationary equation of filtration. The common decision (1.198) will be presented in the form [4]

$$P(t) = P_k + \left[(P_0 - P_k)^{m+1} - (m+1)\int \frac{1}{\varepsilon} b(t)\,dt \right]^{1/(m+1)} \tag{1.199}$$

In particular, if $m = 1$ that is confirmed experimentally for many wells, the solution of equation (1.198) is presented in the form

$$P(t) = P_k - (P_k - P_c)\exp\left[-\int_0^t \frac{1}{\varepsilon} b(t)\,dt \right] \tag{1.200}$$

1.8.4 Oil Filtration and Construction of the Pressure Recovery Curve

For the solution of this task, it can be assumed that in equation (1.115) P_k and P_c are the reservoir and bottom hole pressures, respectively, at the time point t after closing the well.

The choice of structure and type of coefficient $b(t)$ is carried out, proceeding from pilot studies of a curve of restoration of pressure taking into account experience and the researcher's intuition. It should be noted that in the existing literature, experimental data are represented in coordinates $P(t) \sim \ln t$, proceeding from it, and as a first approximation it is possible to accept

$$b(t) = a(\ln t)^m, \quad \int_0^t a(\ln t)^m d(\ln t) = \frac{a}{m+1}(\ln t)^{m+1} \tag{1.201}$$

Then expression (1.200) will be presented in the form

$$P(t) = P_k - (P_k - P_c)\exp\left[-m_0 (\ln t)^{m+1} \right] \tag{1.202}$$

Here, $m_0 = \frac{a}{\varepsilon(m+1)}$. It is important to note that expression (1.202) isn't the only formula for assessment of a curve of restoration of pressure, as also other options of determination of coefficient are possible $b(t)$. The value can depend on properties and the sizes of the well and also time of achievement of the established layer pressure. Figure 1.17 gives the comparison of experimental data from references and own researches of a curve of distribution of pressure in the well with their calculated values on a formula (1.202) below: $P(t) = 0.3 - 0.235\exp[-0.05(\ln t)^3]$, for the first curve; for the second curve constructed on own researches of authors $P(t) = 125 - 28.5\exp[-0.02(\ln t)^3]$, atm.

Advantage of equation (1.200) consists that it describes all curves of restoration of pressure the uniform equation that estimates hydrodynamic properties of the oil layer without additional tangents or approximations of a quasi-linear part of a curve and to calculate efficiency of the well. As shown in Figure 1.17, pressure derivative from time practically doesn't become a constant anywhere that demonstrates lack of the line section of a curve of restoration of pressure. Using this equation and having put expressions of dependence of bottom hole pressure in time point after closing of the well and at the time of closing of the well from properties of oil layer, it is possible to estimate in number hydrodynamic characteristics of layer. We will consider determination of hydrodynamic parameters at the closed well with use of the solution of the equation of filtration for limited layer at

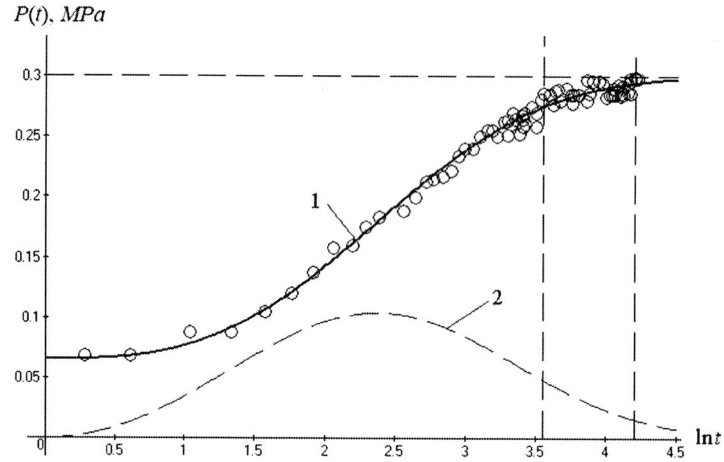

FIGURE 1.17 Comparison of an experimental curve of restoration of pressure with calculated values (1.200) ($m = 2$, $m_0 = 0.05$): (1) curve recovery of pressure; and (2) curve is derivative pressure on time.

Ke>>1. For calculation of hydrodynamic characteristics of layer, it is desirable to use the received analytical decisions (1.181) and (1.184). As a row (1.184) for the specified values quickly meets μ_n, it is enough to be limited to the first member, and we will define that.

$$J_1(2.408) \approx 0.52, \quad A_1 = -\frac{2(P_k - P_c)}{0.52 \times 2.408} = -1.5974(P_k - P_c) \tag{1.203}$$

The decision (1.181) will register in a look

$$\frac{P_k - P}{P_k - P_c} = 1.5974 J_0\left(\mu_1 {R_c}/{R_k}\right) \exp(-5.783 \frac{\chi}{\varepsilon R_k^2} t) \tag{1.204}$$

For the solution of the return problem of assessment of coefficient of a hydraulic diffusivity logarithm, both parts of this equation, having put that $R_c << R_k$ or ${R_c}/{R_k} \approx 10^{-4}$ and $J_0\left(2.408 \times 10^{-4}\right) \approx 1.0$,

$$\ln \frac{0.626(P_k - P)}{(P_k - P_c)} = -5.783 \frac{\chi t}{\varepsilon R_k^2} \tag{1.205}$$

From this expression, we will find assessment for effective coefficient of a piezoconductivity in a look

$$\chi = -\frac{0.1729 \varepsilon R_k^2}{t} \ln \frac{0.626(P_k - P)}{(P_k - P_c)} \tag{1.206}$$

Using dependence of pressure on time (Figure 1.16), look at the following data: $\ln t = 3.5$, $t = 33.115$ hour $= 119214$s, $R_k = 300$m, $P_k = 0.3$ MPa, $P_c = 0.06$ MPa, $P = 0.27$ MPa $\varepsilon \approx 0.2$ From equation (1.159), we will define that the coefficient of hydraulic diffusivity is equal $\chi = 7.181 \times 10^{-2}\ {m^2}/{s}$.

Interpretation of a surface of change of coefficient of a piezoconductivity from time and pressure shows that for a rather big area of a curve of restoration of pressure the coefficient of a piezoconductivity strives for constant value for all values of time and pressure, equal $\chi = 7.181 \times 10^{-2}\ {m^2}/{s}$, though for small values of the specified parameters gives an essential deviation. The sizes of this area are defined from a condition $\frac{\varepsilon R_k^2}{\chi t} > 2.3712$ that corresponds to a condition $\partial P/\partial t < 0$ (Figure 1.17). It means that the formula (1.206) can be suitable for calculation of coefficient of a hydraulic diffusivity for various values of time and pressure of a curve of restoration of pressure in the specified area. When using area $\partial P/\partial t > 0$ it is necessary to consider the second member of a row (1.184). Thus, assessment of coefficient of a hydraulic diffusivity is used for almost all curves of restoration of pressure. The coefficient of a hydraulic diffusivity can be also defined from expression (1.195) with use of the measured values of an output of the well where for various oil layers of value Ke fluctuate within 300–1200

$$\chi = \frac{q}{2\pi h}\frac{R_k}{R_c}\mathrm{Kr}^{-1} \tag{1.207}$$

The coefficient of permeability is defined in a look [4]

$$k = \chi \eta_c \beta^* \tag{1.208}$$

Here, $\beta^* = \varepsilon\beta_s + \beta_c$ is the provided layer for elastic capacity coefficient, and β_s, β_c are the respectively coefficients of an elastic capacity of liquid (oil) and layer. In particular, if that viscosity of oil $\eta_c \approx 3\times10^{-3}\,Pa\,s$, $\beta_c = 10^{-10}\,Pa^{-1}$, $\beta_s \approx 6\times10^{-10}\,Pa^{-1}$, $\beta^* \approx 2.2\times10^{-10}\,Pa^{-1}$, then the coefficient of permeability will be defined as $k = 7.181\times10^{-2}\times3\times10^{-3}\times2.2\times10^{-10} = 47.5\times10^{-15}\,\mathrm{m}^2$.

The coefficient of hydraulic conductivity of layer is defined as follows: $\xi_P = kh/\eta_c = \chi\beta^* h$. With a height of layer equal $h = 10$ the coefficient of hydraulic conductivity is in number equal

$$\xi_p = 18.84\times10^{-11}\frac{m^3}{Pa\,s} \tag{1.209}$$

It is obvious that the given calculations for assessment of hydrodynamic characteristics have conditional character as many macroscopic properties of layer (coefficients of permeability, an hydraulic diffusivity, and hydraulic conductivities) depend on its porosity.

Thus, fundamental laws and the equations of hydrodynamics of liquid and a current of disperse systems with various inclusions of types of firm particles, drops, and bubbles are provided in this section. For solving the different problems of heat and mass transfer, discussed in other sections, the results of this section will be used either in explicit or implicit form by using different similarity criteria.

2 Heat Transfer and Energy Dissipation

Heat transfer processes are an important energy factor that determines the material and energy state of the system. The necessity of presenting the problems associated with heat transfer is determined by the dependence of the transport coefficients and the physical and chemical properties of the flow and particles and the rate of chemical and physical transformations on temperature, the dependence of the phenomena of interphase heat transfer and mass on temperature and heat exchange with the external medium, and energy dissipation during fluid flow and dispersed media. A detailed analysis of heat transfer processes with various mechanisms can be found in many papers, among which it is important to note.

2.1 HEAT TRANSFER EQUATIONS AND BOUNDARY CONDITIONS

There are three mechanisms of heat transfer: (1) thermal conductivity—the process of heat (energy) propagation due to the interaction of structural particles of matter (molecules, atoms, ions); (2) convection (mixing)—the process of heat transfer due to the displacement of relatively large masses of matter in driving media; and (3) radiation—the process of energy transfer in the form of electromagnetic waves through the medium. Consider a fluid characterized by a variable temperature, $T(x, y, z, t)$. The amount of heat flux per unit surface per unit time is determined as follows

$$q = -\lambda\, q\mathrm{rad}\, T \tag{2.1}$$

where q is the heat flux and λ is the coefficient of thermal conductivity, which depends on the physical and chemical properties of the substance and the degree of flow turbulence.

The processes of heat transfer in a moving fluid are in many respects similar to the processes of convective mass transfer, although along with the similarities there are also significant differences. These differences and significant transport difficulties arise in the layer of dispersed particles or in capillary-porous media. The transfer of heat in such media is carried out by: (1) the thermal conductivity of the material of the particles themselves, (2) the thermal conductivity of the gas or liquid filling the pores of the material, (3) the transfer of heat from the thermal conductivity from one particle to another at the points of contact and between the particles and the liquid, and (4) convection of the liquid and radiation from the particle to the particle. The analysis of numerous theoretical sources and experimental studies makes it possible to reveal that the thermal conductivity of a particles material, in comparison with the thermal conductivity of a gas, negatively affects the effective thermal conductivity. Unlike homogeneous media in disperse systems and capillary-porous media, effective thermal conductivity is a complex function of temperature, pressure, chemical composition, porosity, particle and pore sizes and shapes, and other factors. It is efficient to solve most problems in analytical form, since for numerical research the numerical methods have the following disadvantages: (1) the lack of an approximate analytical form of the solution, which is often more convenient than the tabular and graphical representation, the impossibility of interpreting the results, and obtaining dependencies for calculating and analyzing heat transfer coefficients and processes of heat transfer and mass; (2) the lack of local versatility with a change in the geometric shape of the region, the nature of the hydrodynamic flow, the kinetics of the reactions, nonlinear boundary conditions, and the shape of the particles; and (3) the presence of different types of singularities and singular points in the differential equations of heat and mass transfer, which leads to additional analytical investigation and analysis of the difference scheme. At the same time, the basic principles that make it difficult to obtain exact solutions are due to the nonlinearity of the equations and boundary conditions, the dependence of the coefficients of the equations on the coordinates and temperature, and so on.

The compilation, analysis, and solution of equations that take into account all types of heat transfer are very difficult, in connection with which the calculation of heat transfer in such systems is based on simplification of the physical picture of the process under consideration. One of the main assumptions when considering heat transfer processes is the provision on the additivity of various heat transfer mechanisms, which allows one to neglect one or another mode of heat transfer under certain conditions. In the general case, heat transfer, taking into account dissipative heat, external and internal sources, is described by the Fourier–Kirchhoff equation, which in different coordinate systems with constant physicochemical properties of the medium is represented in the form:

1. In Cartesian coordinates:

$$\rho c_P \begin{pmatrix} \dfrac{\partial T}{\partial t} + V_x \dfrac{\partial T}{\partial x} + \\ V_y \dfrac{\partial T}{\partial y} + V_z \dfrac{\partial T}{\partial z} \end{pmatrix} = \lambda \left(\frac{\partial^2 T}{\partial x^2} + \frac{\partial^2 T}{\partial y^2} + \frac{\partial^2 T}{\partial z^2} \right) + 2\eta \left[\left(\frac{\partial V_x}{\partial x} \right)^2 + \left(\frac{\partial V_y}{\partial y} \right)^2 + \left(\frac{\partial V_z}{\partial z} \right)^2 \right]$$

$$+\eta \left[\left(\frac{\partial V_x}{\partial y} + \frac{\partial V_y}{\partial x} \right)^2 + \left(\frac{\partial V_x}{\partial z} + \frac{\partial V_z}{\partial x} \right)^2 + \left(\frac{\partial V_y}{\partial z} + \frac{\partial V_z}{\partial y} \right)^2 \right] + q_V \tag{2.2}$$

The last term in equation (2.2) determines the contribution of energy dissipation, where λ is the coefficient of thermal conductivity, η is the viscosity of the environment, ρ is the density of the environment, q_v is the internal source of heat, and c_P is the thermal capacity.

For a stationary medium on a plane surface without taking into account the dissipative function

$$\frac{\partial T}{\partial t} = a_T \left(\frac{\partial^2 T}{\partial x^2} + \frac{\partial^2 T}{\partial y^2} + \frac{\partial^2 T}{\partial z^2} \right) + q_V / c_P \rho \tag{2.3}$$

2. In cylindrical coordinates:

$$\rho c_P \begin{pmatrix} \dfrac{\partial T}{\partial t} + V_r \dfrac{\partial T}{\partial r} + \\ \dfrac{V_\theta}{r} \dfrac{\partial T}{\partial \theta} + V_z \dfrac{\partial T}{\partial z} \end{pmatrix} = \lambda \begin{pmatrix} \dfrac{1}{r} \dfrac{\partial}{\partial r} \left(r \dfrac{\partial T}{\partial r} \right) + \\ + \dfrac{1}{r^2} \dfrac{\partial^2 T}{\partial \theta^2} + \dfrac{\partial^2 T}{\partial z^2} \end{pmatrix} + 2\eta \begin{bmatrix} \left(\dfrac{\partial V_r}{\partial r} \right)^2 + \dfrac{1}{r} \left(\dfrac{\partial V_\theta}{\partial \theta} + V_r \right)^2 + \\ + \left(\dfrac{\partial V_z}{\partial z} \right)^2 \end{bmatrix} +$$

$$+\eta \left[\left(\frac{\partial V_\theta}{\partial z} + \frac{1}{r} \frac{\partial V_z}{\partial \theta} \right)^2 + \left(\frac{\partial V_z}{\partial r} + \frac{\partial V_r}{\partial z} \right)^2 + \left(\frac{1}{r} \frac{\partial V_r}{\partial \theta} + r \frac{\partial}{\partial r} \left(\frac{V_\theta}{r} \right) \right)^2 \right] + q_V \tag{2.4}$$

In the simplest case on a cylindrical surface

$$\frac{\partial T}{\partial t} = a_T\left(\frac{1}{r}\frac{\partial}{\partial r}\left(r\frac{\partial T}{\partial r}\right) + \frac{1}{r^2}\frac{\partial^2 T}{\partial \theta^2} + \frac{\partial^2 T}{\partial z^2}\right) + q_V / c_P\rho \tag{2.5}$$

where $a_T = \lambda / \rho c_p$ is the coefficient of thermal diffusivity.

3. In spherical coordinates:

$$\rho c_P\left(\frac{\partial T}{\partial t} + V_r\frac{\partial T}{\partial r} + \frac{V_\theta}{r}\frac{\partial T}{\partial \theta} + \frac{V_\varphi}{r\sin\varphi}\frac{\partial T}{\partial \varphi}\right) = \lambda\left[\frac{1}{r^2}\frac{\partial}{\partial r}\left(r^2\frac{\partial T}{\partial r}\right) + \frac{1}{r^2\sin\theta}\frac{\partial}{\partial\theta}\left(\sin\theta\frac{\partial T}{\partial\theta}\right) + \frac{1}{r^2\sin^2\theta}\frac{\partial^2 T}{\partial\varphi^2}\right] +$$

$$+2\eta\left[\left(\frac{\partial V_r}{\partial r}\right)^2 + \left(\frac{1}{r}\frac{\partial V_\theta}{\partial\theta} + \frac{V_r}{r}\right)^2 + \left(\frac{1}{r\sin\theta}\frac{\partial V_\varphi}{\partial\varphi} + \frac{V_r}{r} + \frac{V_\theta ctg\theta}{r}\right)^2\right] + \eta\left[\left(r\frac{\partial}{\partial r}\left(\frac{V_\theta}{r}\right) + \frac{1}{r}\frac{\partial V_r}{\partial\theta}\right)^2\right] + \tag{2.6}$$

$$+\eta\left[\left(\frac{1}{r\sin\theta}\frac{\partial V_r}{\partial\varphi} + r\frac{\partial}{\partial r}\left(\frac{V_\varphi}{r}\right)\right)^2 + \left(\frac{\sin\theta}{r}\frac{\partial}{\partial\theta}\left(\frac{V_\varphi}{\sin\theta}\right) + \frac{1}{r\sin\theta}\frac{\partial V_\theta}{\partial\varphi}\right)^2\right] + q_V$$

In the simplest case on a spherical surface without taking into account the dissipative function

$$\frac{\partial T}{\partial t} = a_T\left[\frac{1}{r^2}\frac{\partial}{\partial r}\left(r^2\frac{\partial T}{\partial r}\right) + \frac{1}{r^2\sin\theta}\frac{\partial}{\partial\theta}\left(\sin\theta\frac{\partial T}{\partial\theta}\right) + \frac{1}{r^2\sin^2\theta}\frac{\partial^2 T}{\partial\varphi^2}\right] + q_V/c_P\rho \tag{2.7}$$

The right-hand side of equations (2.2) through (2.6) characterizes heat transfer due to convection. In the left part, the first term characterizes the diffusion heat transfer due to thermal conductivity, and the subsequent terms reflect the heat source due to the dissipation of the internal energy of motion. It should be noted that the need to take energy dissipation into account in the analysis of heat transfer processes occurs at considerable medium velocities. If the medium is stationary, then the corresponding dissipative terms in the heat transfer equations are zero. The differential equations of thermal conductivity establish a connection between the temporal and spatial changes in the temperature field. To solve a particular problem, these equations are simplified (e.g., the internal energy of dissipation is neglected, symmetry of flow is assumed, etc.).

2.1.1 Coefficient of Thermal Conductivity

The main parameter determining the temperature distribution is the coefficient of molecular and turbulent thermal conductivity. Thus, the coefficient of thermal conductivity λ of a liquid at a temperature $T \approx 30°C$ can be calculated from the formula

$$\lambda = Ac\rho_c\sqrt[3]{\rho_c / M} \tag{2.8}$$

where c is the specific heat of the liquid, ρ is the density of the liquid, M is the molecular mass of the liquid, and is the coefficient equal to the associated liquids $A = 3.58\times10^{-8}$, for non-associated liquids $A = 4.22\times10^{-8}$. The coefficient of thermal conductivity for a liquid at any temperature is defined as

$$\lambda_T = \lambda_{30}\left[1-\zeta\left(T-30\right)\right] \tag{2.9}$$

where ζ is the coefficient is $\zeta = (1.4-2.2)\times10^{-3}$. The coefficient of thermal conductivity of aqueous solutions at a temperature is determined by the formula

$$\lambda_{PT} = \lambda_{P30}\frac{\lambda_{BT}}{\lambda_{B30}} \tag{2.10}$$

where λ_P and λ_B are the coefficients of thermal conductivity of the solution and water, and λ_{P30} is the thermal conductivity of the liquid at $T = 30°C$.

Example 2.1

Calculate the coefficient of thermal conductivity of methyl alcohol at $T = 120°C$ a specific heat capacity of methyl alcohol at $T = 30°C$ equal, $c = 2597.8\ \frac{J}{kg°C}$ density is $\rho = 783\ \frac{kg}{m^3}$, molecular weight of methyl alcohol is $M = 42\ ^{kg}/_{kmol}$. The coefficient of thermal conductivity for $T = 30°C$ $\lambda_{30} = 3.58\times10^{-8}\times2597.8\times783\sqrt[3]{783/42} = 0.193\ ^{W}/_{mK}$.

The coefficient of thermal conductivity at a temperature $T = 120°C$ is

$$\lambda_{120} = 0.193\left[1-1.2\times10^{-3}\left(120-30\right)\right] = 0.172\ ^{W}/_{mK}$$

Example 2.2

Calculate the coefficient of thermal conductivity of 25% aqueous solution of sodium chloride at a temperature 80°C, if the density of the solution $\rho = 1189\ ^{kg}/_{m^3}$, the specific heat capacity at is $c = 3390\ ^{J}/_{kgK}$.

We will determine the mole fraction of sodium chloride and water in the solution if the molecular masses are equal, $M_N = 58.5$ and $M_B = 18$, respectively, as

$$\mu_N = \frac{25/58.5}{\left(25/58.5\right)+\left(75/18\right)} = 0.093$$

$$\mu_B = 1-\mu_N = 0.907$$

The molecular weight of the solution is

$$M = 0.907\times18+0.093\times58.5 = 21.7\ ^{kg}/_{kmole}$$

The coefficient of thermal conductivity of a solution of sodium chloride in water is $T = 30°C$ determined as

$$\lambda_{30} = 3.58\times10^{-8}\times3390\times1189\sqrt[3]{1189/21.7} = 0.548\ ^{W}/_{(mK)}$$

The coefficient of thermal conductivity for $T = 80°C$

$$\lambda_{80} = 0.548\left(0.674/_{0.615}\right) = 0.6\ {}^{W}/_{(mK)}$$

where 0.674 and 0.615 ${}^{W}/_{(mK)}$ are the coefficients of thermal conductivity of water at 80°C and 30°C.

The coefficients of turbulent thermal conductivity, according to the hydrodynamic analogy, are determined identically to the coefficients of turbulent viscosity and turbulent diffusion. For solid particles in a layer, the thermal conductivity coefficient depends on the porosity of the medium. It should be noted that in practice, in the equations given, the coefficients of heat transfer, thermal conductivity, and thermal diffusivity are temperature dependent. The values of the coefficients can be assumed constant if the change in the temperature of the flow from point to point is sufficiently slow. In particular, the dependence of the thermal diffusivity on temperature in the form

$$a_T = \frac{\lambda(T)}{c_P(T)\rho(T)} = \frac{a_0}{1 + 2a_1T + a_2T} \tag{2.11}$$

where a_0, a_1, and a_2 are the coefficients determined on the basis of the experimental data. For turbulent flow, the value of the coefficient of turbulent thermal diffusivity is determined in the form

$$a_T \approx \beta_m U_D y \tag{2.12}$$

where β is the coefficient, determined on the basis of experimental data, $U_D = \sqrt{\tau/_\rho}$ is the dynamic flow velocity, and τ is frictional stress.

By analogy with this equation, the thermal conductivity in the porous layer of particles in the transverse direction is determined in the form

$$\lambda_r = \lambda_0 + BV\rho c_P r \tag{2.13}$$

or in dimensionless form is

$$\frac{\lambda_r}{\lambda_g} = \frac{\lambda_0}{\lambda_g} + B\mathrm{RePr} \tag{2.14}$$

where V is the flow velocity, c_P is heat capacity, λ_0 is the sum of all components of thermal conductivity, not depending on the velocity, $B = B_0\frac{6(1-\zeta)}{4}\varphi$ is coefficient, $\phi-$ is particle shape factor, $\mathrm{Pr} = \frac{v\rho c_P}{\lambda_g}$ is Prandtl number, and λ_g is thermal conductivity of the gas. The coefficient of thermal conductivity in continuous homogeneous media is determined by the thermophysical properties of substances or by the degree of flow turbulence in a turbulent medium.

2.1.2 Boundary Conditions for Equation of Heat Transfer

In order to solve the heat transfer equations, it is necessary to formulate the boundary conditions (initial and boundary conditions).

The initial condition is determined by setting the law of temperature distribution inside the body

$$t = 0, \quad T = T_0(x, y, z) \tag{2.15}$$

Boundary conditions can be specified in various ways:

1. A boundary condition of the first kind consists of assigning the temperature distribution over the surface of the body at any instant of time

$$x = L, \quad T = T(t, L) \tag{2.16}$$

2. A boundary condition of the second kind consists of setting the density of the heat flux for each point of the surface

$$x = L, \quad \lambda\left(\frac{\partial T}{\partial x}\right) = q(x,t) \tag{2.16a}$$

3. A boundary condition of the third kind characterizes the law of convective heat exchange between the surface of the body and the surrounding medium with a constant heat flux. When flowing around a body's surface with a liquid or gas, the transfer of heat from the liquid to the body near its surface is defined as

$$\lambda\left(\frac{\partial T}{\partial x}\right)_{x=L} = \alpha\left(T - T_\infty\right) \tag{2.16b}$$

where α, λ are the coefficients of heat transfer and thermal conductivity, the temperature of the liquid far from the wall. The heat given by the heat carrier is transferred to the heat exchange surface due to the thermal conductivity of the wall layer of the viscous medium. From the equation of convective heat transfer (2.16a), by dividing its right side by the left

$$\frac{\alpha\left(T - T\infty\right)}{\lambda\frac{\left(T - T_\infty\right)}{L}} = \frac{\alpha L}{\lambda} = Nu \tag{2.16c}$$

Thus, the Nusselt number Nu is the ratio of the amount of heat transferred to the surface due to convective heat transfer to the amount of heat that would be transmitted across a fixed layer of coolant thickness.

4. A boundary condition of the fourth kind corresponds to heat exchange on the surface of the contacting bodies. On the contact surface of two bodies, the conditions for the equality of temperatures

$$T_1(x,t) = T_2(x,t) \tag{2.16d}$$

and the equality of heat fluxes

$$\lambda_1\frac{\partial T_1}{\partial x} = -\lambda_2\frac{\partial T_2}{\partial x} \tag{2.16e}$$

When the state of the body changes (melting, evaporation, dissolution, etc.), the condition (2.16e) changes to the form

$$\lambda_1\frac{\partial T_1}{\partial x} = -\lambda_2\frac{\partial T_2}{\partial x} + \psi(T) \tag{2.16f}$$

where $\psi(T)$ is the function characterizing the heat flux due to the phase transformation. When solving certain problems, the boundary conditions (2.15) and (2.16) can be changed according to the conditions of the problem being solved.

2.2 STATIONARY ONE-DIMENSIONAL SOLUTIONS OF HEAT TRANSFER PROBLEMS

The solution of stationary one-dimensional heat transfer problems reduces to simplifying equations (2.2) through (2.6), in the absence of convective heat transfer for small values of the number, to the form

$$\frac{\lambda}{x^n}\frac{\partial}{\partial x}\left(x^n\frac{\partial T}{\partial x}\right)=q_V \tag{2.17}$$

under the boundary conditions (2.16). Here is a volumetric heat source. In this equation, there is a flat surface, a cylindrical surface, and a spherical surface. Below we give stationary solutions of (2.17) for a plane, cylindrical, and spherical surface with different boundary conditions.

For a plane surface, the heat transfer equation in stationary conditions without taking into account the dissipative function can be represented in the following form

$$\lambda\frac{d^2T}{dx^2}=q_V \tag{2.17a}$$

Next we consider the solution (2.17) with different boundary conditions.

1. Suppose that on the surface of a plane surface, the boundary conditions of the first kind are

$$x=0,T=T_{S1};\;\; x=L,\; T=T_{S2} \tag{2.17b}$$

Here are the temperatures T_{s1},T_{s2} on the surfaces of the flat plate, the thickness of the plate, and the thermal conductivity of the plate material.

The temperature distribution and the specific heat flux, as a result of solving equation (2.17), will be presented as

$$\frac{T_{S1}-T}{T_{S1}-T_{S2}}=\frac{x}{L}\left[1-\frac{q_VL^2}{2\lambda(T_{S1}-T_{S2})}\left(1-\frac{x}{L}\right)\right] \tag{2.18}$$

$$q=\frac{\lambda(T_{S1}-T_{S2})}{L}\left[1-\frac{q_VL^2}{\lambda(T_{S1}-T_{S2})}\left(\frac{1}{2}-\frac{x}{L}\right)\right] \tag{2.19}$$

2. Assume that heat transfer on both sides of the plate is carried out conventionally (conditions of the third kind)

$$x=0,\;-\lambda\frac{\partial T}{\partial x}=\alpha_1(T-T_{01});\;\; x=L,\;-\lambda\frac{\partial T}{\partial x}=\alpha_2(T-T_{02}) \tag{2.20}$$

Here, α_1,α_2 are the heat transfer coefficients of the flow to the plate for two liquids. The distribution of temperature and specific heat flux will be determined as

$$\frac{T_{01}-T}{T_{01}-T_{02}}=\frac{1-\dfrac{q_LL^2}{\lambda(T_{01}-T_{02})}\left(\dfrac{1}{2}+\dfrac{1}{\mathrm{Bi}_2}\right)}{1/\mathrm{Bi}_1+1+1/\mathrm{Bi}_2}\left(\frac{x}{L}+\frac{1}{\mathrm{Bi}_1}\right)+\frac{q_VL^2}{2\lambda(T_{01}-T_{02})}\left(\frac{x}{L}\right)^2 \tag{2.21}$$

$$q=\frac{\lambda\left(T_{01}-T_{02}\right)}{L}\left[\frac{1-\dfrac{q_V L^2}{\lambda\left(T_{01}-T_{02}\right)}\left(\dfrac{1}{2}+\dfrac{1}{\mathrm{Bi}_2}\right)}{1/\mathrm{Bi}_1+1+1/\mathrm{Bi}_2}+\frac{q_V L^2}{\lambda\left(T_{01}-T_{02}\right)}\frac{x}{L}\right] \tag{2.22}$$

where $\mathrm{Bi}_1={}^{\alpha_1 L}/_{\lambda}$ and $\mathrm{Bi}_1={}^{\alpha_1 L}/_{\lambda}$ are Bio numbers for both liquids and fluid temperatures away from surfaces.

3. Assume that a heat flow is defined on one surface of the plate (a condition of the second kind), and on the second side the surface temperature

$$x=0,\ q=q_0;\ x=L,\ T=T_{S2} \tag{2.23}$$

The distribution of temperature and specific heat flux will be determined as

$$T-T_{S2}=\frac{q_0 L}{\lambda}\left(1-\frac{x}{L}\right)+\frac{q_V L^2}{2\lambda}\left(1-\frac{x^2}{L^2}\right) \tag{2.24}$$

$$q=q_0+q_V x$$

4. Assume that convection heat transfer is (a condition of the third kind) on one side of the plate and a surface temperature on the other surface (a condition of the first kind)

$$x=0, T=T_{S1};\ x=L \tag{2.25}$$

$$-\lambda\frac{\partial T}{\partial x}=\alpha_2\left(T-T_{02}\right) \tag{2.26}$$

The distribution of temperature and specific heat flux will be determined as

$$\frac{T_{S1}-T}{T_{S1}-T_{02}}=\frac{x}{L}\left[\frac{1-\dfrac{q_V L^2}{\lambda\left(T_{S1}-T_{02}\right)}\left(\dfrac{1}{2}+\dfrac{1}{Bi_2}\right)}{1+1/\mathrm{Bi}_2}+\frac{q_V L^2}{2\lambda\left(T_{S1}-T_{02}\right)}\frac{x}{L}\right] \tag{2.27}$$

$$q=\frac{\lambda\left(T_{S1}-T_{02}\right)}{L}\left[\frac{1-\dfrac{q_V L^2}{\lambda\left(T_{S1}-T_{02}\right)}\left(\dfrac{1}{2}+\dfrac{1}{Bi_2}\right)}{1+1/\mathrm{Bi}_2}+\frac{q_V L^2}{\lambda\left(T_{S1}-T_{02}\right)}\frac{x}{L}\right] \tag{2.28}$$

2.2.1 Heat Transfer in a Cylindrical Solid Particles

For a cylindrical body, the equation of stationary thermal conductivity without taking into account the dissipative function is written in the form

$$\frac{\lambda}{r}\frac{\partial}{\partial r}\left(r\frac{\partial T}{\partial r}\right)=q_V \tag{2.29}$$

Let us consider the solution of the problem of thermal conductivity for a cylindrical body with an internal radius and with an external radius under different boundary conditions.

1. Assume that on the inner and outer surfaces of the cylinder boundary conditions of the first kind

$$r = R_1,\ T = T_{S1};\ r = R_2,\ T = T_{S2} \tag{2.30}$$

The distribution of temperature and specific heat flux will be determined as

$$\frac{T_{S1}-T}{T_{S1}-T_{S2}} = \left[1+\frac{q_V R_1^2}{4\lambda\left(T_{S1}-T_{S2}\right)}\left(1-\frac{R_2^2}{R_1^2}\right)\right]\frac{\ln\frac{r}{R_1}}{\ln\frac{R_2}{R_1}} - \frac{q_V R_1^2}{4\lambda\left(T_{S1}-T_{S2}\right)}\left(1-\frac{r^2}{R_1^2}\right) \tag{2.31}$$

$$Q = -2\pi RL\lambda\frac{\partial T}{\partial r} = \frac{2\pi\lambda L\left(T_{S1}-T_{S2}\right)}{\ln\left(R_2/R_1\right)}\left[1+\frac{q_V R_1^2}{4\lambda\left(T_{S1}-T_{S2}\right)}\left(1-\frac{R_2^2}{R_1^2}\right)\right]+\pi Lq_V R^2 \tag{2.32}$$

2. Assume that heat transfer on both sides of the cylinder is carried out conventionally (conditions of the third kind)

$$r = R_1,\ -\lambda\frac{\partial T}{\partial r} = \alpha_1\left(T-T_{01}\right);\ r = R_2,\ -\lambda\frac{\partial T}{\partial r} = \alpha_2\left(T-T_{02}\right) \tag{2.33}$$

The temperature distribution and the heat flux are represented in the form

$$\begin{aligned}\frac{T_{01}-T}{T_{01}-T_{02}} &= \left[1+\frac{q_V R_1^2}{4\lambda\left(T_{01}-T_{02}\right)}\left(1-\frac{R_2^2}{R_1^2}-\frac{2}{\mathrm{Bi}_2}\frac{R_2^2}{R_1^2}-\frac{2}{\mathrm{Bi}_1}\right)\right]\frac{1/\mathrm{Bi}_1+\ln\left(r/R_1\right)}{1/\mathrm{Bi}_1+\ln\left(R_2/R_1\right)+1/\mathrm{Bi}_2}+\\ &+\frac{q_V R_1^2}{4\lambda\left(T_{01}-T_{02}\right)}\left(\frac{2}{\mathrm{Bi}_1}-1+\frac{r^2}{R_1^2}\right)\end{aligned} \tag{2.34}$$

$$Q = \frac{2\pi L\lambda\left(T_{01}-T_{02}\right)}{1/\mathrm{Bi}_1+\ln\left(R_2/R_1\right)+1/\mathrm{Bi}_2}\left[1+\frac{q_V R_1^2}{4\lambda\left(T_{01}-T_{02}\right)}\left(1-\frac{R_2^2}{R_1^2}-\frac{2}{\mathrm{Bi}_2}\frac{R_2^2}{R_1^2}-\frac{2}{Bi_1}\right)\right]+\pi Lq_V r^2 \tag{2.35}$$

3. Assume that a heat flow is defined on one surface of the cylinder (a condition of the second kind), and on the other surface the temperature (conditions of the first kind)

$$r = R_1,\ q = q_1;\ r = R_2,\ T = T_{S2} \tag{2.36}$$

The temperature distribution and the heat flux will be determined as

$$\begin{aligned}T_{S1}-T &= \frac{q_2R_2}{\lambda}\ln\frac{r}{R_1}-\frac{q_V R_1^2}{4\lambda}\left(2\frac{R_2^2}{R_1^2}\ln\frac{r}{R_1}+1-\frac{r^2}{R_1^2}\right)\\ Q &= 2\pi L\left[q_2R_2-\frac{q_V R_1^2}{2}\left(\frac{R_2^2}{R_1^2}-\frac{r^2}{R_1^2}\right)\right]\end{aligned} \tag{2.37}$$

4. Let us assume that convection heat transfer (a condition of the third kind) on one surface and a temperature on the other surface (a condition of the first kind)

$$r = R_1,\ T = T_{S1};\ r = R_2,\ -\lambda \frac{\partial T}{\partial r} = \alpha_2 \left(T - T_{02}\right) \tag{2.38}$$

The temperature distribution and the heat flux as a result of the solution (2.17b) are determined as

$$\frac{T_{S1} - T}{T_{S1} - T_{02}} = \left[1 + \frac{q_V R_1^2}{4\lambda\left(T_{S1} - T_{02}\right)}\left(1 - \frac{R_2^2}{R_1^2} - \frac{2}{Bi_2}\frac{R_2^2}{R_1^2}\right)\right]\frac{\ln\left(r/R_1\right)}{\frac{1}{\mathrm{Bi}_2} + \ln\left(R_2/R_1\right)} - \frac{q_V R_1^2}{4\lambda\left(T_{S1} - T_{02}\right)}\left(1 - \frac{r^2}{R_1^2}\right) \tag{2.39}$$

$$Q = \frac{2\pi L\lambda\left(T_{S1} - T_{02}\right)}{1/\mathrm{Bi}_2 + \ln\left(R_2/R_1\right)}\left(1 - \frac{R_2^2}{R_1^2} - \frac{2}{\mathrm{Bi}_2}\frac{R_2^2}{R_1^2}\right) + \pi.Lq_V r^2 \tag{2.40}$$

If you accept the conditions $r = R_1,\ -\lambda \frac{\partial T}{\partial r} = \alpha_1\left(T - T_{01}\right);\ r = R_2,\ T = T_{S2}$ then the temperature distribution is defined as

$$\frac{T_{01} - T}{T_{01} - T_{S2}} = \left[1 + \frac{q_V R_1^2}{4\lambda\left(T_{01} - T_{S2}\right)}\left(1 - \frac{R_2^2}{R_1^2} - \frac{2}{\mathrm{Bi}_1}\right)\right]\frac{\ln\left(r/R_1\right) + 1/\mathrm{Bi}_1}{\ln\left(R_2/R_1\right) + 1/\mathrm{Bi}_1} + \frac{q_V R_1^2}{4\lambda\left(T_{01} - T_{S2}\right)}\left(\frac{2}{\mathrm{Bi}_1} - 1 + \frac{r^2}{R_1^2}\right) \tag{2.41}$$

2.2.2 Heat Transfer in a Spherical Solid Particles

Let us consider the solution of the heat conduction problem for a hollow spherical body with an internal radius and with an external radius. The heat conduction equation for a spherical body is represented in the form

$$\frac{\lambda}{r^2}\frac{\partial}{\partial r}\left(r^2 \frac{\partial T}{\partial r}\right) = q_V \tag{2.42}$$

1. Suppose that on both surfaces the temperature is given

$$r = R_1,\ T = T_{S1};\ r = R_2,\ T = T_{S2} \tag{2.43}$$

The temperature distribution and the heat flux are determined in the form

$$\frac{T_{S1} - T}{T_{S1} - T_{S2}} = \frac{1 + \frac{q_V R_1^2}{6\lambda\left(T_{S1} - T_{S2}\right)}\left(1 - \frac{R_2^2}{R_1^2}\right)}{\frac{R_2}{R_1} - 1}\left(\frac{R_2}{R_1} - \frac{R_2}{r}\right) - \frac{q_V R_1^2}{6\lambda\left(T_{S1} - T_{S2}\right)}\left(1 - \frac{r^2}{R_1^2}\right) \tag{2.44}$$

$$Q = -4\pi r^2 \lambda \frac{\partial T}{\partial r} = \frac{4\pi\lambda R_2 \left(T_{S1} - T_{S2}\right)}{R_2/R_1 - 1}\left[1 + \frac{q_V R_1^2}{6\lambda\left(T_{S1} - T_{S2}\right)}\left(1 - {R_2^2}/{R_1^2}\right)\right] + {4}/{3}\,\pi q_V r^3 \tag{2.45}$$

2. Suppose that convective heat transfer occurs on both surfaces

$$r = R_1,\ -\lambda\frac{\partial T}{\partial r} = \alpha_1\left(T - T_{01}\right);\ r = R_2,\ -\lambda\frac{\partial T}{\partial r} = \alpha_2\left(T - T_{02}\right) \tag{2.46}$$

The temperature distribution and the heat flux are defined as

$$\frac{T_{01} - T}{T_{01} - T_{02}} = \left[1 + \frac{q_V R_1^2}{6\lambda\left(T_{01} - T_{02}\right)}\left(1 - \frac{R_2^2}{R_1^2} - \frac{2}{\mathrm{Bi}_1} - \frac{2}{\mathrm{Bi}_2}\frac{R_2^2}{R_1^2}\right)\right] \times$$

$$\times \frac{\dfrac{1}{\mathrm{Bi}_1}\dfrac{R_2}{R_1} + \dfrac{R_2}{R_1} - \dfrac{R_2}{r}}{\dfrac{1}{\mathrm{Bi}_1}\dfrac{R_2}{R_1} + \dfrac{R_2}{R_1} - 1 + \dfrac{1}{\mathrm{Bi}_2}} - \frac{q_V R_1^2}{6\lambda\left(T_{01} - T_{02}\right)}\left(1 - \frac{2}{\mathrm{Bi}_1} - \frac{r^2}{R_1^2}\right) \tag{2.47}$$

$$Q = \frac{4\pi\lambda R_2\left(T_{01} - T_{02}\right)}{\dfrac{1}{\mathrm{Bi}_1}\dfrac{R_2}{R_1} + \dfrac{R_2}{R_1} - 1 - \dfrac{1}{\mathrm{Bi}_2}}\left[1 + \frac{q_V R_1^2}{6\lambda\left(T_{01} - T_{02}\right)}\left(\begin{array}{c} 1 - \dfrac{R_2^2}{R_1^2} - \\ \dfrac{2}{\mathrm{Bi}_1} - \dfrac{2}{\mathrm{Bi}_2}\dfrac{R_2^2}{R_1^2}\end{array}\right)\right] + \frac{4}{3}\pi q_V r^3 \tag{2.48}$$

where $\mathrm{Bi}_1 = \dfrac{\alpha_1 R_1}{\lambda}$ and $\mathrm{Bi}_2 = \dfrac{\alpha_2 R_2}{\lambda}$ are Bio numbers.

3. Assume that the surface temperature is given on one surface and the heat flux is given on the other surface

$$r = R_1,\ q = q_1;\ r = R_2,\ T = T_{S2} \tag{2.49}$$

The temperature distribution and the heat flux will be determined as

$$T - T_{S2} = \frac{q_1 R_1^2}{\lambda R_2}\left(\frac{R_2}{r} - 1\right) + \frac{q_V R_1^2}{6\lambda}\left(\frac{R_2^2}{R_1^2} - 2\frac{R_1}{r} + 2\frac{R_1}{R_2} - \frac{r^2}{R_1^2}\right)$$

$$Q = 4\pi\left[q_1 R_1^2 - q_V R_1^3\left(1 - {r^3}/{R_1^3}\right)/3\right] \tag{2.50}$$

If we accept the condition $r = R_1,\ T = T_{S1};\ r = R_2,\ q = q_2$, then the temperature distribution and the heat flux are determined as

$$T_{S1} - T = \frac{q_2 R_2^2}{\lambda R_1}\left(1 - \frac{R_1}{r}\right) - \frac{q_V R_2^2}{6\lambda}\left(\frac{2R_2}{R_1} + \frac{R_1^2}{R_2^2} - \frac{2R_2}{r} - \frac{r^2}{R_2^2}\right)$$

$$Q = 4\pi\left[q_2 R_2^2 - q_V R_2^3 {\left(1 - r^3/R_2^3\right)}/{3}\right] \tag{2.51}$$

4. Suppose that a temperature is set on one surface of the sphere (a condition of the first kind), and on the other surface convective heat transfer (conditions of the third kind)

$$r = R_1,\ T = T_{S1};\ r = R_2,\ -\lambda \frac{\partial T}{\partial r} = \alpha_2 \left(T - T_{02}\right) \tag{2.52}$$

The temperature distribution and the heat flux are represented in the form

$$\frac{T_{S1}-T}{T_{S1}-T_{02}} = \frac{x}{L}\left[\frac{1-\dfrac{q_V L^2}{\lambda\left(T_{S1}-T_{02}\right)}\left(\dfrac{1}{2}+\dfrac{1}{\mathrm{Bi}_2}\right)}{1+{}^{1}\!/_{\mathrm{Bi}_2}} + \frac{q_V L^2}{2\lambda\left(T_{S1}-T_{02}\right)}\frac{x}{L}\right]$$

$$q = \frac{\lambda\left(T_{S1}-T_{02}\right)}{L}\left[\frac{1-\dfrac{q_V L^2}{\lambda\left(T_{S1}-T_{02}\right)}\left(\dfrac{1}{2}+\dfrac{1}{\mathrm{Bi}_2}\right)}{1+{}^{1}\!/_{\mathrm{Bi}_2}} + \frac{q_V L^2}{\lambda\left(T_{S1}-T_{02}\right)}\frac{x}{L}\right] \tag{2.53}$$

The above solutions make it possible to determine the temperature profiles on the wall under various boundary conditions. These solutions also determine the coefficients of thermal conductivity of the wall material, if the structural dimensions of the wall, the temperature conditions, and the source of internal specific heat are known. In most cases, if the velocity of the fluid is very small, the dissipated term can be neglected, as follows from equations (2.2) through (2.6), and the energy dissipation is proportional to the squares of the derivatives of the velocities.

In this paper, we confine ourselves to solving typical problems with mixed boundary conditions, assuming that the set of solutions of heat transfer problems is given in the literature [14,27,30].

These solutions allow us to calculate the heat transfer in practical applications and can be suitable for the design of heat exchange equipment. For a more detailed study of heat transfer, it is necessary to have more complex solutions, for example, solutions of non-stationary or convective heat transfer.

2.3 NON-STATIONARY ONE-DIMENSIONAL SOLUTIONS OF THERMAL CONDUCTIVITY EQUATIONS

In the practice of chemical technology, the problems of non-stationary thermal conductivity are often encountered in the interphase interaction of a gas or liquid medium with dispersed materials, with a no stationary nature of the course of processes and various dynamic conditions. The presence of no stationary or the time factor in heat transfer equations makes it very difficult to theoretically solve the problem. More complicated are the problems of non-stationary heat exchange, connected with the dependence of the boundary conditions on time. These are external oscillatory phenomena or internal sources of temperature fluctuations.

In chemical technology, heat exchange processes occur in a continuous phase in both cylindrical and spherical particles (spherical particles of a catalyst, an adsorbent, cylindrical Raschig rings, etc.). The study of heat transfer in such particles reduces to solving problems with symmetric boundary conditions, which to some extent simplifies the practical solution.

The differential equation of non-stationary heat conductivity for small numbers with neglect of dissipative terms follows from equations (2.2) through (2.6) with the corresponding boundary conditions:

$$\frac{\partial T}{\partial t}=\frac{1}{x^n}\frac{\partial}{\partial x}\left(\chi^2(T)x^n\frac{\partial T}{\partial x}\right)+q_{V0} \tag{2.54}$$

where $\chi^2=a_T=\lambda/c_P\rho$ is the thermal diffusivity coefficient, c_P is isobaric heat capacity of the medium, λ is the thermal conductivity of the medium, and ρ is density of the medium, $q_{V0}=q_V/c_P\rho$. The thermal diffusivity coefficient characterizes the thermophysical properties of the medium and depends on the temperature. If we introduce a dimensionless temperature $\theta=\frac{T-T_0}{T_S-T_0}$ (where T_0 is the initial temperature of the body, T_S is the temperature of the medium), then equation (2.54) can be written in the form

$$\frac{\partial \theta}{\partial t}=\frac{1}{x^n}\frac{\partial}{\partial x}\left(\chi^2(\theta)x^n\frac{\partial \theta}{\partial x}\right)+q_{V0} \tag{2.55}$$

For different values n we have: a flat problem $(n=0)$, a cylindrical problem $(n=1)$, or a spherical problem $(n=2)$. Depending on the specification of the boundary conditions, the solution of the corresponding heat transfer problem can be obtained in the form of infinite, in most cases, rapidly converging series by the method of separation of variables.

2.3.1 Method of Separation of Variables

In most cases, the practical solution of thermal conductivity problems is obtained by the method of variables separation [31] with subsequent evaluation, according to the boundary conditions, eigenvalues, and coefficients of the series. We consider the solution of the boundary-value problem for the heat equation

$$\frac{\partial \theta}{\partial t}=\chi^2\frac{\partial^2\theta}{\partial x^2} \tag{2.56}$$

with initial conditions

$$t=0,\ \theta(x,0)=\varphi(x) \tag{2.56a}$$

and boundary conditions

$$x=0,\ \theta(0,t)=0;\ x=l,\ \theta(0,t)=0 \tag{2.56b}$$

Here, $\theta=\frac{T-T_0}{T_S-T_0}$ is the dimensionless temperature. We represent the general solution of equation (2.56) in the form

$$\theta(x,t)=\varphi(x)\psi(t) \tag{2.57}$$

Substituting this expression into (2.56) and dividing both sides by $\chi^2\varphi\psi$, we obtain

$$\frac{1}{\chi^2\psi}\frac{d\psi}{dt}=\frac{1}{\varphi}\frac{d\varphi}{dx}=-\lambda \tag{2.58}$$

where $\lambda = \text{const}$ since the left-hand side depends on t, and the right-hand side only on x.
It follows two equations, one of which depends on t, and the other on x.

$$\frac{d\psi}{dt} + \chi^2 \lambda \psi = 0 \tag{2.59}$$

$$\frac{d^2\varphi}{dx^2} + \lambda \varphi = 0 \tag{2.60}$$

The solution of the first equation can be represented as

$$\psi(t) = C_1 \exp\left(-\chi^2 \lambda t\right) \tag{2.61}$$

and the second equation

$$\varphi(x) = C_2 \sin\left(\sqrt{\lambda} x\right) \tag{2.62}$$

where C_1, C_2 are the coefficients of integration, determined from the boundary conditions. Using the boundary conditions (2.56a) or $\varphi(0) = 0$ from the second equation, we have $\sin(\lambda x) = 0$. The roots of this equation are equal $\lambda_n = \left(\frac{\pi.n}{l}\right)^2$. Finally, the general solution (2.56) is represented in the form

$$\theta_n(x,t) = \psi(t)\varphi(x) = C_n \sin\left(\frac{\pi.n}{l} x\right) \exp\left(-\chi^2 \left(\frac{\pi.n}{l}\right)^2 t\right) \tag{2.63}$$

Since $n = 1, 2, 3, \ldots$, then this solution should be written as the sum of particular solutions

$$\theta(x,t) == \sum_{n=1}^{\infty} C_n \sin\left(\frac{\pi.n}{l} x\right) \exp\left(-\chi^2 \left(\frac{\pi.n}{l}\right)^2 t\right) \tag{2.64}$$

Requiring the fulfillment of the initial conditions, we have

$$\theta(x,0) = \phi_0(x) = \sum_{n=1}^{\infty} C_n \sin\left(\frac{\pi.n}{l} x\right) \tag{2.65}$$

Multiplying both sides of this equation by $\sin\left(\frac{\pi.n}{l} x\right)$ and further integrating from 0 to 1, we will get:

$$C_n = \frac{2}{l} \int_0^l \phi_0(\xi) \sin\left(\frac{\pi.n}{l} \xi\right) d\xi \tag{2.66}$$

Without going into the details of the theoretical solution, solutions of the most practical problems of non-stationary heat transfer for various surfaces by the method of variables separation will be subsequently given. These solutions, with the identity of the boundary conditions, can also be used to solve the problems of mass transfer. For a planar surface, the equation of no stationary heat transfer for constant thermophysical properties of the medium is represented in the form

$$\frac{\partial \theta}{\partial t} = \chi^2 \frac{\partial^2 \theta}{\partial x^2} + q_{V0} \tag{2.67}$$

where $q_{V0} = q_V / c_P \rho$, $\chi^2 = a_T$ are coefficients of thermal diffusivity.

Consider the solution of this problem under various boundary conditions:

1. Assume that, in the absence of an internal heat source $q_V = 0$, the plate temperature for is equal to $T = T_0$ (a condition of the first kind), the wall is thermally insulated $x = l$ (a condition of the second kind), and the initial temperature distribution of the body is an arbitrary function of x. Then the boundary conditions of the problem are written as follows

$$t = 0,\ T(x,t) = \phi(x)$$

$$x = 0,\ T(x,t) = 0,\ x = l,\ \frac{\partial T}{\partial x} = 0 \tag{2.68}$$

The solution of problem (2.67) is represented in the form

$$T(x,t) = \sum_{k=0}^{\infty} A_k \exp\left[-\left(\frac{(2k+1)\pi}{2}\right)^2 Fo\right] \sin\frac{(2k+1)\pi}{2l}x \tag{2.69}$$

where A_k is the coefficients of the series, determined from equation

$$A_k = \frac{2}{l}\int_0^l \varphi(x)\sin\frac{(2k+1)\pi}{2l}xdx \tag{2.70}$$

where $\mathrm{Fo} = \frac{a_T t}{l^2}$ is the Fourier number.

The amount of heat transferred per unit surface per unit time is equal to $\left(\cos\frac{2k+1}{2}\pi = \frac{1}{2}\left(1-(-1)^k\right)\right)$

$$q_T = -\lambda\left(\frac{\partial T}{\partial x}\right)_{x=l} = \frac{\lambda\pi}{4l}\sum_{k=0}^{\infty} A_k(2k+1)\exp\left[-\left(\frac{(2k+1)\pi}{2}\right)^2 \mathrm{Fo}\right] \tag{2.71}$$

2. Assume that, in the absence of an internal heat source, the temperatures of both sides of the plate are equal, and the initial temperature distribution is a linear function, that is, boundary conditions of the first kind. The boundary conditions of the problem are written in the form

$$\begin{gathered} t = 0,\ T(x,0) = Ax \\ x = 0,\ T(0,t) = T(l,t) = Ax \end{gathered} \tag{2.72}$$

The solution of problem (2.67) for the indicated boundary conditions is written in the form

$$T(x,t) = \frac{2Al}{\pi}\sum_{k=1}^{\infty}\frac{(-1)^{k+1}}{k}\exp\left[-(k\pi)^2 \mathrm{Fo}\right]\sin\frac{k\pi}{l}x \tag{2.73}$$

The amount of heat transferred per unit surface per unit time is equal to $\left(\cos(k\pi) = (-1)^k\right)$

$$q_T = 2A\lambda\sum_{k=1}^{\infty}(-1)^{2k+1}\exp\left[-(k\pi)^2 \mathrm{Fo}\right] \tag{2.74}$$

TABLE 2.1
Roots of Equation Bi = *p*tg (*p*)

Bi	p_1	p_2	p_3	p_4
0.00	0.000	3.142	6.283	9.424
0.04	0.198	3.154	6.289	9.429
0.10	0.311	3.173	6.299	9.435
0.20	0.433	3.204	6.315	9.446
0.50	0.655	3.292	6.361	9.477
1.00	0.860	3.425	6.437	9.529
2.00	1.077	3.644	6.575	9.629
4.00	1.265	3.935	6.814	9.812
8.00	1.397	4.226	7.126	10.095
∞	1.571	4.712	7.854	10.995

3. Assume that $x = 0$ the surface of the plate is thermally insulated (a condition of the second kind), for $x = l$, the heat exchange between the plate surface and the flow is of a convective nature (conditions of the third kind), and the initial temperature distribution is arbitrary. Then the boundary conditions are written in the form

$$t = 0,\ T(x,0) = \phi(x)$$

$$x = 0,\ \frac{\partial T}{\partial \vartheta} = 0;\ \frac{\partial T}{\partial \vartheta} + \text{Bi}\,\theta(l,t) = 0 \tag{2.75}$$

The solution of equation (2.67) is represented in the form

$$T(x,t) = 2\sum_{k=1}^{\infty}\left[\frac{Bi^2 + \mu_k^2}{l\left(Bi^2 + \mu_k^2\right) + Bi}\int_0^l \varphi(\varsigma)\cos\mu_k\varsigma.d\varsigma\right]\exp\left(-\chi^2\mu_k^2 t\right)\cos\mu_k x \tag{2.76}$$

where μ_k is the eigenvalue determined from the solution of the equation $\mu\,\text{tg}(\mu l) = \text{Bi}$. Here and in the subsequent exposition $\text{Bi} = {}^{\alpha L}\!/\!_{\lambda}$ is the number of Bio, where α is the heat transfer coefficient, and λ is the coefficient of thermal conductivity. The positive roots of the solution to the equation $\text{Bi} = p\,\text{tg}(p)$ are shown in Table 2.1, where $p = \mu.l$.

4. Assume that in the absence of an internal heat source $q_v = 0$ on either side of the plate, convective heat transfer (conditions of the third kind) and the initial temperature of the plate is constant

$$t = 0,\ \theta(x,0) = \theta_M$$

$$x = 0,\ \frac{\partial \theta}{\partial x}(0,t) - B\theta(0,t) = 0,\ x = l \tag{2.77}$$

Solution (2.67) is represented in the form

$$\theta(x,t) = \sum_{k=1}^{\infty} a_k \exp\left(-\chi^2\mu_k^2 t\right).\left[\mu_k\cos\mu_k x + B\sin\mu_k x\right] \tag{2.78}$$

where $a_k = \dfrac{2\theta_M}{l\left(B^2+\mu_k^2\right)}\left(\dfrac{B}{\mu_k}+\dfrac{B^2+\mu_k^2}{2\mu_k^2}\sin\mu_k l\right)$ and μ_k are the positive roots of the equation $ctg(\mu.l) = \dfrac{1}{2}\left(\dfrac{\mu}{B}-\dfrac{B}{\mu}\right)$, $B = \alpha/\lambda$.

5. Let us assume that there is an arbitrary source of heat in the volume of the plate $q_{V0} = f(\theta)$, at $x = 0$, the surface temperature $T = T_0$ (conditions of the first kind), and when $x = l$ there is a constant heat source on the surface q (conditions of the second kind). The boundary conditions are represented in the form

$$t = 0,\ \theta(x,0) = \varphi(x)$$

$$x = 0,\ \theta(0,t) = 0,\ x = l,\ \partial\theta/\partial x = q \tag{2.79}$$

The solution of equation (2.76) is represented in the form

$$\theta(x,t) = \theta_R(x) + \sum_{k=0}^{\infty} a_k \exp\left[-\left(\frac{(2k+1)\pi}{2}\right)^2 \text{Fo}\right]\sin\frac{(k+1)\pi}{2l}x \tag{2.80}$$

where $\theta_R(x) = -\dfrac{1}{\chi^2}\displaystyle\int_0^x\left[\int_0^y f(\xi)d\xi\right]dy + \dfrac{x}{\chi^2}\int_0^l f(\xi)d\xi + qx,$

$$a_k = \frac{2}{l}\int_0^l \left[\varphi(x) - \theta_R(x)\right]\sin\frac{(2k+1)\pi}{2l}xdx$$

6. Let us assume that a negative heat source acts linearly in the volume of a planar body, linearly depending on the temperature $q_{V0} = -\beta\theta$, the temperatures of both surfaces are constant (the boundary condition of the first kind), and the initial temperature of the body is an arbitrary function of x

$$t = 0,\ \theta(x,0) = \varphi(x)$$

$$x = 0,\ \theta(0,t) = 0;\ x = l,\ \theta(l,t) = 0 \tag{2.81}$$

The solution of equation (2.54) with the indicated boundary conditions is represented as

$$\theta(x,t) = \frac{2}{l}\sum_{k=1}^{\infty}\left(\int_0^l \phi(\xi)\sin\left(\frac{k\pi}{l}\xi\right)d\xi\right)\exp\left[-\left(\left(\frac{\chi k\pi}{l}\right)^2 + \beta\right)t\right]\sin\left(\frac{k\pi}{l}x\right) \tag{2.82}$$

7. Assume that in the presence of a linear negative heat source in the volume, the temperature of a flat body on one side is constant (a boundary condition of the first kind). On the other hand, it is insulated (a boundary condition of the second kind), and the initial temperature is an arbitrary function of the coordinate

$$t = 0,\ \theta(t,0) = \phi(x)$$

$$x = 0,\ \theta(0,t) = 0;\ x = 1,\ \partial\theta/\partial x\,(l,t) = 0 \tag{2.83}$$

The solution of the boundary-value problem (2.67) is represented in the form

$$\theta(x,t) = \exp\left[-\left(\frac{\chi^2\pi^2}{4l^2} + \beta\right)t\right]\sin\left(\frac{\pi}{2l}x\right) \tag{2.84}$$

Since the surface is thermally insulated, then, as follows from (2.84), the amount of heat transferred is zero.

Boundary problems with other possible tasks of boundary conditions can also be solved by the method of separation of variables. The solution of the boundary-value problems of semibounded bodies is carried out by transforming equation (2.67) to a single variable, in more detail, as will be described in subsequent chapters on mass transfer. Introducing the new variable $\eta = x/\sqrt{4\chi^2 t}$ equation (2.67) can be reduced to an ordinary differential equation of the second order

$$\frac{d^2\theta}{d\eta^2} + 2\eta\frac{d\theta}{d\eta} = 0 \tag{2.85}$$

Solution (2.85) will be presented in the form

$$\theta(\eta) = A_0\int_0^\eta \exp\left(-\xi^2\right)d\xi + A_1 \tag{2.86}$$

Depending on the boundary conditions, this expression can take the solution of the problem. Suppose that heat transfer on the surface of a flat body is carried out conventionally, and there is no heat exchange away from the surface, that is,

$$x = 0,\ -\frac{\partial\theta}{\partial x} = B\theta,\ x \to \infty,\ \frac{\partial\theta}{\partial x} = 0 \tag{2.87}$$

For a constant initial temperature of the body, the solution of this boundary-value problem is represented in the form

$$\theta(x,t) = \operatorname{erfc}\left(\frac{x}{2\sqrt{\chi^2.t}}\right) - \exp\left(\frac{\alpha}{\lambda}x + \frac{\alpha^2}{\lambda^2}\chi^2 t\right)\operatorname{erfc}\left(\frac{x}{2\sqrt{\chi^2 t}} + \frac{\alpha}{\lambda}\sqrt{\chi^2 t}\right) \tag{2.88}$$

Here, $\operatorname{erfc}(u) = 1 - \operatorname{erf}(u) = 1 - \frac{2}{\sqrt{\pi}}\int_0^u \exp\left(-\xi^2\right)d\xi$, $\operatorname{erf}(u) = \frac{2}{\sqrt{\pi}}\int_0^u \exp\left(-\xi^2\right)d\xi$ are the error function, the heat transfer coefficient and the thermal conductivity.

2.3.2 Cylindrical Heat Transfer Surface

Heat transfer in a cylindrical body at Pe << 1 and with constant thermal and physical properties of the flow is described by an equation in cylindrical coordinates

$$\frac{\partial\theta}{\partial t} = \frac{\chi^2}{r}\frac{\partial}{\partial r}\left(r\frac{\partial\theta}{\partial r}\right) + q_{V0} \tag{2.89}$$

where $q_{V0} = q_V/c_p\rho$. Below are solutions of (2.89) for mixed boundary conditions with limited results.

TABLE 2.2
The Roots of the Equation

Bi	p_1	p_2	p_3	p_4
0,00	0,000	3,381	7,015	10,173
0,04	0,281	3,842	7,021	10,177
0,10	0,442	3,857	7,029	10,183
0,20	0,617	3,883	7,044	10,193
0,40	0,851	3,934	7,072	10,213
1,00	1,256	4,079	7,155	10,271
2,00	1,599	4,291	7,288	10,365
4,00	1,908	4,601	7,520	10,543
6,00	2,049	4,803	7,704	10,969
8,00	2,128	4,986	8,846	11,791
∞	2,405	5,520	8,864	11,791

1. Let us assume that, in the absence of a heat source in the volume, the heat transfer is carried out conventionally on the surface of a cylindrical body (a boundary condition of the third kind), and the initial temperature is a function of the radius

$$t=0,\ \ \theta(r,0)=\varphi(r),\ \ 0\le r<R$$

$$r=R,\ \ \frac{\partial\theta}{\partial r}(R,t)+B\theta(R,t)=0,\ \ |\theta(0,t)|<\infty \tag{2.90}$$

The solution of this boundary-value problem is represented in the form

$$\theta(r,t)=\frac{2}{R^2}\sum_{k=1}^{\infty}\frac{1}{\mathrm{J}_0^2(\mu_k)+\mathrm{J}_1^2(\mu_k)}\left[\int_0^R\rho\phi(\rho)\mathrm{J}_0\left(\frac{\mu_k\rho}{R}\right)d\rho\right]\exp\left[-\mu_k^2\mathrm{Fo}\right]\mathrm{J}_0\left(\frac{\mu_k r}{R}\right) \tag{2.91}$$

where μ_k is the positive roots of equation

$$\mu\mathrm{J}_0'(\mu)+B\mathrm{J}_0(\mu)=0 \tag{2.92}$$

$\mathrm{J}_0(\mu)$ and $\mathrm{J}_1(\mu)$ are Bessel functions of the first kind of zero and first order, R is the radius of the cylinder, and $\mathrm{Fo}=\frac{a_T t}{R^2}$ is the Fourier number. Assuming that $\mathrm{J}_0'(\mu)=-\mathrm{J}_1(\mu)$ we determine the positive roots of the equation $\mathrm{Bi}=p\frac{\mathrm{J}_1(p)}{\mathrm{J}_0(p)}$ given in Table 2.2, where $p=\mu R$.

2. Suppose that the surface of an infinite homogeneous cylindrical body is thermally insulated, and the initial temperature distribution is described by a nonlinear function of the type $\theta(r,0)=\theta_0 r^2$ (where θ_0 is coefficient, θ is temperature). The boundary conditions have the form

$$r=R,\ \ \frac{\partial\theta}{\partial r}(R,t)=0,\ \ |\theta(0,t)|<\infty \tag{2.93}$$

The solution of this boundary-value problem is represented in the form

$$\theta(r,t)=\frac{\theta_0 R^2}{2}+4\theta_0 R^2\sum_{k=1}^{\infty}\frac{1}{\mu_k^2 \mathrm{J}_0(\mu_r)}\exp\left[-\mu_k^2\mathrm{Fo}\right]\mathrm{J}_0\left(\frac{\mu_k r}{R}\right) \tag{2.94}$$

where μ_k is the positive roots of the equation $\mathrm{J}_1(\mu)=0$, $\mathrm{Fo}=\frac{a_T t}{R^2}$.

3. Assume that there is convective heat transfer on the surface of an infinite cylindrical body, and the initial temperature distribution is analogous to the above condition. Then the boundary condition is written in the form

$$r=R,\ \frac{\partial\theta}{\partial r}(R,t)+B\theta(R,t)=0 \tag{2.95}$$

The solution of this boundary-value problem is represented in the form

$$\theta(r,t)=2\theta_0 R^2\sum_{k=1}^{\infty}\frac{4\mathrm{Bi}+(2-Bi)\mu_k^2}{\mu_k^2\left(\mu_k^2+\mathrm{Bi}^2\right)\mathrm{J}_0(\mu_k)}\exp\left(-\mu_k^2\mathrm{Fo}\right)\mathrm{J}_0\left(\frac{\mu_k r}{R}\right) \tag{2.96}$$

Here, μ_k is the positive root of the equation $\mu\mathrm{J}_0'(\mu)+\mathrm{Bi}\,\mathrm{J}_0(\mu)=0$, $\mathrm{Fo}=\frac{a_T t}{R^2}$ is the Fourier number. The amount of heat transferred per unit surface per unit time is equal to $\left(\mathrm{J}_0'(\mu)=-\mathrm{J}_1(\mu)\right)$

$$q_T=2\theta_0 R\lambda\sum_{k=1}^{\infty}\frac{4\mathrm{Bi}+(2-\mathrm{Bi})\mu_k^2}{\mu_k\left(\mu_k^2+\mathrm{Bi}^2\right)}\frac{\mathrm{J}_1(\mu_k)}{\mathrm{J}_0(\mu_k)}\exp\left(-\mu_k^2\mathrm{Fo}\right) \tag{2.97}$$

4. Let us assume that at the initial instant of time, the temperature of the flow and the walls of the pipe $r_0\le r\le R$ (r_0, R is the inner and outer radii of the tube) are constant and equal.
Let us determine the distribution of the temperature of the pipe wall for $t>0$ if its inner surface is thermally insulated (boundary conditions of the second kind), and a constant temperature is maintained on the external surface equal to θ_0 (boundary conditions of the first kind)

$$t=0,\ \theta(r,0)=\theta_0$$

$$r=r_0,\ \frac{\partial\theta}{\partial r}=0,\ r=R,\ \theta(r,t)=\theta_1 \tag{2.98}$$

The solution of this boundary-value problem is represented in the form

$$\theta(r,t)=\theta_1+\pi(\theta_0-\theta_1)\sum_{k=1}^{\infty}a_k\exp\left(-\mu_k^2\chi^2 t\right)\left[\mathrm{N}_0(\mu_k R)\mathrm{J}_0(\mu_k r)-\mathrm{J}_0(\mu_k R)\mathrm{N}_0(\mu_k r)\right] \tag{2.99}$$

where $a_k=\frac{J_0^2(\mu_k r_0)}{J_1^2(\mu_k r_0)-J_0^2(\mu_k R)}$, μ_k are the positive roots of the equation, $\mathrm{N}_0(\mu R)\mathrm{J}_1(\mu r_0)-\mathrm{J}_0(\mu R)\mathrm{N}_1(\mu r_0)=0$, $N_0(\mu), N_1(\mu)$ are Bessel functions of the second kind of zero and first order or Neumann functions.

5. Let us assume that the temperature on the cylindrical body surface is constant (a boundary condition of the first kind), and an arbitrary source of heat acts in the body volume $q_{V0} = f(r,t)$. Let us determine the distribution of body temperature under the following boundary conditions

$$t = 0,\ \ \theta(r,t) = 0$$

$$r = R,\ \theta(R,t) = 0,\ \left|\theta(r,t)\right| < \infty \tag{2.100}$$

The solution of the boundary-value problem is represented in the form

$$\theta(r,t) = \frac{2}{R^2}\sum_{k=1}^{\infty}\frac{1}{J_1^2(\mu_k)}\left[\int_0^t\int_0^R \exp\left[-\left(\frac{\chi\,\mu_k}{R}\right)^2(t-\tau)\right]\xi\, f(\xi,\tau)J_0\left(\frac{\mu_k\xi}{R}\right)d\xi\, d\tau\right]J_0\left(\frac{\mu_k r}{R}\right) \tag{2.101}$$

where μ_k is the positive root of the equation.

6. Suppose that the outer surface of a cylindrical body is thermally insulated, and a negative heat source acts in the volume $q_{V0} = -\beta\theta$, that is, at

$$t = 0,\ \ \theta(r,0) = \phi(r)$$

$$r = R,\ \frac{\partial\theta}{\partial r} = 0,\ \left|\theta(r,t)\right| < \infty \tag{2.102}$$

The solution of this boundary-value problem is represented in the form

$$\theta(r,t) = \exp(-\beta t)\sum_{k=0}^{\infty} c_k \exp\left(-\mu_k^2\mathrm{Fo}\right)J_0\left(\frac{\mu_k r}{R}\right) \tag{2.103}$$

Here, $c_k = \frac{2}{R^2 J_0^2(\mu_k)}\int_0^R \xi.\phi(r)J_0\left(\frac{\mu_k\,\xi}{R}\right)d\xi$ and μ_k are the positive roots of the equation $J_0'(\mu) = -J_1(\mu) = 0$, with $\mu_0 = 0$, $\mu_1 = 3.8317$, $\mu_2 = 7.0156$, etc. With these eigenvalues, the series (2.103), with allowance for $\mathrm{Fo} \geq 0{,}3$ convergence, already converges on the second term.

7. Suppose that the temperature on the surface of a homogeneous cylinder obeys an exponential law $\theta(R,t) = A\exp\left(-b^2 t\right)$, and the initial temperature is constant $\theta(r,0) = 0$. The solution of the boundary-value problem (2.94) is represented as

$$\theta(r,t) = \frac{A\exp\left(-b^2 t\right)}{J_0\left(\dfrac{bR}{\chi}\right)}J_0\left(\frac{br}{\chi}\right) - 2\chi^2 A\sum_{k=1}^{\infty}\frac{\mu_k}{\chi^2\mu_k^2 - b^2R^2}\exp\left(-\mu_k^2\mathrm{Fo}\right)J_0\left(\frac{\mu_k r}{R}\right) \tag{2.104}$$

where μ_k is the positive root of the equation $J_0(\mu) = 0$.

8. We consider the problem of heat transfer in an infinite cylindrical body having a certain initial temperature, which is an arbitrary function, and a constant surface temperature. The heat transfer equation can be written in the form

$$\frac{\partial\theta}{\partial t}=\chi^2\left(\frac{\partial^2\theta}{\partial r^2}+\frac{1}{r}\frac{\partial\theta}{\partial r}+\frac{1}{r^2}\frac{\partial^2\theta}{\partial\phi^2}\right) \tag{2.105}$$

$$t=0,\ \theta(r,\phi,0)=f(r,\phi),\ r=R,\ \theta(R,\phi,t)=0$$

The solution of this boundary-value problem by the method of separation of variables is represented in the form

$$\theta(r,\phi,t)=\sum_{n=0}^{\infty}\sum_{m=1}^{\infty}\left(A_{nm}\cos n\phi+B_{nm}\sin n\phi\right)\mathrm{J}_n\left(\frac{\mu_{nm}r}{R}\right)\exp\left(-\mu_{nm}^2\mathrm{Fo}\right) \tag{2.106}$$

where A_{nm}, B_{nm} are the coefficients of the series and defined as follows

$$A_{nm}=\frac{\int_0^R\int_0^{2\pi}f(r,\phi)\mathrm{J}_n\left(\frac{\mu_{nm}r}{R}\right)r\cos n\phi\,d\phi\,dr}{\frac{\pi R^2}{2}\alpha_n\left[\mathrm{J}_n'(\mu_{nm})\right]^2}$$

$$B_{nm}=\frac{\int_0^R\int_0^{2\pi}f(r,\phi)\mathrm{J}_n\left(\frac{\mu_{nm}r}{R}\right)r\sin n\phi\,d\phi\,dr}{\frac{\pi R^2}{2}\alpha_n\left[\mathrm{J}_n'(\mu_{nm})\right]^2} \tag{2.107}$$

where $\alpha_n=1$, if and $n\neq 0$ and $\alpha_n=2$, if $n=0$; and μ_{nm} are eigenvalues or root of the equation $J_n(\mu)=0$. If the initial temperature depends only on r or $\theta(r,\varphi,0)=f(r)$, then the solution is somewhat simplified

$$\theta(r,t)=\sum_{m=1}^{\infty}C_m\mathrm{J}_0\left(\frac{\mu_{nm}r}{R_0}\right)\exp\left(-\mu_{0m}\mathrm{Fo}\right) \tag{2.108}$$

where $C_m=\frac{2\int_0^R f(r)\mathrm{J}_0\left(\frac{\mu_{0m}r}{R}\right)r\,dr}{R^2\left[\mathrm{J}_1(\mu_{0m})\right]^2}$ and μ_{0m} are the roots of the equation $J_0(\mu)=0$. These are the following values: $\mu_{01}=2.4048$ ($\mathrm{J}_1(\mu_{01})=0.5191$); $\mu_{02}=5.5201$ ($\mathrm{J}_1(\mu_{02})=-0.33403$); $\mu_{03}=$ 8.6537 (μ_{03})= 02.2715); etc. If the initial temperature is a constant value $\theta(r,\varphi,0)=\theta_0$, then the solution becomes simpler to the form

$$\theta(r,t)=\theta_0\sum_{m=1}^{\infty}\frac{2\mathrm{J}_0\left(\frac{\mu_{0m}r}{R}\right)}{\mu_{0m}\mathrm{J}_1(\mu_{0m})}\exp\left(-\mu_{0m}^2\mathrm{Fo}\right) \tag{2.109}$$

This series converges for values $\mathrm{Fo} \geq 0.3$, so we can restrict ourselves to the first terms. For example, on the axis of the cylinder, using the appropriate values of μ_{01} and $J_1(\mu_{01})$, we can write

$$\theta(t) = \theta_0 \frac{2}{2.40 \times 0.52} \exp\left(-(2.40)^2 \mathrm{Fo}\right) = 1.6\theta_0 e^{-5.76 Fo} \tag{2.110}$$

2.3.3 Spherical Heat Transfer Surface

Heat transfer in a spherical body for $\mathrm{Pe} << 1$ asymmetric flow in the general case is described by equations

$$\frac{\partial \theta}{\partial t} = \frac{\chi^2}{r^2}\frac{\partial}{\partial r}\left(r^2 \frac{\partial \theta}{\partial r}\right) + \frac{\chi^2}{r^2 \sin\psi}\frac{\partial}{\partial \psi}\left(\sin\psi \frac{\partial \theta}{\partial \psi}\right) + \frac{\chi^2}{r^2 \sin^2\psi}\frac{\partial^2 \theta}{\partial \varphi^2} + q_{V0} \tag{2.111}$$

with the corresponding boundary conditions (where φ and ψ are polar angles). The solution of the nonlinear equation (2.111) for various inhomogeneous boundary conditions presents certain difficulties. Below are various solutions (2.111) for mixed boundary-value problems.

1. Suppose that in the absence of an internal heat source $q_{v0} = 0$, the temperature on the surface of the sphere is constant $\theta(R,\psi,\varphi,t) = 0$ (a boundary condition of the first kind), and the initial surface temperature is an arbitrary function $\theta(r,\psi,\varphi,0) = f(r,\psi,\varphi)$, where $0 \leq r \leq R$, $0 \leq \psi \leq \pi$, and $0 \leq \varphi \leq 2\pi$. The general solution of this boundary-value problem is represented in spherical coordinates in the form

$$\theta(r,\psi,\phi,t) = \sum_{n=0}^{\infty}\sum_{m=1}^{\infty}\sum_{k=0}^{n} \exp\left(-\mu_{nm}^2 \mathrm{Fo}\right)\frac{1}{\sqrt{r}} \mathrm{J}_{n+1/2}\left(\frac{\mu_{n,m} r}{R}\right) \times \left(A_{mnk}\cos k\phi + B_{mnk}\sin k\phi\right)\mathrm{P}_n^k(\cos\psi) \tag{2.112}$$

where $\mu_{n,m}$ are the positive roots of the equation, $\mathrm{J}_{n+1/2}(\mu) = 0$

$$A_{mnk} = \frac{(2n+1)(n-k)!}{\pi\alpha_k (n+k)!\left[\mathrm{J}_{n+1/2}'(\mu_{n,m})\right]^2 R^2}\int_0^R\int_0^\pi\int_0^{2\pi} r^{3/2} f(r,\psi,\phi)\mathrm{J}_{n+1/2}\left(\frac{\mu_{n,m} r}{R}\right) \times \cos k\phi\, \mathrm{P}_n^k(\cos\psi)\sin\psi\, dr\, d\psi\, d\phi \tag{2.113}$$

$$B_{mnk} = \frac{(2n+1)(n-k)!}{\pi\alpha_k (n+k)!\left[\mathrm{J}_{n+1/2}'(\mu_{n,m})\right]^2 R^2}\int_0^R\int_0^\pi\int_0^{2\pi} r^{3/2} f(r,\psi,\phi)\mathrm{J}_{n+1/2}\left(\frac{\mu_{n,m} r}{R}\right) \times \sin k\phi\, P_n^k(\cos\psi)\sin\psi\, dr\, d\psi\, d\phi \tag{2.114}$$

where $\alpha_k = 2$, if $k = 0$, $\alpha_k = 1$ if $k \neq 0$, and $\mathrm{P}_n^k(\cos\psi)$ are adjoin Legendre functions. This problem is of interest in the inhomogeneous distribution of the initial temperature on the surface, in the case of a no-uniform concentration distribution during mass transfer, in coagulation and heterocoagulation of dispersed particles, coalescence of droplets, bubbles, etc.

2. Under the conditions of the preceding problem, we give a somewhat simplified solution of the problem (2.112), assuming that there is no temperature change with respect to the coordinate, that is, there is no third term. The solution of the boundary-value problem is represented in the form

$$\theta(r,\psi,t)=\sum_{n=0}^{\infty}\sum_{m=0}^{\infty}\frac{a_{nm}}{\sqrt{r}}\exp\left(-\mu_{nm}^2\mathrm{Fo}\right)\mathrm{P_n}(\cos\psi)\mathrm{J}_{n+1/2}\left(\frac{\mu_{nm}r}{R}\right) \tag{2.115}$$

Here, $a_{nm}=\frac{2n+1}{R^2\left[\mathrm{J}'_{n+1/2}(\mu_{nm})\right]^2}\int_0^R\int_0^{\pi} r^{3/2}f(r,\psi)\mathrm{J}_{n+1/2}\left(\frac{\mu_{nm}r}{R}\right)\mathrm{P}_n(\cos\psi)\sin\psi\,dr\,d\psi$, μ_{nm} are the positive roots of the equation $\mathrm{J}_{n+1/2}(\mu)=0$.

3. Suppose that the distribution of temperature in the body of a spherical shape is symmetrical, that is, there is no temperature dependence on the coordinates φ and ψ. In this case, (2.69) for a homogeneous spherical body introduces itself as

$$\frac{\partial\theta}{\partial t}=a_T\left(\frac{\partial^2\theta}{\partial r^2}+\frac{2}{r}\frac{\partial\theta}{\partial r}\right) \tag{2.116}$$

Let us assume that the distribution of the temperature of the ball at the initial instant of time is assumed not to be uniform, but convective heat transfer occurs on the surface of the ball

$$\theta(0,t)\neq 0,\ r=R,\ -\lambda\frac{\partial\theta}{\partial r}=\alpha\theta$$

$$\theta(r,0)=\theta_0(r),\ r=0,\ \frac{\partial\theta}{\partial r}=0 \tag{2.117}$$

The latter condition means that the temperature is bounded at the center of the sphere. The solution of this boundary-value problem is represented in the form

$$\theta(r,t)=\frac{2}{R}\sum_{n=0}^{\infty}a_n\frac{\mu_n}{\mu_n-\sin\mu_n\cos\mu_n}\frac{\sin\left(\frac{\mu_n r}{R}\right)}{r}\exp\left(-\mu_n^2\mathrm{Fo}\right) \tag{2.118}$$

where $a_n=\int_0^R r\theta_0(r)\sin\left(\mu_n\frac{r}{R}\right)dr$ μ_n are the eigenvalues of the problem, defined as the positive roots of the equation, $\mathrm{tg}\mu=\frac{\mu}{1-\mathrm{Bi}}$, $\mathrm{Bi}={}^{\alpha R}\!/_{\lambda}$ are the numbers of Bio.

The roots of this equation are presented in Table 2.3.

If the initial temperature of the ball is uniformly distributed $\theta_0=1$, this solution is transformed to the form

$$\theta(r,t)=2\sum_{n=1}^{\infty}\frac{\sin\mu_n-\mu_n\cos\mu_n}{\mu_n-\sin\mu_n\cos\mu_n}\frac{\sin\left(\mu_n\frac{r}{R}\right)}{\mu_n\frac{r}{R}}\exp\left(-\mu_n^2\mathrm{Fo}\right) \tag{2.119}$$

TABLE 2.3
The Roots of Equation $t_g\mu = \frac{\mu}{1-\mathrm{Bi}}$

Bi	μ_1	μ_2	μ_3	μ_4
0.00	0.000	4.493	7.725	10.904
0.04	0.345	4.502	7.730	10.907
0.10	0.542	4.516	7.738	10.913
0.40	1.052	4.582	7.777	10.940
1.00	1.571	4.712	7.854	10.966
2.00	2.028	4.913	7.978	11.086
4.00	2.455	5.233	8.204	11.256
6.00	2.654	5.454	8.391	11.408
8.00	2.765	5.607	8.540	11.540
∞	3.140	6.283	9.424	12.566

The average temperature of the sphere is

$$\bar{\theta}(t) = \sum_{n=1}^{\infty} \frac{6\mathrm{Bi}^2}{\mu_n^2\left(\mu_n^2 + \mathrm{Bi}^2 - \mathrm{Bi}\right)} \exp\left(-\mu_n^2 \mathrm{Fo}\right) \tag{2.120}$$

In the limiting case, $Bi \to \infty$, this expression simplifies to a simpler form

$$\bar{\theta}(t) = 6\sum_{n=1}^{\infty} \frac{1}{\mu_n^2} \exp\left(-\mu_n^2 \mathrm{Fo}\right) \tag{2.121}$$

This problem is of interest both in the transfer of heat in a spherical grain and in the diffusion transfer of mass in the grain of a catalyst, an adsorbent, etc.

4. Assume that the initial temperature of a homogeneous sphere is constant, and the surface temperature is also maintained at a constant temperature (a boundary condition of the first kind)

$$\frac{\partial \theta}{\partial t} = \frac{\chi^2}{r^2} \frac{\partial}{\partial r}\left(r^2 \frac{\partial \theta}{\partial r}\right) \tag{2.122}$$

$$\left|\theta(0,t)\right| < \infty,\ r = R,\ \theta(R,t) = \theta_P,\ t = 0,\ \theta(r,0) = \theta_0$$

The solution of this boundary-value problem is represented in the form

$$\theta(r,t) = \theta_P + \frac{2R(\theta_0 - \theta_P)}{\pi r} \sum_{k=1}^{\infty} \frac{(-1)^{k+1}}{k} \exp\left[-k^2\pi^2 \mathrm{Fo}\right] \sin\frac{k\pi r}{R} \tag{2.123}$$

5. Suppose that a constant heat source acts Q in a homogeneous spherical body, the initial temperature of the body is equal T_0, and the temperature of the medium is equal to T_∞. For a number Pe << 1, let us determine the temperature distribution in the sphere if convective heat exchange with the surrounding medium occurs on the surface of the body (a boundary condition of the third kind). The heat transfer equations can be represented in the form

$$\frac{\partial T}{\partial t} = \frac{\chi^2}{r^2} \frac{\partial}{\partial r}\left(r^2 \frac{\partial T}{\partial r}\right) + \frac{Q}{c_P \rho} \tag{2.124}$$

$$t = 0,\ T(r,0) = T_0,\ r = R,\ \lambda \frac{\partial T}{\partial r} = -\alpha \left(T_\infty - T(R,t)\right)$$

The solution of this problem can be represented in the form

$$T(r,t) = T_\infty + \frac{QR}{3\alpha} + \frac{Q\left(R^2 - r^2\right)}{6\lambda} + \\ + \frac{2\mathrm{Bi}}{r} \sum_{k=1}^{\infty} \left(T_0 - T_\infty - \frac{Q}{\lambda \mu_k^2}\right) \frac{\cos R\mu_k \sin \mu_k r}{\mu_k \left(1 - \mathrm{Bi} - \cos^2 R\mu_k\right)} \exp\left(-\chi^2 \mu_k^2 t\right) \tag{2.125}$$

Here, α is the coefficient of heat transfer, λ is the coefficient of thermal conductivity of the body, and μ_k is the positive root of the equation $\mathrm{tg}\, R\mu = \frac{R\mu}{1-\mathrm{Bi}}$.

Example 2.3

Zeolite is subjected to desorption—heating with hot air in a device with a suspended layer of particles. Average air temperature $T_\infty = 200°\mathrm{C}$. Zeolite enters the apparatus with temperature $T_0 = 20°\ \mathrm{C}$. The diameter of the zeolite grain $d = 4\ \mathrm{mm}$, its density $\rho_T = 1100\ \mathrm{kg/m^3}$, specific heat capacity $c_T = 870\ \mathrm{J/(kgK)}$, and thermal conductivity coefficient $\lambda_T = 0.24\ \mathrm{W/(mK)}$. Determine the amount of heat transferred to the spherical zeolite grain, if the air velocity $V = 4.36\ \mathrm{m/s}$ and the porosity of the layer $\xi = 0.67$. As a solution, we use equation (2.125), setting $Q = 0$

$$T(r,t) = T_\infty + \frac{2\mathrm{Bi}\left(T_0 - T_\infty\right)}{r} \sum_{k=1}^{\infty} \frac{\cos \mu_k R \sin \mu_k r}{\mu_k \left(1 - \mathrm{Bi} - \cos^2 \mu_k R\right)} \exp\left(-\mu_k^2 R^2 \mathrm{Fo}\right) \tag{2.126}$$

The amount of heat transferred will be determined as

$$Q_P = -4\pi R^2 \lambda \frac{\partial T}{\partial r} \uparrow_{r=R} \tag{2.127}$$

Define the derivative

$$\frac{\partial T}{\partial r} \uparrow_{r=R} = -\frac{2Bi\left(T_0 - T_\infty\right)}{R^2} \sum_{k=1}^{\infty} \frac{\cos^2 \mu_k R}{1 - \mathrm{Bi} - \cos^2 \mu_k R} \exp\left(-\mu_k^2 R^2 \mathrm{Fo}\right) \tag{2.128}$$

Then we have

$$Q_P = 8\pi Bi\, \lambda_T \left(T_0 - T_\infty\right) \sum_{k=1}^{\infty} \frac{\cos^2 \mu_k R}{1 - \mathrm{Bi} - \cos^2 \mu_k R} \exp\left(-\mu_k^2 R^2 \mathrm{Fo}\right) \tag{2.129}$$

Let us determine the coefficient of heat transfer from air to a solid particle by the formula

$$\mathrm{Nu} = 0.4 \mathrm{Re}^{0.67} \mathrm{Pr}^{0.33}$$

where $\mathrm{Re} = \frac{V d \rho_C}{\xi \eta_C} = \frac{4.36 \times 0.004 \times 0.736}{0.67 \times 25.7 \times 10^{-6}} = 773$ and $\mathrm{Pr} = \frac{c_C \eta_C}{\lambda_C} = \frac{1020 \times 25.7 \times 10^{-6}}{0.0385} = 0.681$, the air density is $\rho_C = 0.763\ \mathrm{kg/m^3}$, and the dynamic viscosity of air $\eta_C = 25.7 \times 10^{-6}\mathrm{Pas}$, at temperature 200°C. Hence, we have

$$\mathrm{Nu} = 0.4 \times 773^{0.67}\ 0.681^{0.33} = 30.2$$

The heat transfer coefficient is

$$\alpha = \frac{Nu\,\lambda_C}{d} = \frac{30.2\times 0.0385}{0.004} = 290\ \mathrm{W}/\mathrm{m^2K}$$

The number of Bio is defined as

$$\mathrm{Bi} = \frac{\alpha R}{\lambda_T} = \frac{290\times 0.002}{0.24} = 2.42$$

Then the values of the eigenvalues are determined from the solution of equation

$$\mathrm{tg}\,\mu R = \frac{\mu R}{1-\mathrm{Bi}} = -\frac{\mu R}{1.42} \tag{2.130}$$

Since the series (2.129) converges rapidly, one of the first solutions of this equation is $\mu R \approx 2.154$. We define the Fourier number

$$\mathrm{Fo} = \frac{\lambda . t}{c_T \rho_T R^2} = \frac{0.24\times 10}{870\times 1100\times (0.002)^2} \approx 0.627$$

The amount of heat transferred to one particle is determined from equation (2.129)

$$Q_P = -8\times 3.14\times 2.42\times 0.24\times 180\frac{\cos^2\frac{180}{3.14}2.154}{1-2.154-\cos^2\frac{180}{3.14}2.154}\exp\left[-(2.154)^2\,0.627\right] = \tag{2.131}$$

$$= 25.1W$$

Above we considered various solutions of the equations of stationary and no stationary heat conduction with mixed boundary conditions.

Since the heat transfer equations and the convective diffusion equations with similar boundary conditions have a great similarity (although these conditions are not always observed), this allows us to use the above solutions of heat transfer to mass diffusion problems. The difference in the problems of heat transfer and mass transfer is determined for the most part by the presence of a dissipative term in the heat transfer equations that is absent in the convective diffusion equations, the discrepancy due to the nature of the problem, the boundary conditions, and also the different degrees of the thermal diffusivity and diffusion dependence on temperature.

2.4 CONVECTIVE HEAT TRANSFER ON A FLAT SURFACE

The basis of convective heat transfer is the calculation of the flow velocity distribution in the heat transfer equations at moderate values. Let us consider heat transfer when flowing over a solid plane surface by a stream. At the same time, far from the surface of the body, heat transfer will be carried out mainly by convection, and as we approach the interphase boundary, the velocity of the medium decreases, and accordingly the role of convection in heat transfer and the role of thermal conductivity increases. Near the surface of a solid there are regions in which the heat fluctuates due to the convection and thermal conductivity. In this region, the temperature of the environment rapidly changes in the direction perpendicular to the surface of the solid. This region, by analogy with the hydrodynamic layer, is called the thermal or temperature layer. The equations of convective transport for a plane surface are represented in the form

$$V_x\frac{\partial V_x}{\partial x} + V_y\frac{\partial V_x}{\partial y} = \nu\frac{\partial^2 V_x}{\partial y^2} \tag{2.132}$$

$$\frac{\partial V_x}{\partial x}+\frac{\partial V_y}{\partial y}=0 \tag{2.133}$$

$$V_x\frac{\partial T}{\partial x}+V_y\frac{\partial T}{\partial y}=a_T\frac{\partial^2 T}{\partial y^2} \tag{2.134}$$

with boundary conditions

$$y=0,\quad V_x=0,\quad V_y=0,\quad T=T_S \tag{2.135}$$

$$y\to\infty, V_x=V_\infty, T=T_\infty \tag{2.136}$$

Introducing functions analogous to the preceding expressions

$$\eta=y\sqrt{V_\infty/_{v.x}},\qquad \psi=\sqrt{v.xV_\infty}f(\eta),\qquad \theta=\frac{T_S-T}{T_S-T_\infty} \tag{2.137}$$

Equation (2.134) can be represented in the form

$$\theta''(\eta)+\frac{\Pr}{2}f(\eta)\theta'(\eta)=0 \tag{2.138}$$

with boundary conditions

$$\eta=0,\theta=0$$

$$\eta\to\infty,\theta=1 \tag{2.139}$$

where $\Pr={}^{v}/_{a_T}$ is the Prandtl number. Integrating equation (2.138) twice, we obtain

$$\theta=C_1\int_0^\eta\exp\left[-\frac{\Pr}{2}\int_0^\eta f(\eta)d\eta\right]d\eta+C_2 \tag{2.140}$$

Using the boundary conditions, we define the coefficients in the form

$$C_2=0$$

$$C_1=\frac{1}{\int_0^\infty\exp\left[-\frac{\Pr}{2}\int_0^\eta f(\eta)d\eta\right]d\eta} \tag{2.141}$$

Then the solution of (2.139) can be represented in the form

$$\theta[\eta]=\frac{\int_0^\eta\exp\left[-\frac{\Pr}{2}\int_0^\eta f(\eta)d\eta\right]d\eta}{\int_0^\infty\exp\left[-\frac{\Pr}{2}\int_0^\eta f(\eta)d\eta\right]d\eta} \tag{2.142}$$

From the boundary layer equation we have

$$f = -{}^{2f'''}\!/_{f''} \tag{2.143}$$

integrating which we have

$$\int_0^{\eta} f(\eta)\,d\eta = -2\int_0^{\eta} \frac{f'''(\eta)}{f''(\eta)}\,d\eta = -2\ln\frac{f''(\eta)}{f''(0)} \tag{2.144}$$

Then

$$\exp\left[-\frac{\Pr}{2}\int_0^{\eta} f(\eta)\,d\eta\right] = \left[\frac{f''(\eta)}{f''(0)}\right]^{\Pr} \tag{2.145}$$

In view of the foregoing, the solution (2.142) is represented in the form

$$\theta = \frac{\int_0^{\eta}\left[f''(\eta)\right]^{\Pr} d\eta}{\int_0^{\infty}\left[f''(\eta)\right]^{\Pr} d\eta} \tag{2.146}$$

On the surface of the body, the conditions for convective heat transfer appear in the form

$$\alpha\left(T_S - T_\infty\right) = -\lambda\left(\frac{\partial T}{\partial y}\right)_{y=0} \tag{2.147}$$

where α and λ are the coefficients of heat transfer and thermal conductivity. From this equation (2.147), we determine the heat transfer coefficient in the form

$$\alpha = -\frac{\lambda}{T_S - T_\infty}\left(\frac{\partial T}{\partial y}\right)_{y=0} = -\frac{\lambda}{T_S - T_\infty}\sqrt{\frac{V_\infty}{\nu x}}\left(\frac{\partial T}{\partial \eta}\right)_{\eta=0} \tag{2.148}$$

We define the derivative on the surface of the body as

$$\left(\frac{\partial T}{\partial \eta}\right)_{\eta=0} = -\left(T_S - T_\infty\right)\left(\frac{\partial \theta}{\partial \eta}\right)_{\eta=0} \tag{2.149}$$

and from equation (2.146)

$$\left(\frac{\partial \theta}{\partial \eta}\right)_{\eta=0} = \frac{\left[f''(0)\right]^{\Pr}}{\int_0^{\infty}\left[f''(\eta)\right]^{\Pr} d\eta} = m(\Pr) \tag{2.150}$$

The numerical values of the function $m(\Pr)$ as a function of the number Pr are given in Table 2.4.

TABLE 2.4
Values *m*(Pr)

Pr	0.6	0.7	0.8	0.9	1.0	1.1	7.0	10.0	15.0
m	0.276	0.293	0.307	0.320	0.332	0.334	0.645	0.730	0.835

These values can be approximated by the following expression $m(\mathrm{Pr}) = 0.332\,\mathrm{Pr}^{1/3}$. Given this expression, you can write

$$\left(\frac{\partial \theta}{\partial \eta}\right)_{\eta=0} = 0.332\,\mathrm{Pr}^{1/3} \tag{2.151}$$

$$\left(\frac{\partial T}{\partial \eta}\right)_{\eta=0} = -(T_S - T_\infty)0.332\,\mathrm{Pr}^{1/3} \tag{2.152}$$

The average value of the heat transfer coefficient over the length of the layer is

$$\alpha_S = \frac{1}{L}\int_0^L \alpha\, dx = 0.664\,\mathrm{Pr}^{1/3}\sqrt{\frac{V_\infty}{\nu x}} \tag{2.153}$$

Putting $\mathrm{Nu} = \frac{\alpha_S L}{\lambda}$ and $\mathrm{Re} = \frac{V_\infty L}{\nu}$ equation (2.153) is represented in the form

$$\mathrm{Nu} = 0.664\,\mathrm{Pr}^{1/3}\,\mathrm{Re}^{1/2} \tag{2.154}$$

Equation (2.154) is the heat transfer equation for a laminar flow past a plane surface. The thickness of the thermal boundary layer can be expressed as

$$\frac{\delta_T}{\delta_D} \sim \frac{1}{\mathrm{Pr}^{1/2}} \tag{2.155}$$

where δ_D is the thickness of the dynamic layer. Thus, depending on the number, the thickness of the thermal layer can be greater or less than the thickness of the hydrodynamic layer. For gases, the thickness of the thermal layer is comparable to the thickness of the hydrodynamic boundary layer. For liquids, when $\mathrm{Pr} \approx 10^3$, the thermal boundary layer is much thinner than the hydrodynamic layer. For very small values of the number $\mathrm{Pr} << 1$ (for gases), the thickness of the hydrodynamic layer is small and therefore $V_y << V_x$. Then the convective heat transfer equations can be represented in the form

$$V_x \frac{\partial T}{\partial x} = a_T \frac{\partial^2 T}{\partial y^2} \tag{2.156}$$

$$y = 0,\quad T = T_w,\quad y \to \infty,\quad T = T_\infty$$

$$x = 0,\quad T = T_\infty$$

The solution of equation (2.156) for given boundary conditions is represented in the form

$$\frac{T - T_w}{T_\infty - T_w} = \mathrm{erf}\,\frac{y\sqrt{V_x x}}{2\sqrt{a_T x}} \tag{2.157}$$

The local Nusselt number is represented in the form

$$\mathrm{Nu} = \frac{\alpha . x}{\lambda} = \frac{x}{T_\infty - T_w}\left(\frac{\partial T}{\partial y}\right)_{y=0} = \frac{1}{\pi}\mathrm{Pr}^{1/2}\,\mathrm{Re}^{1/2} \tag{2.158}$$

where $\mathrm{Re} = \frac{V_x L}{v}$ is the Reynolds number. We define the Stanton number as

$$\mathrm{St_T} = \frac{q_S}{c_P V_\infty \rho (T_\infty - T_w)} = \frac{\mathrm{Nu}}{\mathrm{RePr}} = \frac{\alpha}{c_P \rho V_\infty} \tag{2.159}$$

In practical applications, the heat transfer during flow along a flat horizontal surface, depending on the number, is calculated by a formula similar to (2.159):

1. $\mathrm{Re} < 5\times10^5$, $\mathrm{Nu} = 0.664\,\mathrm{Re}^{1/2}\,\mathrm{Pr}^{1/3}\left(\mathrm{Pr}/\mathrm{Pr}_w\right)^{0.25}$ (2.160)

2. $\mathrm{Re} > 5\times10^5, \mathrm{Nu} = 0.037\,\mathrm{Re}^{0.8}\,\mathrm{Pr}^{0.43}\left(\mathrm{Pr}/\mathrm{Pr}_w\right)^{0.25}$ (2.161)

The determining temperature is the average temperature of the fluid, which determines the size and length of the streamlined surface along the direction of flow. For gases, in particular for air, equations (2.160 and 2.161) simplify to form

$$\mathrm{Nu} = 0.032\,\mathrm{Re}^{0.8} \tag{2.162}$$

Heat transfer during the flow of a liquid film along a vertical surface is determined in the form

1. with turbulent flow of the film ($\mathrm{Re} > 2000$)

$$\mathrm{Nu} = 0.01\left(\mathrm{Ga\,Pr\,Re}\right)^{1/3} \tag{2.163}$$

2. at laminar flow of film ($\mathrm{Re} < 2000$)

$$\mathrm{Nu} = 0.67\left(\mathrm{Ga}^2\,\mathrm{Pr}^3\,\mathrm{Re}\right)^{1/3} \tag{2.164}$$

where $\mathrm{Nu} = \alpha H/\lambda$ is the Nusselt number, $\mathrm{Ga} = H^3\rho^2 g/\eta^2$ is Galileo number, H is surface height, $\mathrm{Re} = V d_e \rho/\eta$ is Reynolds number, $d_e = 4f/P$ is equivalent film diameter, f is cross-sectional area of film, and P is film-washed perimeter. The average temperature of the boundary layer is taken as the determining temperature.

2.5 CONVECTIVE HEAT TRANSFER IN ENGINEERING APPLICATIONS

In engineering calculations of convective heat transfer processes, empirical equations are mainly used, which are a function of the following parameters: (1) Nusselt number, $\mathrm{Nu} = \alpha L/\lambda$; (2) Prandtl number, $\mathrm{Pr} = v/a_T$; (3) the Reynolds number, $\mathrm{Re} = VL/v$; (4) Galileo number, $\mathrm{Ga} = gL^3/v^2$; (5) Grashof number, $\mathrm{Gr} = \mathrm{Ga}\beta\,\Delta T = \frac{gL^3}{v^2}\beta\Delta T$; and (6) the Peclet number, $\mathrm{Pe} = \mathrm{Re\,Pr} = VL/a_T$, where $\beta = -\frac{1}{\bar\rho}\left(\frac{\partial\rho}{\partial T}\right)_P$ is the coefficient of volume expansion, $\bar\rho$ is the average density of the medium, α is the heat transfer coefficient, λ is the thermal conductivity coefficient, η is the dynamic coefficient of viscosity, v is kinematic viscosity coefficient, ρ is density, $a_T = \lambda/(c_P\rho)$ is thermal diffusivity coefficient, c_P is specific heat capacity at constant pressure, V is velocity, L is determining the size, g is acceleration of free fall, and ΔT is temperature difference.

These numbers take into account the influence of the physical properties of the coolant and the features of the hydromechanics of its movement on the intensity of heat transfer. The physicochemical properties of the liquid (gas) entering into these equations must be taken at the determining temperature. Many factorial equations of convective heat transfer include a factor $(\mathrm{Pr}/\mathrm{Pr}_w)^{0.25}$, where Pr_w is the value of the number calculated at the wall temperature. For gases $\mathrm{Pr}/\mathrm{Pr}_w = 1$ with both heating and cooling, since for gases the number is approximately constant, independent of temperature and pressure. At present, there are a lot of formulas in the literature for calculating heat transfer coefficients, both for forced and for natural convection. Below are some of the criteria equations used in practical heat transfer calculations for forced (laminar and turbulent flows) and natural convection. The heat transfer in the turbulent flow is mainly determined by the intensity of the movement of chaotically pulsating volumes of liquid. The difficulty in analyzing turbulent heat exchange is that the intensity of turbulent exchange depends on the distance to the solid wall. It is generally recognized that a purely turbulent heat transfer prevails far from the solid wall, whereas in the immediate vicinity of the wall, the molecular nature of the heat transfer is paramount. The ratio of turbulent heat transfer coefficients is determined by the numerical value of the turbulent Prandtl number. According to many experimental data, Pr_T it can have values from 0.5 to 1 and remain constant or vary in the transverse direction of the turbulent flow. For practical calculations of the heat transfer intensity in turbulent flow, the relationships correlating the experimental results are applied.

2.5.1 Heat Transfer in Forced Convection

When flowing around different bodies, the expressions for calculating the heat transfer coefficient for forced convection are obtained theoretically and on the basis of experimental studies, which makes it possible to write the following formulas [27,29,30] for a sphere with $3.5 < \mathrm{Re} < 7.6 \times 10^4$ and $0.7 < \mathrm{Pr} < 380$

$$\mathrm{Nu} = 2 + \left(0.4\,\mathrm{Re}^{0.5} + 0.06\,\mathrm{Re}^{2/3}\right)\mathrm{Pr}^{0.4}\left(\frac{\eta}{\eta_w}\right)^{0.25} \tag{2.165}$$

and $\mathrm{Re} < 200$

$$\mathrm{Nu} \approx 2 + 0.61\,\mathrm{Re}^{1/2}\,\mathrm{Pr}^{1/3} \tag{2.166}$$

The experimental data for the heat exchange of single spherical particles confirm that the value of the Nusselt criterion tends to a value equal to 2. For a cylindrical body with $3.5 < \mathrm{Re} < 7.6 \times 10^4$ and $0.7 < \mathrm{Pr} < 300$

$$\mathrm{Nu} = \left(0.4\,\mathrm{Re}^{0.5} + 0.06\,\mathrm{Re}^{2/3}\right)\mathrm{Pr}^{0.4}\left(\frac{\eta}{\eta_w}\right)^{0.25} \tag{2.167}$$

Experimental studies of heat transfer in cylindrical bodies made it possible to obtain a number of approximation formulas for determining the Nusselt number

1. When $5 \le \mathrm{Re} \le 10^3$

$$\mathrm{Nu} = 0.5\,\mathrm{Re}^{0.5}\,\mathrm{Pr}^{0.38}\left(\mathrm{Pr}/\mathrm{Pr}_w\right)^{0.25} \tag{2.168}$$

2. When $10^3 \le \mathrm{Re} \le 2 \times 10^5$

$$\mathrm{Nu} = 0.25\,\mathrm{Re}^{0.6}\,\mathrm{Pr}^{0.38}\left(\mathrm{Pr}/\mathrm{Pr}_w\right)^{0.25} \tag{2.169}$$

3. When $\mathrm{Re} = 3\times10^5 - 2\times10^6$

$$\mathrm{Nu} = 0.023\,\mathrm{Re}^{0.8}\,\mathrm{Pr}^{0.37}\left(\mathrm{Pr}/\mathrm{Pr}_w\right)^{0.25} \tag{2.170}$$

1. Thermal efficiency with developed turbulent flow in straight pipes and channels ($\mathrm{Re} > 10^4$) can be calculated from the formula

$$\mathrm{Nu} = 0.021\vartheta_i\,\mathrm{Re}^{0.8}\,\mathrm{Pr}^{0.43}\left(\mathrm{Pr}/\mathrm{Pr}_w\right)^{0.25} \tag{2.171}$$

The average temperature of the liquid is taken as the determining temperature. The value of the correction factor ϑ_i can be assumed to be approximately equal to the ratio of the length of the pipe to its diameter $10 < L/d < 50$ equal $\vartheta_i = 1.0 - 1.23$. In addition, for a developed turbulent flow in smooth tubes, a number of empirical expressions can be used to calculate the heat transfer. If $0.5 < \mathrm{Pr} < 100$ and $10^4 < \mathrm{Re} < 5\times10^6$, then the heat transfer can be calculated from formula

$$\mathrm{Nu} = \frac{\left(\xi_T/8\right)\mathrm{Re}\,\mathrm{Pr}}{1.07 + 9\left(\xi_T/8\right)\left(\mathrm{Pr}-1\right)\mathrm{Pr}^{1/4}} \tag{2.172}$$

If $0.5 < \mathrm{Pr} < 200$ and $10^4 < \mathrm{Re} < 5\times10^6$, then the heat transfer can be calculated from formula

$$\mathrm{Nu} = \frac{\left(\xi_T/8\right)\mathrm{Re}\,\mathrm{Pr}}{1.07 + 12.7\left(\xi_T/8\right)^{1/2}\left(\mathrm{Pr}^{2/3}-1\right)}\left(\frac{\mathrm{Pr}}{\mathrm{Pr}_w}\right)^{0.11} \tag{2.173}$$

where ξ_T is the coefficient of resistance in the pipes.

2. Thermal efficiency in straight pipes and channels at $(\mathrm{Gr\,Pr}) < 8\times10^5$ and $\mathrm{Re} < 10^4$ is calculated as

   a. If $\mathrm{Re} < 2300$ and $\mathrm{Pe}\dfrac{d}{L} \geq 20$, then

$$\mathrm{Nu} = 1.55\vartheta_i\left(\mathrm{Re}\frac{d}{L}\right)^{1/3}\left(\frac{\eta}{\eta_w}\right)^{0.14} \tag{2.174}$$

   b. For values $\left(\mathrm{Pe}\dfrac{d}{L}\right) < 20$ the value asymptotically tends to the limiting value $\mathrm{Nu} \approx 3.66$.

3. Heat transfer in straight pipes and channels at $(\mathrm{GrPr}) > 8\times10^5$, $\mathrm{Re} < 10^4$, and horizontal arrangement of pipes

   a. If $\mathrm{Re} < 3500$ and $20 \leq \left(\mathrm{Pe}\dfrac{d}{L}\right) \leq 120$ can be calculated from formula

$$\mathrm{Nu} = 0.8\left(\mathrm{Pe}\frac{d}{L}\right)^{0.4}\left(\mathrm{Gr\,Pr}\right)^{0.1}\left(\eta/\eta_w\right)^{0.14} \tag{2.175}$$

b. If $\left(\mathrm{Pe}\frac{d}{L}\right) \le 10$, then

$$\mathrm{Nu} = 0.5\left(\mathrm{Pe}\frac{d}{L}\right) \tag{2.176}$$

c. If $\mathrm{Re} > 3500$, then the heat transfer coefficient is

$$\mathrm{Nu} = 0.022\,\mathrm{Re}^{0.8}\,\mathrm{Pr}^{0.4}\left(\eta/\eta_W\right)^n \tag{2.177}$$

where $n = 0.11$ with heating and $n = 0.25$ with cooling.

4. In the vertical arrangement of pipes and in the case of non-coincidence of free and forced convection, the heat transfer is calculated by formula

$$\mathrm{Nu} = 0.037\,\mathrm{Re}^{0.75}\,\mathrm{Pr}^{0.4}\left(\eta/\eta_w\right)^n \tag{2.178}$$

where $n = 0.11$ with heating and $n = 0.25$ with cooling. This formula is effective for $n = 0.25$ and $1.5\times10^6 < (\mathrm{Gr\,Pr}) < 12\times10^6$.

5. In the case of a transverse flow around a beam of smooth tubes, the heat transfer is calculated by the formulas:

at $\mathrm{Re} < 1000$, for corridor and chess beams

$$\mathrm{Nu} = 0.56\xi_\varphi\,\mathrm{Re}^{0.5}\,\mathrm{Pr}^{0.36}\left(\mathrm{Pr}/\mathrm{Pr}_w\right)^{0.25} \tag{2.179}$$

at $\mathrm{Re} > 1000$, for corridor bundles

$$\mathrm{Nu} = 0.22\xi_\varphi\,\mathrm{Re}^{0.65}\,\mathrm{Pr}^{0.36}\left(\mathrm{Pr}/\mathrm{Pr}_w\right)^{0.25} \tag{2.180}$$

but for chess beams

$$\mathrm{Nu} = 0.4\xi_\varphi\,\mathrm{Re}^{0.6}\,\mathrm{Pr}^{0.36}\left(\mathrm{Pr}/\mathrm{Pr}_w\right)^{0.25} \tag{2.181}$$

where ξ_ϕ is the coefficient, taking into account the influence of the angle of attack of the surface by the flow, for, $10 < \varphi < 90$, $0.42 < \xi_\varphi < 1$. For gases, these formulas are simplified, since $\left(\mathrm{Pr}/\mathrm{Pr}_w\right) = 1$. In particular, for air with a staggered arrangement of pipes, one can write

$$\mathrm{Nu} = 0.356\xi_\varphi\,\mathrm{Re}^{0.6} \tag{2.182}$$

2.5.2 Heat Transfer with Natural Convection

The fluid motion caused by the difference in density in the field of external forces is called natural convection. Such external forces are gravity forces, and the difference in densities in the simplest case is caused by a temperature drop between the surface of a solid and a liquid. The equations of momentum and heat transfer for the case of heat transfer between a vertical plate and a liquid can be represented in the form.

$$V_x\frac{\partial V_x}{\partial x} + V_y\frac{\partial V_x}{\partial y} = \nu\frac{\partial^2 V_x}{\partial y^2} + \beta g\left(T - T_\infty\right)$$

$$V_x \frac{\partial T}{\partial x} + V_y \frac{\partial T}{\partial y} = a_T \frac{\partial^2 T}{\partial y^2} \tag{2.183}$$

$$\frac{\partial V_x}{\partial x} + \frac{\partial V_y}{\partial y} = 0$$

The boundary conditions are

$$y = 0, \quad V_x = 0, \quad V_y = 0, \quad T = T_w$$

$$y \to \infty, \quad V_x = 0, \quad T = T_\infty \tag{2.184}$$

Introducing the following notation

$$\psi = 4\nu \left(\frac{\text{Gr}}{4} \right)^{1/4} \varphi(\zeta), \tag{2.185}$$

$$\zeta = \frac{y}{x} \left(\frac{1}{4} \text{Gr} \right)^{1/4}, \tag{2.186}$$

$$V_x = \left[\frac{\beta g (T_w - T_\infty)}{4\nu^2} \right]^{1/4} 4\nu x^{1/2} \varphi', \tag{2.187}$$

$$V_y = \nu x^{-1/4} \left[\frac{\beta g (T_w - T_\infty)}{4\nu^2} \right]^{1/4} (\zeta \phi' - 3\phi) \tag{2.188}$$

Equation (2.183) can be written in the form

$$\phi'' + 3\phi\phi' - 2\phi'^2 + \theta = 0 \tag{2.189}$$

$$\theta'' + 3\text{Pr}\,\phi\theta' = 0 \tag{2.190}$$

where $\theta = \frac{T - T_\infty}{T_w - T_\infty}$ is the dimensionless temperature. An approximate solution of the nonlinear equations (2.189 and 2.190) for a parabolic temperature distribution along the normal to the surface can be represented in the form

$$\text{Nu} = 0.508\,\text{Pr}^{0.5} (0.952 + \text{Pr})^{-0.25} \text{Gr}^{0.25} \tag{2.191}$$

In practical calculations, the following formulas are used to calculate the heat transfer under natural convection:

1. Heat transfer outside horizontal pipes at $10^3 < \text{Ra} < 10^9$ (where $\text{Ra} = \text{GrPr}$ is the Rayleigh number)

$$\text{Nu} = 0.5\text{Ra}^{0.25} (\text{Pr}/\text{Pr}_w)^{0.25} \tag{2.192}$$

2. For vertical surfaces, flat and cylindrical

   a. When $10^3 < \mathrm{Ra} < 10^9$

$$\mathrm{Nu} = 0.76\mathrm{Ra}^{0.25}\left(\mathrm{Pr}/\mathrm{Pr}_w\right)^{0.25} \tag{2.193}$$

   b. When $\mathrm{Ra} > 10^9$

$$\mathrm{Nu} = 0.15\mathrm{Ra}^{0.33}\left(\mathrm{Pr}/\mathrm{Pr}_w\right)^{0.25} \tag{2.194}$$

The ambient temperature is taken as the determining temperature. For particles of spherical shape, heat transfer at natural convection can be calculated from formulas $(\mathrm{Pr} = 1)$

$$\mathrm{Nu} = 2 + 0.43\mathrm{Ra}^{1/4}, \quad 1 < \mathrm{Ra} < 10^5 \tag{2.195}$$

$$\mathrm{Nu} = 2 + 0.5\mathrm{Ra}^{1/4}, \quad 3\cdot 10^5 < \mathrm{Ra} < 8\cdot 10^8 \tag{2.196}$$

$$\mathrm{Nu} = 2 + \frac{0.589\mathrm{Ra}^{1/4}}{\left[1+\left(\dfrac{0.469}{\mathrm{Pr}}\right)^{9/16}\right]^{4/9}}, \quad \mathrm{Ra} \le 10^4 \tag{2.197}$$

In addition to these expressions, in the literature one can find more complicated formulas for calculating the number for free convection. In conclusion, we note that the phenomenon of natural convection consists of the motion of the medium in the field of gravity, which arises from the spatial inhomogeneity of the density of the medium, which is a consequence of the inhomogeneity of the temperature field.

Example 2.4

Toluene is heated in the tube space of the heat exchanger. The internal diameter of the pipes is $d = 0.021\,\mathrm{m}$, the length of the pipes is $L = 4\,\mathrm{m}$, and the velocity of toluene is $V = 0.05\,\mathrm{m/s}$, the temperature of toluene. The temperature of the wall surface in contact with toluene is $T_W = 50°\mathrm{C}$. Determine the heat transfer coefficient of toluene.

At an average temperature $T_S = \frac{30+50}{2} = 40°\mathrm{C}$ and toluene density at 40°C $\rho = 847\,\mathrm{kg/m^3}$ is the coefficient of volumetric expansion $\beta = 1.11\times 10^{-3}\mathrm{K}^{-1}$, the dynamic viscosity of toluene $\eta = 0.466\times 10^{-3}$ Pas, we calculate the values of the numbers Gr, Pr, and Re

$$\mathrm{Gr} = \frac{d^3\rho^2\beta\Delta T g}{\eta^2} = \frac{0.021^3\times 847^2\times 1.11\times 10^{-3}(50-40)9.81}{0.466^2\times 10^{-6}} = 3.33\times 10^6$$

$$\mathrm{Pr} = \frac{c_p\eta}{\lambda} = \frac{1718\times 0.466\times 10^{-3}}{0.14} = 5.72$$

where $c_p = 1718\,\mathrm{J/(kgK)}$ is the specific heat capacity of toluene, and $\lambda = 0.14$ W/mK is the thermal conductivity of toluene at an average temperature

$$\mathrm{Re} = \frac{Vd\rho}{\eta} = \frac{0.05\times 0.021\times 847}{0.466\times 10^{-3}} = 1900$$

The Rayleigh number is

$$\mathrm{Ra} = \mathrm{GrPr} = 3.33\times 10^6\times 5.72 = 19\times 10^6 > 8\times 10^5.$$

1. For horizontal pipes when Re < 3500, we use the formula (2.34b)

$$\mathrm{Nu} = 0.8\left(\mathrm{Pe}\frac{d}{L}\right)^{0.4}\mathrm{Ra}^{0.1}\left(\frac{\eta}{\eta_w}\right)^{0.14} = 0.8\left(1900\times 5.72\frac{0.021}{4}\right)\left(19\times 10^6\right)^{0.1}\left(\frac{0.466}{0.42}\right)^{0.14} = 21.75$$

Here, $\mathrm{Pe} = \mathrm{RePr}$ $\eta_w = 0.42\times 10^{-3}$Pas is the dynamic viscosity of toluene at the wall temperature. The heat transfer coefficient for a horizontal wall is

$$\alpha_g = \frac{\mathrm{Nu}\lambda}{d} = \frac{21.75\times 0.14}{0.021} = 145\ \mathrm{W/m^2K}$$

2. For vertical pipes, if the free and forced convection do not coincide, we use formula (2.35)

$$\mathrm{Nu} = 0.037\mathrm{Re}^{0.75}\mathrm{Pr}^{0.4}\left(\eta/\eta_w\right)^{0.11} = 0.037\times 1900^{0.75}\times 5.72^{0.4}\left(0.466/0.42\right)^{0.11} = 21.17$$

The coefficient of heat transfer of toluene to the vertical wall of the pipes is determined as

$$\alpha_b = \frac{\mathrm{Nu}\lambda}{d} = \frac{21.17\times 0,14}{0.021} = 141\mathrm{BM/m^2K}$$

Example 2.5

Through the tube space of the tubular heat exchanger, calcium chloride brine with a concentration of 24.7% (mass.) is pumped at an average temperature of $T_p = -20°\mathrm{C}$ and velocity $V = 0.1\ \mathrm{m/s}$. The internal diameter of the pipes is $d = 0.021\ \mathrm{m}$, and the length of the pipes is $L = 3\ \mathrm{m}$. The average temperature of the wall surface in contact with the brine is $T_w = -10°\mathrm{C}$. Determine the heat transfer coefficient of calcium chloride.
We calculate the criterion Re

$$\mathrm{Re} = \frac{Vd\rho}{\eta} = \frac{0.1\times 0.021\times 1248}{99.96\times 10^{-4}} = 262$$

where $\rho = 1246\ \mathrm{kg/m^3}$ is the density of the brine, and $\eta = 99.96\times 10^{-4}$ Pas is the dynamic viscosity of the brine at a temperature of $T_P = -20°\mathrm{C}$. At a determining temperature equal to $T_S = 0.5\left[(-10)+(-20)\right] = -15°\mathrm{C}$ and $\mathrm{Re} < 10^4$, we calculate the numbers Gr;Pr and Re

$$\mathrm{Gr} = \frac{d^3\rho^2\beta\Delta Tg}{\eta^2} = \frac{0.021^3\times 1246^2\times 0.00361\times 9.81}{81.32^2\times 10^{-8}} = 0.77\times 10^4$$

$$\mathrm{Pr} = \frac{c_p\eta}{\lambda} = \frac{2861\times 81.32\times 10^{-4}}{0.467} = 49.8,\ \mathrm{Re} = \frac{0,1.0,021.1246}{81,32.10^{-4}} = 322$$

Here, $\rho = 1246\ \mathrm{kg/m^3}$ is the density of the brine at $T_S = -15°\mathrm{C}$, $\eta = 81.32\times 10^{-4}$ Pas is the dynamic viscosity of the brine at $T_S = -15°\mathrm{C}$, $c = 2861\ \mathrm{J/(kg\,K)}$ is the specific heat of the brine, and $\lambda = 0.467\ \mathrm{W/mK}$ is the thermal conductivity of the brine. Let us determine the number $\mathrm{Ra} = \mathrm{GrPr} = 0.77\times 10^4\times 49.8 = 3.84\times 10^5 < 8\times 10^5$. Under these conditions, for both horizontal and vertical pipes we take the formula (2.34a)

$$\mathrm{Nu} = 1.55\left(\mathrm{Pe}\frac{d}{L}\right)^{1/3}\left(\frac{\eta}{\eta_w}\right)^{0.14} = 1.55\left(322\times 49.8\frac{0.021}{3}\right)^{1/3}\left(\frac{81.32}{62.69}\right)^{0.14} = 7.24$$

The coefficient of heat transfer of the brine is

$$\alpha = \frac{\mathrm{Nu}\,\lambda}{d} = \frac{7.24 \times 0.467}{0.021} = 161\,\mathrm{BM/m^2K}$$

2.5.3 Convective Heat Transfer in Rough Pipes

The roughness of the surface of pipes and channels has a significant effect on the coefficient of resistance and heat transfer. The use of rough surfaces to certain roughness values (the height of the protrusions) intensifies heat exchange due to the turbulence of the flow near the wall, which changes the structure of the turbulent boundary layer, although at high roughness values friction (hydraulic resistance) increases and, accordingly, mechanical energy consumption for pumping the flow. Investigation of the influence of pipe roughness on heat transfer is considered in papers and in many other works.

It is known that the use of rough surfaces is one of the ways to intensify the processes of heat exchange. In some cases, the intensity of heat transfer can be increased several times due to the creation of artificially rough surfaces. The presence of a roughness under certain conditions also increases the intensity of mass-transfer processes due to turbulization of the diffusion boundary layer.

To estimate the effect of roughness on heat transfer, the following dependence

$$\eta_e = \frac{\mathrm{Nu}/\mathrm{Nu}_0}{\xi_T/\xi_{T0}} \tag{2.198}$$

where η_e is the coefficient of effective use of rough surfaces, and Nu_0, ξ_{T0} are Nusselt number and resistance coefficient for smooth pipes, respectively. In some papers, equation (2.198) is represented in the form

$$\eta_e = \frac{\mathrm{St}/\mathrm{St}_0}{\xi_T/\xi_{T0}} \tag{2.199}$$

where St is the Stanton number, defined for rough surfaces in the form

$$\mathrm{St} = \frac{\xi_T/8}{(m-8.48)\sqrt{\xi_T/8}+1} \tag{2.200}$$

where $m = 4.5(y_+)^{0.24}\,\mathrm{Pr}^{0.44}$. Since the efficiency coefficient depends on the roughness and the values of the numbers Pr and Re, then with increasing numbers Pr, Re and by decreasing the roughness, the area of effective use of rough surfaces increases. Using the experimental data, for variable values of Re and Pr, an empirical dependence was obtained for the coefficient of effective utilization of rough surfaces

$$\eta_e = \frac{1.15\,\mathrm{Pr}^{1/7}}{1+0.82\times10^{-5}\,\mathrm{Pr}^2}\left(1-0.106\,K_+^{1/4}\right) \tag{2.201}$$

where K_+ is the dimensionless roughness parameter, defined as

$$K_+ = \frac{U_D\varepsilon}{\nu} = \zeta\,\mathrm{Re}\left(\frac{\xi_T}{8}\right)^{1/2} \tag{2.202}$$

ε is the height of roughness ridges, $\zeta = \varepsilon/D$ is the dimensionless height of roughness, and D is pipe diameter. Equation (2.201) agrees satisfactorily with the experimental data for $1 \le K_+ \le 1000$ and $0.7 \le \mathrm{Pr} \le 200$. For small values of the number $\mathrm{Pr} < 50$, equation (2.201) simplifies to the form

$$\eta_e = 1.15\,\mathrm{Pr}^{1/7}\left(1 - 0.106K_+^{1/4}\right)\frac{\xi_T}{\xi_{T0}} \tag{2.203}$$

Using (2.201) and (2.203), we can write

$$\mathrm{Nu} = 1.15\mathrm{Nu}_0\,\mathrm{Pr}^{1/7}\left(1 - 0.106K_+^{1/4}\right)\frac{\xi_T}{\xi_{T0}} \tag{2.204}$$

Using equation (2.203) and the preceding equations for calculating heat transfer, one can express the number Nu from the roughness. If $\zeta\,\mathrm{Re} << 2.3\times10^4\,\xi_{t0}^{-0.5}$, then equation (2.204) is transformed to the form

$$\mathrm{Nu} = \mathrm{Nu}_0\,\mathrm{Pr}^{1/7}\,\frac{\xi_T}{\xi_{T0}} \tag{2.205}$$

For the coefficient of effective utilization of rough surfaces, an empirical equation

$$\eta_e = 0.9\,\mathrm{Pr}^{1/40} - \alpha\,\mathrm{Pr}^{-1/6}\,K_+ f^{-1} + 0.106\,\mathrm{Pr}^{2/3}\exp\left[-\alpha\,\mathrm{Pr}^{-1/5}\left(K_+ - 60f\right)^2\right] \tag{2.206}$$

where $\alpha \approx 8.75\times10^{-4}$ is the empirical constant, and f is the factor characterizing the shape of the roughness protrusions. In particular, for pyramidal three-dimensional roughness $f = 3/2$. For numbers $\mathrm{Pr} < 1$, equation (2.206) is simplified to the form

$$\eta_e = 0.9\,\mathrm{Pr}^{1/40} - \alpha\,\mathrm{Pr}^{-1/6}\,K_+ f^{-1} \tag{2.207}$$

As shown from formulas (2.206) and (2.207) and research results [32–35], the presence of roughness of a certain shape and dimensions can intensify the process of heat transfer, if $\eta > 1$.

2.6 HEAT TRANSFER AND ENERGY DISSIPATION IN STRUCTURED DISPERSE MEDIA

The rheology of disperse systems is based on the formation of coagulation structures that depend on the phenomena of coagulation, crushing, collision frequency, and particle concentration, which significantly affect the structural viscosity of the dispersed medium. Moreover, the coagulation structures can be in the form of aggregates, clusters of aggregates, which subsequently form a definite framework or a continuous mesh, which has a fractal structure and corresponds to the maximum viscosity of the system and minimum energy dissipation and to nullify the dissipative function $\Phi_D = 0$. The structure-forming elements of a disperse system are its particles, and sometimes certain groups of particles, for example, aggregates, which determine the disordered fixation of the mutual arrangement of its elements. Thus, if a set of particles are interconnected in a framework, then for the flow of the disperse system it is necessary to provide a condition for the destruction of the framework, at which the shear stress exceeds the yield stress corresponding to the difference between the dissipative function and zero. As a result of structural failure, the viscosity of such a system is sharply reduced.

The phenomena of coagulation, coalescence, and fragmentation of particles, the result of which are the formation of coagulation structures, representing very complex physical processes, are characterized by stochastic nature and conceal in themselves a very large number of problems unclear in their structure, sometimes beyond description and description.

The transfer of heat and mass in disperse media is determined by the conditions of interaction of dispersed particles and the flow, interactions between the particles themselves, leading to friction with energy dissipation and their coagulation with the formation of coagulation structures. Energy dissipation, which forms the basis of all dissipative systems, is one of the important parameters of the turbulent flow and forms the basis for calculating the parameters (turbulent diffusion, turbulent thermal conductivity, turbulent viscosity, etc.) of all transport phenomena occurring in the processes of chemical technology. The energy dissipated by the flow of a viscous fluid in a unit volume is expressed by the formula

$$-\frac{dE}{dt} = \int \frac{\eta_c}{2}\left(\frac{\partial U_i}{\partial x_k} + \frac{\partial U_k}{\partial x_i}\right)^2 d\upsilon \tag{2.208}$$

where U_i, U_k are the components of the flow velocity. From a practical point of view, the solution of this equation, connected with the solution of a system of hydrodynamic nonlinear partial differential equations, presents certain difficulties. In principle, the solution of these equations is possible only in the simplest cases of laminar flow, which is not possible with the turbulent flow of dispersed flows. In this connection, in the evaluation of energy dissipation in the turbulent flow of disperse systems, they tend to simplify and, in most cases, use semiempirical or empirical expressions.

2.6.1 Heat Transfer on a Flat Surface with Allowance for Dissipation Energy

For a plane surface, the dissipative function is defined as

$$\Phi_D = 2\eta\left(\frac{\partial V_x}{\partial x}\right)^2 + 2\eta\left(\frac{\partial V_y}{\partial y}\right)^2 + \eta\left(\frac{\partial V_x}{\partial y} + \frac{\partial V_y}{\partial x}\right)^2 \tag{2.209}$$

Having estimated the terms of this expression for a plane flow, it is established that $\partial V_x/\partial y >> \partial V_y/\partial x$, in connection with which one can write

$$\Phi_D \approx \eta\left(\frac{\partial V_x}{\partial y}\right)^2 \tag{2.210}$$

Then the heat transfer equations taking into account the dissipative term will be simplified to the form

$$\rho c_P\left(V_x\frac{\partial T}{\partial x} + V_y\frac{\partial T}{\partial y}\right) = \lambda\frac{\partial^2 T}{\partial y^2} + \eta\left(\frac{\partial V_x}{\partial y}\right)^2 \tag{2.211}$$

Let us consider the flow in a plane-parallel channel, assuming that the cause of fluid motion is the entrainment of its plate (the upper one), moving at a constant velocity V_1. The equations of motion in a plane channel are simplified $(V_z = V_y = 0)$ and under stationary conditions and assuming that $\partial V_x/\partial x = \partial^2 V_x/\partial x^2 = 0$ they take the form

$$\partial^2 V_x / \partial y^2 = 0 \tag{2.212}$$

$$y = 0, \quad V_x = 0, \quad y = \Delta, \quad V_x = V_1 \tag{2.213}$$

The solution of this equation can be represented in the form: $V_x = V_1 \frac{y}{\Delta}$, where Δ is the width of the channel. Taking this expression into account, the heat transfer equation in a laminar flow, taking into account the dissipative term, is represented in the form

$$\frac{d^2T}{dy^2} + \frac{\eta}{\lambda}\left(\frac{V_1}{\Delta}\right)^2 = 0 \tag{2.214}$$

$$y = 0, \quad T = T_0, \quad y = \Delta, \quad T = T_1 \tag{2.215}$$

The solution of this equation is represented as

$$\theta = \frac{T - T_0}{T_1 - T_0} = \xi + \Pr \theta_R \xi (1 - \xi) \tag{2.216}$$

Here, $\xi = y/\Delta$ is the dimensionless coordinate. A dimensionless complex $\theta_R = \frac{V_1^2}{2c_P(T_1 - T_0)}$ is the temperature of adiabatic flow braking, which the flow would accept if all of its kinetic energy of motion turned into heat. The linearity of the flow temperature, in connection with the presence of the second nonlinear term in solution (2.132), is violated depending on the term $\Pr\theta_R$. At large values $\Pr\theta_R$, the temperature profile becomes a parabola with a shifted maximum.

2.6.2 Heat Transfer in Pipes Taking into Account Energy Dissipation

In laminar flow in tubes, the heat transfer equations, taking energy dissipation into account, are described by equations of the form

$$\frac{a_T}{r}\frac{d}{dr}\left(r\frac{dT}{dr}\right) = -\frac{\eta}{c_P \rho}\left(\frac{dV_x}{dr}\right)^2 + q_{V0} \tag{2.217}$$

Assuming that the distribution of the flow velocity over the pipe cross section obeys equation

$$V_x = 2V_m\left[1 - \left(\frac{r}{R}\right)^2\right] \tag{2.218}$$

(where V_m is the average flow velocity) and, having determined the derivative as $\frac{dV_x}{dr} = -\frac{4V_m r}{R^2}$, we rewrite equation (2.217) in the form

$$\frac{1}{r}\frac{d}{dr}\left(r\frac{dT}{dr}\right) = -\frac{\Pr}{c_P}\frac{16V_m^2}{R^4}r^2 + q_{V0} \tag{2.219}$$

Integrating this equation twice, we have

$$T = -\frac{\Pr V_m^2}{c_P}\frac{r^4}{R^4} + \frac{q_{V0} r^2}{4} + C_1 \tag{2.220}$$

Using the boundary condition $r = 0, T = T_0$ and $r = R, T = T_1$ we have

$$\theta = \theta_R \Pr . \left[1 - \left(\frac{r}{R}\right)^4\right] + \frac{q_{V0} R^2}{4}\left[1 - \left(\frac{r}{R}\right)^2\right] \tag{2.221}$$

where $\theta = \frac{T-T_0}{T_1-T_0}$ is the dimensionless temperature, $\theta_R = \frac{V_m^2}{c_P(T_1-T_0)}$.

For gases $\Pr \approx 1$, the dissipation energy is insignificant, for liquids $\Pr \approx 10^3$, and for concentrated liquids $\Pr \approx 10^6$ the value of dissipation energy becomes significant at high flow rates, since the quantity θ_R is proportional to the square of the velocity.

2.6.3 Dissipation of Energy in the System of a Solid Particle–Liquid at Small Reynolds Numbers

Suppose that in a laminar flow $\mathrm{Re}_d << 1$ there is a solid spherical particle of radius R and surface temperature T_0.

As a result of the flow around the spherical particle by the flow, friction between it and the liquid arises on its surface, which turns into heat.

The equation of heat transfer in a solid particle–liquid system is described as

$$\frac{a_T}{r^2}\frac{\partial}{\partial r}\left(r^2\frac{\partial T}{\partial r}\right)+\frac{a_T}{r^2\sin\phi}\frac{\partial}{\partial\phi}\left(\sin\phi\frac{\partial T}{\partial\phi}\right)=-\frac{\eta}{2c_P\rho}\left(r\frac{\partial}{\partial r}\left(\frac{V_\theta}{r}\right)+\frac{1}{r}\frac{\partial V_r}{\partial\phi}\right)^2 \tag{2.222}$$

Here, $a_T = \lambda/(c_P\rho)$ is the coefficient of thermal diffusivity.

Using the expressions for the stream functions (2.217) represented in the form

$$V_T = V_m\cos\phi\left(1-\frac{3R}{2r}+\frac{R^3}{2r^3}\right)$$
$$V_\theta = -V_m\sin\phi\left(1-\frac{3R}{4r}-\frac{R^3}{4r^3}\right) \tag{2.223}$$

Equation (2.222) is represented as

$$\frac{1}{r^2}\frac{\partial}{\partial r}\left(r^2\frac{\partial T}{\partial r}\right)+\frac{1}{r^2\sin\phi}\frac{\partial}{\partial\phi}\left(\sin\frac{\partial T}{\partial\phi}\right)=\frac{9}{4}\frac{V_m^2\Pr}{c_P}\frac{R^4}{r^2}\left[\cos^2\phi.\left(\begin{matrix}3-\frac{6R^2}{r^2}+\\+\frac{2R^4}{r^4}\end{matrix}\right)+\frac{R^4}{r^4}\right] \tag{2.224}$$

with the boundary conditions: $r = R, T = T_1$ and $r \to \infty, T = T_0$.

The solutions of this equation are sought in the form

$$T = f(r)\cos^2\phi+\psi(r) \tag{2.225}$$

substitution, which in (2.222), taking into account the separation of variables, allows us to obtain two independent equations

$$r^2f''+2rf'-6f = -\theta_m\left(\frac{3R^2}{r^2}-\frac{6R^4}{r^4}+\frac{2R^6}{r^6}\right)$$
$$r^2\psi''+2r\psi'+2f = -\theta_m\frac{R^6}{r^6} \tag{2.226}$$

where $\theta_m = \frac{9}{4}\frac{V_m^2 \Pr}{c_p}$. The final solution (2.222), taking into account (2.225) and solution (2.226), is represented as

$$T - T_0 = \frac{\theta_m R}{r}\left[\begin{array}{l} \frac{2}{3} - \frac{3}{4}\frac{R}{r}\sin^2\phi - \frac{5}{3}\frac{R^2}{r^2}\left(\cos^2\phi - \frac{1}{3}\right) + \frac{R^3}{r^3}\left(\cos^2\phi - \frac{1}{6}\right) \\ -\frac{R^5}{12r^5}\left(\cos^2\phi + \frac{1}{3}\right) \end{array}\right] \tag{2.227}$$

At the point of flow $\phi = 0$ and on the surface of the particle $r = R$, the difference in temperature at the surface and in the flow will be equal $T_1 - T_0 = \frac{5}{8}\theta_m$.

Let us determine the average surface temperature of a solid sphere

$$T_m(R) = \frac{1}{R^2}\int_0^{\pi} T_c(r,\theta) r^2 \sin\theta d\theta \big|_{r=R} \tag{2.228}$$

Using equation (2.227), after simple transformations, we obtain

$$\frac{T_m - T_{c0}}{T_{d0}} = 1 + \frac{5}{2}\Pr_c \frac{U^2}{T_{d0}c_{pc}} \tag{2.229}$$

The amount of heat transferred from a unit surface of a solid particle to an external flux using (2.227) is determined in the form

$$q_c \approx 10\lambda_c \Pr_c \frac{U^2 R}{c_{pc}} + \frac{\lambda_c R(T_{d0} - T_{c0})}{2} \tag{2.230}$$

This problem is of interest for the estimation of energy dissipation during the laminar flow of disperse systems, such as a solid particle-liquid.

Obviously, in this case the value of the converted energy will depend on the volume fraction of particles φ in the flow. At large values φ, the following types of energy dissipation resulting from the presence of friction should be taken into account: between the layers of the liquid, between the liquid and the particles and between the particles themselves and the channel wall, since the particle velocity on the channel surface is not zero. It is important to note that with an increase in the volume fraction of particles, a natural decrease in the velocity and turbulence of the flow can reduce the amount of dissipative energy.

2.6.4 Dissipation of Energy in a Liquid–Liquid System with Small Reynolds Numbers

Let us consider the flow of a single spherical liquid drop around a small Reynolds number, on the surface of which, as a result of friction, processes of mechanical energy dissipation into thermal energy arise, which changes the temperature of the surface and surrounding the droplet of the medium. The difficulty in calculating the energy of dissipation and surface temperature is determined by the presence of a circulation flow inside the drop, depending on the viscosity of the droplet and the medium.

Let us assume that there is a spherical drop or bubble in the laminar flow of a liquid, the radius R, and the surface temperature T_0. The heat transfer equation in a liquid with allowance for energy dissipation is analogous to equation (2.222). Using the current functions, taking into account the circulation of the internal liquid in the drop (1.175) and the external flow, we determine the components of the flow velocity in the form

$$V_r = V_m \cos\varphi \left(1 - \frac{3R\psi_1}{2r} + \frac{R^3\psi_2}{2r^3}\right)$$
$$V_\theta = -V_m \sin\varphi \left(1 - \frac{3R\psi_1}{4r} - \frac{R^3\psi_2}{4r^3}\right) \qquad (2.231)$$

Equation (2.222), taking (2.231) into account, is represented as

$$\frac{1}{r^2}\frac{\partial}{\partial r}\left(r^2\frac{\partial T}{\partial r}\right) + \frac{1}{r^2 \sin\phi}\frac{\partial}{\partial\phi}\left(\sin\frac{\partial T}{\partial\phi}\right)$$
$$= \frac{9}{4}\frac{V_m^2 \Pr}{c_P}\frac{R^2}{r^4}\left[\cos^2\phi\left(3\psi_1^2 - 6\psi_1\psi_2\frac{R^2}{r^2} + 2\psi_2^2\frac{R^4}{r^4}\right)\right] + \psi_2^2\frac{R^4}{r^4} \qquad (2.232)$$

where $\psi_1 = \frac{\gamma+2/3}{1+\gamma}$, $\psi_2 = \frac{\gamma}{1+\gamma}$, $\gamma = \eta_d/\eta_C$, η_C and η_d are the dynamic viscosity of the medium and the droplets. For the solution of (2.232), similar to the solution for a solid particle, we can write

$$T = f(r)\cos^2\phi + \varphi(r) \qquad (2.233)$$

and the final decision is presented in the form

$$T - T_0 = \frac{\theta_m R}{r}\left[\begin{array}{l}\frac{2}{3}\psi_1 - \frac{3}{4}\psi_1^2\frac{R}{r}\sin^2\varphi - \frac{5}{3}\psi_2\frac{R^2}{r^2}\left(\cos^2\varphi - 1/3\right)\\ + \frac{R^3}{r^3}\psi_1\psi_2\left(\cos^2\varphi - 1/6\right) - \frac{1}{12}\psi_2^2\frac{R^5}{r^5}\left(\cos^2\varphi + 1/3\right)\end{array}\right] \qquad (2.234)$$

At the flow point $\varphi = 0$ of the flow, we have

$$T_1 - T_0 = \theta_m\left(\frac{2}{3}\psi_1 - \frac{10}{9}\psi_2 + \frac{5}{6}\psi_1\psi_2 - \frac{1}{9}\psi_2^2\right) \qquad (2.235)$$

For bubbles in a liquid $\gamma = \eta_d/\eta_C \to 0$ and $\psi_1 = 2/3$, $\psi_2 = 0$ the temperature distribution is described by the equation

$$T - T_0 = \frac{\theta_m R}{3r}\left(\frac{4}{3} - \frac{R}{r}\sin^2\phi\right) \qquad (2.236)$$

and the difference in temperature on the surface at the point of flow and the medium is defined as

$$T_1 - T_0 = \frac{4}{9}\theta_m \qquad (2.237)$$

The amount of heat transported inside the drop from the entire surface using (2.234) is determined from equation

$$q_d = -\lambda_d \int_0^{\pi} \frac{\partial T_d}{\partial r} r^2 \sin\theta d\theta \Big|_{r=R} = \frac{\lambda_d A_d R}{3(1+\gamma)^2} \left[\frac{11}{3} + \frac{\lambda_d}{\lambda_c} \frac{(4+16\gamma+33\gamma^2)}{2\gamma} + \frac{3(T_{c0}-T_{d0})(1+\gamma)^2}{2A_d} \right] \tag{2.238}$$

The amount of heat transferred to the external medium from the entire surface of the drop is determined as

$$q_c = -\lambda_c \int_0^{\pi} \frac{\partial T_c}{\partial r} r^2 \sin\theta d\theta \Big|_{r=R} = \frac{\lambda_c A_c R}{3(1+\gamma)^2} \left[\frac{\lambda_c}{\lambda_d} \frac{\gamma}{3} + (4+24\gamma+41\gamma^2) + \frac{3(T_{d0}-T_{c0})(1+\gamma)^2}{2A_c} \right] \tag{2.239}$$

Here, $A_c = \frac{3}{4} U^2 \frac{\mathrm{Pr}_c}{c_{pc}}$, $\mathrm{Pr}_c = \frac{\nu_c c_{pc} \rho_c}{\lambda_c}$ is the Prandtl number for the external environment, $\psi_3 = \frac{1}{1+\gamma}$, $A_d = \frac{3}{4} U^2 \frac{\mathrm{Pr}_d}{c_{pd}}$, $\mathrm{Pr}_d = \frac{\eta_d c_{pd}}{\lambda_d}$ is the Prandtl number for the inner medium of the drop.

For $\gamma \to \infty$, that is, for a solid particle, $\psi_1 = \psi_2 = 1$, solution (2.237) tends to solution (2.227). Usually for gases the number u, as follows from equations (2.236) and (2.237) dissipated energy, has little effect on the temperature difference between the gas bubble and the medium.

2.6.5 Relation of the Drag Coefficient of Particles to Dissipation Energy

The relationship between the phenomena of coalescence, coagulation, and fragmentation of particles with energy dissipation in a turbulent flow was previously cited. In particular, for a laminar flow, the dissipative function in spherical coordinates for spherical particles is defined as

$$\Phi_D = 2\eta_c \left[\left(\frac{\partial V_r}{\partial r}\right)^2 + \left(\frac{1}{r}\frac{\partial V_\theta}{\partial \theta} + \frac{V_r}{r}\right)^2 + \left(\frac{V_\theta}{r} ctg\theta + \frac{V_r}{r}\right)^2 \right] + \eta_c \left[r\frac{\partial}{\partial r}\left(\frac{\partial V_\theta}{\partial r}\right) + \frac{1}{r}\frac{\partial V_r}{\partial \theta} \right]^2 \tag{2.240}$$

The velocities for the internal and external fluid flows in spherical coordinates can be determined from the formulas given in Section 1.7, respectively

$$V_{r1} = \frac{U}{2(1+\gamma)}\left(1-\frac{r^2}{R^2}\right)\cos\theta, \quad V_{\theta 1} = -\frac{U}{2(1+\gamma)}\left(1-\frac{2r^2}{R^2}\right)\sin\theta \tag{2.241}$$

$$V_{r1} = U\left[1 - \frac{2+3\gamma}{2(1+\gamma)}\frac{R}{r} + \frac{\gamma}{2(1+\gamma)}\frac{R^3}{r^3}\right]\cos\theta,$$
$$V_{\theta 2} = -U\left[1 - \frac{2+3\gamma}{2(1+\gamma)}\frac{R}{r} - \frac{\gamma}{4(1+\gamma)}\frac{R^3}{r^3}\right]\sin\theta \tag{2.242}$$

where U is the flow velocity and is far from the droplet, R is the droplet radius, $\gamma = \frac{\eta_d}{\eta_c}$, η_d, η_d are the dynamic viscosities of the droplet and the medium. Using the last formulas for the external flow of a fluid, the dissipative function Φ_D for flow around a drop after labor-intensive but complicated transformations is defined as

$$\Phi_{DC} = \frac{3\eta_c}{4}\frac{U^2}{R^2}\left[\cos^2\theta\left(\psi_1^2 - 6\psi_1\psi_2\frac{R^2}{r^2} + 6\psi_2^2\frac{R^4}{r^4}\right) + 3\frac{R^4}{r^4}\psi_2^2\right] \tag{2.243}$$

Here, it is $\psi_1 = \frac{2+3\gamma}{1+\gamma}$, $\psi_2 = \frac{\gamma}{1+\gamma}$.

The drag force for a spherical particle in a stream is defined in the form, using the above formulas

$$E_D = \iint \Phi_{DC} 2\pi r^2 dr \Big|_{r=R} \sin^2\theta d\theta = \pi\eta_c R U^2 f(\psi_1, \psi_2)$$
$$E_D = -\frac{\partial E}{\partial t} = -\pi\eta_c R U^2 f(\psi_1, \psi_2) \tag{2.244}$$

The drag coefficient of a deformable particle can be determined from the following expression

$$C_D = \frac{F_T}{\pi R^2 \frac{1}{2}\rho_c U^2} = \frac{\pi\eta_c U R}{\pi R^2 \frac{1}{2}\rho_c U^2} f(\psi_1, \psi_2) = \frac{4}{\mathrm{Re}_d} f(\psi_1, \psi_2) \tag{2.245}$$

where $\mathrm{Re}_d = \frac{Ud}{\nu_c}$ is the Reynolds number, determined by the particle diameter, $\psi_1 = \frac{2+3\gamma}{1+\gamma}$, $\psi_2 = \frac{\gamma}{1+\gamma}$, $f(\psi_1, \psi_2) = \psi_1^2 - 6\psi_1\psi_2 + 3\psi_2^2$

1. In particular, if we assume that for very small numbers the droplets and bubbles behave like solid particles $(\gamma >> 1, \psi_1 = 3, \psi_2 = 1)$, then we can write $f(\psi_1, \psi_2) = -6$

$$-\frac{dE}{dt} = -6\pi R\eta_c U^2 \tag{2.246}$$

We define the particle resistance force according to formula (2.245) as

$$F_T = 6\pi R\eta_c U \tag{2.247}$$

which agrees with the Stokes law for a solid particle. Accordingly, the coefficient of resistance is defined as

$$C_D = \frac{F_T}{\pi R^2 \frac{1}{2}\rho_c U^2} = \frac{6\pi\eta_c U R}{\pi R^2 \frac{1}{2}\rho_c U^2} = \frac{24}{\mathrm{Re}_d} \tag{2.248}$$

From the literature, it follows that expression (2.248) agrees very well with the resistance curve up to $\mathrm{Re}_d < 1$.

2. For gas bubbles in a liquid, we can assume that, $\gamma \to 0$, $\psi_1 = 2, \psi_2 = 0$ and, respectively $f(\psi_1, \psi_2) = 4$. The drag force for gas bubbles is determined according to equation (2.244) in the form

$$F_T = 4\pi\eta_c U R \tag{2.249}$$

Then the coefficient of resistance, taking into account (2.249) in analogy with (2.248), is defined in the form

$$C_{DG} = \frac{16}{\mathrm{Re}_d} \tag{2.250}$$

This equation is consistent with the results of the work up to $0.4 \le \mathrm{Re}_d \le 1.385$. With increasing number Re_d of bubbles begin to deform and lose spherical shape, as a result of which the drag coefficient is significantly influenced by the numbers Mo and We [4,6,9,21].

2.6.6 Dissipation of Energy in a Turbulent Flow in the Presence of Interaction of Dispersed Particles

Structured systems correspond to the state of minimum energy dissipation. According to Kolmogorov's theory of isotropic turbulence, the pulsation velocity through energy dissipation is expressed in the form $V_\lambda = (\varepsilon_R \lambda)^{1/3}$, from which we have $\varepsilon_R = \frac{V_\lambda^3}{\lambda}$. In the review paper [23,24] on the basis of the statistical theory of isotropic turbulence, the expression for determining the energy dissipation per unit mass is proposed in the form: $\varepsilon_R = C_m U^3 / L$ where L is the integral length characterizing the energy transfer by turbulent vortices $C_m = \text{const}$. Subsequent studies have shown that it C_m is a function, that is, $C_m = f(\mathrm{Re}) \sim \mathrm{Re}^{-m}$. In particular, taking the scale of turbulent pulsations equal to the diameter of the tube, the dissipation of energy in the turbulent flow of liquid in the pipes can be determined as [23]: $\varepsilon_R = f(\mathrm{Re}) U^3 / d_T$ where $f(\mathrm{Re})$ is the drag coefficient in the pipe, which can be determined by Blasius's formula $f(\mathrm{Re}) = 0.3164 / \mathrm{Re}^{1/4}$, $\mathrm{Re} > 10^4$. At the constant value $C_m = \text{const}$, the turbulent kinematic viscosity is proposed to identify as: $\nu_T \sim C_m U^4 / \varepsilon_R$. Using experimental data, the following empirical formula can be obtained for the drag coefficient in a turbulent flow for large-scale vortices

$$C_D = \left(\frac{40}{\mathrm{Re}_\lambda}\right)^5 + 0.49 \tag{2.251}$$

Here, $\mathrm{Re}_\lambda = V_\lambda \lambda / \nu_C$ is the pulsation velocity of scale λ. In Figure 2.1 shows the correspondence of the formula to the experimental data. The relationship between the drag coefficient and the dissipation of energy in an isotropic turbulent flow can also be represented in the form: $C_D \sim \varepsilon_R \lambda / V_\lambda^3$.

The phenomena of coalescence and crushing depend on the dissipation of energy in a unit of mass, the determination of which for a dispersed stream is of great complexity. At the same time, the total dissipation of energy in a dispersed stream is determined by the concentration (number) of particles and the rate of their flow, which obviously vary depending on the ongoing phenomena of coalescence and crushing. In the same paper [24], the following dependence of the dissipation of the energy of the dispersed and continuous phase on their viscosity for highly concentrated disperse systems is given

$$\frac{\varepsilon_{Rd}}{\varepsilon_{Rc}} = \left(\frac{\nu_C}{\nu_d}\right)^3 \tag{2.252}$$

Here, $\varepsilon_{Rd}, \varepsilon_{Rc}$ are the dissipation of energy in the dispersed and continuous phases. If we consider the total dissipation of energy as an additive sum of energy dissipation in the continuous and dispersed phase, then we can write

$$\varepsilon_{R0} = \varepsilon_{Rd} + \varepsilon_{Rc} = \varepsilon_{Rc}\left[1 + \gamma^{-3}\left(\frac{\rho_d}{\rho_C}\right)^3\right] \tag{2.253}$$

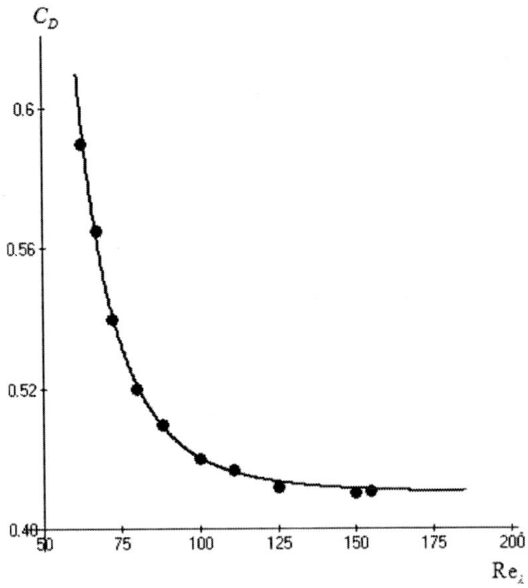

FIGURE 2.1 Dependence of the drag coefficient on the number, $Re_\lambda = V_\lambda \lambda / \nu_C$.

where ε_{R0} is the total energy dissipation, and $\gamma = \eta_d / \eta_c$ is the degree of mobility of the droplet surface. Thus, as the degree of mobility of the droplet surface increases, the total energy dissipation decreases. The paper [24] presents experimental data on the dependence of the volume fraction of particles on energy dissipation for different rotational velocities of the stirrer. These data are satisfactorily approximated by the following empirical expression

$$\varphi = b_0 \varepsilon_{R0} \exp\left(-b_1 \varepsilon_{R0}\right) \tag{2.254}$$

Here, b_0, b_1 are the coefficients that depend on the rotational velocity of the mixer (with; $n_m = 400\,\text{min}^{-1}$, $b_0 = 2.85 \times 10^{-5}$, $b_1 = 3.5 \times 10^{-4}$ and when $n_m = 800\ \text{min}^{-1}$, $b_0 = 2.55 \times 10^{-6}$, $b_1 = 3.5 \times 10^{-5}$) (Figure 2.2).

Several other data are presented in the literature on the dependence of energy dissipation on the volume fraction of particles, which are approximated by the following expression:

$$\varepsilon_{R0} = 340 - 2.96 \times 10^6 \varphi + 3.75 \times 10^{10} \varphi^2 - 10^{14} \varphi^3,\ m^2/c^3,\ \ 4.5 \times 10^{-6} < \varphi \le 1.5 \times 10^{-4} \tag{2.255}$$

It should be noted that such a difference in data is associated with a complex structure and mechanism of dissipation. At a certain content of particles in the flow, the total dissipation of energy is composed of the additive sum of many terms: (1) for the flow of the carrier and the dispersed phase in the gradient field; (2) as a result of the interaction of particles among themselves (frictional forces, adhesion forces, etc.); (3) as a result of the interaction of the carrier and disperse phase; and (4) due to irreversible processes of multicomponent diffusion (thermal diffusion, barodiffusion) in dispersed and carrier phases, etc. Obviously, the determination of each component of energy dissipation in a dispersed turbulent flow presents certain difficulties associated with the formalization of the mechanism of the flow of individual stages. For a constant number of particles per unit volume, the energy dissipation can be

$$\varepsilon_R = \varepsilon_{R0} f(N) \tag{2.256}$$

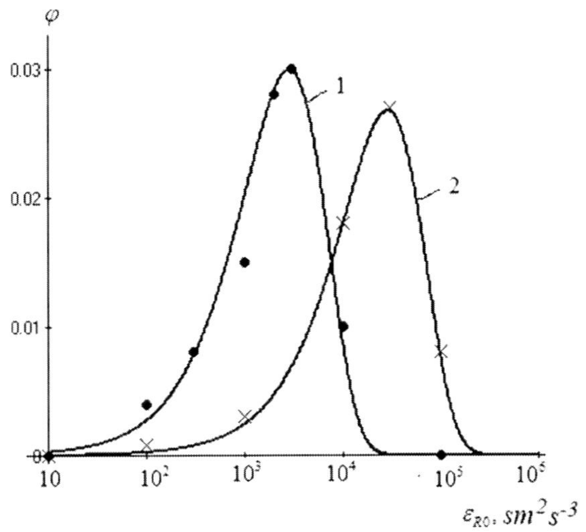

FIGURE 2.2 The dependence of energy dissipation on the volume fraction of particles at rotation frequencies (1) $n_m = 400\,\text{min}^{-1}$ and (2) $n_m = 800\,\text{min}^{-1}$.

where $f(N)$ is the function that determines the nature of the change in the number of particles.

It should be noted that since structure formation in dispersed systems leads to a decrease in the flow velocity to almost zero and the system becomes stationary, the dissipation of energy also decreases and, in the limiting case, approaches zero.

Significant differences in the experimental data are associated with a complex structure and mechanism of dissipation energy. At a certain content of particles in the flow, the total dissipation of energy is composed of the additive sum of many terms: (1) for the flow of the carrier and the dispersed phase in the gradient field; (2) as a result of the interaction of the particles with each other (frictional forces, adhesion forces, etc.); (3) as a result of the interaction of the carrier and disperse phase; and (4) due to irreversible processes of multicomponent diffusion (thermal diffusion, barodiffusion) in dispersed and carrier phases, etc. Obviously, the determination of each component of energy dissipation in a dispersed turbulent flow presents certain difficulties associated with the formalization of the mechanism of the flow of individual stages.

2.7 CHARACTERISTIC FEATURES OF HEAT AND MASS TRANSFER IN CHEMICAL ENGINEERING

The phenomena of heat, mass, and momentum transfer are usually described by complex nonlinear differential equations complicated by the dependence of the coefficients (viscosity, thermal conductivity, diffusion and heat transfer coefficients, etc.) on the temperature distribution. When calculating the technological process in order to simplify the solution of problems, it is usually necessary to assume various assumptions inherent in this process. For example, in the theory of chemical reactors, it is customary to assume that, instead of the complex chemical, diffusion, and thermal interaction of various transport phenomena, limit variants of these processes can be analyzed: (1) the rate of chemical reaction is much less than the rate of supply of reagents to the apparatus, that is, the limiting stage is the kinetic region of the process; (2) the rate of chemical reaction is high and the rate of chemical transformation is determined by the rate of supply of reagents to the reaction zone—the diffusion region; (3) the reaction rate depends on the reaction kinetics and diffusion transport—a mixed region. It is important to note that in chemical reactor processes occur in most cases under turbulent conditions at relatively high temperatures, and therefore, according to different values of the number, one can judge the diffusion or convective nature of mass and heat transfer. In such processes, the reactions are carried out in the gas–solid particle system, which have small values of the number, so that the magnitude of the dissipative transformation has an insignificant effect on the heat balance.

Possible assumptions can be made when calculating heat exchange equipment, where diffusion heat transfer can be neglected. In the processes of separation and purification of liquids and gases and in a number of other mass-exchange processes, heat and mass transfer occurs in thin interphase films and are realized in such systems as: (1) gas–liquid (rectification, absorption), (2) gas–solid (adsorption, drying, and moistening of solids); (3) liquid–liquid (coalescence and crushing of droplets in emulsion, liquid extraction); and (4) liquid–solid (extraction of matter from solids by solvent). The phenomena of heat and mass transfer in thin fluid films are also characteristic of gas–liquid reactors, where heat and mass-transfer processes flow through the interface and can be complicated by phase transformations (condensation, evaporation, dissolution). Usually in such devices the residence time of one of the phases is small, and the processes of heat and mass transfer proceed with high intensity.

In the processes of crystallization of solid particles from the liquid phase, the phenomena of mass and heat transfer at the interphase boundary are closely intertwined, since the processes of crystallization of solid inclusions from solutions proceed in two ways: (1) cooling solution, with decreasing heat, and (2) when the solution is heated and evaporated, that is, with the growth of heat. In the general case, the rate of transformation of the liquid phase of the solution into a crystalline solid depends on the kinetics of nucleation of the solid phase, the rate of their diffusion growth, and the intensity of heat transfer in the crystallization zone. The formulation and solution of the problem of moving the boundary of a phase transition is possible only within the framework of making certain assumptions and simplifications. In connection with the difficulties of obtaining analytical solutions, various approximate solutions are widely used, the advantage of which is the possibility of analyzing the crystallization process for bodies of finite dimensions and geometric shapes.

Certain assumptions make it possible to simplify the heat and mass-transfer equations, although the simultaneous solution of these equations for nonlinear kinetics and variable thermophysical flow properties for exothermic and endothermic processes with phase transformations is fraught with many difficulties. In these cases, the nonlinear terms in the transport equations should be considered as a dissipative function that determines the temperature and mass of particles on the surface, the flow characteristic, the size and geometric shape of the particles, and many other factors that complicate the problem.

The solutions presented in this chapter are of value for studying the processes of heat transfer in order to reveal the mechanism of phenomena and determine the transport coefficients and the empirical equations for their calculation under various conditions for the realization of the process. The theoretical solution of the transport problem allows us to obtain an analytical formula suitable for analysis and establishment of a truer mechanism of the process.

3 Introduction to Mass Transfer

This basis of all chemical and many other (food, pharmacological, etc.) processes makes a mass transfer, that is, the irreversible transfer of weight within one phase or of one phase in another. The process of a mass transfer is carried out as a result of chaotic motion of the molecules of the medium (molecular diffusion), as a result of the macroscopic movement of the environment (convective transfer), and in turbulent streams, as a result of the chaotic movement of the pulsation whirlwinds of various sizes. In this chapter, we will consider the basic concepts of various mass-transfer processes and dimensionless criteria for interphase convective mass transfer, the laws of Henry and Raoult, the determination of mass-transfer coefficients for two-phase systems, etc.

3.1 GENERAL INFORMATION ABOUT PROCESSES OF A MASS TRANSFER IN THE PROCESSES OF CHEMICAL TECHNOLOGY

Mass transfer includes: (1) mass output, where mass transfer occurs from the interface boundary in the volume of flow; (2) mass transfer, where mass transfer from one phase to other occurs through the surface of interface phase; and (3) mass conductivity, where mass transfer occurs in the volume of porous body [4]. The transfer of heat proceeds only at a deviation from the equilibrium state, that is, in the presence of the difference of temperatures between heat transfers, and the transition of a substance in another comes from one phase in the absence of a balance between phases. Respectively, at a deviation from the equilibrium state there is a transition of substance from a phase in which the content is a higher equilibrium, in a phase where the content of this substance is lower than the equilibrium. The rate of transition of a substance is proportional to the deviation degree from its balance, which can be expressed as the difference of concentration of a substance in one of the phases and the equilibrium concentration in this substance and the interphase surface of the contact of phases. This difference of concentration is the driving force of the process of a mass transfer. Therefore, the substance transition velocity in a unit of time from one phase to another can be determined as

$$M = KF\Delta c \tag{3.1}$$

where M – is the amount of the substance that has passed from one of a phase into another, K is mass-transfer coefficient, F – is the surface of the contact of phases, $\Delta c = (c^* - c)$ is the driving force of a mass transfer, and c^* is the equilibrium concentration of the substance. It is important to note that if the concentration of a substance in this phase is above the equilibrium concentration, then it passes the substance from this phase into another. Two main types of transfers of mass can be distinguished: (1) a mass transfer between liquid and gas or between two immiscible liquids (distillation, absorption, liquid extraction, separation of emulsions, transfer of mass in disperse systems, etc.); and (2) a mass transfer between a solid and liquid, gas or steam (adsorption, catalytic processes, drying, moistening, etc.). Three substances participate in the majority of processes of a mass transfer: (1) the distributing substance making the first phase, (2) the distributing substance making the second phase, and (3) the distributed substance passing from one phase into another. In all solvable problems of a mass transfer, it is required to determine generally three parameters: (1) distribution of the concentration in an interphase layer, (2) calculation and assessment of the amount of transferable substance from one phase in another, and (3) calculation of the coefficient of a mass transfer. The last task in the majority, considering hydrodynamic and thermal factors, it is presented in the form of the criteria equations. Among the main dimensionless criteria used in calculations of a mass exchange, it is important to note the following: (1) the Reynolds number expressing itself a measure of a ratio of forces of inertia and viscosity, $\mathrm{Re} = VL/\nu$; (2) diffusive number of Prandtl or Schmidt's number characterizing by itself a measure of a ratio of viscous and concentration properties of a stream, $\mathrm{Sc} = \nu/D$; (3) diffusive Nusselt number or Sherwood number characterizing by itself the dimensionless number of transferable weight $\mathrm{Sh} = \beta L/D$; (4) the diffusive Peclet number expressing the relation of the number of transferable mass convection to the number of the mass postponed in the diffusive way, $\mathrm{Pe_D} = VL/D$; (5) Fourier's number characterizing change in time of rate of transfer of substance at a non-stationary mass transfer, $\mathrm{Fo} = Dt/L^2$; and (6) Weber's number characterizing a measure of a ratio of forces of inertia and a superficial tension $\mathrm{We} = V^2 d\rho/\sigma$. Here, V is stream velocity, $\mathrm{m/s}$; L is characteristic size, m , D is diffusion coefficient, $\mathrm{m/s^2}$; β is mass-transfer coefficient, $\mathrm{m/s}$; ν is viscosity, $\mathrm{m^2/s}$; and σ is coefficient of a superficial tension, $\mathrm{N/m}$.

From the point of view of disperse systems, interfacial mass transfer can be classified as follows [4]:

- Mass transfer from the solid phase to the liquid phase (dissolution of solid particles in the liquid, melting of the solid phase, separation of the substance from the solid phase by the liquid solvent by extraction) and vice versa, from the liquid phase to the solid phase (crystallization of solids from the solution, wetting of the powders, processes of sedimentation of solid particles from a liquid);
- Mass transfer from the solid phase to the gas phase (drying of materials, desorption of products from the surface, sublimation or evaporation of the solid phase), and vice versa, from the gas phase to the solid phase (adsorption, chemical processes occurring on the surface of porous solids, sedimentation processes of solid particles);
- Transfer of mass from one liquid phase to another liquid phase (liquid extraction, separation of emulsions, crushing and coalescence of droplets, spraying);
- Mass transfer from the gas phase to the liquid phase (condensation, absorption, dissolution of gas bubbles in liquids, gas–liquid reactors); and
- Transfer of mass from the liquid phase to the gas phase (rectification, distillation, evaporation, evaporation).

It is important to note that the presence of the polydispersity of particles participating in the mass-transfer processes imposes additional conditions on the solution of transport problems. This is primarily due to the time evolution of the particle size distribution function, the determination of the average volume, the average interfacial surface, and the average particle size. In addition, the presence of polydisperse systems requires the need to take into account the phenomena of coagulation (coalescence) and their crushing deformation in the calculation of the distribution function and time-varying interfacial surface. The presence of phenomena of coagulation and fragmentation of particles and their deformation contributes to a change in the surface boundary conditions and the concentration of particles per unit volume, thereby affecting dissipative processes.

In chemical technology, there is a class of mass-exchange processes, where the main driving force of the process is the rate of mass transfer from one phase to another. Let's briefly dwell on these processes:

- Selective absorption of gases or vapors by a liquid solvent (absorbent) is called the absorption process, where the substance is transferred from the gas phase to the liquid by diffusion. The use of absorption in the technique for separating and purifying gases, separating vapors from vapor–gas mixtures, is based on the difference in the solubility of gases and vapors in liquids. The process inverse to absorption is called desorption; it is used to separate the absorbed gas from the solution and to regenerate the absorbent. Absorption processes in practice are carried out in packed or tray columns (countercurrent), and therefore the interfacial surface (specific contact surface) is determined by the design of the plates and the size and shape of the nozzle particles. Physical absorption is carried out at ambient temperature (20°C–40°C) or at low temperatures, because the solubility of highly soluble gases increases with decreasing temperature. In addition, as the temperature decreases, the solubility of poorly soluble gases decreases, that is, the selectivity increases and the losses of the poorly soluble component and the contamination of the extracted gas are reduced, as well as the vapor pressure of the absorbent and its loss. With chemical absorption, an increase in temperature leads to a significant increase in the mass-transfer coefficient and, in addition, to an increase in the solubility of many absorbents in diluents, and hence to an increase to a certain limit of the total absorbent capacity of the absorbent;

- Selective absorption of various components of gases and substances dissolved in liquid substances by the surface of a porous solid (adsorbent) is called the adsorption process. In the industrial practice of chemical technology, absorption and adsorption processes are used to separate gas mixtures into separate components, to purify and dry gases from unwanted elements, and to isolate the most valuable elements or components. At the same time, these processes differ from each other by the presence of different phases and the nature of the hydrodynamic flow, which has a significant effect on the processes of mass and heat transfer, as well as structural design. The reverse process of adsorption is the desorption process, where mass transfer is carried out from the surface of a solid to a volume of gas or liquid (desorption of the absorbent). Knowledge of diffusion rates is important not only for the theory of adsorption but also for calculating industrial adsorption processes. In this case, they usually do not deal with individual grains of the adsorbent, but with their layers. The kinetics of the process in the layer is expressed by very complicated dependences. At each point of the layer at a given instant of time, the amount of adsorption is determined not only by the form of the isotherm equation and by the laws of the process kinetics, but also by the aerodynamic or hydrodynamic conditions;
- Separation of a liquid mixture with different boiling points into individual components by repeated evaporation and condensation and countercurrent interaction of vapor and liquid flows is called the distillation process. In the distillation process, a continuous transfer of matter from the liquid phase to the vapor phase and, conversely, from the vapor phase to the liquid phase on the contact plates or on the nozzles of the column apparatuses, called distillation columns, is carried out. The driving force of the rectification process is the difference between the actual (operating) and equilibrium concentrations of the components in the vapor phase that correspond to a given composition of the liquid phase. The vapor–liquid system tends to achieve an equilibrium state, as a result of which the vapor, when in contact with the liquid, is enriched with volatile (low boiling) components, and the liquid by highly volatile (low boiling) components. Since liquid and vapor tend to move countercurrent (vapor-up, liquid-down), at sufficiently high column height, a sufficiently pure component can be obtained. The quality and quantity of the carried substance, which depends on the hydrodynamics of the flows and the heat transfer, completely determines the degree of purity of the individual substances obtained during rectification of the liquid mixtures. Rectification processes can be used to separate gas mixtures after their liquefaction (separation of air into nitrogen and oxygen, separation of hydrocarbon gases, etc.);
- Selective extraction of one or more of one liquid phase into a liquid that is in contact with and immiscible with it, containing a selective solvent or from a solid with another liquid (solvent), is called an extraction process. Being one of mass-exchange processes, extraction is used for extraction and concentration of dissolved substances from a liquid (liquid–liquid) and solid medium (liquid–solid). The recovery phase includes only the extractant or is a solution of the extractant in the solvent. The main stages of liquid extraction are: (1) contacting and dispersing phases, (2) separation or delamination of

phases into extract (extracting phase) and raffeinate (exhaust phase), and (3) the isolation of the target components from the extract and the regeneration of the extractant. The mechanism of the extraction process from solids differs from the mechanism of liquid extraction, the presence of phenomena occurring in a porous medium. During the extraction process, diffusion transfer of the mass from the solid or liquid phase to another liquid phase (solvent) takes place. Extraction processes are carried out in tray columns or in apparatus with mixing devices, and the quality of mass transfer, in addition to other parameters, is determined by the design of plates and mixing devices, as well as by the selectivity of the solvent extractant;

- Removal of moisture or liquid (most often moisture–water, less often various liquids–solvents) from porous solid wet materials by its evaporation is called the drying process. The drying process is carried out by evaporation of liquid and removal of the formed vapors in heat is supplied to the material being dried, most often by means of so-called drying agents of heat (heated air, flue gases and their mixtures with air, inert gases, superheated steam). Drying is carried out by wet bodies: solid-colloid particles, granular, powdery, lumpy, granular, leafy, woven, etc. In the drying process, the moisture mass is transferred from the pores of the solid to the gas or vapor phase, which is determined by the properties of the porous bodies, the temperature of the medium, and nature of the current. Depending on the properties of the dried material, the drying processes are carried out in devices of different design, which determines the nature of the mass transfer.
- The process of moistening of solid particles and powdery materials where transfer of moisture in the volume of a solid body is carried out is called the reverse drying of solid material. All calculations concerning the process of drying are valid also in moistening processes only with the return sign. The process of moistening of an external and internal surface of porous material is often used in the food industry and in chemical technology (granulation of powdery materials, in processes of a desorption, etc.);
- The gradual transfer of mass of a solid phase from a surface of particles in the volume at their direct contact and at a certain temperature and concentration of a surface layer equal to the saturation concentration is called dissolution. Dissolution of firm particles is widely used in chemical, food, and the pharmaceutical industry and carried out in mixers and in a fluidized layer. Dissolution of gas bubbles in liquid by nature differs from dissolution of firm particles, first of all in existence of an internal hydrodynamic current in bubbles. In certain cases, dissolution of a firm phase makes a basis of process of extraction of substance of a solid body liquid solvent or extractions in system liquid—liquid, dissolution of gas bubbles makes a basis of processes of absorption. In certain cases, dissolution or absorption by a solid body of liquid or its vapors brings to their swelling maintaining of property of not fluidity by it (i.e., the form and geometry of a body doesn't change), though it leads to an increase in volume and mass of a solid body. Swelling is purely a diffusive transfer with prevalence of capillary penetration of liquid during a time of a solid body.

3.1.1 Membrane Processes

Membrane processes are selective extractions of components of mix or their concoction by means of a semipermeable partition—a membrane. These processes represent substance transition (or substances) from one phase in another through the membrane dividing them. Gas and liquid mixes are applied to division, sewage treatment, and gas emissions.

Unlike a heat transfer, which occurs usually through a wall, the mass transfer is carried out, as a rule, at direct contact of disperse phases (except for membrane processes). At the same time, the border of contact or the surface of contact of phases can be mobile (gas–liquid, vapors–liquid, liquid) or motionless (gas–solid, liquid–solid). By this principle, mass-exchanged processes subdivide on mass-transfer systems with free limit of the section of phases, in systems with the motionless surface of contact of phases, and through semipermeable partitions (membranes).

Many thermal processes, such as evaporation, condensation evaporation, and also hydromechanical processes—flotation, sedimentation, hashing—are also followed by a mass exchange between phases. It is important to note that mass-exchanged processes are physically reversible, that is, the distributed substance can pass from one phase into another depending on the concentration of this substance in both phases and balance conditions. In the equilibrium state, there is certain dependence between concentrations of the distributed substance in both phases, which is subsequently discussed. In chemical technology, all mass-exchanged processes are carried out in devices of various designs determining the maximum velocity of a mass exchange. These devices are designed so that they create the maximum mass transfer surface, and the minimum hydraulic resistance at the maximum interfacial transfer intensity is created on it. The principle of formation of an interphase surface is the basis for classification of mass-exchanged devices that can be presented as follows: (1) devices with the fixed surface of phase contact, nozzle and film devices, and devices where interaction of gas with a firm surface is carried out belong to this type; (2) devices with the surface of contact formed in the course of the movement of streams (dish-shaped columns) of which discrete interaction of phases on device height is characteristic; and (3) devices with an external supply of energy—devices with mixers, rotor, pulsation devices, etc.

The condition of interaction of phases in mass transfer processes is determined by the purpose of the main process. For example, during the rectification process, there is saturated steam and boiling liquid in direct contact, which facilitates the transfer of relatively heavy components from steam to liquid and light components from liquid to steam. In adsorption processes, the gas mix is divided as a result of selective sorption of one of components into the surfaces of the solid adsorbent.

The traditional approach to the solution of problems of a mass exchange comes down to a research of the equations of convective transfer in which components of velocity are defined from the considered hydrodynamic task. In many cases, chemical and phase transformations happen in streams, and their kinetic features (process velocity) are influenced by the convective transfer of weight. The presence of phase and chemical transformations in the flow (chemisorption, evaporation, condensation) can significantly change the hydrodynamic velocity fields as a result of the appearance of various convective flows (due to changes in surface tension, differences in temperature and pressure) and changes in the properties of the medium, density and viscosity of the current environment, which depends on the composition of phases. The estimated task is very difficult and doesn't give in to the theoretical analysis and decision. Consideration of problems of a mass exchange makes an assumption that the hydrodynamics influences a mass transfer, and diffusive streams influence the current a little. At the same time, such an approach to a research of problems of a mass exchange doesn't save one from the mathematical difficulties connected, taking into account difficult hydrodynamic conditions both at laminar and at a turbulent flow.

3.2 WAYS OF EXPRESSING THE CONCENTRATION OF PHASES

When calculating processes of a mass exchange between various phases it is necessary to distinguish a condition of mix of these phases under set conditions. If there are several immiscible and insoluble particles in other phases in the mixture (liquid, gas, solid particles), then such systems (one of the phases is carrier) form the interfacial separation region and are called multiphase systems. If such systems contain only two phases, then such systems are called disperse systems. If mix contains several mutually dissolved components, then such systems form multicomponent solutions. Solutions in which the number of components is equal to two are called binary. An important question when calculating processes of a mass exchange is, what are the ways of expression of concentration of phases in these mixes? In processes of a mass exchange, mass and molar concentration are generally used.

1. If a mix with total amount v contains N components $A, B, C, \ldots$ and volumes in mix respectively $v_A, v_B, v_C, \ldots$, then the volume fraction of each component ϕ_i can be calculated by formulas

$$\phi_A = {v_A}/{v},\ \phi_B = {v_B}/{v},\ \phi_C = {v_C}/{v},\ v = v_A + v_B + v_C = \sum_{i=1}^{N} v_i,\ \phi_A + \phi_B + \phi_C = \sum_{i=1}^{N} \phi_i. \tag{3.2}$$

2. If a mix with lump G contains N components $A, B, C, \ldots$ and their mass in mix respectively $G_A, G_B, G_C, \ldots$, and then the mass fraction of each component is defined as

$$x_A = {G_A}/{M_A},\ x_B = {G_B}/{M_B},\ x_C = {G_C}/{M_C} \tag{3.3}$$

3. If there are N components $A, B, C, \ldots$ in a mixture and their mass respectively G_A, G_B, G_C and the molecular masses M_A, M_B, M_C, then the number of moles of each component in the mix will be defined as

$$n_A = {G_A}/{M_A},\ n_B = {G_B}/{M_B},\ n_C = {G_C}/{M_C} \tag{3.4}$$

Total number of moles will be defined as

$$n = n_A + n_B + n_C = \sum_{i=1}^{N} n_i = {G_A}/{M_A} + {G_B}/{M_B} + {G_C}/{M_C} = \sum_{i=1}^{N} {G_i}/{M_i} \tag{3.5}$$

4. If there are N components $A, B, C, \ldots$ in a mixture with the number of moles in the mix equals n_A, n_B, n_C, then the molar fraction for each component will be defined as

$$x_A = {n_A}/{n},\ x_B = {n_B}/{n},\ x_C = {n_C}/{n},\ x_A + x_B + x_C = \sum_{i=1}^{N} x_i = 1 \tag{3.6}$$

Using these expressions, it is possible to write down

$$x_A + x_B + x_C = \sum_{i=1}^{N} x_i = 1,\ x_A = {n_A}/{n} = \frac{{G_A}/{M_A}}{\sum_{i=1}^{N} {G_i}/{M_i}},\ x_B = {n_B}/{n} = \frac{{G_B}/{M_B}}{\sum_{i=1}^{N} {G_i}/{M_i}},$$
$$x_C = {n_C}/{n} = \frac{{G_C}/{M_C}}{\sum_{i=1}^{N} {G_i}/{M_i}} \tag{3.7}$$

Or in a general view, it is possible to write down

$$x_j = \frac{{G_j}/{M_j}}{\sum_{i=1}^{N} {G_i}/{M_i}} = \frac{{m_j}/{M_j}}{\sum_{i=1}^{N} {m_i}/{M_i}} \tag{3.8}$$

For $n = 1$ *mole* mix, it is possible to write $x_A = {G_A}/{M_A}$, $x_B = {G_B}/{M_B}$, $x_C = {G_C}/{M_C}$ or $G_A = x_A M_A$, $G_B = x_B M_B$, and $G_C = x_C M_C$. If to consider that lump 1 *mole* to asking mixes it is equal

$$G = G_A + G_B + G_C = x_A M_A + x_B M_B + x_C M_C = \sum_{i=1}^{N} x_i M_i \tag{3.9}$$

that the mass fraction of components will be defined respectively

$$m_A = \frac{x_A M_A}{\sum_{i=1}^{N} x_i M_i}, \quad m_B = \frac{x_B M_B}{\sum_{i=1}^{N} x_i M_i}, \quad m_C = \frac{x_C M_C}{\sum_{i=1}^{N} x_i M_i} \tag{3.10}$$

or in a general view

$$m_j = \frac{x_j M_j}{\sum_{i=1}^{N} x_i M_i} \tag{3.11}$$

These expressions connect mass and molar fractions for 1 to asking mixes. We will determine the general molecular mass of mix as

$$M = M_A x_A + M_B x_B + M_C x_C = \sum_{i=1}^{N} M_i x_i \tag{3.12}$$

3.2.1 Dalton's Law

With a constant pressure and temperature, the general pressure of mix is equal in the closed system to the sum of partial pressure of the components of mix

$$P = p_1 + p_2 + \dots + p_n = \sum_{i=1}^{N} p_i \tag{3.13}$$

At a constant temperature, the volume of the mixture v and the pressure P for any component i of the gas can be written as

$$p_i v = P v_i \tag{3.14}$$

From here $v_i = \frac{p_i}{P} v$ the volume fraction will be defined as

$$\varphi_i = \frac{v_i}{v} = \frac{p_i}{P} \tag{3.15}$$

For 1 *mole* mixes, they can be written as $\varphi_i = x_i$. Thus, the partial pressure of that component i is equal in mix $p_i = \varphi_i P$. Using the equation of the gas law, we will determine the volume of gas mix at a temperature T

$$v = \frac{m_i}{M_i} \frac{RT}{p_i} \tag{3.16}$$

where m_i is mass, i is that component of gas, M_i is molecular mass of that, and i is the component of gas. From this equation, we will define mass and molar concentration

$$\rho_i = \frac{m_i}{v} = \frac{M_i p_i}{RT}, \left[\text{kg/m}^3\right] \tag{3.17}$$

$$C_i = \frac{m_i}{M_i v} = \frac{p_i}{RT}, \left[\text{kmole/m}^3\right] \tag{3.18}$$

where $R = 8314$ J/kmole, and K is gas constant.

The density of mix of liquids can be determined in a formula

$$\frac{1}{\rho} = \frac{m_1}{\rho_1} + \frac{m_2}{\rho_2} + \ldots.. = \sum_{i=1}^{N} \frac{m_i}{\rho_i} \tag{3.19}$$

where m_i are mass fractions of components to sweep away liquids.

Example 3.1

In air mixes with gas CO_2, the content of carbon dioxide is about 15%. At pressure $P = 20$ *atm* and temperature $T = 20°C$ to define mass fractions, the mix of mass is $M_{CO_2} = 44$ and $M_B = 29$ molar concentration of components.

For 1 mole CO_2, the weight is equal to the mix, and $G_{CO_2} = \phi_{CO_2} M_{CO_2} = 0{,}15.44 = 6{,}6$ kg the mass of air is equal to $G_B = (1 - \phi_{CO_2}) M_B = (1 - 0{,}15).29 = 24{,}65$ kg.

Molar shares CO_2 and air are respectively equal

$$m_{CO_2} = \frac{G_{CO_2}}{G_{CO_2} + G_B} = \frac{6.6}{6.6 + 24.65} = 0.212,\; m_B = (1 - 0.212) = 0.788$$

Partial pressure CO_2 is equal $p_{CO_2} = \phi_{CO_2} P = 0.15/20 = 3$ atm $= 3.10^5$ $^{N}/_{m^2}$, and air is $p_B = (1 - \varphi_{CO2}) P = (1 - 0.15).20 = 17$ atm $= 17.10^5$ $^{N}/_{m^2}$. Mass concentration CO_2 will be defined as follows

$$\rho_{CO_2} = \frac{M_{CO_2} p_{CO_2}}{RT} = \frac{44.3.10^5}{8314.(273 + 25)} = 5.327\ ^{kg}/_{m^3}$$

and air $\rho_B = \frac{M_B p_B}{RT} = \frac{29.17.10^5}{8314.(273 + 25)} = 19.89$ $^{kq}/_{m^3}$. Molar concentration CO_2 and air will be defined as

$$C_{CO_2} = \frac{p_{CO_2}}{RT} = \frac{3.10^5}{8314.(273 + 25)} = 0.121\ ^{kmole}/_{m^3}$$

$$C_A = \frac{p_B}{RT} = \frac{17.10^5}{8314.(273 + 25)} = 0.686\ ^{kmole}/_{m^3}$$

3.2.2 Expression for the Mass and Molar Rate of a Stream

We will express molar and mass velocity through these concentrations. Average mass velocity will be defined as

$$V = \frac{\sum_{i=1}^{N} \rho_i V_i}{\rho} \tag{3.20}$$

where ρ_i, V_i is the mass concentration and mass rate of components of mix, and total $\rho = \sum_{i=1}^{N} \rho_i$ is mass concentration of mix. Average molar rate of mix will be defined in a look

$$V^* = \frac{\sum_{i=1}^{N} C_i V_i}{C} \tag{3.21}$$

where C_i, V_i is molar concentration and molar rate of components of mix, and $C = \sum_{i=1}^{N} C_i$ are total molar concentrations of mix. Between mass and molar rate communication is expressed by the following formulas

$$V - V^* = \sum_{i=1}^{N} m_i \left(V_i - V^*\right) \tag{3.22}$$

$$V^* - V = \sum_{i=1}^{N} x_i \left(V_i - V\right) \tag{3.23}$$

Generally, the number of transferable weights through surface unit of area for a unit of time from one phase to another can be defined as mass $\left(\text{kq}/\text{m}^2\text{s}\right)$ or molar $\left(\text{kmole}/\text{m}^2\text{s}\right)$ streams. The expressions defining dependence of transferable weight through various units of concentration and various velocities of a stream (3.20) and (3.1) are given in Table 3.1.

We will consider the binary mix containing two components A and B.

1. The mass diffusive stream concerning an average of mass velocity will be defined with equation (3.20) as

$$j_A + j_B = \rho_A\left(V_A - V\right) + \rho_B\left(V_B - V\right) = \rho_A V_A + \rho_B V - V\left(\rho_A + \rho_B\right) = \rho_A V_A + \rho_B V_B - V\rho \tag{3.24}$$

Here, $\rho = \rho_A + \rho_B$ are total mass concentration, and V is average mass velocity. Considering these ratios, $V = \frac{V_A\rho_A + V_B\rho_B}{\rho}$ we will receive

$$j_A + j_B = \rho_A V_A + \rho_B V_B - \frac{V_A\rho_A + V_B\rho_B}{\rho}\rho = 0 \tag{3.25}$$

That is, the mass diffusive stream of rather average mass velocity is equal to zero.

TABLE 3.1
Diffusive Streams

Ratio \ Parameters	Mass Rate, $\text{kg}/\text{m}^2\text{s}$	Molar Velocity, $\text{kmol}/\text{m}^2\text{s}$
Rather stationary axes	$\vec{V}_m = \sum_{i=1}^{N} \rho_i \vec{V}_i$	$\vec{V}_C = \sum_{i=1}^{N} C_i \vec{V}_i$
Rather average mass velocity	$\vec{j} = \sum_{i=1}^{N} \rho_i\left(\vec{V}_i - \vec{V}_m\right)$	$\vec{J} = \sum_{i=1}^{N} C_i\left(\vec{V}_i - \vec{V}_C\right)$
Rather average molar velocity	$\vec{j}^* = \sum_{i=1}^{N} \rho_i\left(\vec{V}_i - \vec{V}^*\right)$	$\vec{J}^* = \sum_{i=1}^{N} C_i\left(\vec{V}_i - \vec{V}^*\right)$

2. We will define a molar diffusive stream of rather average mass velocity

$$J_A + J_B = C_A(V_A - V) + C_B(V_B - V) = C_A V_A + C_B V_B - V(C_A + C_B) = C(V^* - V) \tag{3.26}$$

   Here, $CV^* = C_A V_A + C_B V_B$, $C = C_A + C_B$.
3. We will define a molar stream of rather average molar rate

$$J_A^* + J_B^* = C_A(V_A - V^*) + C_B(V_B - V^*) = C_A V_A + C_B V_B - V^* C \tag{3.27}$$

   Considering that $V^* = \frac{C_A V_A + C_B V_B}{C}$, $C = C_A + C_B$, we have

$$J_A^* + J_B^* = C_A V_A + C_B V_B - \frac{C_A V_A + C_B V_B}{C} C = 0 \tag{3.28}$$

4. The mass stream of rather average molar rate is equal

$$j_A^* + j_B^* = \rho_A(V_A - V^*) + \rho_B(V_B - V^*) = \rho_A V_A + \rho_B V_B - V^*(\rho_A + \rho_B) = \rho(V - V^*) \tag{3.29}$$

   Here, $\rho V = \rho_A V_A + \rho_B V_B$, $\rho = \rho_A + \rho_B$.
5. For multicomponent mix for that component in a stationary system, we will define a molar stream of rather average molar rate

$$\begin{aligned} J_i^* &= C_i(V_i - V^*) = C_i V_i - C_i \frac{\sum_{j=1}^{n} C_j V_j}{C} = C_i V_i - x_i \sum_{j=1}^{n} C_j V_j \\ &= C_i V_i - x_i \sum_{j=1}^{N} N_j = N_i - x_i \sum_{j=1}^{n} N_j \end{aligned} \tag{3.30}$$

From this equation, the molar stream of a substance will be defined as

$$N_i = x_i \sum_{j=1}^{n} N_j + J_i^* \tag{3.31}$$

Subsequently, it will be shown that it represents the full stream of the substance consisting of convective, diffusive, and other components of transfer. For a binary system, equation (3.31) can be written down in a look

$$\begin{aligned} N_A &= x_A(N_A + N_B) + J_A^* \\ N_B &= x_B(N_A + N_B) + J_B^* \end{aligned} \tag{3.32}$$

We will note that in further statements, x is the concentration of substance in a liquid phase, and y is the concentration of the same substance in a gas phase.

Thus, the certain ratios connecting mass and molar streams are received, and these ratios cover all types of streams, which are found in the theory of a mass exchange and specify a way of transition from one stream to others. Furthermore, for monotony of statement and convenience of the application, we will use only one form of definition of a stream J. However, using the ratios given in this section it is possible to pass at the solution of the corresponding task to this or that stream.

3.3 MASS TRANSFER BY CONVECTION AND DIFFUSION

Transfer of weight between two phases (binary systems) is carried out in the diffusive and convective way, though it is important to note the influence of effects of thermal diffusion, a barodiffusion, and some other the transfers connected with a hydrodynamic flow. At the same time, an important role in transfer of weight is played by the hydrodynamic current defining the nature of diffusive and convective transfers of substance. In heterogeneous processes, the transfer of weight is carried out in a thin interphase layer, formed between various phases. If in a stream of liquid or in volume of an interphase layer there are physical and chemical transformations, then distribution of concentration on coordinate and on time depend on the velocity of these transformations.

Thus, the transfer of mass caused by concentration change is carried out in two ways:

1. The convection caused macroscopic movement of the media. In this case, the players of each moving volume aren't changed, but each point of space concentration of the component, which is in this place of liquid, will change;
2. By diffusion, resulting in a change in the concentration of the substance of the mixture by molecular and turbulent transfer.

Alignment of concentration by direct change of composition of liquid is called diffusion. Diffusion, an irreversible process (dissipative system, Section 1.7), also represents one of the sources of dissipation of energy in a liquid mix. Characteristic of diffusive process, the coefficient of diffusion defines a diffusive stream in the presence of a concentration gradient. The diffusive stream caused by a gradient of temperature and pressure is defined, respectively, by the coefficient of thermal diffusion and a bar diffusion.

One of the main models of a mass transfer, which are widely used in numerous works, is the film model offered by Uitman and Luis. According to this theory, in a stream kernel concentration constant and process of transfer is described by the stationary equation of molecular diffusion in thin interphase films on condition of phase balance on limit of the section of phases. Process of transfer of weight of one phase, according to the film theory, can be presented to another as follows. Let the concentration of a transferable substance in a gas phase y be a higher equilibrium, and the substance passes into a liquid phase where its concentration is x. The transfer of substance in both phases is carried out by molecular diffusion and convective diffusion. In each phase, distinguish two areas: a stream kernel where transfer of substance is carried out by convective diffusion and the interface, which is formed on limit of the section of phases where transfer of substance is carried out by molecular diffusion. Owing to intensive hashing in a stream kernel concentration is distributed equally. The interface is an area of sharp change of concentration of transferable substance. In approaching an interface of phases, there is an attenuation of convective streams and increase of a role of molecular diffusion. Concentration of transferable substance in a gas phase decreases from size up x to y_P, the size on the limit of the section of phases. In a liquid phase, the concentration of transferable substance decreases from size on x_P, the limit of the section of phases up to the size x in a stream kernel. At the established process on limit of the section of phases, according to the film model of a mass transfer, balance is observed, that is, x_P and y_P are in equilibrium (Figure 3.1).

The rate of transition of a substance from one phase to another is proportional to the degree of deviation from equilibrium, which can be expressed as the difference in concentration of the working concentration of the substance in one of the phases and the equilibrium concentration of the substance in the other.

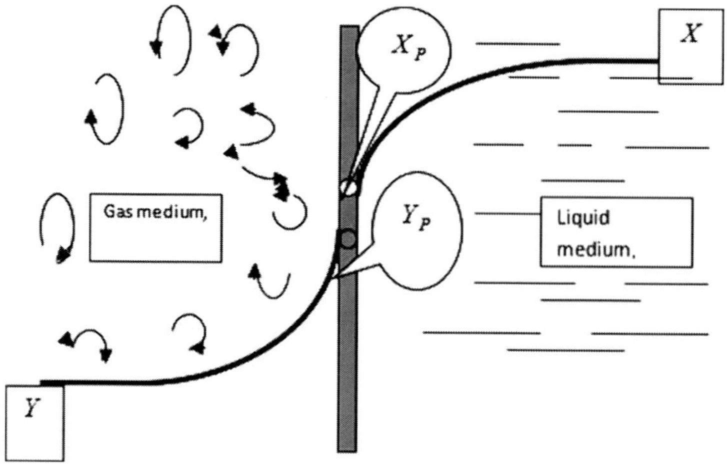

FIGURE 3.1 The scheme of the movement of streams of mass between gas and liquid in a film model.

This difference of concentration is the driving force of process of a mass exchange. The amount of transferable substance from one phase to another for a unit of time is proportional to the difference of concentration at an interface of phases and in the volume of a stream and is defined as

$$N = K_y F \Delta C_y \tag{3.33}$$

$$N = K_x F \Delta C_x \tag{3.33a}$$

where K_y is the mass-transfer coefficient carried to driving force $\Delta y = \Delta C_y$, expressed through molar shares of a component in a gas phase, K_x is the mass-transfer coefficient $\Delta x = \Delta C_x$, F is the area of interphase surface, m^2, and N is the amount of substance carried per unit of time. The dimension of coefficient of a mass transfer is defined by the dimension of the driving force. If the dimension of driving force in mass concentration is [kq/m^3], then the dimension of coefficient of a mass transfer is equal to $K_C = [\mathrm{m/s}]$. If the dimension of driving force is expressed through partial pressure $\Delta P = P - P^* \left[{}^{\mathrm{N}}\!/\!_{\mathrm{m}^2}\right]$, then the dimension of coefficient of a mass transfer is $K_P = \left[\frac{\mathrm{kg}}{\mathrm{m^2 s N/m^2}}\right] = \left[\frac{\mathrm{s}}{\mathrm{m}}\right]$. Between coefficients K_P and K_C there is also a communication $K_P = \frac{M_K}{RT} K_C$ (where M_K is the molecular mass of transferable substance and T is temperature.) If the driving force is expressed as the difference of relative mass concentrations $\left[{}^{\mathrm{kg}}\!/\!_{\mathrm{kg}}\right]$, then the dimension of the mass-transfer coefficient can be determined as follows: $K \approx \frac{M_K}{M_H} P K_P$ (where M_H is the molecular mass of the carrier phase). If the motive power is expressed in molar fractions, then the mass-transfer coefficient is measured in $\left[\frac{\mathrm{kg}}{\mathrm{m^2 s}}\right]$.

3.3.1 Driving Force of Process

The driving force of transfer of any component from one phase to another is expressed through the difference of chemical potentials of this component in the interacting phases. However, in practice, the driving force of a mass exchange is usually expressed through a gradient of concentration that considerably simplifies communication between the velocity of process and structure of moving streams. As appears from formulas (3.33), the driving force of process is a major factor, the defining amount of transferable substance from one phase to another. Generally, the value of the driving force changes with the change of concentration in a stream. For practical calculations, it is convenient to use the average driving force determined in a look

$$\Delta y_c = \frac{\bar{y}_h - \bar{y}_b}{\int\limits_{\bar{y}_b}^{\bar{y}_h} \frac{d\bar{y}}{\bar{y} - \bar{y}^*}} \tag{3.34}$$

where $\bar{y}_h$ and $\bar{y}_b$ are both initial final contents (in molar shares) a component in a gas phase, and $\bar{y}^*$ is equilibrium value of concentration of a component. Equation (3.34) is suitable for the nonlinear character of an equilibrium curve. If the equilibrium curve of substance is linear, then equation (3.34) becomes simpler, to a look

$$\Delta y_c = \frac{\Delta y_h - \Delta y_b}{2{,}3\lg\left(\Delta\bar{y}_h / \Delta\bar{y}_b\right)} \tag{3.35}$$

where $\Delta\bar{y}_h = \bar{y}_h - \bar{y}_h^*$ and $\Delta\bar{y}_b = \bar{y}_b - \bar{y}_b^*$. If $0{,}5 \le \Delta\bar{y}_h / \Delta\bar{y}_b \le 2$, then it is possible to write down

$$\Delta y_c = \frac{\Delta\bar{y}_h + \Delta\bar{y}_b}{2} \tag{3.36}$$

Determination of the average driving force on equation (3.34) is connected with numerical integration of the expression, and it is possible to carry out by a method of trapezes

$$F_0 = \int\limits_{\bar{y}_b}^{\bar{y}_h} \frac{dy}{y - y^*} = \frac{\bar{y}_h - \bar{y}_b}{n}\left(\frac{\phi_0 + \phi_n}{2} + \sum_{i=1}^{i=n-1} \phi_i\right) \tag{3.36a}$$

where φ_i is the value of the subintegral function.

3.3.2 Balance in the System of Couples (Gas–Liquid)

Such processes of chemical technology as rectification, distillation, extraction, and absorption are carried out in conditions when the multicomponent system is in a condition of one or several solutions capable of more or less noticeable evaporation. When a liquid evaporates from the surface, a heterogeneous phase is formed and an equilibrium state is established between the liquid and its vapor, depending on temperature and pressure. The heterogeneous phase balances are called the balances, which are established in physical processes of transition of substance of one phase to other phases. Usually such balances are observed when boiling liquid with a constant pressure (liquid–steam), melting of firm particles (firm–liquid), at allocation of crystals from saturated solution (liquid–firm–steam), etc. Thermodynamic balance in heterogeneous systems is characterized by as much as long coexistence of several phases in the conditions of constancy of pressure and temperature. Based on the number of components, systems are distinguished as follows: single-component, two-component, and multicomponent; and based on the number of phases, one-, two-, and multiphase. Similarly, based on the number of degrees of freedom, without degrees of freedom ($n = 0$), with one degree of freedom, with two degrees of freedom, etc. Let's say some system consists of the K components and f phases, which are in stable equilibrium, the determined constant temperature, pressure, and structure. The structure of a phase is defined by the concentration $(K-1)$ of components as the sum of molar shares of all components is equal to unit. Then the number of independent concentrations for all f phases is equal $f(K-1)$. If considering temperature and pressure, then the total number of variables will be equal $f(K-1)+2$. Obviously for each component is available the $(f-1)$ equations, and for all components K of the $K(f-1)$ equations. The difference between the total number of the variables and number connecting their equations it is equal to the number of degrees of freedom

$$n = f(K-1)+2-K(f-1) = K+2-f \tag{3.37}$$

Equation (3.37) defines the number of degrees of freedom and makes a basis of the rule of Gibbs. If we use Gibbs's rule to some capacity filled with water, the number of phases $f=1$ and number of components $K=1$, the number of degrees of freedom is equal $n=2$. Thus, the condition of system is defined by two parameters, which can be temperature and pressure. Now we will assume that some heat at which water begins to evaporate is brought to capacity. In this case, the number of phases is equal $f=2$ (water–steam), and the number of components $K=1$. The number of degrees of freedom on equation (3.37) is equal $n=1$, that is, the condition of the system can be operated one variable, and we will allow temperature. Between steam and liquid there can be several steady conditions of balance depending on temperature. In the equilibrium state, vapor pressure is called pressure of saturation or pressure of saturated steam. The dependence of pressure of saturated steam on temperature can be calculated on the empirical equation

$$\ln P_D = A_1 + \frac{A_2}{A_3+T} + A_4T + A_5T^2 + A_6\ln T \tag{3.38}$$

where T is expressed ^0K and P_D is pressure of saturated steam, atm. If $A_4 = A_5 = A_6 = 0$, then this equation comes down to Anteon's equation (Table 3.2)

$$\ln P_D = A_1 + \frac{A_2}{T+A_3} \tag{3.39}$$

$$\lg P_D = A_1 - \frac{A_2}{T+A_3} \tag{3.40}$$

In the presence of experimental data about dependences of pressure of saturated steam on temperature, coefficients A_1, A_2, and A_3 can be also calculated by methods of nonlinear programming. Usually the coefficient A_3 is defined as a function of temperature of boiling of liquid; therefore, with dependence $A_3 = f(T)$, other coefficients A_1 and A_2 can be calculated by the method of the smallest squares.

TABLE 3.2
Value of the Coefficients, Entering of Anteon's Equation

Substance	A_1	A_2	A_3
Methyl chloride	7.09349	948.582	349.336
Chloroform	6.95465	1170.966	226.253
Acetone	6.11714	1210.595	229.664
Ethyl oxide	6.92032	1064.066	228.799
Methanol	8.08097	1582.271	239.736
Ethanol	8.11220	1592.864	226.184
Benzene	6.89272	1203.531	219.888
Toluene	6.95805	1346.773	219.693
Water	7.96681	1668.210	228.000

Partial vapor pressure of any component of an ideal solution is connected with its concentration in liquid in the equation

$$p^* = Px \tag{3.41}$$

where p^* is the partial pressure of a component in steam–gas mix over the liquid in the conditions of balance; P is pressure of saturated steam of a clean component, which is unambiguous function of temperature; and x is a molar share of a component in liquid. Equation (3.41) is the expression of Raoult's law: at a constant temperature, the equilibrium partial vapor pressure of any component is equal to the product of the saturated vapor pressure of this component at a given temperature and its mole fraction in the liquid phase. Equation (3.41) calls such solutions ideal, which submits to Raul's law in all areas of change of concentration from $x=0$ to $x=1$. For extremely diluted solutions where the concentration of the dissolved substance is small, instead of Raul's law we will apply Henry's law

$$p^* = H(T)x \tag{3.42}$$

where $H(T)$ are Henry's coefficients depending on temperature and the nature of gas and liquid.

As per this law, at a constant temperature, the partial vapor pressure of a dissolved liquid substance is proportional to its mole fraction. Having divided both members of the equation into the general pressure, we will receive

$$y^* = mx \tag{3.43}$$

where y is a molar share of a component in a gas phase, equilibrium with liquid; and $m = {}^{H(T)}\!/_{P}$ is the dimensionless coefficient, gas, constant for this system—liquid at a constant temperature and pressure.

We will consider the two-component (A, B) and two-phase (liquid–steam) systems. According to Raul's law, having put $y_A + y_B = 1$, it is possible to write

$$\begin{aligned} y_A^* &= p_A x_A \\ y_B^* &= 1 - y_A^* = p_B(1 - x_B) \end{aligned} \tag{3.44}$$

Having divided the first equation into the second, we will receive

$$\frac{y_A^*}{1 - y_A^*} = \frac{p_A x_A}{p_B(1 - x_B)} \tag{3.45}$$

After simple transformations, this equation can be presented in the form

$$y_A^* = \frac{\alpha\, x_A}{1 + (\alpha - 1)x_A} \tag{3.46}$$

where $\alpha = {}^{p_A}\!/_{p_B} < 1$ are coefficients of relative volatility, p_A is the pressure of saturated steam of more flying components at the same temperature, and p_B is pressure of saturated steam of less flying components at the same temperature.

It is important to note that in comparison with the working line of process, the curve of equation (3.46) is located above the working line. If the mass of the substance passes from liquid into a gas phase (distillation), and vice versa if the substance passes from a gas phase in liquid, then the curve of balance is located below the working line.

Equation (3.46) is the equation of an equilibrium curve for binary mix. For many two-component systems, the liquids given about equilibrium structures and couple for various solutions is given in a tabular or graphic style on the basis of numerous experimental data.

3.4 MASS-TRANSFER COEFFICIENTS

Mass-transfer coefficients in binary systems are key parameters of calculation of a mass exchange between various phases, and finally define the number of transferable weights in various hydrodynamic and temperature conditions for surfaces of different geometry. It should be noted that coefficients of a mass transfer are functions of coefficients of a mass transfer, which are defined from statements of the problem at a certain current of a stream in a thin interphase layer in which sharp falling of a gradient of concentration on a normal limit of the section of phases is observed. As it will be shown below, the solution of this task is possible only for a certain circle of tasks at certain simplifications.

Let's say that from a gas phase to a liquid phase the convective transfer of the weight (Figure 3.1) is carried out. The number of transferable weights for a gas and liquid phase will be defined in the form of equations (3.33) [1]

$$N = \beta_y F\left(y - y_P\right) = K_y F\left(y - y^*\right) \tag{3.47}$$

$$N = \beta_x F\left(x_P - x\right) = K_x F\left(x^* - x\right) \tag{3.47a}$$

where β_y, β_x are mass-transfer coefficients for a gas and liquid phase, respectively; y_P, x_P are the concentrations of transferable substances on limit of the section of phases in a gas and liquid phase; y^*, x^* are equilibrium concentrations of substances in a gas and liquid phase; K_y, K_x are total coefficients of mass transfer for a gas and liquid phase; and F is an interphase surface. Considering Henry's law $y^* = mx$, it is also possible to write $y_P = mx_P$. Then from these equations we have $x = y^*/m$ and $x_P = y_P/m$. Substituting these expressions in (3.47a), we will receive

$$N = \beta_x F\left(\frac{y_P}{m} - \frac{y^*}{m}\right) = \frac{\beta_x F}{m}\left(y_P - y^*\right) \tag{3.48}$$

From equations (3.47) and (3.48) it is possible to write

$$\begin{aligned} y - y_P &= \frac{N}{\beta_y F} \\ y_P - y^* &= \frac{Nm}{\beta_x F} \end{aligned} \tag{3.49}$$

Putting these two equations together, we will receive

$$\left(y - y^*\right) = \frac{N}{F}\left(\frac{1}{\beta_y} + \frac{m}{\beta_x}\right) \tag{3.50}$$

The number of transferable mass from one phase to another is defined by equation (3.33), and for a gas phase will register in a look

$$N = K_y F\left(y - y^*\right) \tag{3.51}$$

from where having $(y-y^*)$ defined and equating (3.50), we will receive

$$\left(y-y^*\right)=\frac{N}{K_yF}=\frac{N}{F}\left(\frac{1}{\beta_y}+\frac{m}{\beta_x}\right) \tag{3.52}$$

From this expression, we will receive a formula for calculation of total coefficient of a mass transfer for a gas phase in a look

$$\frac{1}{K_y}=\frac{1}{\beta_y}+\frac{m}{\beta_x} \tag{3.53}$$

Here, $1/\beta_y$ are the resistances to transfer of weight for a gas phase, and m/β_x are the resistances to transfer of weight for a liquid phase. At great values β_y or $1/\beta_y \to 0$, the mass-transfer coefficient for a gas phase is defined by mass-transfer coefficient for a liquid phase $K_y \approx \beta_x/m$, and at great values β_x or $m/\beta_x \to 0$ we have $K_y \approx \beta_y$, or the mass transfer entirely is defined by the mass-transfer coefficient for a gas phase.

We will similarly define the mass-transfer coefficient for a liquid phase. From equation (3.47), it is possible to write

$$N=\beta_yF\left(mx^*-mx_P\right)=m\beta_yF\left(x^*-x_P\right) \tag{3.54}$$

From equations (3.47a) and (3.54) we have

$$\begin{aligned} x_P-x&=\frac{N}{\beta_xF}\\ x^*-x_P&=\frac{N}{m\beta_yF} \end{aligned} \tag{3.55}$$

Putting these two equations together, we will receive

$$x^*-x=\frac{N}{F}\left(\frac{1}{m\beta_y}+\frac{1}{\beta_x}\right) \tag{3.56}$$

For a liquid phase from equation (3.33a), it is possible to write

$$N=K_xF\left(x^*-x\right) \tag{3.57}$$

from which, having $\left(x^*-x\right)$ defined and comparing to the previous expression, we will receive

$$x^*-x=\frac{N}{K_xF}=\frac{N}{F}\left(\frac{1}{m\beta_y}+\frac{1}{\beta_x}\right) \tag{3.58}$$

From this equality we have

$$\frac{1}{K_x}=\frac{1}{m\beta_y}+\frac{1}{\beta_x} \tag{3.59}$$

If β_x is a very big size $1/\beta_x \to 0$, we have $K_x \approx m\beta_y$, and if β_y is a very big size $1/m\beta_y \to 0$, we have $K_x \approx \beta_x$. Inverse values to mass-transfer coefficients make sense of resistance to transfer of substance in the corresponding phases and are called phase resistance to substance transfers. Equations (3.53) and (3.59) are called the equations of additivity of phase resistance. It should be noted that expressions for calculation of coefficient of a mass transfer, both for a liquid phase and for a gas phase, are received for small concentrations of the transferable substance submitting to Henry's law. The last condition is always observed for local coefficients of a mass transfer at a small change of the driving force. At a curvature of the line of balance, it is necessary to consider change with concentration. As appears from equations (3.47) and (3.47a), total coefficients of mass transfer for a liquid and gas phase are connected by a ratio

$$\frac{1}{K_x} = \frac{1}{\alpha_P K_y}, \quad \alpha_P = \left(y - y^*\right) \Big/ \left(x^* - x\right) \tag{3.60}$$

The restrictions connected with curvature of the equilibrium line becomes insignificant in cases when the process is controlled by resistance of a gas phase $1/\beta_y >> m/\beta_x$ or liquid phase $1/\beta_x >> m\beta_y$. The principle of additivity of phase resistances cannot be reliably used until all resistances are properly defined. It is important to note that convective streams on a surface in the form of regular structures appear, owing to the emergence of local gradients of a superficial tension (Marangoni's effect). This is because of natural convection of a stream, owing to a difference in density or pressure at the limit of the section and in volume of a stream (a Stefanov's stream) and many other phenomena that influence mass-transfer coefficients. Increase β_x (sometimes several times) can happen under the influence of the superficial convection caused by local gradients of a superficial tension, which arise in some cases as a result of a mass transfer, especially at simultaneous courses of reactions. Besides, near the section of phases there can be rather big gradients of temperature and pressure leading to essential changes of true phase boundary concentration, owing to the emergence of thermal diffusion and a bar diffusion. For high concentrations of transferable substances, in most cases it is impossible to receive theoretical expressions; therefore, in such cases, use the empirical expressions received on the basis of pilot studies and the theory of similarity.

Example 3.2

In an absorbing column, the general pressure is equal $P = 3.1 \times 10^5$Pa. In balance conditions between liquid and gas, Henry's law is observed $p^* = 108.8 \times 10^5 x$. Mass-transfer coefficients for a gas phase is equal $\beta_y = 1.07 \frac{\text{kmol}}{\text{m}^2\text{s}}$, and for a liquid phase $\beta_x = 22 \frac{\text{kmol}}{\text{m}^2\text{s}}$ is to define the total coefficient of heat transfer for a liquid and gas phase.

First of all, having divided Henry's equation into the general pressure, we will receive

$$y^* = \frac{p^*}{P} = \frac{108.8 \times 10^5}{3.1 \times 10^5} x = 35.096x, \ m = 35.096$$

Then the mass-transfer coefficient for a gas phase will be defined as

$$K_y = \frac{1}{\frac{1}{\beta_y} + \frac{m}{\beta_x}} = \frac{1}{\frac{1}{1.07} + \frac{35.096}{22}} = 0.396 \frac{\text{kmol}}{\text{m}^2\text{s}}$$

and for a liquid phase

$$K_x = \frac{1}{\frac{1}{m\beta_y} + \frac{1}{\beta_x}} = \frac{1}{\frac{1}{35.096 \times 1.07} + \frac{1}{22}} = 13.878 \frac{\text{kmole}}{\text{m}^2\text{s}}$$

3.4.1 Mass-Transfer Coefficient with Chemical Reaction

The majority of chemical processes are characterized by the existence of a chemical reaction on a solid surface. Let's say that on a surface of a solid body reaction of the first order $A \xrightarrow{K_1} B$ and to surfaces of a spherical particle concentration of substance proceeds, it is A equal C_S, and in the medium is C_0. Convective transfer of a substance on a surface of a firm particle is equal

$$N = \beta(C_0 - C_S)\pi.a^2 \tag{3.61}$$

Here, β is mass-transfer coefficient, and a is particle size. As a result of the chemical reaction, the amount of substance is equal

$$N = K_1(C_S - C^*)\pi a^2 \tag{3.62}$$

where C^* is the equilibrium concentration of substance. From these two equations, it is possible to write

$$C_0 - C_S = \frac{N}{\beta\pi a^2}$$
$$C_S - C^* = \frac{N}{K_1\pi a^2} \tag{3.63}$$

Putting which, we will receive

$$C_0 - C^* = \frac{N}{\pi a^2}\left(\frac{1}{\beta} + \frac{1}{K_1}\right) \tag{3.64}$$

On the other hand, the amount of transferable substance is equal

$$N = K(C_0 - C^*)\pi a^2 \tag{3.65}$$

where K is total coefficient of mass transfer. From these two equations, it is possible to write

$$\frac{1}{K} = \frac{1}{\beta} + \frac{1}{K_1} \tag{3.66}$$

In the case under consideration, there are inverse values of a constant of rate of reaction and coefficient of a mass transfer representing kinetic and diffusive resistance. If $\beta \to \infty$ or $1/K_1 >> 1/\beta$, then $K \approx K_1$ or the rate of process is defined by the velocity of chemical reaction. If $K_1 \to \infty$ or $1/\beta >> 1/K_1$, then $K \approx \beta$ or the rate of process is defined by the mass-transfer rate. For reactions whose rate does not satisfy the first order, one can formally introduce chemical resistance, defining it as the ratio of the concentration at the surface to the reaction rate [36].

4 The Basic Laws of Mass Transfer and Diffusion

In this chapter, we will consider the basic laws of convective and diffusion mass transfer (Fick's first and second laws), the determination of the coefficients of molecular, Knudsen and turbulent diffusion, the calculations of the one-dimensional diffusion, and Stefan flows and the character of the equimolar and non-equimolar mass transfers and the consequences arising from them. The types of diffusion transfers in liquids and in multicomponent media of laminar and turbulent flow is analyzed, and various boundary conditions for solving the problems of mass transfer in disperse media will be considered.

4.1 FICK'S FIRST LAW OF THE MASS TRANSFER

The diffusive stream of a substance arises in the presence of liquid of gradients of concentration, and in certain cases from a gradient of temperature and a gradient of pressure. Thus, the diffusive stream of a substance through a unit of an interphase surface area for a unit of time is proportional to a gradient of concentration and is equal

$$J_i^* = -D_i \frac{\partial C}{\partial y} \tag{4.1}$$

where D_i is the proportionality coefficient called by the coefficient of diffusion i of that component in mixes, and y is the direction of mass transfer.

Expression (4.1) is the equation of Fick's first law. In this equation, the minus sign means that the amount of substance is sent from a surface to stream volume, that is, $\partial C/\partial y < 0$. If the amount of substance is directed from volume to a surface, then in this case before the derivative there will be a plus and $\partial C/\partial y > 0$. If units of measure of concentration of a component [kmole/m^3] or [kq/m^3], then, respectively, units of measure of a stream of substance are [kmole/m^{2}s] or [kq/m^{2}s], and a unit of measure of coefficient of molecular diffusion will be [m^2/s]. In mass concentration, equation (4.1) will register as

$$j_i = -D_i \frac{\partial \rho_i}{\partial y} \tag{4.2}$$

and in molar and mass fractions

$$J_i^* = -CD_i \frac{\partial x_i}{\partial y}$$
$$j_i = -\rho D_i \frac{\partial m_i}{\partial y} \tag{4.3}$$

For a two-component mix (A, B) the diffusive stream of substance is equal A

$$J_A^* = -CD_{AB}\frac{\partial x_A}{\partial y}$$
$$j_A = -\rho D_{AB}\frac{\partial m_A}{\partial y} \tag{4.4}$$

and for a stationary system the molar stream is equal

$$N_A = (N_A + N_B)x_A - CD_{AB}\frac{\partial x_A}{\partial y} \tag{4.5}$$

For two components we can write $J_A^* + J_B^* = 0$ or $J_A^* = -J_B^*$, According to Fick's first law

$$-CD_{AB}\frac{dx_A}{dy} = -\left(-CD_{BA}\frac{dx_B}{dy}\right) \tag{4.6}$$

If $\sum x_i = x_A + x_B = 1$ or $x_A = 1 - x_B$, $dx_A = -dx_B$ then from this equation it is possible to write $D_{AB} = D_{BA}$. For a multicomponent mix equality to zero sum, all of the available streams on components is fair $\sum_{i=1}^{N} J_i = 0$.

Generally, the diffusive stream is defined as the additive sum of separate components of molecular, thermal diffusion, and bar diffusion transfers

$$J = -\left(\left(\nabla DC + k_T \nabla \ln T + k_P \nabla \ln P\right)\right) \tag{4.7}$$

If coefficients of diffusion don't depend on coordinates, then from (4.7) we will receive

$$J = -D\left(\nabla C + \frac{k_T}{T}\nabla T + \frac{k_P}{P}\nabla P\right) \tag{4.8}$$

where $\nabla = \frac{\partial}{\partial x} + \frac{\partial}{\partial y} + \frac{\partial}{\partial z} = \text{grad}$. Equation (4.8) gives the complete idea of Fick's first law for the diffusive transfer of a substance. The diffusive stream caused by a temperature gradient is defined by the thermal diffusion coefficient $k_T D$ where the coefficient k_T is called "the thermal diffusion relation." The diffusive stream caused by the pressure gradient is defined by the pressure diffusion coefficient $k_P D$. It should be noted that coefficients also are entirely defined only by thermodynamic properties of liquid. It should be noted that at small concentration of mix, coefficients of thermal diffusion and pressure diffusion are very small; therefore in equation (4.8), the two last members can be neglected. Brownian particles are more prone to displacement due to thermal diffusion and pressure diffusion, although these effects are also observed in highly dispersed systems, which determine the movement of particles.

Except other types of transfer in disperse systems, movement of particles with sizes more than the size of molecules of the environment in a stream are possible thanks to some other effects connected by heterogeneity of temperature and pressure. If the characteristic extent of nonuniformity is less or is comparable with the sizes of particles, then there is a force proportional to a gradient of temperature acting in the direction of reduction of temperature. This phenomenon has received the name *thermophoresis*. It is caused by the fact that the boost received by a particle from sites of the environment with more high temperature isn't compensated by a boost, which is received by a particle from colder sites of the environment. For determination of thermophoresis force, a number of theoretical formulas among which it is important to note is offered:

1. Formula Cawood [37]

$$F_T = -\frac{1}{2}\left(\pi a_r^2\right)\left(P\lambda_0/T\right)dT/dy,\ \mathrm{Kn} >> 1 \tag{4.9}$$

2. Formula Waldmann [38]

$$F_T = -4a_r^2\left(P\lambda_0/T\right)dT/dy,\ \mathrm{Kn} >> 1 \tag{4.10}$$

thus, the first formula gives underestimation of thermophoresis force in $8/\pi$;

3. Deryagin and Bakanov's formula [39]

$$F_T = -\frac{3\pi\eta^2 a_r}{\rho T}\left[\frac{8\lambda_c + \lambda_d + 2C_t Kn\lambda_d}{2\lambda_c + \lambda_d + 2C_t Kn\lambda_d}\right]\frac{dT}{dy},\ \mathrm{Kn} << 1 \tag{4.11}$$

where P is the pressure, ρ is the gas density, λ_0 is the length of a free run of gas molecules, C_t is the-coefficient of temperature jump, λ_c, λ_d are the coefficients of heat conductivity of gas and particles, a_r is the radius of particles, and $\mathrm{Kn} = \lambda_0/a_r$ is the Knudsen's number. It follows from the formulas that the size of the thermophoresis force in all cases is proportional to a gradient of temperature of gas and is inversely proportional to its absolute value. The size of particles, coefficients of heat conductivity of gas and particles, and the value of number of Knudsen also exert an impact on the value of the thermophoresis force. The velocity of a thermophoresis of small particles can be determined by a formula [12]

$$V_T = -\frac{\lambda_c}{5P\left(1+\pi\alpha/8\right)}\frac{dT}{dy} \tag{4.12}$$

If $\mathrm{Kn} >> 1$, α a share of the gas molecules disseminated by a particle. For a case $\mathrm{Kn} << 1$, that is, for larger particles, the velocity of a thermophoresis will be defined by a formula

$$V_T = -\frac{3\lambda_c\eta}{2\left(2\lambda_c + \lambda_d\right)\rho_C}\frac{dT}{dy} \tag{4.13}$$

The order of magnitude of the thermophoresis rate of high-disperse particles is small, and with an increase in the sizes of particles it sharply decreases and is considerable at high gradients of temperatures.

Similarly, with emergence in the environment of a gradient of total or partial pressure, any components affect particles in the same way and for the reason, as the temperature gradient. This phenomenon is called diffusophoresis. For Brownian particles, the gradients of temperature (pressure) in the environment lead to thermal diffusion and a bar diffusion, that is, to migration of particles to the area with a smaller temperature (pressure).

A stream influences the migration of drops, also physical phenomena proceeding on a surface. For example, two evaporating drops, with a certain stream of mass of steam from a surface, mutually make a start, growing—are mutually attracted (effect of Fasi) [40].

4.2 COEFFICIENTS OF MOLECULAR AND TURBULENT DIFFUSION

It has been previously shown that the diffusion coefficient entering the equation of Fick's first law is the additive sum molecular, turbulent, and many other types of diffusion (thermal diffusions, bar diffusion). Except for molecular diffusion, coefficients of other types of diffusion in most cases decide by an empirical way on the use of pilot studies. The coefficient of molecular (Brownian) diffusion is defined in a look [12]

$$D=\left(k_BT/6\pi\eta a\right)\left(1+\left(10^{-4}/Pa\right)\left(6.32+2.01\exp\left(-2190\,Pa\right)\right)\right) \tag{4.14}$$

where $k_B = 1.38 \times 10^{-16}$ (constant of Boltzmann), T is the absolute temperature, P is the pressure, and a is the size of particles. Now values of molecular diffusion for binary mixes (in water and air) are given in various reference books and monographs and in various situations, perhaps to determine the following formulas. In Figure 4.1, the scheme of main types of diffusion of gases is given: molecular, Knudsen, total Knudsen, and molecular and diffusions in a granular layer.

4.2.1 Coefficients of Molecular Diffusion in Binary Mix

Now there is a set of empirical formulas on which it is possible to calculate diffusion coefficients in various binary systems. For gases with a low density, the coefficient of mutual molecular diffusion can be determined by a formula Chapman–Ensco [1]

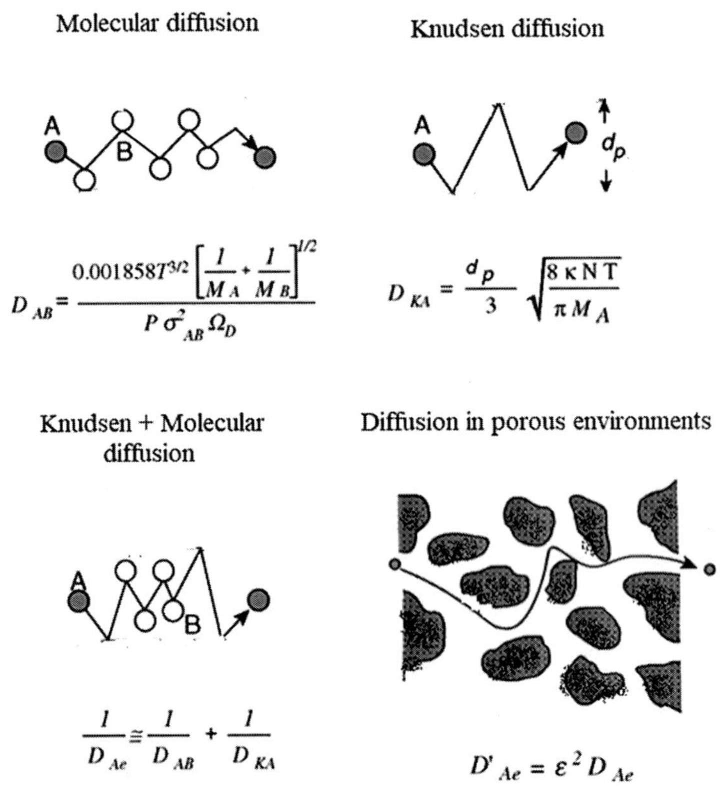

FIGURE 4.1 Characteristic scheme of diffusion of gases.

$$D_{AB} = \frac{1.858\times10^{-27}\sqrt{T^3\left(\frac{1}{M_A}+\frac{1}{M_B}\right)}}{P\sigma_{AB}^2\Omega_D} \quad (4.15)$$

where T is the temperature, K; P is the pressure, atm; σ_{AB} is the diameter of collisions; Ω_D is the integral of impacts (σ, ε are the constants in potential function of Lennard-Jones of Figure 4.2), and M_A, M_B are the molecular mass of components A and B. For determination of coefficient molecular diffusion gas, there is also another formula

$$D_{AB} = \frac{4.3\times10^{-7}T^{3/2}}{P\left(v_A^{1/3}+v_B^{1/3}\right)}\sqrt{\frac{1}{M_A}+\frac{1}{M_B}} \quad (4.16)$$

where v_A, v_B are the molar volumes of gases A and B, defined as the sum of atomic volumes of the elements that are a part $v_{C_nH_m} = nv_{AC} + mv_{AH}$.

Example 4.1

To determine the molar volume of gas CO_2. Using tabular data, we will determine the molar volume of gas in a look

$$v_{CO_2} = 1\times v_{AC} + 2\times v_{AO} = 1\times14.8 + 2\times7.4 = 29.6\,\mathrm{sm^3/mol}$$

For two temperatures T_1 and T_2 also two P_1 and P_2 pressure, it is also possible to write

$$D_1 = \frac{4.3\times10^{-7}T_1^{3/2}}{P_1\left(v_A^{1/3}+v_B^{1/3}\right)}\sqrt{\frac{1}{M_A}+\frac{1}{M_B}}$$
$$D_2 = \frac{4.3\times10^{-7}T_2^{3/2}}{P_2\left(v_A^{1/3}+v_B^{1/3}\right)}\sqrt{\frac{1}{M_A}+\frac{1}{M_B}} \quad (4.17)$$

Having divided the second equation into the first, we will receive

$$D_2 = D_1\frac{P_1}{P_2}\left(\frac{T_2}{T_1}\right)^{3/2} \quad (4.18)$$

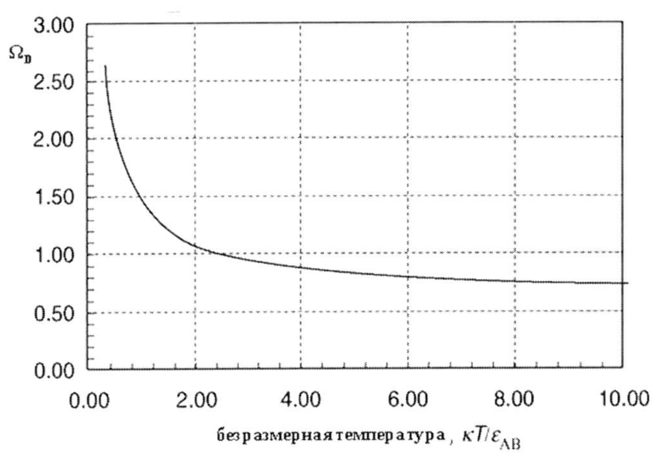

FIGURE 4.2 Integral of impacts of Lennard-Jones.

with a constant pressure we have

$$D_2 = D_1\left(\frac{T_2}{T_1}\right)^{3/2} \tag{4.19}$$

Example 4.2

To determine diffusion coefficient SO_2 in air with a constant pressure $P = 1$ atm and temperature $T = 20°C$ by formulas (4.15) and (4.16).

1. For determination of coefficient of diffusion on a formula (4.15) from the corresponding tables we will find sizes

$$\sigma_A = 4.11\times 10^{-10}\,m,\ \sigma_B = 3.711\times 10^{-11}\,m,\ \sigma_{AB} = \frac{1}{2}(\sigma_A+\sigma_B) = 3.9115\times 10^{-10},$$
$$\varepsilon_D = 1.2,\ M_A = 64,\ M_B = 29.$$

The coefficient of diffusion is equal

$$D_{AB} = \frac{1.858\times 10^{-27}}{1\times\left(3.9115.10^{-10}\right)\times 1.2}\sqrt{293^3\left(\frac{1}{64}+\frac{1}{29}\right)} = 1.13\times 10^{-5}\mathrm{m}^2/\mathrm{s}$$

2. For determination of coefficient of diffusion on a formula (4.16), we will calculate molar volumes of gases

$$v_{SO_2} = 1\times v_{AS} + 2\times v_{AO} = 1\times 25.6 + 2\times 7.4 = 40.4$$

$$v_{air} = 29.9$$

The coefficient of diffusion is equal (Table 4.1)

$$D_{AB} = \frac{4.3\times 10^{-7}(273+20)^{3/2}\sqrt{\frac{1}{64}+\frac{1}{29}}}{1\times\left(40.4^{1/3}+29.9^{1/3}\right)^2} = 1.01\times 10^{-5}\mathrm{m}^2/\mathrm{s}$$

In Table 4.2, some expressions for calculation of coefficients of diffusion of gases at low pressures, borrowed from work, are given below [19].

Here, δ is the effective thickness of a layer, $\rho_r = \rho_A/\rho_{Ac}$ is the relative density, ρ_A is the gas density A, ρ_{Ac} is the critical density of gas A, V_C is the molar volume, $\beta = M_A^{1/2}P^{1/3}T^{5/8}$, and P is the pressure.

TABLE 4.1
Atomic Volumes of Some Elements

Atomic Volumes		Atomic Volumes	
Hydrogen, H	3.7	Carbon, C	14.8
Oxygen, O:	9.1	Sulfur, S	25.6
in esters;	9.9	Chlorine,	24.6
in simple air;		Nitrogen:	15.6
with two saturated communications;	7.4	with two saturated communications;	10.5
in connections with S, P, N:	8.3	air	29.9

TABLE 4.2
Formulas for Determination of Coefficients of Diffusion for Gases

Authors	Expression
Binary mixes with low pressure	
1. Chapman–Enskog	$D_{AB} = \dfrac{0.001858T^{3/2}M_{AB}^{1/2}}{P\sigma_{AB}^2\Omega_D}$
2. Wilke–Lee	$D_{AB} = \dfrac{\left(0.00217 - 0.0005M_{AB}^{1/2}\right)T^{3/2}M_{AB}^{1/2}}{P\sigma_{AB}^2\Omega_D}$
3. Fuller–Schetler–Giddings	$D_{AB} = \dfrac{0.001T^{1.75}M_{AB}^{1/2}}{P\left(\upsilon_A^{1/3} + \upsilon_B^{1/3}\right)}$
Self-diffusion	
4. Mathur–Todos	$D_{AA} = \dfrac{10.7\times10^{-3}T}{\beta\rho_r}, \quad \rho_r \le 1.5,$
5. Lee–Todos	$D_{AA} = \dfrac{0.77\times10^{-5}T}{\delta\rho_r}, \quad \rho_r \le 1,$
Supercritical mixes	
6. Sun and Chen	$D_{AB} = \dfrac{1.28\times10^{-10}T}{\eta^{0.799}V_C^{0.49}}$

Source: *Perry's Chemical Engineers Handbook*, Section 5: Heat and Mass Transfer, 8th Edition, McGraw-Hill Companies, 2008.

The coefficient of diffusion of gases in liquids can be determined by a formula Wilke–Chang [1,16], though there are also other empirical correlations

$$D_{AB} = \frac{1.17\times10^{-3}\sqrt{\psi_B M_B}}{\nu V_A^{0.6}}T \tag{4.20}$$

where M_B is the molecular mass of liquid of solvent, T is the temperature, V_A is the molar volume of gas, ν is the viscosity of liquid, and ψ_B is the parameter characterizing liquid, in particular for water $\psi_B = 2{,}6$; for methanol is 1.9; for ethanol is 1.5; for benzene 1.0. Other expressions for the calculation of coefficient of diffusion in liquids at a temperature, equal $T = 20°C$, is the formula [1,16]

$$D_{20} = \frac{10^{-6}}{AB\sqrt{\nu}\left(\nu_A^{1/3} + \nu_B^{1/3}\right)^2}\sqrt{\frac{1}{M_A} + \frac{1}{M_B}} \tag{4.21}$$

where D_{20} is the diffusion coefficient, m²/s; ν is the coefficient of dynamic viscosity of liquid, MPa × s; ν_A, ν_B are the molar volumes of the dissolved substance and solvent; and M_A and M_B are the molar mass of the dissolved substance and solvent; the A and B coefficients depend on properties of the dissolved substance and solvent. The value of coefficient A for some substances dissolved in water: for gases—1.0; for ethyl alcohol—1.24; methyl alcohol—1.19, etc. For coefficient B: for water—4.7; for ethyl alcohol—2.0; for methyl alcohol—2.0; for acetone—1.15, etc. The coefficient of diffusion of gases in liquid at a temperature T is the connected with coefficient D_{20} in the following ratio

$$D_T = D_{20}\left[1 + b(T - 20)\right] \tag{4.22}$$

in which the temperature coefficient b can be determined by an empirical formula

$$b = 0.2\sqrt{\nu}\Big/\sqrt[3]{\rho} \tag{4.23}$$

TABLE 4.3
Diffusion Coefficients in Liquids

Authors	Expression	Mistake,%
1. Wilke–Chang	$D_{AB} = \frac{7.4\times10^{-8}(\varphi_B M_B)^{1/2} T}{\eta_B V_A^{0.6}}$	20
2. Tyn–Calus	$D_{AB} = \frac{8.93\times10^{-8}\left(V_A/V_B^2\right)^{1/6}\left(\psi_A/\psi_B\right)^{0.6} T}{\eta_B}$	10
3. Siddiqi–Lucas	$D_{AB} = \frac{9.89\times10^{-8} V_B^{0.265} T}{V_A^{0.45}\eta_B^{0.907}}$	13
Gases in low-viscous liquids		
4. Sridhar–Potter	$D_{AB} = D_{BB}\left(\frac{V_{Bc}}{V_{Ac}}\right)^{2/3}\left(\frac{V_B}{V_l}\right)$	18
Water solutions		
5. Siddiqi– Lucas	$D_{AW} = 2.98\times10^{-7} V_A^{-0.5473}\eta_W^{-1.026} T$	13

Source: *Perry's Chemical Engineers Handbook*, Section 5: Heat and Mass Transfer, 8th Edition, McGraw-Hill Companies, 2008.

where ν is the dynamic coefficient of viscosity of liquid at $T = 20°C$, MPa × s; and ρ is the liquid density kq/m³. Except this formula, in Table 4.3, some expressions for determination of coefficients of diffusion in liquids are given below.

There, ψ is the an empirical parameter used in processes of absorption and rectification, $25 < \psi < 190$; and V_{ml} is the molar volume of a liquid phase.

Example 4.3

At a temperature $T = 50°C$ to calculate ammonia gas diffusion coefficient in water $\left(\rho_B = 1000 \, {}^{kg}/_{m^3},\ \nu = 1.0\ cp\right)$, we will define the diffusion coefficient at a temperature $T = 20°C$, having put $A = 1,\ B = 4.7$ also molar volumes

$$v_{NH_3} = 1\times15.6 + 3\times3.7 = 26.7,\ v_B = 2\times3.7 + 1\times7.4 = 14.8$$

$$D_{20} = \frac{10^{-6}}{1\times4.7\sqrt{1}\left(26.7^{1/3} + 14.8^{1/3}\right)^2}\sqrt{\frac{1}{17} + \frac{1}{18}} = 0.00243\times10^{-6}\text{m}^2/\text{s}$$

At $T = 50°C$ the coefficient of diffusion is equal

$$b = {}^{0.2\sqrt{1}}\!\Big/_{\sqrt[3]{1000}} = 0.02$$

$$D_{50} = 0.00243\times10^{-6}\left[1 + 0.02\times(50-20)\right] = 0.0039\times10^{-6}\,{}^{\text{m}^2}\!/_{\text{s}}$$

In the liquid environment, the coefficient of diffusion is connected with viscosity, and for many liquids with an increase in viscosity the coefficient of diffusion decreases $D \sim {}^{1}/_{\text{v}}$.

4.2.2 Coefficient of Knudsen Diffusion

Knudsen diffusion is in narrow channels and a time that it is most characteristic of processes of adsorption, a desorption, drying, catalytic processes, etc. If gas density or diameter pore are small, then the impact of molecules happens with walls of a time considerably more often than the impact of molecules with each other. The type of diffusion, typical for these conditions, is known as the Knudsen diffusion. The molecules that have reached a time wall are instantly adsorbed, and after a desorption move in any direction. Knudsen diffusion isn't observed in liquids. Knudsen's number is usually defined as the relation of length of a free run of molecules to diameter of a time, that is, the coefficient of Knudsen diffusion for gases in a time is defined in a look

$$D_{KA}=\frac{d_p}{3}\sqrt{\frac{8k_BNT}{\pi M_A}} \tag{4.24}$$

Spherical particles of the correct form length of a free run are defined as

$$\lambda=\frac{\eta}{P}\sqrt{\frac{\pi k_BT}{2m_C}} \tag{4.25}$$

where m_C is the mass of molecules of the environment, and P is the pressure. The coefficient of Knudsen diffusion of a component A is defined as

$$D_{KA}=97r\left(\frac{T}{M_A}\right)^{1/2} \tag{4.26}$$

where T is the temperature, K; r is the average radius of a time; and M_A is the molecular mass of component A. For a cylindrical time length L, it is possible to write

$$\frac{V_P}{S}=\frac{\pi r^2L}{2\pi rL}=\frac{r}{2} \tag{4.27}$$

where V_P is the specific volume of a time, m^3/kq; and s is the specific surface of a time, m^2/kq. The average radius of a time is equal $r=\frac{V_P}{S/2}$. If we determine the specific volume of a pore through porosity $V_P=\varepsilon/\rho_P$, where ρ_P is the particle density, then we can write the expression for the average radius of a time in a look

$$r=\frac{2\varepsilon}{\rho_PS} \tag{4.28}$$

Having entered a factor of tortuosity ζ of a time-effective coefficient of the Knudsen diffusion, it is possible to define in a look

$$D_{Keff}=D_K\frac{\varepsilon}{\zeta} \tag{4.29}$$

and effective coefficient of molecular diffusion

$$D_{Meff}=D_M\frac{\varepsilon}{\zeta} \tag{4.30}$$

The general coefficient of molecular and Knudsen diffusion will be defined as

$$\frac{1}{D_{eff}} = \frac{1}{D_{Meff}} + \frac{1}{D_{Keff}} \tag{4.31}$$

Example 4.4

At $T = 20°C$ sulphur dioxide gas from air is adsorbed on the surface of absorbent carbon. The specific surface of absorbent carbon is equal $S = 700 \times 10^3 m^2/Kg$, density $\rho_P = 1200\ kg/m^3$, tortuosity of a time is accepted, equal $\zeta = 2$ and coefficient of molecular diffusion $D_M = 1.22 \times 10^{-5} m^2/s$. To define the effective coefficient of diffusion of sulphur dioxide gas, the molecular mass of gas is equal $M_A = 64$.

We will determine the time radius by a formula (4.28)

$$r = \frac{2 \times 0.5}{700 \times 10^5 \times 1200} = 1.19 \times 10^{-9} m$$

The coefficient of the Knudsen diffusion is equal (4.26)

$$D_K = 97 \times 1.19 \times 10^{-9} \sqrt{\frac{293}{64}} = 2.47 \times 10^{-7} m^2/s$$

We will determine the effective coefficient of molecular diffusion by a formula (4.30)

$$D_{Keff} = D_K \frac{\varepsilon}{\zeta} = 2.47 \times 10^{-7} \frac{0.5}{2} = 6.177 \times 10^{-8} m^2/s$$

The effective coefficient of molecular diffusion is equal

$$D_{Meff} = 1.22.10^{-5} \frac{0.5}{2} = 3.10^{-6} m^2/s$$

The general coefficient of diffusion of gas will be defined in a look

$$D_{eff} = \frac{1}{\frac{1}{3 \times 10^{-6}} + \frac{1}{6.177 \times 10^{-8}}} = 6.05 \times 10^{-8} m^2/s$$

For disperse (colloidal and aerosol) particles, Knudsen's number is equal to the relation of length of a free run of molecules of the environment to the size of a spherical particle $Kn = \lambda/a$. For determination of the coefficient of diffusion, the formula is used [40]

$$D_{eff} = D\xi^{-1}(Kn) \tag{4.32}$$

and at where $\xi(Kn)$ is the some function depending on Knudsen's number $Kn \to 0$, $\xi(Kn) \to 1$. In particular, for $Kn << 1$ dependence $\xi(Kn)$ it is linear, and it is possible to use the expression $\xi(Kn) \approx 1 - 1.137Kn$.

To obtain true information on the behavior of function $\xi(Kn)$ for any numbers Kn, it is necessary to solve a kinetic problem about transformation of function of distribution of molecules of gas in the neighborhood of a firm spherical particle (Table 4.4). The numerical solution of this task is received by *Cercignani*, and approximation of this decision $0.08 \le Kn \le 20$ is by Millikan.

TABLE 4.4
Dependence of Coefficient of Diffusion on the Sizes of Particles

R, mkm	3.34×10^{-3}	5.91×10^{-2}	0.117	0.40	0.80	2.0
D, sm²/s	1.02×10^{-3}	6.28×10^{-6}	1.66×10^{-6}	3.43×10^{-7}	1.6×10^{-7}	0.6×10^{-7}

$$\xi^{-1}(\mathrm{Kn}) \approx 1 + 1.234\mathrm{Kn} + 0.414\mathrm{Kn}\exp\left(-\frac{0.876}{\mathrm{Kn}}\right) \tag{4.32a}$$

This expression gives satisfactory consent with experimental data within 2%. We will give some values of the coefficient of diffusion of the water drops in air at $T = 273°K$, $P = 1atm$ calculated by formulas (4.32) and (4.32a).

As appears from this table, with an increase in the sizes of particles, the coefficient of effective diffusion significantly falls.

4.2.3 Diffusion Coefficient in Multicomponent Mix

According to the molecular-kinetic theory of gases, the diffusive stream for n component system is expressed by the following expressions

$$\begin{aligned} j_1 &= D_{11}\nabla y_1 + D_{12}\nabla y_2 + D_{13}\nabla y_3 + \ldots + D_{1n}\nabla y_n \\ j_2 &= D_{21}\nabla y_1 + D_{22}\nabla y_2 + D_{23}\nabla y_3 + \ldots + D_{2n}\nabla y_n \\ &\ldots\ldots\ldots\ldots\ldots\ldots\ldots\ldots\ldots\ldots \\ j_1 &= D_{11}\nabla y_1 + D_{12}\nabla y_2 + D_{13}\nabla y_3 + \ldots + D_{1n}\nabla y_n \end{aligned} \tag{4.33}$$

where ∇c_i is a gradient of concentration of i component, D_{ij} is the coefficient of multicomponent diffusion, and y_i is the molar shares of i component in mix. *Wilkes* [16,17] enters a concept of effective coefficient of diffusion, which in a multicomponent mix for a separate component in other gases is expressed by the following formula

$$D_A = \frac{1 - y_A}{\sum_{j=2}^{N} \frac{y_j}{D_{Aj}}} \tag{4.34}$$

where y_i are the molar shares of i components. Let us say what mix is available (A, B, C, D) components and are known coefficients of diffusion of the A component in other components D_{AB}, D_{AC}, D_{AD}. Then the coefficient of diffusion of an A component in mix will be defined as

$$D_A = \frac{1 - y_A}{y_B/D_{AB} + y_C/D_{AC} + y_D/D_{AD}} \tag{4.35}$$

Such method allows defining coefficients of diffusion of separate components in a multicomponent mix. In practice of calculation of processes of a mass exchange, it is usually supposed that coefficients of diffusion do not depend on the concentration of diffusing substance and other parameters of process and is a constant for the set temperature and pressure. However, pilot studies show [41] that in many processes (in particular at dissolution, drying) the diffusion coefficient linearly decreases at increase in concentration of a diffusing substance. Various dependences of coefficients of diffusion on concentration are known

$$D/D_0 = 1 - \alpha\sqrt{c}\,,\ D/D_0 = (1-c)^{1/2} \tag{4.36}$$

Similar cases are observed when drying materials, when in the course of drying, materials are exposed to different types of deformation (shift, lengthening, a bend, etc.). In this case, the coefficient of diffusion depends on a deformation factor, on time of deformation, and on the current shape of a body. For drying of the particles having the extended form, the coefficient of deformation depends on the deformation of a bend and lengthening and can be presented in the form

$$D/D_0 = (1+\alpha\xi)^{-1/2} \tag{4.37}$$

where α is the coefficient defined experimentally, and ξ is the parameter characterizing deformation of a body.

Obviously, the given formulas aren't the final expression for calculation of coefficient of diffusion as other dependences in practice can meet. Except for molecular diffusion in processes of chemical technology, the coefficient of turbulent diffusion and viscosity is very important. Generally when calculating various processes, the general effective coefficient of diffusion is considered as the additive sum of molecular and turbulent diffusion

$$D = D_M + D_T \tag{4.38}$$

It is important to note that for the developed turbulence, the coefficient of turbulent diffusion is much more a coefficient of molecular diffusion $D_T >> D_M$. We will note that an order of coefficient of molecular diffusion for gases and for liquids significantly differ: for gases—$5\times10^{-6} - 10^{-5}\,\mathrm{m^2/sec}$, for liquids—$10^{-9} - 10^{-6}\,\mathrm{m^2/sec}$. This distinction significantly affects criteria of similarity and especially on Schmidt's number.

4.3 TURBULENT DIFFUSION AND TURBULENT VISCOSITY

Coefficients of turbulent diffusion in an isotropic stream, according to Levich [6], in the field of the pulsations exceeding the internal Kolmogorov scale of turbulence $\lambda > \lambda_0$ will be defined in a look

$$D_T \approx V_\lambda \lambda = \alpha(\varepsilon_R \lambda)^{1/3}\lambda,\ \ \lambda > \lambda_0 \tag{4.39}$$

with an increase in scale of turbulence, D_T increases in proportion $\lambda^{4/3}$, and this value of coefficient of turbulent diffusion much more exceeds the value of coefficient of molecular diffusion. For the area of pulsations of small scales, $\lambda < \lambda_0$ the coefficient of turbulent diffusion is defined in a look

$$D_T \approx \beta\left(\frac{\varepsilon_R}{\nu}\right)^{1/2}\lambda^2 \tag{4.40}$$

where α and β are the proportionality coefficients. In this area, the value of the coefficient of turbulent diffusion decreases with reduction of scale of turbulent pulsations under the square law and with increase in the viscosity proportional $\nu^{-1/2}$.

According to the statistical theory of turbulent diffusion, the size of the average square shift of particles of the environment from the initial situation in cross section to a stream is defined by dependence

$$\bar{y}^2 = 2\bar{V}'^2 \int_{t_0}^{t} dt \int_{t_0}^{t} R_L(\vartheta) d\vartheta \tag{4.41}$$

where $R_L(\vartheta)$ is the Lagrangian correlation function. For big intervals of time $t \to \infty$, when $\int_{t_0}^{t} R_L(\vartheta) d\vartheta \to \vartheta_L = const$ from (4.41) it is possible to write down

$$\bar{y}^2 = 2\bar{V}'^2 \vartheta_L (t - t_0) \tag{4.42}$$

Here, $D_T = \bar{V}'^2 \vartheta_L$ is the size coefficient of turbulent diffusion. The coefficient of turbulent diffusion is a function of stream turbulent properties, cross coordinate, and also depends on coefficient of molecular diffusion, physical properties of the environment and, in some problems of a mass transfer, on the concentration of any component. Mechanisms of turbulent transfer numbers of the movement and weight are in most cases identical, though at the solution of some practical problems of flow of a surface of not streamline shape (a sphere, the cylinder), such analogy is broken. The hydrodynamic picture of such surface flows is much more difficult than a flat surface. In this case, profiles of liquid velocity are deformed when they move to each point and have a different character that complicates the transfer of the substance.

We will consider flow with a stream of a sphere or the cylinder whose axis is perpendicular to the direction of a stream. The front of the ball and the cylinder smoothly circulate with the liquid flow and a boundary layer formed on it, which differs little from the boundary layer on a flat surface. A forward part of a body liquid in an interface moves in the direction of the pressure gradient, and in a back part, against the pressure gradient. In some back point, counterpressure completely slows down the liquid near a surface. Furthermore, there is a returnable movement of liquid where liquid layers are pushed aside in the volume of liquid, and the interface comes off the surface of a body.

At large numbers of Re, the movement of liquid behind a point of a separation turbulizes, which leads to formation of a turbulent trace whose length significantly depends on flow velocity. Such a non-uniform picture of flow of bodies leads to essential violation of analogy between transfer of an impulse and transfers of heat and weight. So, for example, near a forward critical point of run-ups of a stream, the density of a diffusive and thermal stream are final, though resistance of friction addresses in zero. Therefore, to this point, there is no analogy between processes of transfer of substance and transfer of an impulse. Lavish [6] notes that at numbers $\Pr >> 1$ and $Sc >> 1$, not only is there similarity in distribution of temperature, concentration, and velocity but also transfer of weight is broken, and heat and an impulse have various mechanisms. These circumstances significantly reduce expediency of use of quantitative analogy between convective diffusion and friction. However, it is important to note that when it can be received concrete, at least the approximate solution of a problem of transfer of substance is more preferable to use in this decision than an imperfect analogy between transfer of an impulse and transfer of a substance. Therefore, the coefficient of diffusion is accepted turbulent viscosity equal to kinematic coefficient $D_T = \nu_T$. In literature, there is a set of the empirical correlations estimating turbulent viscosity. So, in work of Lin [42], the following expressions are given:

$$\nu_T / \nu_c = \left(y_+ / 14.5 \right)^3, \; 0 < y_+ \le 5 \tag{4.43}$$

$$\nu_T / \nu_c = 0.2 y_+ - 0.959, \; 5 < y_+ \le 30 \tag{4.43a}$$

$$\nu_T/\nu = 0.001y_+^3, \qquad 0 < y_+ \le 5$$

in the work of Owen [43]: $$\nu_T/\nu_c = 0.012(y_+ - 1.6)^2, \qquad 5 < y_+ \le 20 \qquad (4.43b)$$

$$\frac{\nu_T}{\nu_c} = 0.4(y_+ - 10), \qquad y_+ > 20$$

in the work of Mizushina [44]:

$$\nu_T/\nu_c = Ay_+^2,\ 0 < y_+ \le y_{1k},$$

$$\nu_T/\nu_c = 0.4y_+\left[1 - \left(y_+/\mathrm{Re}_+\right)\right] - 1,\ y_{1k} < y_+ \le y_{2k}, \qquad (4.43c)$$

$$\nu_T/\nu_c = 0.07\,\mathrm{Re}_+,\ y_{1k} < y_+ \le \mathrm{Re}_+$$

where $A = (4.37 - 5.23) \times 10^{-4}$, $y_{1k} = 26.3$, $y_{2k} = (0.25 - 0.30)\mathrm{Re}_+, \mathrm{Re}_+ = 2RU_D/\nu_c$

At a turbulent flow in pipes, using various experimental data of Laufer [45], we have received the expression for distribution of a cross component of the pulsation velocity of a stream

$$V'/U_D = 0.72\left[1 - \exp(-6.25r_+) + 4.46r_+^{1/2}\exp(-7.49r_+)\right] \qquad (4.44)$$

where $r_+ = y/R$, $4 \times 10^4 < \mathrm{Re} < 5 \times 10^5$, and U_D is the dynamic velocity of a stream. The characteristic curve of change of cross velocity on a radius is given in Figure 4.3. It follows from the dependence that at the $r_+ \approx 0.15$ cross-pulsation velocity of a turbulent stream makes $V' \approx 0.85U_D$, and at $r \to 1$, that is, in the center of a pipe $V' \approx 0.72U_D$. On the basis of experimental data of Laufer [45], Reichardt [46], and Nunner [47] at $3 \times 10^4 < \mathrm{Re} < 5 \times 10^5$ distribution of coefficient of turbulent diffusion on the cross section of a pipe in the form of empirical dependence is received [4,48]

$$\frac{D_T}{\nu_c\,\mathrm{Re}_+} = 0.275\,\mathrm{Re}^{-1/15}\left(\frac{V'}{U_D}\right)\left[1 - \exp(-3.78r_+) + 0.95r_+\exp(-9.765r_+^3)\right] \qquad (4.45)$$

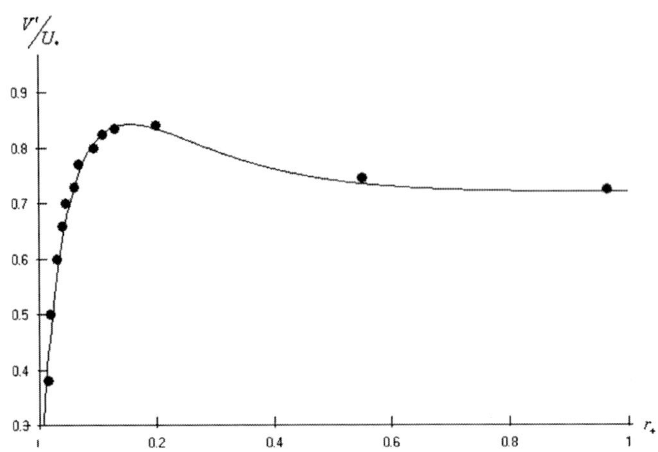

FIGURE 4.3 Comparison of a calculated value of distribution of cross velocities (4.44) with experimental data. (From Mizushina,T. and Ogino, F. *J. Chem. Eng. Jap.*, 3, 166–170, 1970.)

As it appears from (4.45), the coefficient of turbulent diffusion, at $r_+ \to 1$, that is, in the center of a pipe, can be defined in a look

$$D_T \approx 0.275 v_c \,\mathrm{Re}_+ \mathrm{Re}^{-1/15}\left(\frac{V'}{U_D}\right) \tag{4.46}$$

This equation, $V' \approx 0.72$ and $U_D \approx 0.2 U_m \mathrm{Re}^{-1/8}$ is taken into account and also represented as

$$D_T \approx 0.0285 D \mathrm{Sc}\,\mathrm{Re}^{4/5},\ 3\times10^4 < \mathrm{Re} < 5\times10^5 \tag{4.47}$$

where $\mathrm{Sc} = {}^{\nu}\!/_{D}$ is the Schmidt's number, D is the coefficient of molecular diffusion, and U_m is the average velocity of a stream. The nature of change of the coefficient of turbulent diffusion on the pipe radius depends on the number and the cross velocity of a stream as given in Figure 4.3.

As appears from Figure 4.4, the coefficient of turbulent diffusion passes through a maximum and is established in the center of a pipe. In work [49], the determination of coefficient of turbulent diffusion is presented by formulas

$$D_T/D = \frac{\mathrm{Sc}\left(y_+\right)^{4-y_+^{0.08}}}{1000\left(\frac{2.5\times10^7}{\mathrm{Re}}\right)^{y_+(400+y_+)}},\ y_+ \ge 0.05 \tag{4.48}$$

$$D_T/D = \left(\frac{y_+}{14.5}\right)^3 \mathrm{Sc},\ y_+ < 0.05 \tag{4.49}$$

Numerous pilot studies and calculations show that coefficients of turbulent viscosity and diffusion (4.45) are distributed in the section of pipes and channels similar to distribution of values of a cross component of velocity of pulsations (4.44). A viscous underlay growth of turbulent viscosity happens significantly quicker than beyond its limits. According to one researcher, the change of turbulent viscosity can be defined as ${}^{\nu_T}\!/_{\nu_c} \sim y_+^4$, according to others $-{}^{\nu_T}\!/_{\nu_c} \sim y_+^3$. The last detailed researches [12] show that the wall ${}^{\nu_T}\!/_{\nu_c} \sim y_+^3$ has a further steady recession of an exponent up to zero.

The turbulent diffusion coefficient is proportional to the macroscopic scale of turbulent pulsations and in a stream when the increased turbulence exceeds the molecular diffusion coefficient.

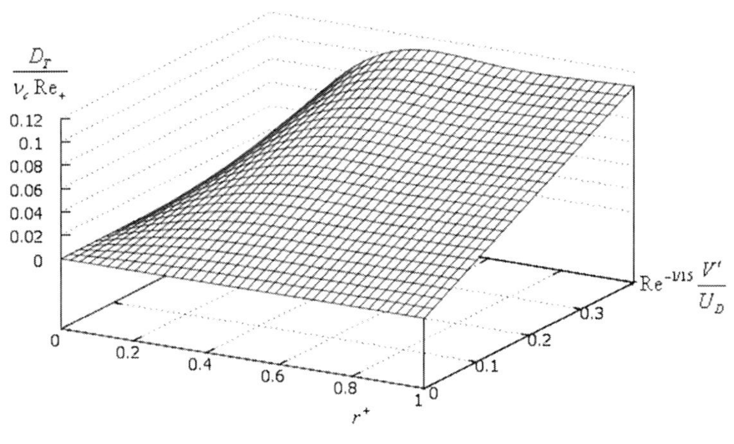

FIGURE 4.4 Change of coefficient of turbulent diffusion on radius pipes.

4.3.1 Turbulent Diffusion of Particles in a Stream

The movement of the weighed particles in a turbulent stream of gas and liquid differs more than in a laminar stream in complexity and intensity in all directions. In dispersed systems, if the particle sizes are small, then they will be completely taken away by turbulent pulsations and their motion in the liquid will be chaotic. With an increase in the sizes of particles, the last will lag behind the movement of liquid and, in this case extent of hobby of particles for turbulent pulsations, will decrease. It is important to note that for large particles in a turbulent stream, the coefficient of turbulent diffusion of particles will be defined not only by the velocity of a stream but also by their sedimentation. At a turbulent flow of disperse systems, the coefficient of turbulent diffusion of particles will be defined as [12]

$$D_{TP} = \mu_p^2 D_T \tag{4.50}$$

where D_{TP} is the coefficient of turbulent diffusion of particles, D_T is the coefficient of turbulent diffusion of liquid, and μ_p^2 is the extent of hobby of particles for the pulsing environment, determined by a formula

$$\mu_p^2 = \frac{1}{1+\omega_E^2 \tau_P^2} \tag{4.51}$$

where ω_E is the frequency of turbulent pulsations, and τ_P is the time of a relaxation of particles. The value μ changes in $0 \le \mu \le 1$ limits. For the small sizes of particles μ value it is close to zero, and for the big sizes of particles, aspire to the unit. Relaxation time for fine Stokes particles is defined by expression

$$\tau_P = \frac{1}{18}\frac{\rho_d}{\eta_c}a^2 \tag{4.52}$$

and for the coarsely dispersed particles in a look [49]

$$\tau_P = \frac{\rho_d a^2}{18\eta_c\left(1+\dfrac{\mathrm{Re}^{2/3}}{6}\right)} \tag{4.52a}$$

The extent of the hobby of fine particles will be defined as

$$\mu = \frac{1}{\left[1+\left(\dfrac{\rho_d a^2 \omega_E}{18\eta_c}\right)\right]^{1/2}} \tag{4.52b}$$

In a turbulent stream for assessment of coefficients of turbulent diffusion, using various pilot studies, it is possible to receive a number of empirical dependences for vertical and horizontal channels, type $D_{TP}/D_T = \mu^2 = f\left(U_D, V_S\right)$. So, using experimental data on a current of water drops of various size in an air stream in a vertical pipe diameter equal to 200 mm [12], in work [48], the following formulas are received

$$\frac{D_{TP}}{D_T} = \mu^2 = 0.023\psi\left(U_D\right)\left(\frac{U_D^3}{V_S}\right)^{1/4}, \mathrm{Re}_d = \frac{V_S a}{\nu} \le 5 \tag{4.53}$$

where V_S is the velocity of sedimentation of particles $\psi(U_D) = 1 + 0.786.10^{-6}U_D^4$. In the case of $\mathrm{Re}_d > 5$, then we receive the following formula

$$\frac{D_{TP}}{D_T} = \mu^2 = 0.054\left(\frac{U_D}{V_S}\right)^{1/4} \tag{4.54}$$

It follows from these formulas that the coefficient of turbulent diffusion of particles at a current in vertical channels is directly proportional to the dynamic velocity of a stream and is inversely proportional to the velocities of particle sedimentation. A comparison of turbulent diffusion coefficient of particle calculated values (4.53) and (4.54) with experimental values is given in Table 4.5.

For horizontal channels, using experimental data on a current of firm particles and oil drops [50] in an air stream for coefficients of turbulent diffusion, the following correlations are received [48]

$$\frac{D_{TP}}{D_T} = \mu^2 = 0.24\left(\frac{V_S}{U_D^3}\right)^{1/4}, \ \mathrm{Re}_d \geq 2.5 \tag{4.55}$$

$$\frac{D_{TP}}{D_T} = \mu^2 = k(V_S)\left(\frac{V_S}{U_D}\right)^{1/4}, \ \mathrm{Re}_d < 2.5 \tag{4.56}$$

where $k(V_S)$ is the parameter equal $k(V_S) = \frac{2.16}{V_S^{1/8}}$. As it appears from (4.55), unlike vertical channels, in horizontal channels the coefficient of turbulent diffusion is directly proportional to the velocity of sedimentation and is inversely proportional to dynamic velocity. Thus, both for horizontal and for vertical channels, the coefficient of diffusion depends on sedimentation velocity, and this dependence increases with the growth of the sizes of particles. The dependence of coefficient of turbulent diffusion of particles on the velocity of sedimentation defines the influence of mass of particles on coefficient diffusion. If in vertical channels the increase in mass of particles generally influences lag of particles from the velocity of the bearing environment, then in horizontal channels an increase in mass of a particle directly influences the coefficient of their cross diffusion. Comparisons of calculated values of coefficients of turbulent diffusion in horizontal channels (4.55) and (4.56) with experimental data are given in Tables 4.6 and 4.7.

TABLE 4.5
Coefficients of Turbulent Diffusion of Particles at a Current

U_m (m/s)	U_m (sm/s)	a (mkm)	Vs (sm/s)	D_T (sm²/s)	D_{TP} (sm²/s)	D_{TP}/D_T	μ^2	Re_d
1.55	9.0	80	17	6.3	0.370	0.059	0.059	<5
3.44	18.0	80	17	12.6	1.35	0.107	0.107	<5
7.60	36.0	80	17	25.2	10.1	0.400	0.386	<5
1.55	9.0	150	50	6.3	0.26	0.042	0.045	=5
3.44	18.0	150	50	12.6	1.05	0.083	0.082	= 5
7.60	36.0	150	50	25.2	6.20	0.246	0.295	=5
1.55	9.0	200	69	6.3	0.20	0.032	0.032	>5
3.44	18.0	200	69	12.6	0.44	0.035	0.038	>5
7.60	36.0	200	69	25.2	1.22	0.048	0.046	>5

TABLE 4.6
Experimental Data and Calculated Values Coefficient of Turbulent Diffusion (4.55) for Firm Particles in Air (a = 100–200 mkm, $Re_d > 2.5$) in the Square Channel of Section 76 × 76 mm

U_m (m/s)	U_D (sm/s)	V_s (sm/s)	D_T (sm²/s)	D_{TP} (sm²/s)	D_{TP}/D_T	μ^2
7.6	40.7	41	23.5	0.9	0.038	0.038
16.7	81.0	41	63.7	1.5	0.024	0.023
25.9	119.0	41	103.0	1.8	0.017	0.017
7.6	40.7	104	23.5	1.36	0.058	0.048
16.7	81.0	104	63.7	1.8	0.028	0.027
25.9	119.0	104	103.0	2.3	0.022	0.022

Source: Dasvies, C.N., Deposition from moving aerosols, In *Aerosol Science*, London, UK, Academic Press, 1966.

TABLE 4.7
Experimental Data and Settlement Values of Coefficient of Turbulent Diffusion (4.56) for Oil Drops in an Air Stream (a = 45 mkm, $Re_d > 2.5$) in a Horizontal Pipe (D = 152 mm, V_S = 3.5 sm/s)

U_m (m/s)	U_D (sm/s)	D_T (sm²/s)	D_{TP} (sm²/s)	D_{TP}/D_T	μ^2
58	241	358	229	0.640	0.641
87.2	344	511	298	0.583	0.585
122	461	686	386	0.563	0.545
148	546	812	460	0.566	0.522

Source: Dasvies, C.N., Deposition from moving aerosols, In *Aerosol Science*, London, UK, Academic Press, 1966.

The roughness of a surface of pipes has significant effect on the coefficient of turbulence. In work [48], the formula for calculation of coefficient of resistance of rough pipes is received

$$\xi_{TS} = \xi_S + \left[0.79 \ln \mathrm{Re} - 1.64 + \left(\zeta \, \mathrm{Re}\right)^{1/2}\right]^{-2} \tag{4.57}$$

where ξ_{TS} is the resistance coefficient for rough pipes, ξ_S is the established value of coefficient of resistance, $\xi_S \approx 0.22\left(\zeta/_D\right)^{3/8}, \zeta = {}^{\varepsilon_s}/_D$, ε_s is the height of ledges of roughnesses.

In literature [40], it is possible to encounter other formula for calculation of coefficient resistance in rough pipes

$$\frac{1}{\sqrt{\xi_{TS}}} = -2\lg\left[\frac{\varsigma}{3.7} + \left(\frac{6.81}{\mathrm{Re}}\right)^{0,9}\right] \tag{4.58}$$

though the formula (4.57) is more exact. If in formulas (4.55) and (4.56) use the expression and $U_D = U_m\sqrt{\xi/8}$ and $V_S = 6\times10^{-4} U_D \tau_+^2$ ($\tau_+ = \frac{\tau_p U_D^2}{\nu}$ the dimensionless time of relaxation), then equation (4.55) for $\mathrm{Re}_d > 5, \tau_+ < 20$ in vertical channels will be presented as

$$\frac{D_{TP}}{D_T} = 0.58\left(\frac{\nu_c}{\tau_p U_m}\right)^{1/2} \xi_{TS}^{-1/4} \tag{4.59}$$

For high values of roughness $\left(\varepsilon_s/D > \frac{2000}{\text{Re}}\right)$, equation (4.59) will be presented as [4,48]

$$\frac{D_{TP}}{D_T} = 0.846\left(\frac{\nu_c}{\tau_p U_m}\right)^{1/2}\left(\frac{\varepsilon_s}{D}\right)^{-0.094} \tag{4.60}$$

In horizontal channels, the coefficient of turbulent diffusion is estimated as

$$\frac{D_{TP}}{D_T} = 0.0155\left(\frac{\tau_p U_m}{\nu}\right)^{1/2}\left(\frac{\varepsilon_s}{D}\right)^{0.094}, \tau_+ \le 20 \tag{4.61}$$

$$\frac{D_{TP}}{D_T} = 0.142 U_m^{-1/2}\left(\frac{\varepsilon_s}{D}\right)^{-0.094}, \tau_+ > 20 \tag{4.62}$$

Obviously, these formulas are confidants, in view of a lack of pilot studies of influence of roughness on the coefficient of turbulent diffusion.

4.4 THE EFFECT OF THE COMPACTION OF A LAYER OF PARTICLES ON THE DIFFUSION COEFFICIENTS AND THERMAL CONDUCTIVITY

In devices to the big thickness of a layer of particles, there is a consolidation of a layer under the influence of the external deforming forces and a body weight of a layer over time that significantly influences transfer coefficients. Consolidation of a layer of particles is characterized by generally technological devices with a motionless layer of the catalyst, adsorbent, and other particles fillers. Non-stationary consolidation of a layer of particles promotes, first of all, the narrowing of channels between layer particles and the reduction of porosity of a layer and a natural change of fields of velocity, temperature, and concentration. On the other hand, as a result of the non-stationary consolidation of particles, there is a change of coefficients of heat conductivity and diffusion on time, depending on change of porosity. With change of porosity, it is a connected pressure difference in a layer of porous particles and some other parameters of a layer and process.

If the distance between particles is big in comparison with the sizes of particles, that is, at a big porosity of a layer, Maxwell has offered the following formula for definition of effective heat conductivity of a layer [14,29]

$$\lambda_{eff} = \lambda_c\left[\frac{\lambda_d + 2\lambda_c - 2(1-\varepsilon)(\lambda_c - \lambda_d)}{\lambda_d + 2\lambda_c + (1-\varepsilon)(\lambda_c - \lambda_d)}\right] \tag{4.63}$$

where λ_d and λ_c are the heat conductivity according to dispersed and a continuous phase, and ε is the porosity. The geometry of the porous environment has a significant effect on the value of effective heat conductivity of a layer. So, for dense hexagonal packing of particles for determination of coefficient of heat conductivity [14,29] in works, the formula of the following look is offered

$$\lambda_{eff} = 3\pi\lambda_d \ln\frac{0.43 + 0.31\varepsilon}{\varepsilon - 0.26} \tag{4.64}$$

If the layer of particles is located perpendicular to the direction of a stream, then heat conductivity of such system will be minimum and will be defined by expression

$$\lambda_{eff}^{\min} = \frac{\lambda_d\lambda_c}{\varepsilon\lambda_d + (1-\varepsilon)\lambda_c} \tag{4.65}$$

If the layer of particles is located parallel to the direction of a stream, then in this case heat conductivity of system will be maximum

$$\lambda_{eff}^{max} = \varepsilon\lambda_c + (1-\varepsilon)\lambda_d \tag{4.66}$$

The geometry of porous structure and, first of all, tortuosity and variable sections of channels in a time, significantly influences the coefficient of effective diffusion. Using a set of experimental data for a layer from various particles (glass balls, sand, chloride sodium, friable soil, mica), it is possible to put that for values of porosity the coefficient of effective diffusion linearly increases with increase in porosity. At great values of porosity ε <0.65, such linearity is broken. The correlation dependence of effective coefficient of diffusion according to average experimental data can be presented in the form of Figure 4.5.

$$D_{eff}/D_0 = 0.62\varepsilon + 0.28\varepsilon^{4.4} \tag{4.67}$$

where D_0 is the diffusion coefficient in free liquid.

Comparison of average experimental values of the coefficient of effective diffusion in a hydrogen–air system for layers of various particles (glass balls, particles of chloride sodium, mica, soil, etc.) with calculated values on a formula (4.67) is given in Table 4.8. As it appears from experimental data, with reduction of porosity of a layer, the coefficient of effective diffusion also decreases.

As it appears from this table, the relative error of calculation makes 8%–9%, and coefficient of correlation $r^2 = 0.926$.

Though the problem of consolidation of particles is non-stationary, the simplest problem of consolidation of two particles can be considered as a hydrodynamic problem of expression of the liquid concluded between two particles. We will put that two parallel round plates of radius R are located one on another at small distance h from each other, and the space between them is filled with liquid.

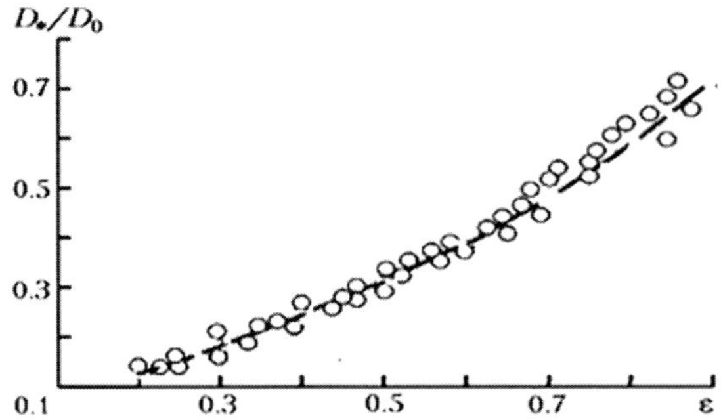

FIGURE 4.5 Change of effective coefficient of diffusion in dependence of porosity (points—experimental data).

TABLE 4.8
Dependence of Coefficient of Effective Diffusion on Porosity of the Media

Porosity ε	D_{eff}/D_0 (exp.)	D_{eff}/D_0 (4.35)	Porosity ε	D_{eff}/D_0 (exp.)	D_{eff}/D_0 (4.35)
0.25	0.140	0.155	0.70	0.49	0.48
0.30	0.185	0.187	0.80	0.60	0.60
0.40	0.253	0.253	0.90	0.73	0.76
0.60	0.400	0.400	1.00	0.90	0.90

Under the influence of external forces of a plate, it approaches the constant velocity of U, forcing out liquid. Having put a current of liquid laminar and in view of a subtlety of a layer of liquid $V_z << V_r$ and $\partial V_r/\partial r << \partial V_r/\partial z$, equation (1.7) will be presented as [5]

$$\eta_c \frac{\partial^2 V_r}{\partial z^2} = \frac{\partial P}{\partial r}, \frac{\partial P}{\partial z} = 0 \quad (4.68)$$

$$\frac{1}{r}\frac{\partial (rV_r)}{\partial r} + \frac{\partial V_z}{\partial z} = 0 \quad (4.69)$$

with boundary conditions

$$z = 0, V_r = V_z = 0$$

$$z = h, V_r = 0, V_z = -U \quad (4.70)$$

$$r = R, P = P_0$$

where P_0 is the external pressure. From equation (4.68) we find

$$V_r = \frac{1}{2\eta_c}\frac{\partial P}{\partial r} z(z-h) \quad (4.71)$$

Integrating equation (4.69) on dz we will receive distribution of velocity of a current on liquid layer thickness

$$U = \frac{1}{r}\frac{d}{dr}\int_0^h rV_r dz = -\frac{h^3}{12\eta_c r}\frac{d}{dr}\left(r\frac{dP}{dr}\right) \quad (4.72)$$

from where we will determine distribution of pressure by liquid layer thickness

$$P = P_0 + \frac{3\eta_c U}{h^3}\left(R^2 - r^2\right) \quad (4.73)$$

4.4.1 Non-Stationary Compaction of a Layer of Particles

Consolidation of a layer of particles can be considered as the task connected with questions of deformation. By analogy with the theory of elasticity, it can be assumed that the elastic volume deformations of a porous environment are associated with a deforming pressure correlation

$$\frac{\Delta V_s}{V_s} = -\frac{\sigma_D}{\eta_s}\Delta t \quad (4.74)$$

where $V_s = 1/\rho_P$ is the specific volume of the porous environment, $\rho_p = \rho_d(1-\varepsilon)$ is the bulk density of a layer, ρ_d is the density of material of a firm phase, and $\sigma_D = -\Delta P_f + g(\rho_d - \rho)z$ is the deforming tension developing the pressure difference on length of a layer and body weight of a layer. Differentiating (4.74) on t, we will receive

$$\frac{1}{1-\varepsilon}\frac{d\varepsilon}{dt} = -\frac{\sigma_D}{\eta_s} \quad (4.75)$$

where η_s is the volume viscosity of the porous environment. Passing from a substantive derivative to local, we will receive

$$\frac{\partial \varepsilon}{\partial t} + div(\varepsilon U_s) = -(1-\varepsilon)\eta_s^{-1}\sigma_D \tag{4.76}$$

Expression (4.76) is the equation of porosity variation under the action of deforming stresses. In practice when calculating a current porous layer, in view of the limitation of the last the surface of the device, it is possible to put $U_{sx} << U_{sz}$ and $U_{sy} << U_{sz}$. Then equation (4.76) will register in a look

$$\begin{aligned} &\frac{\partial \varepsilon}{\partial t} + U_s \frac{\partial \varepsilon}{\partial z} = -(1-\varepsilon)\eta_s^{-1}\sigma_D \\ &\varepsilon(t,z)_{t=0} = \varepsilon_0(z) \end{aligned} \tag{4.77}$$

The decision (4.77) will be presented in the form

$$\frac{1-\varepsilon}{1-\varepsilon_0} = \exp\left[-\eta_s^{-1}\sigma_D\left(t + \frac{z}{U_s}\right)\right] \tag{4.78}$$

Having expressed volume viscosity from shift viscosity as ξ_s

$$\eta_s = \frac{4}{3}\xi_s \frac{1-\varepsilon}{\varepsilon} \tag{4.79}$$

equation (4.77) can be presented in the form

$$\frac{\partial \varepsilon}{\partial t} + U_s \frac{\partial \varepsilon}{\partial z} = -\frac{3}{4}\xi_s^{-1}\varepsilon\sigma_D \tag{4.80}$$

with the decision

$$\varepsilon = \varepsilon_0 \exp\left[-\frac{3}{4}\xi_s^{-1}\sigma_D\left(t + \frac{z}{U_s}\right)\right] \tag{4.81}$$

For average porosity on a layer, it is possible to write down

$$\varepsilon = \varepsilon_0 \exp\left(-\frac{3}{4}\xi_s^{-1}\sigma_D t\right) \tag{4.82}$$

Expressions (4.81) and (4.82) make it possible to estimate the value of the porosity of a layer at different points in time in its compacted state and take their values into account in determining transfer coefficients.

4.5 ONE-DIMENSIONAL DIFFUSIVE TRANSFER AND STEFAN FLOW

Although the research of one-dimensional diffusion problems is the simplest task, it allows us to analyze diffusive streams more deeply. Because the diffusive process is slow, this phenomenon is significantly shown in motionless liquid or at a weak laminar current.

Let's assume that in a slowly moving stream we choose an elementary volume with a length dz (z is the coordinate in the direction of diffusion, calculated from the surface of the liquid and taken as the diffusion path) and with a constant cross-sectional area S. The stream enters this volume at the left J_{AZ} and leaves this volume $J_{A(Z+dZ)}$ (z is the direction of the movement of a stream). In the absence of chemical and phase transformations, it is possible to write down

$$SJ_{AZ} = SJ_{A(Z+dZ)} \tag{4.83}$$

Having divided both parts $S\Delta z$ on and under a condition $\Delta z \to 0$, we will receive

$$\lim_{\Delta z \to 0} \frac{J_{AZ} - J_{A(Z+dZ)}}{\Delta z} = 0 \tag{4.84}$$

In a limit case, we can write

$$-\frac{dJ_A}{dz} = 0 \quad \text{or} \quad J_A = const \tag{4.85}$$

Thus, at the absence in a stream of chemical and phase transformations, change of a diffusive stream doesn't happen. The condition (4.85) is fair in binary mixes when diffusion of both components happens to an identical velocity. If velocities of diffusion of components are different, then besides the diffusive movement, there is an apparent motion of all mix caused by a gradient of the general pressure. Let's say that from the surface of motionless liquid in the closed vessel there is an evaporation in air. At evaporation of liquid, there is a mutual diffusion of steam in air and, back, air in liquid. And if steam freely diffuses in air, then for the last the surface of liquid is a barrier, and as a result the amount of air at the surface of liquid continuously increases, causing an increase in the general pressure. As pressure is constant, performance of this condition requires the movement of all mix in the form of convective transfer with some velocity V_{St}. This stream connected with excessive pressure is sent from the surface of liquid to the volume of air and is called a Stefan stream. We will define the value of this stream, and having put that on the surface of liquid, the following conditions are met

$$\begin{aligned} z &= z_1,\ y_A = y_{AC} \\ z &= z_2,\ y_A = y_{Aw} \end{aligned} \tag{4.86}$$

where y_A is the concentration of steam, $y_B = 1 - y_A$ is the concentration of air in binary mix, and z_1, z_2 are the coordinates of a surface of liquid and a surface of pipe.

According to Fick's first law, the general stream from a surface is equal

$$J_A = (J_A + J_B)\,y_A - CD_{AB}\frac{dy_A}{dz} \tag{4.87}$$

If ρ is the to put that the air stream at the surface of liquid is equal to zero $J_B = 0$, then from this equation we will receive

$$J_A = \frac{CD_{AB}}{(1 - y_A)}\frac{dy_A}{dz} \tag{4.88}$$

In principle, J_A represents a stream from the surface of liquid with a velocity

$$V_{St} = {J_A}/{\rho} \tag{4.88a}$$

(where ρ is the mix density), that is, the velocity of convective transfer is equal to steam density, divided into density of the steam–gas mix. Substituting (4.88) in equation (4.85), we will receive expression

$$-\frac{d}{dz}\left[-\frac{CD_{AB}}{1-y_A}\frac{dy_A}{dz}\right]=0 \tag{4.89}$$

integrating twice at constancy CD_{AB}, we have

$$-\ln(1-y_A)=A_1 z+A_2 \tag{4.90}$$

Using conditions (4.86), we will define integration constants

$$A_1=\frac{1}{z_2-z_1}\ln\frac{1-y_{AC}}{1-y_{AW}}=\frac{1}{\Delta z}\ln\frac{y_{BC}}{y_{BW}} \tag{4.91}$$

$$A_2=-\ln y_{BC}-\frac{\ln\left(y_{BC}/y_{BW}\right)}{\Delta z}z_1 \tag{4.92}$$

where $\Delta z=z_2-z_1$. Substituting these values in equation (4.90), after its simple transformations, we will receive

$$\frac{y_B}{y_{BC}}=\left(\frac{y_{BW}}{y_{BC}}\right)^{\frac{z-z_1}{\Delta z}} \tag{4.93}$$

Using expression $y_A=1-y_B$ and (4.93), we will receive the expression for a diffusive stream as

$$J_A=-\frac{CD_{AB}}{1-y_A}\frac{dy_A}{dz}=-\frac{CD_{AB}}{\Delta z}\ln\frac{y_{BW}}{y_{BC}}=\frac{CD_{AB}}{\Delta z}\ln\frac{y_{BC}}{y_{BW}} \tag{4.94}$$

Having defined molar concentration as $C=\frac{P}{RT}$, equation (4.94) can be written down in a look

$$J_A=\frac{P}{RT}\frac{D_{AB}}{\Delta z}\ln\frac{y_{BC}}{y_{BW}}=\frac{P}{RT}\frac{D_{AB}}{\Delta z}\ln\frac{1-y_{AC}}{1-y_{AW}} \tag{4.95}$$

Expression (4.95) bears the name of the equation of Stefan and allows us to calculate the velocity of the steam–gas mix from the surface of the liquid. Having increased and having divided the right member of equation (4.95) on $(y_{BW}-y_{BC})$, after the corresponding transformations, we will define

$$J_A=\frac{P}{RT}\frac{D_{AB}}{\Delta z}(y_{BW}-y_{BC})\frac{\ln\left(y_{BC}/y_{BW}\right)}{(y_{BW}-y_{BC})} \tag{4.95a}$$

Having entered the logarithmic average or driving force of process

$$\Delta_y=\frac{y_{BW}-y_{BC}}{\ln\left(y_{BC}/y_{BW}\right)} \tag{4.95b}$$

we will finally receive $\left(y_{BW}-y_{BC}=\left(1-y_{AW}\right)-\left(1-y_{AC}\right)=y_{AC}-y_{AW}\right)$

$$J_A=\frac{P}{RT}\frac{D_{AB}}{\Delta z}\frac{y_{AC}-y_{AW}}{\Delta_y} \tag{4.95c}$$

It should be noted that, in the derivation of these formulas, detection of thermal and pressure diffusion was neglected. In principle, the evaporation process, as well as any process of a mass exchange, is non-stationary, characterized by a certain time of relaxation of the established state. Change of a molar stream depending on time can be presented in the form

$$J_A=\frac{\rho_A}{M_A}\frac{dz}{dt} \tag{4.95d}$$

Comparing this expression to equation (4.95c), we will receive the equation

$$\frac{\rho_A}{M_A}\frac{dz}{dt}=\frac{CD_{AB}\left(y_{AC}-y_{Aw}\right)}{z\Delta_y} \tag{4.96}$$

and integrating from 0 to t and from z_0 to z_t, we have

$$t=\frac{\rho_A\Delta_y}{CD_{AB}M_A\left(y_{AC}-y_{Aw}\right)}\left(\frac{z_t^2-z_0^2}{2}\right) \tag{4.97}$$

From this equation in the established state, we will define diffusion coefficient as

$$D_{AB}=\frac{\rho_A\Delta_y}{M_AC\left(y_{AC}-y_{Ab}\right)t}\left(\frac{z_t^2-z_0^2}{2}\right) \tag{4.98}$$

Example 4.5

In a vertical glass pipe where the open end evaporates toluene, where depth to the surface of liquid at the initial moment it is equal to 20 mm, pressure in a pipe is atmospheric and system temperature toluene—steam is equal $T=39.4°C$. After $t=278$ hours, the depth to the surface of liquid are equal 80 mm. To define the coefficient of diffusion of steams toluene. Pressure of saturated steam of toluene at a temperature T=39.4°C is equal $p_T=7.64\ \text{kH}/\text{m}^2$, density of toluene $\rho_A=850\ {}^{\text{kg}}\!/_{\text{n}}$ and molecular weight of toluene is equal $M_A=92$.

For determination of the established value of coefficient of diffusion, we will use a formula (4.98). We will determine the driving force of process, believing

$$y_{AC}={}^{p_T}\!/_{P}=\frac{7.64\times10^3}{101.3\times10^3}=0.0754,\ y_{B1}=1-y_{Ac}=1-0.0754=0.9246,$$

$$y_{B2}=1-y_{AW}=1-0=1,$$

$$y_{Aw}=0,\ \Delta_y=\frac{1-0.9246}{\ln\dfrac{1}{0.9246}}=0.9618$$

Molar concentration will be defined as

$$C=\frac{P}{RT}=\frac{1.013\times10^5}{8314(273+39.4)}=0.039\,{}^{\text{kmole}}\!/_{\text{m}^3}$$

and the coefficient of diffusion is equal

$$D_{AB} = \frac{850 \times 0.9618}{92 \times 0.039(0.0754 - 0)275 \times 3600}\left(\frac{0.08^2 - 0.02^2}{2}\right) = 9.1572 \times 10^{-6}\,\mathrm{m^2/sec}$$

The Stefan stream is shown in the processes connected with contact of a gas phase with a firm surface (adsorption, catalytic processes) and with a liquid surface (absorption, dissolution of gases in liquid), proceeding with the change of volume of a stream and stoichiometry. In close proximity to a firm surface on which chemical reaction proceeds the velocity of a current, normal to a surface, it doesn't depend on hydrodynamic conditions and entirely is defined by conditions of diffusion and a stoichiometry of streams. For one-dimensional diffusion, the Stefan stream is defined by expression (4.95), which for the interface formed at contact of two phases is equal [36]

$$J_{AB} = \frac{\mathrm{Sh}D}{d}\frac{P}{\gamma RT}\ln\frac{1-\gamma\, y_{AC}}{1-\gamma\, y_{AW}} \tag{4.99}$$

where $\gamma = \sum {}^{v_i}\!/_{v_i}$ is the coefficient characterizing change of volume of mix, v_i is the stoichiometry coefficients, Sh is the Sherwood number, and d is the linear size. As it has been noted, if one of the initial substances is limiting, then the reversible reactions inflow with a reduction of volume. If one of the reaction products is limiting, then $\gamma < 0$ for the reactions proceeds with an increase in volume. At a $\gamma > 0$ the Stefan stream is sent to an interface of phases and at $\gamma < 0$ from a surface to a stream. A condition of lack of a Stefan stream is $\gamma = 0$. It should be noted that in catalytic and adsorptive processes, the size of a Stefan stream is insignificant in comparison with the main convective stream.

4.6 EQUIMOLAR AND NONEQUIMOLAR MASS TRANSFER

The binary systems of a mass transfer distinguish equimolar and nonequimolar transfer of weight between two phases. Equimolar transfer of weight is carried out in processes when the number of transferable weight between phases is equal, that is, $J_A = -J_B$ and whose example is the rectification process where between a steam and liquid phase there is such equality. The nonequimolar systems of a mass transfer are valid in case the amount of transferable substance between phases is unequal, that is, $J_B = -\alpha J_A (\alpha < 1)$. The nonequimolar of systems is an example of processes of absorption, adsorption, etc.

1. **Equimolar** transfer of weight from one phase in another is described by the equation

$$J_A = -D_{AB}\frac{dC_A}{dz} + C_A\left(J_A + J_B\right) \tag{4.100}$$

Having put that $J_A = -J_B$, we will receive

$$J_A = -D_{AB}\frac{dC_A}{dz} \tag{4.101}$$

Integrating this equation under boundary conditions

$$z = z_1, C_A = C_{A1}, z = z_2, C_A = C_{A2} \tag{4.102}$$

we will receive

$$J_A = \frac{D_{AB}}{z_2 - z_1}\left(C_{A1} - C_{A2}\right) \tag{4.103}$$

Having put $C_A = \frac{p_A}{RT}$, we will express a weight stream through partial pressure

$$J_A = \frac{D_{AB}}{RT(z_2 - z_1)}(p_{A1} - p_{A2}) \tag{4.104}$$

Expression (4.104) is the equation of equimolar diffusive transfer of weight. For definition of a concentration profile, we use expression $dJ_A/dz = 0$. Using the expression for Fick's first law, this equation is presenting as

$$\frac{d}{dz}\left(-D_{AB}\frac{dC_A}{dz}\right) = 0 \tag{4.105}$$

The solution of this equation will be presented in the equation form

$$\frac{C_A - C_{A1}}{C_{A1} - C_{A2}} = \frac{z - z_1}{z_1 - z_2} \tag{4.106}$$

the presenting concentration profile of equimolar transfer.

2. **Nonequimolar** transfer of weight is carried out if $J_B = -\alpha J_A$, and it is described by the equation

$$J_A = -CD_{AB}\frac{dy_A}{dz} + y_A(J_A - \alpha J_A) = -CD_{AB}\frac{dy_A}{dz} + y_A J_A(1-\alpha) \tag{4.107}$$

Having presented this equation in the form

$$J_A[1 - y_A(1-\alpha)] = -CD_{AB}\frac{dy_A}{dz} \tag{4.108}$$

and integrating, we will receive expression

$$J_A = \left(\frac{1}{1-\alpha}\right)\frac{CD_{AB}}{z_2 - z_1}\ln\left[\frac{1-(1-\alpha)y_{A2}}{1-(1-\alpha)y_{A1}}\right] \tag{4.109}$$

the presenting equation of nonequimolar transfer of number of weight.

Example 4.6

The vertical pipe with a diameter of 1 cm and 20 cm long is filled with a mix of CO_2 and H_2 gases with a pressure of 2 atm and temperature 0°C. The diffusion coefficient of CO_2 in H_2 is equal $D_{AB} = 0.275\ \mathrm{sm^2/sec}$. The partial pressure CO_2 at the beginning $z = z_1$ equals 1.5 atm and at the end $z = z_2$ −0.5 atm. To define a diffusive stream: (1) equimolar transfer $J_A = -J_B$; and (2) for nonequimolar transfer of weight $J_B = -0.75 J_A$ or $\alpha = 0.75$.

1. The diffusive stream for equimolar transfer will be determined by equation (4.104), under a condition $D_{AB} = 0.275\times10^{-4}\ \mathrm{m^2/Sec}, T = 273°K$:

$$J_A = \frac{0.275\times10^{-4}}{8314\times273\times0.2}(1.5\times1.013\times10^5 - 0.5\times1.013\times10^5) = 6.138\times10^{-6}\ \mathrm{kmole/m^2sec}$$

The general diffusive stream will be defined by a surface of section as

$$J = J_A S = J_A \pi r^2 = 6.138\times10^{-6}\times3.14\times\left(0.5\times10^{-2}\right)^2 = 4.821\times10^{-10}\,\mathrm{kmole}/\mathrm{sec}$$

$$= 1.735\times10^{-3}\,\mathrm{mole}/\mathrm{hour}$$

2. The diffusive stream for nonequimolar transfer will be determined by equation (4.109). To do this, with respect to the specified conditions of the problem, determine the concentration

$$C = \frac{P}{RT} = \frac{2\times1.013\times10^5}{8314\times273} = 0.0893\,\mathrm{kmole}/\mathrm{m}^3$$

$$y_{A1} = \frac{p_{A1}}{P} = \frac{1.5}{2} = 0.75,\; y_{A2} = \frac{p_{A2}}{P} = \frac{0.5}{2} = 0.25$$

$$J_A = \frac{4CD_{AB}}{z}\ln\left(\frac{1-0.25y_{A2}}{1-0.25y_{A1}}\right) = \frac{4\times0.0893\times0.275.10^{-4}}{0.2}\ln\frac{1-0.25\times0.25}{1-0.25\times0.75}$$

$$= 7.028\times10^{-6}\,\mathrm{kmole}/\mathrm{m^2sec}$$

$$J = J_A S = 7.028\times10^{-6}\times3.14\times\left(0.5\times10^{-2}\right)^2 = 5.52\times10^{-10}\,\mathrm{kmole}/\mathrm{sec} = 1.987\times10^{-3}\,\mathrm{mole}/\mathrm{hour}$$

4.6.1 Transfer of Weight at Evaporation of a Drop

We often meet the phenomena of evaporation of a liquid drop in processes of dispersion at high temperatures, rectification, etc. The amount of the evaporated liquid from a surface of a spherical drop is defined by Fick's first law, on the condition of lack of the return diffusion of steam in drop volume, $J_B = 0$

$$J_A = -\frac{CD_{AB}}{1-y_A}\frac{dy_A}{dr} \tag{4.110}$$

It is supposed that the surface of a drop evaporates evenly and the molar stream from a surface of a drop decides by drop geometry on variable radius. For any distance $r+\Delta r$ lectured from the center of a drop, believing constancy of a stream of steam it is possible to write down

$$4\pi r^2 J_A\left(r+\Delta r\right) - 4\pi r^2 J_A\left(r\right) = 0 \tag{4.111}$$

In a limit case, this equation can be written down in a differential form

$$\frac{d}{dr}\left(r^2 J_A\right) = 0 \tag{4.112}$$

Integrating this equation, we have $r^2 J_A = \mathrm{const}$ or $r^2 J_A = r_0^2 J_{A0}$, where r_0, J_{A0} are the radius of the drop and the mass flow corresponding to it. Substituting this expression in equation (4.110), we will receive $\left(J_A = \frac{r_0^2 J_{A0}}{r^2}\right)$

$$-\frac{r^2 CD_{AB}}{1-y_A}\frac{dy_A}{dr} = r_0^2 J_{A0} \tag{4.113}$$

We will make the solution of this equation under entry conditions

$$r = r_0,\ y_A = y_{AS}$$
$$r \to \infty,\ y_A = y_{A\infty} \tag{4.114}$$

As a result we have

$$J_{A0} = \frac{CD_{AB}}{r_0} \ln\left(\frac{1 - y_{A\infty}}{1 - y_{AS}}\right) \tag{4.115}$$

On the other hand, we will define change of volume of a drop in a look

$$4\pi r_0^2 J_{A0} = -\frac{d}{dt}\left(\frac{4}{3}\pi r_0^3 \frac{\rho_k}{M_A}\right) = -4\pi r_0^2 \frac{\rho_k}{M_A}\frac{dr_0}{dt} \tag{4.116a}$$

From this equation, it is possible to write down $J_{A0} = -\frac{\rho_k}{M_A}\frac{dr_0}{dt}$, substituting which in (4.115), we will receive

$$\frac{CD_{AB}}{r_0} \ln\left(\frac{1 - y_{A\infty}}{1 - y_{AS}}\right) = \frac{\rho_k}{M_A}\frac{dr_0}{dt} \tag{4.116b}$$

Integrating this equation under entry conditions, we will receive

$$t = \frac{\rho_k}{2M_A CD_{AB}} \frac{R_0^2}{\ln\left(\frac{1 - y_{A\infty}}{1 - y_{AS}}\right)} \tag{4.117}$$

Equation (4.117) defines time of full evaporation of a spherical drop with an initial radius R_0. Obviously, this decision doesn't consider influence of a hydrodynamic current of a stream, temperature, and many other factors and considers evaporation of a drop in ideal conditions.

4.7 FICK'S SECOND LAW OF MASS TRANSFER

Fick's second law of a mass transfer is based on weight conservation law, and its output with the use of the equation of continuity of a stream allows us to define concentration profiles for various situations of a mass transfer.

The equations of continuity of a stream removed in the first section for some component will be presented A in the form

$$\frac{\partial \rho}{\partial t} + \frac{\partial q_{Ax}}{\partial x} + \frac{\partial q_{Ay}}{\partial y} + \frac{\partial q_{Az}}{\partial z} = 0 \tag{4.118}$$

where q_A is a specific mass A consumption of the substance postponed through unit of area of an interphase surface for a unit of time. The mass expense consists of convective and diffusive members

$$q_A = \rho_A V_A + J_A \tag{4.119}$$

Substituting (4.119) in equation (4.118), and taking into account the velocity of chemical and phase transformations r_A, we will receive

$$\frac{\partial\rho_A}{\partial t}+\frac{\partial\rho_A V_x}{\partial x}+\frac{\partial\rho_A V_y}{\partial y}+\frac{\partial\rho_A V_z}{\partial z}+\frac{\partial J_{Ax}}{\partial x}+\frac{\partial J_{Ay}}{\partial y}+\frac{\partial J_{Az}}{\partial z}=r_A \tag{4.120}$$

where V_x, V_y, V_z are the being substance velocities on coordinates, and J_{Ax}, J_{Ay}, J_{Az} are the components of diffusive streams of substance A on coordinates. At constancy of velocity of a stream, considering equation of a diffusive stream according to Fick's first law (4.1), equation (4.120) will register as

$$\frac{\partial\rho_A}{\partial t}+V_x\frac{\partial\rho_A}{\partial x}+V_y\frac{\partial\rho_A}{\partial y}+V_z\frac{\partial\rho_A}{\partial z}=D_A\left(\frac{\partial^2\rho_A}{\partial x^2}+\frac{\partial^2\rho_A}{\partial y^2}+\frac{\partial^2\rho_A}{\partial z^2}\right)+r_A \tag{4.121}$$

This equation expresses itself as the Fick's second law. The left part of this equation represents convective transfer of substance, and the right part the diffusive transfer. If the velocity of phase and chemical transformations and also velocity of a stream is equal to zero $V_x = V_y = V_z = 0$, then we have

$$\frac{\partial\rho_A}{\partial t}=D_A\left(\frac{\partial^2\rho_A}{\partial x^2}+\frac{\partial^2\rho_A}{\partial y^2}+\frac{\partial^2\rho_A}{\partial z^2}\right) \tag{4.122}$$

This expression is the equation of non-stationary diffusion in a motionless liquid. Subsequently, it will be visible that this equation is also fair in moving a liquid at small numbers Pe. The equation (4.122) in molar concentration $C_A = \rho_A / M_A$ will register in a look

$$\frac{\partial C_A}{\partial t}=D_A\left(\frac{\partial^2 C_A}{\partial x^2}+\frac{\partial^2 C_A}{\partial y^2}+\frac{\partial^2 C_A}{\partial z^2}\right) \tag{4.123}$$

In a turbulent stream, the coefficient of turbulent diffusion depends on coordinates, and therefore equations (4.121) through (4.123) will register in a look

$$\frac{\partial C_A}{\partial t}=\frac{\partial}{\partial x}\left(D_{SX}\frac{\partial C_A}{\partial x}\right)+\frac{\partial}{\partial y}\left(D_{SY}\frac{\partial C_A}{\partial y}\right)+\frac{\partial}{\partial z}\left(D_{SZ}\frac{\partial C_A}{\partial z}\right) \tag{4.124}$$

where D_S is the total coefficient of turbulent and molecular diffusion.

Fick's second law of a mass transfer in various coordinates will be submitted as:

1. In rectangular coordinates:

$$\frac{\partial C_A}{\partial t}+V_x\frac{\partial C_A}{\partial x}+V_y\frac{\partial C_A}{\partial y}+V_z\frac{\partial C_A}{\partial z}=D_A\left(\frac{\partial^2 C_A}{\partial x^2}+\frac{\partial^2 C_A}{\partial y^2}+\frac{\partial^2 C_A}{\partial z^2}\right) \tag{4.125}$$

In case of motionless liquid we have

$$\frac{\partial C_A}{\partial t}=D_A\left(\frac{\partial^2 C_A}{\partial x^2}+\frac{\partial^2 C_A}{\partial y^2}+\frac{\partial^2 C_A}{\partial z^2}\right) \tag{4.125a}$$

In a stationary case, diffusive transfer will be presented as

$$\left(\frac{\partial^2 C_A}{\partial x^2}+\frac{\partial^2 C_A}{\partial y^2}+\frac{\partial^2 C_A}{\partial z^2}\right)=0 \tag{4.125b}$$

2. In cylindrical coordinates:

$$\frac{\partial C_A}{\partial t}+V_r\frac{\partial C_A}{\partial r}+\frac{V_\theta}{r}\frac{\partial C_A}{\partial r}+V_z\frac{\partial C_A}{\partial z}=D_A\left[\frac{1}{r}\frac{\partial}{\partial r}\left(r\frac{\partial C_A}{\partial r}\right)+\frac{1}{r^2}\frac{\partial^2 C_A}{\partial \theta^2}+\frac{\partial^2 C_A}{\partial z^2}\right] \tag{4.126}$$

For motionless liquid we have

$$\frac{\partial C_A}{\partial t}=D_A\left[\frac{1}{r}\frac{\partial}{\partial r}\left(r\frac{\partial C_A}{\partial r}\right)+\frac{1}{r^2}\frac{\partial^2 C_A}{\partial \theta^2}+\frac{\partial^2 C_A}{\partial z^2}\right] \tag{4.126a}$$

For a stationary and symmetric current we have

$$\left[\frac{1}{r}\frac{\partial}{\partial r}\left(r\frac{\partial C_A}{\partial r}\right)\right]=0 \tag{4.126b}$$

3. In spherical coordinates:

$$\frac{\partial C_A}{\partial t}+V_r\frac{\partial C_A}{\partial r}+\frac{V_\theta}{r}\frac{\partial C_A}{\partial r}+\frac{V_\phi}{r\sin\phi}\frac{\partial C_A}{\partial \phi}=$$
$$=D_A\left[\frac{1}{r^2}\frac{\partial}{\partial r}\left(r^2\frac{\partial C_A}{\partial r}\right)+\frac{1}{r^2\sin\theta}\frac{\partial}{\partial \theta}\left(\sin\theta\frac{\partial C_A}{\partial \theta}\right)+\frac{1}{r^2\sin^2\theta}\frac{\partial^2 C_A}{\partial \phi^2}\right] \tag{4.127}$$

For motionless liquid we have

$$\frac{\partial C_A}{\partial t}=D_A\left[\frac{1}{r^2}\frac{\partial}{\partial r}\left(r^2\frac{\partial C_A}{\partial r}\right)+\frac{1}{r^2\sin\theta}\frac{\partial}{\partial \theta}\left(\sin\theta\frac{\partial C_A}{\partial \theta}\right)+\frac{1}{r^2\sin^2\theta}\frac{\partial^2 C_A}{\partial \varphi^2}\right] \tag{4.127a}$$

For a symmetric current we will receive

$$\frac{\partial C_A}{\partial t}=D_A\left[\frac{1}{r^2}\frac{\partial}{\partial r}\left(r^2\frac{\partial C_A}{\partial r}\right)\right] \tag{4.127b}$$

For a stationary current we have

$$\left[\frac{1}{r^2}\frac{\partial}{\partial r}\left(r^2\frac{\partial C_A}{\partial r}\right)\right]=0 \tag{4.127c}$$

We will make assessment of the importance of members of the equation of convective diffusion (4.125) for a stationary case, having entered dimensionless variables and $U_x={}^{V_x}\!/_{V_0}$, $U_y={}^{V_y}\!/_{V_0}$, $U_z={}^{V_z}\!/_{V_0}$, $\bar{x}={}^{x}\!/_{L}$, $\bar{y}={}^{y}\!/_{L}$, $\bar{z}={}^{z}\!/_{L}$, $\bar{c}={}^{C}\!/_{C_0}$

Here, V_0 is the average velocity of a stream, L is the characteristic size, and C_0 is the concentration of substance in a stream. Considering these sizes, equation (4.125) will register as

$$U_x \frac{\partial \overline{c}}{\partial \overline{x}} + U_y \frac{\partial \overline{c}}{\partial \overline{y}} + U_z \frac{\partial \overline{c}}{\partial \overline{z}} = \frac{1}{\text{Pe}_D} \left(\frac{\partial^2 \overline{c}}{\partial \overline{x}^2} + \frac{\partial^2 \overline{c}}{\partial \overline{y}^2} + \frac{\partial^2 \overline{c}}{\partial \overline{z}^2} \right) \tag{4.128}$$

Here, $\text{Pe}_D = \frac{V_0 L}{D}$ is the Peclet's number, expressing the relation of convective transfer to diffusive. Thus, if $\text{Pe}_D >> 1$, then convective transfer prevails over molecular transfer of weight, and the last can be neglected, that is, the equation of convective diffusion will be transformed to a look

$$V_x \frac{\partial C}{\partial x} + V_y \frac{\partial C}{\partial y} + V_z \frac{\partial C}{\partial z} = 0 \tag{4.129}$$

The solution of this equation will be presented in the form

$$C(x, y, z) = \text{const} \tag{4.130}$$

This result means that in a stream kernel where convective transfer of weight dominates, concentration of substance is constant. If $\text{Pe} << 1$, then molecular transfer of weight prevails over convective transfer, and the last can be neglected and the equation of convective diffusion will be presented in the form (4.125b). Peclet's number Pe in processes of a mass exchange plays the same role, as the number in transfer of an impulse. The relation of numbers Pe_D and Re also defines the so-called diffusive criterion of Prandtl or Schmidt's number Sc, that is,

$$\frac{\text{Pe}}{\text{Re}} = \frac{V_0 L / D}{V_0 L / \nu} = \frac{\nu}{D} = \text{Sc} \tag{4.131}$$

For gases, the diffusive number of Schmidt has a unit order where as for liquid $\text{Sc} >> 1$, in particular for water $\text{Sc} \approx 10^3$, and for viscous liquids can exceed value $\text{Sc} \geq 10^6$. This circumstance essentially affects distinction of numbers for liquids and gases at identical velocities of the movement of the environment as Pe=ScRe. From this relationship, it follows that even with small numbers of Re, the number of Pe for liquids remains significantly greater than 1 due to large values of Sc, and mass transfer is carried out using a convective mechanism. For gases, the condition Pe >>1 will be observed only within Re >>1. However, the condition of interphases exchange is the existence of the moving force of the process, that is, concentration differences in the core of stream, in connection with which should be a zone with a big concentration gradient at or near the interface boundary. This area, where to come essential change of concentration of substance from superficial value to value in a stream kernel, and carries the name of a diffusion boundary layer. The main role in the course of a mass exchange is played by thickness of the diffusion boundary layer depending on thickness of a dynamic layer.

In model of a "motionless" film, it is accepted that the limit of the section of phases has a thin layer of liquid (gas) in which all gradients of concentration is concentrated, and transfer through this layer happens only owing to molecular diffusion. For assessment of thickness of a diffusion boundary layer, we will use the equation of laminar flow of a flat surface taking into account a mass transfer of some substance. For this purpose, we will use equation (4.119) for a stationary case, having put that the thickness of a dynamic boundary layer is a δ, and the thickness diffusion boundary layer δ_D. For some characteristic length $x = l$, we will estimate the derivatives entering this equation as

$$\frac{\partial C}{\partial x} \approx \frac{\Delta C}{l}, \frac{\partial C}{\partial y} \approx \frac{\Delta C}{\delta_D}, \frac{\partial^2 C}{\partial x^2} \approx \frac{\Delta C}{l^2}, \frac{\partial^2 C}{\partial y^2} \approx \frac{\Delta C}{\delta_D^2} \tag{4.132}$$

Having put that $\delta_D << l$, from these expressions we have

$$\frac{\partial^2 C}{\partial y^2} >> \frac{\partial^2 C}{\partial x^2}, \frac{\partial C}{\partial y} >> \frac{\partial C}{\partial x} \tag{4.133}$$

Under these conditions, the equation of convective diffusion will register as

$$V_y \frac{\partial C}{\partial y} = D \frac{\partial^2 C}{\partial y^2} \tag{4.134}$$

As a result of assessment of members of this equation, we have

$$\delta_D \approx \left(\frac{D}{\nu}\right)^{1/3} \delta \ = \frac{\delta_0}{\mathrm{Sc}^{1/3}} \tag{4.135}$$

For a turbulent flow thickness of a diffusion, boundary layer is equal

$$\delta_D \approx \frac{\delta_0}{\mathrm{Sc}^{1/4}} \tag{4.136}$$

It follows from these expressions that for gases as $\mathrm{Sc} \approx 1$, it is possible to write $\delta \ = \delta_D$. For liquids $\mathrm{Sc} \approx 10^3$, the thickness of a diffusive layer is $\delta_D \approx 0{,}1\delta$. Having put that for a flat surface thickness of a dynamic boundary layer is defined from equation (1.19), for thickness of a diffusion boundary layer in case of a gas stream we will receive $\delta_D \approx 5.2\sqrt{\frac{\nu_c x}{V_0}}\left(\frac{D}{\nu_c}\right)^{1/3} = 5.2 D^{1/3} \nu_c{}^{1/6} \sqrt{\frac{x}{V_0}}$.

As it appears from this equation, the thickness of a diffusion boundary layer with an increase in velocity of a stream decreases, it is proportional $V_0{}^{-1/2}$.

Having put that $\mathrm{Re} = {}^{V_0 x}\!/_{\nu_c}$ this expression can be rewritten in a look $\delta_D \approx \frac{5.2x}{\mathrm{Re}^{1/2}\,\mathrm{Sc}^{1/3}}$.

Development of the theory of a diffusion boundary layer and experimental data have shown that thickness of this layer depends on hydrodynamic conditions. The amount of the substance postponed through a diffusion boundary layer and coefficient of a mass transfer will be defined as

$$J = \frac{D C_0}{\delta_D}, \beta = \frac{D}{\delta_D} \tag{4.137}$$

and thickness is selected so as to receive experimentally observed values of a stream of substance and coefficient of a mass transfer.

4.7.1 Boundary Conditions for Mass Transfer Equations

As well as in problems of heat transfer, at the solution of problems of a mass transfer it is necessary to set regional conditions for concentration similar to temperature. Despite the analogy of the equations and similarity of the phenomena of heat transfer and a mass transfer, boundary conditions for these two tasks can differ somewhat, due to the different values of the coefficients of thermal conductivity and diffusion. First of all, it belongs to processes of a mass transfer with phase transformations and chemical reactions in volume and on a surface, to a mass transfer in porous bodies, the complicated capillary phenomena, Knudsen diffusion, etc. The solution of the equation of convective diffusion (4.110), except entry and boundary conditions for concentration, demands a task of distribution of the field of velocities. The last requirements in practice are seldom fulfilled, and therefore strict solutions of this equation turn out only for separate simple cases. Now we will consider the main types of regional conditions that have to participate at the solution of problems of transfer of weight in the boundary layer. Since the differential equations of transfer of mass are second-order equations, then for their decision it is necessary to set one initial condition and at least two boundary conditions. The initial conditions for the mass-transfer equations are

$$t = 0, C(X,t) = C(X,0) \tag{4.138}$$

where X is a set of the coordinates characterizing the size of a body and space. Boundary conditions are generally connected with features of transfer of weight in a hydrodynamic stream (laminar, turbulent), with geometry of a surface and the existence of phase and chemical transformations. Boundary conditions for the equations of transfer of weight are subdivided into the following childbirth:

1. The condition of the first sort defines a task of values of concentration for the surfaces of a body and in a stream kernel

$$X = B, C(X,t) = C(B,t) = C_0 \tag{4.139}$$

$$X = L, C(X,t) = C(L,t) = C_L \tag{4.139a}$$

   where B is the coordinate (size) of a surface of a body whose value in some decisions accept equal to zero, and L is the distance far from the surface of a body, value which in some decisions accept equal to infinity. It should be noted that at the solution of the majority of tasks, the concentration of transferable substance in a kernel of a stream is known, though on a surface it isn't always set. Superficial concentration is set in case process of a mass exchange (evaporation, dissolution) and proceeds directly on the surface of a body.
2. The second boundary condition defines a task from the surface of a body of a diffusive stream of substance and registers in a look

$$X = B, D\frac{\partial C}{\partial X} = Q \tag{4.139b}$$

   where Q is the amount of transferable substance, which is taken away at the expense of the diffusion mechanism from the interphase border to the boundary layer. Size Q can be also the velocity of phase and chemical transformation. If on a surface the chemical reaction with a velocity proceeds $r = kC^n$, then this condition can be written down in a look

$$X = B, D\frac{\partial C}{\partial X} = kC^n \tag{4.139c}$$

   If at the chemical interaction on limit of the section of phases the target component is reagent, and the velocity of the chemical reaction considerably surpasses the velocity of a supply of a target component, that is, $D/(kBC_0^{n-1}) << 1$, then as a first approximation it is possible to consider that the concentration on the limit of the section of phases is equal to a zero

$$X = B, C = 0 \tag{4.139d}$$

Otherwise, this boundary $D/(kBC_0^{n-1}) >> 1$ condition will be transformed to a look

$$X = B, \partial C/\partial X = 0 \tag{4.139e}$$

3. The boundary condition of the third sort proceeds from equality on the surface of a body of diffusive and convective streams and is expressed in a look

$$X = B, -D\frac{\partial C}{\partial X} = \beta(C - C_L) \tag{139.f}$$

The ratio (4.139f) reflects the fact that the substance stream on the interphase border is continuous and is defined only by molecular diffusion. From this ratio $\mathrm{Sh} = \beta L/D$ – the Sherwood number is easy to receive a dimensionless parameter.

4. At contact of two liquids or two phases on the interphase border, conditions of equality of streams of substances on both sides will be met

$$D_1\frac{\partial C}{\partial x_1} = D_2\frac{\partial C}{\partial x_2} \tag{4.139g}$$

If in the interphase border any phase (dissolution, evaporation, chemical reaction) occurs, then these conditions will register in a look

$$D_1\frac{\partial C}{\partial x_1} = D_2\frac{\partial C}{\partial x_2} + \omega(C) \tag{4.139h}$$

where $\omega(C)$ is the velocity of phase and of chemical transformations.

Except these conditions, at contact of two liquids with various superficial tension on the limit of the section, there can be additional conditions of a current of liquids concerning each other (Marangoni's effect) caused by uneven distribution of a superficial tension or concentration of a substance. It causes distribution of concentration of a substance along the line of contact and changes the existing boundary conditions. Here it should be noted that the relative movement of two liquids promotes energy dissipation, though the definition of volume dissipation of energy and dissipation of energy in an interphase layer theoretically is represented a difficult task. Existence of dissipation of energy can cause the thermocapillary movements of liquid that also promote distribution of concentration of substance and a superficial tension to interfaces.

In spite of the fact that the previously stated boundary conditions are characteristic of many problems of transfer of weight, in actual practice they can be complemented with boundary conditions of various character. Now, for the solution of problems of a mass exchange with the specified regional conditions, there are several theoretical and practical methods: (1) exact methods in which the mathematical methods allowing to receive required decisions in an analytical form without simplification of an initial task are used. In the theory of a mass exchange, exact methods are used for the solution of some nonlinear tasks described by the equations in private derivatives and for the bodies having simple and final geometry; (2) the asymptotic methods based on the solution of the approximate equations boundary dynamic, thermal and diffusive layers by means of asymptotic decomposition and methods of merging; (3) the numerical methods having big universality and based on replacement of any differential equations with the corresponding confidants of course—differential approximations with use of various numerical algorithms. Despite obvious usefulness and community of numerical methods, they don't allow us to receive approximate analytical decisions problems of a mass exchange; and (4) the approximate methods based on informal understanding of physical essence of the phenomenon of a mass exchange, though these methods quite often allow us to receive the required dependences, which are successfully competing with results of the other relevant decisions. Along with these methods, it is necessary to stop on the principles of the theory of similarity, thanks to which using a set of dimensionless criteria of similarity, it is possible to receive some regularities in processes of transfer of heat and mass transfer.

5 Convective Mass Transfer

Convective mass transfer is the most common form of transport, but also a complex form of transport in connection with which there are great difficulties in determining the mass-transfer coefficients β. When considering the flow as a whole, it is necessary to distinguish it in two main directions: longitudinal, coinciding with the flow direction and transverse, and perpendicular to the flow direction, the transfer components. The main convective transfer is carried out in the longitudinal direction, and the diffusion transfer takes place in the transverse direction. However, in the processes of mass transfer, matter passes from phase to phase in a direction transverse to the motion of the phases, that is, cross-section diffusion transport is important. In this case, turbulent mixing equalizes the concentrations along the cross section, and they speak of a much larger quantity than the molecular diffusion of turbulent diffusion. But the particle cannot reach the wall, since a thin laminar boundary layer is formed near it, where the transfer is carried out predominantly by the molecular diffusion. Therefore, within the boundaries of a viscous boundary layer, the equation includes the molecular diffusion coefficient D. The mass-transfer coefficient in complex processes of film flow, drops, bubbles, jets, etc. with allowance for the Marangoni effect is determined using series of simplified theoretical models of matter transfer.

5.1 MODELS AND MECHANISMS OF MASS TRANSFER

To date, models of a "fixed film" are known-film models or the Lewis and Whitman model, the turbulent boundary layer, the permeability or penetration model, the Higby model, and the infinite surface renewal model. In the fixed-film model, it is assumed that a thin layer of liquid or gas exists at the interface between phases, in which the entire concentration gradient of the substance is concentrated, and the transfer through this layer is accomplished only by molecular diffusion. The basic positions of the fixed film are as follows: (1) at the boundary of the two phases (gas–liquid, vapor–liquid, liquid–liquid), boundary films (gas film, liquid film, etc.) form on the side of each phase, the transition of matter from one phase to another; (2) at the interface of two phases, respectively, at the interface between the films, the conditions of mobile equilibrium are created, that is, stationary conditions of mass transfer are achieved; and (3) the diffusion flux of the component within each phase is proportional to the difference in concentration or the difference in the partial pressure of the component in the bulk and near the boundary. With this method, the thickness of the boundary diffusion layer cannot be predicted. However, the model makes it possible to reliably calculate the mass-transfer rate with a simultaneous chemical reaction.

The theories of the boundary diffusion layer and experimental studies have shown that the thickness of this layer depends on the hydrodynamic conditions and is related to the thickness of the dynamic layer. According to this model, the concentration of matter, constant in the core of the flow, in the turbulent sublayer gradually decreases as the boundary layer (i.e., in the buffer sublayer) approaches, in which the molecular and turbulent viscosity forces are commensurable, that is, the model of a turbulent boundary layer describes the transfer of a mass between a fixed interface and a turbulent flow of a liquid. According to the known value of the flow and the difference between the average concentration and the concentration of matter near the wall, the mass-transfer coefficient is determined, neglecting convective transfer. The boundary layer model best describes mass-transfer processes on solid fixed surfaces. The models of non-stationary mass transfer are more applicable to the interface between gas and liquid. In contrast to conventional films, the diffusion boundary layer takes into account: (1) the motion of the liquid and the convective transport of matter; (2) molecular and convective diffusion in the transverse and tangential directions; and (3) absence of a distinct layer boundary.

In the permeation model, the process of unsteady diffusion is considered for the time interval of contact of two phases characteristic for a given system. For turbulent flows, the theory of permeation is supplemented by the theory of surface renewal, where it is assumed that turbulent pulsations constantly bring fresh liquid to the phase interface and wash off portions of the liquid that have already reacted with the gas (vapor). In the surface renewal model, each element of the liquid surface interacts with the gas (steam) for some time τ (the time of contact or renewal), after which the element is updated, that is, represent a set of updated surface elements, each of which is characterized by its lifetime ranging from zero to infinity. In this work, a complete analysis of the models and mass-transfer mechanism has been made, and it is noted that the most effective of these models is the boundary layer model. In many cases of mass transfer, the most significant drop in the concentration gradient along the normal to the interface is concentrated in a thin boundary layer adjacent to this boundary, at which phase equilibrium exists. Taking this fact into account, in order to calculate the processes of matter transport within this phase from the interface into the interior of the flow or from the core of the flow to the interfacial surface, it is convenient to introduce a mass-transfer coefficient in the form of the ratio of the density of the diffusion flux to the characteristic difference in concentration. The quantitative dependences, both theoretical and experimental, for the mass-transfer rate in a given phase are written in generalized variables that have the meaning of criteria. Analysis of the equations of convective mass transfer make it possible to obtain a diffusion Fourier number (where the characteristic linear dimension), which characterizes the change in the flux of the diffusing mass in time and necessary for the characterization of nonstationary processes, as well as the Peclet diffusion number, which is transformed to the form where the Schmidt number. To characterize the mass transfer, the Sherwood number is introduced, equal to the mass-transfer coefficient. The obtained similarity criteria make it possible to find the equation for the similarity of convective diffusion: $f\left(\mathrm{Sh},\mathrm{Re},\mathrm{Sc},\mathrm{Fo},\mathrm{Gr}\right)=0$, where Gr is the Grashof number that determines free convection of the mass, in particular, in vertical channels. Of all the similarity numbers included in this equation, only the number Sh does not consist entirely of uniqueness conditions, and cause it is a definable criterion. On this basis, this equation can be written in the form $\mathrm{Sh}=f\left(\mathrm{Re},\mathrm{Sc},\mathrm{Fo},\mathrm{Gr}\right)=f\left(\mathrm{Pe},\mathrm{Fo},\mathrm{Gr}\right)$. When considering stationary mass-transfer processes, the number Fo is excluded from this equation, and the given equation is represented as $\mathrm{Sh}=f\left(\mathrm{Re},\mathrm{Sc},\mathrm{Gr}\right)=f\left(\mathrm{Pe},\ \mathrm{Gr}\right)$. If the flow of the phase is forced, natural convection can be neglected, then the number u drops out of the equation Gr and $\mathrm{Sh}=f\left(\mathrm{Re},\mathrm{Sc}\right)=f\left(\mathrm{Pe}\right)$.

It should be noted that the preceding models can be used for the calculation of processes only in special cases, since, due to the extreme complexity of turbulent disperse flows, it is practically impossible to determine in them the phase contact surface, the distribution of concentrations in phases, the thickness of the interphase film, and other parameters necessary for calculation.

5.2 ANALYSIS OF THE EQUATIONS OF CONVECTIVE MASS TRANSFER

Taking into account the foregoing, let us analyze the equation of convective diffusion within the boundary layer, which for a planar surface is represented in the form

$$\frac{\partial C}{\partial t}+V_x\frac{\partial C}{\partial x}+V_y\frac{\partial C}{\partial y\partial}+V_z\frac{\partial V_z}{\partial z}=D\left(\frac{\partial^2 C}{\partial x^2}+\frac{\partial^2 C}{\partial y^2}+\frac{\partial^2 C}{\partial z^2}\right)+w(C) \tag{5.1}$$

Introducing the dimensionless quantities

$$\bar{V}_x=\frac{V_x}{V_\infty},\quad \bar{V}_y=\frac{V_y}{V_\infty},\quad \bar{V}_z=\frac{V_z}{V_\infty},\quad \bar{C}=\frac{C}{C_\infty},\quad \bar{x}=\frac{x}{L},\quad \bar{y}=\frac{y}{L},\quad \bar{z}=\frac{z}{L},\quad \bar{t}=\frac{t}{t_\infty} \tag{5.2}$$

we transform equation (5.1) to the dimensionless form

$$\frac{\partial \overline{C}}{\partial \overline{t}}+\overline{V}_x\frac{\partial \overline{C}}{\partial \overline{x}}+\overline{V}_y\frac{\partial \overline{C}}{\partial \overline{y}}+\overline{V}_z\frac{\partial \overline{C}}{\partial \overline{z}}=\frac{1}{Pe}\left(\frac{\partial^2 \overline{C}}{\partial \overline{x}^2}+\frac{\partial^2 \overline{C}}{\partial \overline{y}^2}+\frac{\partial^2 \overline{C}}{\partial \overline{z}^2}\right)+w\left(\overline{C}\right)\frac{L}{V_\infty C_\infty} \quad (5.3)$$

$$1 \quad 1 \quad 1/1 \quad \delta_* \quad 1/\delta_*^2 \quad 0 \quad \delta_*^2 \quad 1 \quad 1/\delta_*^2 \quad 0$$

Here, $\mathrm{Pe} = V_\infty l/D$, $\mathrm{Sc} = \nu/D$ are the Peclet and Schmidt numbers. Introducing the dimensionless thickness of the dynamic boundary layer $\delta_* = \frac{\delta V_\infty t_\infty}{l^2}$, we estimate the terms of (5.3). Since Re is of order $1/\delta_*^2$ and Pe = ReSc has the order $1/\delta_{D*}^2$,

$$\frac{\delta_{D*}}{\delta_*}=\frac{\delta_D}{\delta}\sim\frac{1}{\mathrm{Sc}^{1/2}} \quad (5.4)$$

Thus, if the number $\mathrm{Sc} \geq 1$, which is typical for liquids, then the thickness of the diffusion layer is less than the thickness of the dynamic layer $\delta_D \leq \delta$. As a result of estimating the terms of equation (5.3), we obtain that $\partial^2\overline{C}/\partial\overline{x}^2 << \partial^2\overline{C}/\partial\overline{y}^2$, $\partial^2\overline{C}/\partial\overline{z}^2 << \partial^2\overline{C}/\partial\overline{y}^2$, as a result of which the equation of mass transfer on a plane surface within the boundary layer simplifies to the form

$$\frac{\partial C}{\partial t}+V_x\frac{\partial C}{\partial x}+V_y\frac{\partial C}{\partial y}=\frac{\partial^2 C}{\partial y^2} \quad (5.5)$$

Similarly, we can estimate the terms for the equation for cylindrical and spherical surfaces. Obviously, to determine the mass-transfer coefficient, it is necessary to solve the transport equations in analytical form, which is possible under certain assumptions only for laminar flow around the surface. Obviously, the boundary layer creates the main resistance to the process of mass transfer of matter. As we approach the laminar regime, the boundary layer grows strongly, as if filling the entire cross section of the flow. Under these conditions, the convective transfer proceeds in a direction parallel to the motion of the flow. The mass transfer to the interface is mainly determined by molecular diffusion.

Let us consider the equation of longitudinal non-stationary convective diffusion in the case of flow past a plane surface of length with a chemical reaction of the first order and an arbitrary distribution of the initial concentration

$$\frac{\partial C}{\partial t}+V_x\frac{\partial C}{\partial x}=D\frac{\partial^2 C}{\partial x^2}-kC$$

$$C(x,t)\big|_{x=0}=C_0;\ C(x,t)\big|_{x=L}=0;\ C(x,t)\big|_{t=0}=\varphi(x) \quad (5.5a)$$

By introducing new variables

$$C=U\exp(\mu x+\lambda t),\quad \mu=\frac{V_x}{D},\quad \lambda=-k-\frac{V_x^2}{4D^2} \quad (5.6)$$

we transform equation (5.5a) to the form

$$\frac{\partial U}{\partial t}=D\frac{\partial^2 U}{\partial x^2} \quad (5.7)$$

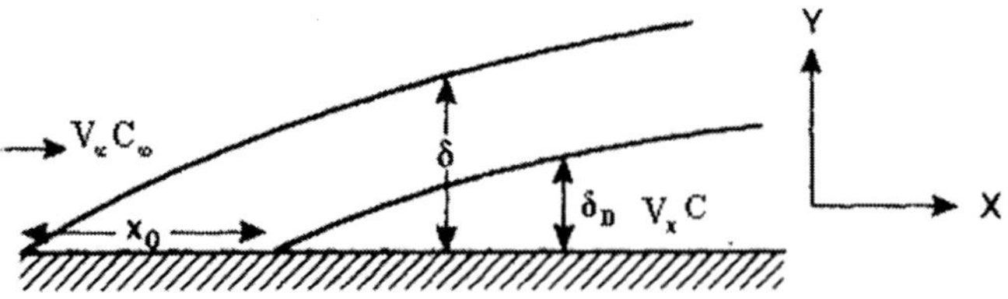

FIGURE 5.1 Structure of the diffusion boundary layer.

The solution of this equation with a linear initial concentration distribution

$$C(x,t)\big|_{t=0}=\varphi(x)=\mu x \tag{5.8}$$

will be presented in the form

$$U(x,t)=\frac{2L\mu}{\pi}\sum_{n=1}^{\infty}\frac{(-1)^{n+1}}{n}\exp\left[-\left(\frac{\sqrt{D}n\pi}{L}\right)^2 t\right]\sin\frac{n\pi}{L}x \tag{5.9}$$

Then the solution of equation (5.5a) can be represented in the form

$$C(x,t)=\frac{2\text{Pe}}{\pi}\exp\left[\text{Pe}\frac{x}{L}-\left(k+\frac{V_x^2}{4D^2}\right)t\right]\sum_{n=1}^{\infty}\frac{(-1)^{n+1}}{n}\exp\left[-n^2\pi^2\text{Fo}\right]\sin\frac{n\pi}{L}x \tag{5.10}$$

Here, $\text{Pe}=\frac{V_xL}{D}$ is the Peclet's number, and $Fo=\frac{Dt}{L^2}$ is the Fourier number.

In most cases, the longitudinal mass transfer compared to the transverse transport is very small, so that the mass-transfer solutions discussed below are related to transverse diffusion.

5.3 CONVECTIVE MASS TRANSFER IN THE DIFFUSION BOUNDARY LAYER

Let us consider the steady-state transport of a mass in the flow of a liquid on a plane surface (5.5), described by the following equation:

$$V_x\frac{\partial C}{\partial x}+V_y\frac{\partial C}{\partial y}=D\frac{\partial^2 C}{\partial y^2} \tag{5.11}$$

The derivation of this equation takes into account the following assumptions, obtained as a result of estimating all the terms of the complete equation $\partial^2 C/\partial x^2 << \partial^2 C/\partial y^2$, $\partial C/\partial t=0$, $\partial C/\partial z=0$, $V_z=0$. In addition, when a fluid flows past a plane surface with a liquid, its concentration varies from the value on the surface to a concentration far from its surface, C_∞, that is, outside the diffusion boundary layer. Consequently, such a change in concentration $(C_\infty - C_w)$ occurs in a small layer characterizing the thickness of the diffusion boundary layer (Figure 5.1).

5.3.1 Laminar Flow Around a Flat Surface

Let us consider the simplest problem of stationary convective mass transfer of a substance under laminar flow around an infinite flat plate (Figure 5.1). The equations of convective transfer, starting from (5.11), can be written in the form

$$V_x\frac{\partial C}{\partial x}=D\frac{\partial^2 C}{\partial y^2} \tag{5.12}$$

with boundary conditions

$$x = 0,\ C_A = C_{A0};\quad y = 0,\ C_A = C_{A1};\ y \to \infty,\ C_A = C_{A0} \tag{5.12a}$$

We introduce the dimensionless function and the variable

$$f = \frac{C_A - C_{A0}}{C_{A1} - C_{A0}},\quad \eta = \frac{y}{2\sqrt{D_A \,x/V_x}} \tag{5.13}$$

Then, taking into account the derivatives

$$\frac{\partial f}{\partial y} = \frac{\partial f}{\partial \eta}\frac{\partial \eta}{\partial y},\ \frac{\partial^2 f}{\partial y^2} = \frac{\partial}{\partial y}\left(\frac{\partial f}{\partial y}\right) = \frac{\partial^2 f}{\partial \eta^2}\left(\frac{\partial \eta}{\partial y}\right)^2,\ \frac{\partial f}{\partial \eta}\frac{\partial \eta}{\partial\left(x/V_x\right)} = D_A \frac{\partial^2 f}{\partial \eta^2}\left(\frac{\partial \eta}{\partial x}\right)^2,$$

$$\frac{\partial \eta}{\partial\left(x/V_x\right)} = -\frac{y}{4\sqrt{D_A \,x/V_x}},\ \frac{\partial \eta}{\partial x} = \frac{1}{2\sqrt{D_A x/V_x}} = \frac{\eta}{y},\ \frac{\partial f}{\partial \eta}\left(-\frac{\eta}{2x/V_x}\right) = D_A \frac{\partial^2 f}{\partial \eta^2}\frac{\eta^2}{x^2} \tag{5.14}$$

equation (5.12) can be rewritten in the form

$$\frac{\partial^2 f}{\partial \eta^2} + 2\eta \frac{\partial f}{\partial \eta} = 0$$
$$\eta \to \infty,\ f = 0,\ \eta = 0,\ f = 1 \tag{5.15}$$

Introducing the new variable $\psi = {}^{\partial f}\!/\!{}_{\partial \eta}$, we transform (5.15) to an equation of the form

$$\frac{d\psi}{d\eta} = -2\psi\eta \tag{5.16}$$

the solution of which is presented in the form

$$\psi = A_1 \exp\left(-\eta^2\right) \tag{5.17}$$

or, passing to the old variables $\psi = {}^{\partial f}\!/\!{}_{\partial \eta}$, we get

$$f = A\int_0^{\eta} \exp\left(-\eta^2\right) d\eta + A_2 \tag{5.18}$$

Taking boundary conditions into account, we finally obtain

$$\frac{C_A - C_{A0}}{C_{A1} - C_{A0}} = 1 - \mathrm{erf}\left(\frac{y}{2\sqrt{D_A x/V_x}}\right) \tag{5.19}$$

The flow of mass per unit time is defined as

$$j = -D_A \frac{\partial C_A}{\partial y} = \sqrt{\frac{D_A V_x}{\pi x}} \exp\left(-\frac{V_x y^2}{4 D_A x}\right)\left(C_{A1} - C_{A0}\right) \tag{5.20}$$

On the surface of the plate we have

$$j\big|_{y=0} = \sqrt{\frac{D_A V_x}{\pi x}}\left(C_{A1} - C_{A0}\right) \tag{5.21}$$

The mass-transfer coefficient is defined in the form

$$\beta_L = \left(\frac{D_A V_x}{\pi x}\right)^{1/2} \tag{5.22}$$

As follows from (5.22), if we assume that there is a contact time $\tau_k = x/V_x$, then it can be asserted that for the penetration model of mass transfer, the mass-transfer coefficient is proportional to the square root of the diffusion coefficient, that is, $\beta_L \sim D_A^{1/2}$.

5.3.2 The Integral Equation of the Diffusion Boundary Layer

Integral equations are derived from the expression (5.11) by integrating over y in the range from 0 to δ_D, taking into account the following relations

$$\begin{aligned}
&\frac{\partial}{\partial x}\int_0^{\delta_D} V_x C dy = \int_0^{\delta_D} C\frac{\partial V_x}{\partial x}dy + \int_0^{\delta_D} V_x\frac{\partial C}{\partial x}dy \\
&\frac{\partial}{\partial y}\int_0^{\delta_D} V_y C dy = \int_0^{\delta_D} V_y\frac{\partial C}{\partial y}dy - \int_0^{\delta_D} C\frac{\partial V_y}{\partial x}dy
\end{aligned} \tag{5.23}$$

Taking into account the boundary conditions

$$y = 0,\ \ V_x = V_y = 0;\ \ y \to \infty,\ \ V_x = V_\infty,\ \left(\frac{\partial C}{\partial y}\right)\Big|_{y\to\infty} \to 0 \tag{5.24}$$

Equation (5.11) is represented in the form

$$\frac{\partial}{\partial x}\int_0^{\delta_D} (C_\infty - C)V_x dy = D\left(\frac{\partial C}{\partial y}\right)\Big|_{y=0} \tag{5.25}$$

The local mass-transfer coefficient, determined from relation

$$\beta_x = \frac{j_w}{(C_\infty - C_w)} \tag{5.26}$$

will be

$$\beta_w = -\frac{D\left(\partial C/\partial y\right)\big|_{y=0}}{C_\infty - C_w} = \frac{1}{C_\infty - C_w}\int_0^{\delta_D} (C_\infty - C)V_x dy \tag{5.27}$$

where β_w is the mass-transfer coefficient, and j_w is the mass flow density per unit time.

We consider the general case when the distribution curves $V_x(y)$ and $C(y)$ do not coincide, while the initial section of the planar surface does not participate in mass transfer. The hydrodynamic boundary layer begins at the anterior edge of the plane, and the diffusion boundary layer at the distance (Figure 5.1). Assuming boundary value boundary conditions

$$y=0,\ V_x=0,\ C=C_w;\ x=0,\ V_x=V_\infty,\ C=C_\infty;\ y\to\infty,\ V_x=V_\infty,\ C=C_\infty \tag{5.28}$$

the concentration profile is assumed to be in the form of a cubic parabola

$$\frac{C-C_\infty}{C_\infty-C_w}=\frac{3}{2}\frac{y}{\delta_D}-\frac{1}{2}\left(\frac{y}{\delta_D}\right)^3 \tag{5.29}$$

An analogous equation can be written for the velocity profile

$$\frac{V_x}{V_\infty}=\frac{3}{2}\frac{y}{\delta}-\frac{1}{2}\left(\frac{y}{\delta}\right)^3 \tag{5.30}$$

where δ is the thickness of the hydrodynamic boundary layer. Taking into account (5.29) and (5.30), we define the integral entering into equation (5.25)

$$\begin{aligned}\int_0^{\delta_D}(C-C_\infty)V_x dy &= C_\infty V_\infty\int_0^{\delta_D}\left[1-\frac{3}{2}\frac{y}{\delta_D}+\left(\frac{y}{\delta_D}\right)^3\right]\left[\frac{3}{2}\frac{y}{\delta}+\left(\frac{y}{\delta}\right)^3\right]dy\\ &= C_\infty V_\infty\delta\left(\frac{3}{20}\Delta^2-\frac{3}{280}\Delta^4\right)\end{aligned} \tag{5.31}$$

where $\Delta = {}^{\delta_D}\!/_{\delta}$. Assuming that $\delta_D<\delta$, the second term in (5.31) can be neglected. Then we have

$$\int_0^{\delta_D}(C-C_\infty)V_x dy \approx \frac{3}{20}C_\infty V_\infty\delta\Delta^2 \tag{5.32}$$

Substituting (5.32) $\left({}^{\partial C}\!/_{\partial y}\right)\big|_{y=0}={}^{3C_\infty}\!/_{2\delta_D}$ and the condition into equation (5.25), we have

$$\frac{\partial}{\partial x}\left(\frac{3}{20}C_\infty V_\infty\delta\,\Delta^2\right)=D\frac{3C_\infty}{2\delta_D}=D\frac{3C_\infty}{2\delta\,\Delta} \tag{5.33}$$

We transform (5.33) to the form

$$\frac{1}{10}V_\infty\left(\delta\,\Delta^3\frac{\partial\delta}{\partial x}+2\delta^2\Delta^2\frac{\partial\Delta}{\partial x}\right)=D \tag{5.34}$$

Setting for the dynamic boundary layer $\delta^2=\frac{280\eta x}{13\rho V_\infty}$, we obtain

$$\delta\frac{\partial\delta}{\partial x}=\frac{140}{13}\frac{\eta}{\rho V_\infty} \tag{5.35}$$

Substituting (5.35) into (5.34), after simple transformations, we obtain

$$\Delta^3+\frac{4x}{3}\frac{\partial\Delta^3}{\partial x}=\frac{13}{14}\mathrm{Sc}^{-1} \tag{5.36}$$

where $\mathrm{Sc}={\eta}/{\rho D}$ is the Schmidt number. Integrating (5.36), we obtain

$$\Delta^3=\frac{13}{14}\mathrm{Sc}^{-1}+B_0x^{-3/4} \tag{5.37}$$

We define the integration constant from the condition: $x=x_0,\ \Delta=0$. Then from expression (5.37) we obtain

$$\Delta^3=\frac{13}{14}\mathrm{Sc}^{-1}\left[1-\left(\frac{x_0}{x}\right)^{3/4}\right]=0.976\mathrm{Sc}^{-1}\left[1-\left(\frac{x_0}{x}\right)^{3/4}\right] \tag{5.38}$$

If $x_0=0$, then we get

$$\Delta^3=0.976\mathrm{Sc}^{-1/3} \tag{5.39}$$

Using the conditions $\left({\partial C}/{\partial y}\right)\big|_{y=0}={3C_\infty}/{2\delta_D}$ in expression

$$\beta C_\infty=D\left({\partial C}/{\partial y}\right)\bigg|_{y=0}=\frac{3}{2}\frac{D}{\delta_D}=\frac{3}{2}\frac{D}{\delta\,\Delta} \tag{5.40}$$

and the equation for the thickness of the hydrodynamic layer $\delta=4.64\sqrt{\frac{\eta_c x}{\rho_c V_\infty}}$, with allowance for (5.38), we find the mass-transfer coefficient in the form

$$\beta_L=\frac{3D}{2}\frac{1}{4.64}\sqrt{\frac{\rho V_\infty}{\eta x}}\frac{\mathrm{Sc}}{0.976\left[1-\left(\frac{x_0}{x}\right)^{3/4}\right]^{1/3}} \tag{5.41}$$

We transform this equation to the form

$$\mathrm{Sh}=0.332\mathrm{Sc}^{1/3}\mathrm{Re}_x^{1/2}\frac{1}{\left[1-\left(\frac{x_0}{x}\right)^{3/4}\right]^{1/3}} \tag{5.42}$$

where $\mathrm{Sh}=\frac{V_\infty\beta_L}{D}$ is the Sherwood number for a plane surface, and $\mathrm{Re}_x=\frac{V_\infty x\rho}{\eta_c}$ is the Reynolds number for a plane surface. If $x_0=0$, then from (5.42) we obtain

$$\mathrm{Sh}=0.332\mathrm{Sc}^{1/3}\mathrm{Re}_x^{1/2} \tag{5.43}$$

The mass-transfer coefficient is defined as

$$\beta_L=\frac{\mathrm{Sh}\,x}{D} \tag{5.43a}$$

Let us determine the average value of the length of the flat surface of the mass-transfer coefficient in the form

$$\bar{\beta}_L = \frac{\int_0^L \beta dx}{\int_0^L dx} = \frac{\int_0^L 0.332\left(\frac{V_\infty x}{\nu}\right)^{1/2} \mathrm{Sc}^{1/3}\frac{D}{x}dx}{\int_0^L dx} = \frac{0.332\left(\frac{V_\infty}{\nu}\right)^{1/2} D\,\mathrm{Sc}^{1/3}\int_0^L \frac{1}{x^{1/2}}dx}{L} \tag{5.43b}$$

$$= \frac{0.332\left(\frac{V_\infty}{\nu}\right)^{1/2} D\,\mathrm{Sc}^{1/3}}{\frac{1}{2}L}\left[x^{1/2}\right]_0^L = \frac{0.332\left(\frac{V_\infty}{\nu}\right)^{1/2} D\,\mathrm{Sc}^{1/3}L^{1/2}}{\frac{1}{2}L}$$

Hence, the mean value of the Sherwood number is determined in the form (Figure 5.2)

$$\overline{\mathrm{Sh}} = \frac{\bar{\beta}_L L}{D} = 0.664\,\mathrm{Re}_x^{1/2}\,\mathrm{Sc}^{1/3} \tag{5.43c}$$

Equations (5.42) and (5.43) make it possible to theoretically determine the mass-transfer coefficient for a flat surface under certain assumptions.

Example 5.1

Air with a temperature $T = 32°\mathrm{C}$ passes through water thickness $L = 1.2$ m with a temperature $T = 20°C$ with and a velocity is $V_\infty = 0.15\,\mathrm{m/s}$. The coefficient of diffusion of air in water is equal $\mathrm{D} = 2.77\times10^{-5}\,\mathrm{m^2/s}$. Determine the coefficient of convective mass transfer.

We will determine average temperature

$$T_c = \frac{32+20}{2} = 26°\mathrm{C}$$

at which to make viscosity of air: $\nu = 1.51\mathrm{x}10^{-5}\,\mathrm{m^2/s}$. Reynolds number is equal

$$\mathrm{Re} = \frac{V_\infty L}{\nu_c} = \frac{0.15\times1.2}{1.51\times10^{-5}} = 1.192\times10^4$$

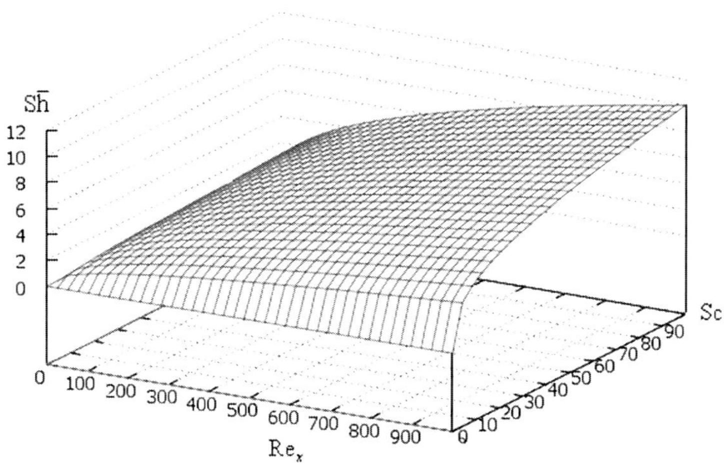

FIGURE 5.2 Change of number $\overline{\mathrm{Sh}}$ depending on numbers Re_x and Sc.

Using formula (5.43c), we will define

$$\overline{\mathrm{Sh}} = 0.664\left(1.192 \times 10^4\right)^{1/2}\left(\frac{1.51}{2.77}\right)^{1/3} = 59.22$$

The coefficient of a convective mass transfer is equal

$$\beta_L = \overline{\mathrm{Sh}}\frac{D}{L} = 59.22\frac{2.77 \times 10^{-5}}{1.2}1.367 \times 10^{-3}\ \mathrm{m/s}.$$

5.4 CONVECTIVE MASS TRANSFER IN TURBULENT FLOW

Convective mass transfer in a turbulent flow is a complex case of mass transfer, and in most cases, this problem is solved numerically or using empirical or semi-empirical relationships. This problem is somewhat simplified if we assume that turbulence is homogeneous or isotropic, as will be described in the next chapter.

Let us consider the case of mass transfer when the direction of convective and turbulent diffusion transport coincide, that is, $\partial j/\partial y = 0$, where $j = \tilde{V}(y)\tilde{C} + \left(D + D_T(y)\partial\tilde{C}/\partial y\right)$. The equation of mass transfer for a plane surface is represented in the form

$$\frac{\partial j}{\partial y} = \frac{\partial}{\partial y}\left[\tilde{V}(y)\tilde{C} + \left(D + D_T(y)\frac{\partial\tilde{C}}{\partial y}\right)\right] = 0,$$
$$y = 0,\ \tilde{V}(y) = 0,\ \tilde{C} = 0;\quad y \to \infty,\ \tilde{C} = \tilde{C}_0,\ \frac{\partial\tilde{C}}{\partial y} = 0 \tag{5.44}$$

where $D_T(y)$ is the coefficient of turbulent diffusion, and $\tilde{C}$ is the mean concentration value. Integrating equation (5.44), we obtain

$$\left(D + D_T(y)\right)\frac{\partial\tilde{C}}{\partial y} = \tilde{V}(y)\tilde{C} + A_0 \tag{5.45}$$

Using the boundary conditions (5.44), we define $A_0 = 0$. Integrating (5.45) repeatedly with the boundary conditions, we have

$$\tilde{C}(y) = \tilde{C}_0\exp\left[-\int\frac{\tilde{V}(y)dy}{D + D(y)}\right] \tag{5.46}$$

In Section 5.1 for the viscous region $0 \le y_+ = \frac{yU_*}{\nu} \le 5$, the coefficient of turbulent diffusion is defined as

$$D_T = D\left(\frac{y_+}{14,5}\right)^3 = D\left(\frac{U_*}{14,5\nu}\right)^3 y^3 = D\gamma y^3 \tag{5.47}$$

Here, $\gamma = \left(U_*/14.5\nu\right)^3$ and $U_* = \tilde{V}\sqrt{\lambda/8}$ are the dynamic flow velocity, and λ is the coefficient of friction resistance. Putting $x = y/L$, $\tilde{V}(y) = \tilde{V}$, we represent the solution of (5.46) in the form

$$\tilde{C}(y)=\tilde{C}_0\exp\left[-\frac{\tilde{V}L}{D}\int_0^x\frac{dx}{1+\gamma_0 x^3}\right]$$
$$=\tilde{C}_0\exp\left[-\mathrm{Pe}\gamma_0^{-1/3}\left(\frac{1}{6}\ln\frac{(1+z)^2}{1-z+z^2}+\frac{1}{\sqrt{3}}\mathrm{arctg}\frac{2z-1}{\sqrt{3}}+\frac{\pi}{6\sqrt{3}}\right)\right] \quad (5.48)$$

Here, $z=\gamma_0^{1/3}x$, $\gamma_0=(U_*L/14{,}5\nu)^3=\left(\mathrm{Re}_D/14{,}5\right)^3$, $\mathrm{Re}_D=U_*L/\nu$, $\mathrm{Pe}=VL/D$ is the Peclet number. Equation (5.47) determines the concentration profile in the viscous sublayer of the turbulent boundary layer (Figure 5.3).

5.4.1 Mass Transfer in an Isotropic Turbulent Flow with Chemical Reaction

Consider mass transfer in an isotropic turbulent flow $\lambda>\lambda_0$ on the surface of a spherical particle, in the volume of which a first-order reaction occurs. For small numbers Pe, the mass-transfer equation is represented in the form

$$\frac{\partial\tilde{C}}{\partial t}=\frac{1}{r^2}\frac{\partial}{\partial r}\left(D_T r^2\frac{\partial\tilde{C}}{\partial r}\right)-k(\tilde{C}-\tilde{C}_P) \quad (5.49)$$

where k is the rate constant of reaction, and $\tilde{C}_p$ is the equilibrium concentration of matter. Using formula (4.39) for turbulent diffusion, we transform (5.49) to the form

$$\frac{\partial\tilde{C}}{\partial t}=\frac{1}{r^2}\frac{\partial}{\partial r}\left(\varepsilon_R^{1/3}r^{10/3}\frac{\partial\tilde{C}}{\partial r}\right)-k\tilde{C} \quad (5.50)$$

with boundary conditions

$$t=0,\ r>R,\ \tilde{C}=\tilde{C}_0;\ t>0,\ r=R,\ \tilde{C}=0;\ r\to\infty,\tilde{C}=\tilde{C}_p \quad (5.51)$$

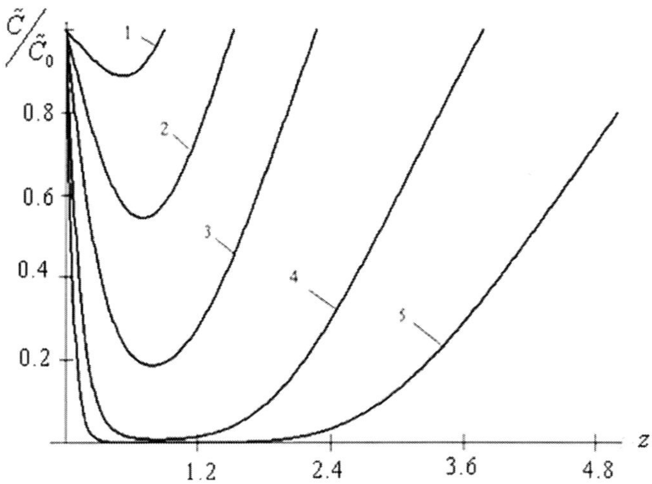

FIGURE 5.3 The concentration profile in a viscous sublayer for different values of the number Pe: (1) 5; (2) 10; (3) 20; (4) 50; (5) 100.

We seek the solution of (5.50) by the method of separation of variables, for which we set

$$\tilde{C}(r,t)=\phi(t)\psi(r) \tag{5.52}$$

Substituting (5.52) into equation (5.50) and dividing the variables, we obtain two equations

$$\frac{\partial\phi}{\partial t}=-\mu^2 k\phi, \quad \phi=A_1\exp\left(-\mu^2 kt\right) \tag{5.53}$$

$$r^{4/3}\frac{d^2\psi}{dr^2}+\frac{10}{3}r^{1/3}\frac{d\psi}{dr}+m_R\psi=0 \tag{5.53a}$$

where $m_R=\frac{(\mu^2-1)k}{\varepsilon_R^{1/3}}$. Introducing a new variable $\xi^2=2r^{2/3}$, equation (5.53a) can be represented in the form

$$\frac{d^2\psi}{d\xi^2}+\frac{b}{\xi}\frac{d\psi}{d\xi}+m_R\psi=0, \quad b=10/3 \tag{5.54}$$

If to accept $b\approx 3$, then the solution of equation (5.54) is represented in the form

$$\psi(\xi)=A_2 J_2\left(\sqrt{m_R}\xi\right) \tag{5.55}$$

Using the boundary condition $r=R,\ C(R,t)=0,\ \psi(\xi)=0$, we obtain from (5.50)

$$J_2\left(\sqrt{m_R}\xi\right)=J_2(q_n)=0 \tag{5.56}$$

The roots of this equation are as follows: $q_1=5.05;\ q_2=8.45;\ q_3=11.8$; and so on. The eigenvalues of the equation are defined as

$$\mu_n^2=1+\frac{q_n^2\varepsilon_R^{1/3}}{3k\,R^{2/3}} \tag{5.57}$$

Finally, the solution (5.50) is represented in the form

$$\tilde{C}(r,t)\approx\tilde{C}_p+\sum_{n=0}^{\infty}A_n J_2\left(\sqrt{m_R}\xi\right)\exp\left(-\mu_n^2 kt\right) \tag{5.58}$$

Here, $A_n=\frac{2(\tilde{C}_0-\tilde{C}_p)\int_o^R J_2\left[q_n\left(r/R\right)^{1/3}\right]rdr}{R^2\left[J_2(q_n)\right]^2}$. Since the series converges rapidly, we can restrict ourselves to the first term and $A_n=\frac{2(\tilde{C}_0-\tilde{C}_p)J_3(q_n)}{q_n J_1^2(q_n)}$. Using the properties of the Bessel functions $J_3(q_n)=-J_1(q_n)$ and $J_1(q_1)=J_1(5.05)=-0.33$, we obtain $A_1\approx-\frac{2(\tilde{C}_0-\tilde{C}_p)}{q_1 J_1(q_1)}=1.2(\tilde{C}_0-\tilde{C}_p)$.

Solution (5.58) is represented in the form

$$\tilde{C}(r,t)\approx\tilde{C}_p+1.2\left(\tilde{C}_0-\tilde{C}_p\right)J_2\left(B_0\frac{k^{1/2}}{\varepsilon_R^{1/6}}r^{1/3}\right)\exp\left(-B_1\frac{\varepsilon_R^{1/6}k^{1/2}}{R^{1/3}}t\right) \tag{5.59}$$

Here, $B_0 = \left(\mu_1^2 - 1\right)\sqrt{3}$, $B_1 = q_1/\sqrt{3}$. The flux of mass to the surface of a spherical particle per unit time will be determined as

$$j = -D_T \frac{\partial \tilde{C}}{\partial r}\Big|_{r=R} \approx 0.132 B_0 \left(\tilde{C}_p - \tilde{C}_0\right) k^{1/2} \varepsilon_R^{1/6} R^{2/3} \exp\left(-B_1 \frac{\varepsilon_R^{1/6} k^{1/2}}{R^{1/3}} t\right) \tag{5.60}$$

Introducing the Fourier number for an isotropic turbulent flow $\mathrm{Fo_T} = \frac{\varepsilon_R^{1/6} k^{1/2} t}{R^{1/3}}$, we obtain

$$j = -D_T \frac{\partial \tilde{C}}{\partial r}\Big|_{r=R} \approx 0.132 B_0 \left(\tilde{C}_p - \tilde{C}_0\right) k^{1/2} \varepsilon_R^{1/6} R^{2/3} \exp\left(-B_1 \mathrm{Fo_T}\right) \tag{5.61}$$

Under the steady-state regime, we have

$$j = 0.132 B_0 \left(\tilde{C}_p - \tilde{C}_0\right) k^{1/2} \varepsilon_R^{1/6} R^{2/3} \tag{5.62}$$

As follows from expressions (5.60) and (5.62), the mass flux to the particle surface depends on the difference in concentration, the specific dissipation of the energy of the isotropic turbulent flow, and the rate constant of the chemical reaction. The mass-transfer coefficient is defined as

$$\beta_L = \frac{j}{C_p - C_0} = 0.132 B_0 k^{1/2} \varepsilon_R^{1/6} R^{2/3} \tag{5.63}$$

As follows from equation (5.63), the mass-transfer coefficient in an isotropic turbulent flow involving a chemical reaction depends on the turbulence parameter and on the rate constant of the chemical reaction.

5.5 DETERMINATION OF THE SHERWOOD NUMBER FOR CONVECTIVE MASS TRANSFER

To describe the processes of convective mass transfer, an important parameter for calculating the mass-transfer coefficient is the determination of the mass flow to the surface or its dimensionless quantity, called the Sherwood number, that is, $\beta_L = \frac{D}{L}\mathrm{Sh}$ (where L is the characteristic particle size). In most cases, $\mathrm{Sh} = f\left(\mathrm{Re}, \mathrm{Sc}\right)$, the dependence is very complicated except for small numbers Re, and its definition is due to the presence of qualitative experimental data and a successful choice of the structure of the function $f\left(\mathrm{Re}\right)$.

For dispersed systems of great interest is the transfer of mass to a moving solid particle or drop, which is considered in the paper [6]. We give some relations connected with the determination of the mass flow to the surface of the particle and the mass-transfer coefficient. In particular, the mass flow of matter per unit surface per unit time to the incident solid particle is defined as

$$j = D \frac{\partial C}{\partial y}\Big|_{y=0} = 0.652 \frac{D C_0}{a} \mathrm{Pe}^{1/3} \frac{\sin\theta}{\left(\theta - \frac{\sin 2\theta}{2}\right)^{1/2}} \tag{5.64}$$

1. On the surface of a moving drop in the form

$$j = D \frac{\partial C}{\partial y}\Big|_{y=0} = \sqrt{\frac{3}{\pi}} \frac{D \Delta C}{a} \mathrm{Pe}^{1/2} \frac{1 + \cos\theta}{\sqrt{2 + \cos\theta}} \tag{5.65}$$

2. The dimensionless mass flow to the surface of the streamlined plate is equal to

$$\mathrm{Sh} = 0.68\,\mathrm{Re}^{1/2}\,\mathrm{Sc}^{1/3}\,{}^{b}\!/\!_{h} \tag{5.66}$$

Here, $\mathrm{Pe} = {}^{V_\infty a}\!/\!_{D}$ is the Peclet number, a is the particle size, and b, h are the width and length of the plate.

3. On the inner surface of the pipe is defined as

$$j = D\frac{\partial C}{\partial y}\bigg|_{y=0} = 0.67\frac{DC_0}{x^{1/3}R^{2/3}}\mathrm{Pe}^{1/3} \tag{5.67}$$

For large values of the Peclet number, which characterizes such systems as liquid–liquid, liquid–gas (these are processes of absorption, liquid-phase extraction), the solution of the mass-transfer equations is possible only by approximate methods, such as the introduction of new variables of *Prandtl-Mises* type [21]. In particular, the mass transfer to the surface of a spherical particle for large Peclet numbers, provided that the tangential velocity component on the drop surface is distributed according to a sinusoidal law, is described by equation [21]

$$\mathrm{Sh} = \left(\frac{8V_\theta}{3\pi}\right)^{1/2}\mathrm{Pe}^{1/2} \tag{5.68}$$

where V_θ is the dimensionless velocity of the liquid at the equator of the drop $\left(\theta = {}^{\pi}\!/\!_{2}\right)$.

At $\mathrm{Re} \le 100$ and $\gamma = {}^{\eta_d}\!/\!_{\eta_c} < 0.3$ the flow around the drop is non-discontinuous, although for a solid sphere a detached flow occurs at $\mathrm{Re} \ge 20$, which is important to take into account when determining the mass flow to the surface.

It is important to note that the above expressions for calculating mass transfer correspond to small numbers Re, that is, laminar flow past particles. An increase in the number Re leads to a deviation from the laminar flow and is accompanied by such phenomena as separation of the boundary layer from the surface of the particle, turbulence of the boundary layer, and many other phenomena, which in most cases cannot be taken into account and described analytically. To this end, various interpolations and extrapolations are used in the literature to obtain empirical and semi-empirical relationships. Thus, theoretical methods and models allow us to determine the mass-transfer coefficients on surfaces of a given simple configuration. An analogy between mass transfer, heat transfer, and momentum transfer is also used to determine the mass-transfer coefficients. No less interesting is the method of model analogies presented in the paper [41].

5.5.1 The Method of Model Equations and Analogies

The essence of this method is the use of non-traditional methods of constructing approximate relationships that make it possible to obtain engineering formulas for calculating the mass-transfer coefficients for a large range of parameters. The essence of the method consists of asymptotic splicing of individual solutions using interpolation and extrapolation to obtain a single formula for a large range of number variations [41]. Assume that the average number is known for a single-term asymptotic for large numbers and a two-term asymptotic for small numbers

$$\mathrm{Sh} = B\mathrm{Pe}^{m}, \quad \mathrm{Pe} \to \infty; \tag{5.69}$$

$$\mathrm{Sh} = 1 + A\mathrm{Pe}, \quad \mathrm{Pe} \to 0 \tag{5.70}$$

In the formula (5.69), we make certain increments, as a result of which we obtain

$$\mathrm{Sh} = B\left(\mathrm{Pe} + \Delta_0\right)^{m} + \Delta_1 \tag{5.71}$$

asymptotic, which for large numbers Pe coincides with (5.69). For small numbers Pe from (5.71) we obtain

$$\mathrm{Sh} = \Delta_1 + B\Delta_0^m + mB\Delta_0^{m-1}\mathrm{Pe} + ... \tag{5.72}$$

The conditions for the correspondence of the approximate asymptotic (5.72) to the exact solution (5.69) allows us to determine constants and from expressions

$$\Delta_1 + B\Delta_0^m = 1, \quad mB\Delta_0^{m-1} = A \tag{5.73}$$

As a result of solving these equations, we obtain

$$\Delta_0 = \left(\frac{mB}{A}\right)^{1/1-m}, \quad \Delta_1 = 1 - B\left(\frac{mB}{A}\right)^{m/1-m} \tag{5.74}$$

Substituting these expressions into (5.71), we obtain

$$\mathrm{Sh} = B\left[\mathrm{Pe} + (mB/A)^{1/1-m}\right]^m + 1 - B\left(\frac{mB}{A}\right)^{m/1-m} \tag{5.75}$$

Let us consider diffusion to a bubble in a flow where mass transfer is described by equations $\mathrm{Sh} = 0.461\mathrm{Pe}^{0.5}$, $\mathrm{Pe} \to \infty$; $\mathrm{Sh} = 1 = 0.5\mathrm{Pe}$, $\mathrm{Pe} \to 0$; those $m = 0.5$, $B = 0.461$, $A = 0.5$. From (5.75) we obtain the general expression for determining the number

$$\mathrm{Sh} = 0.461(\mathrm{Pe} + 0.21)^{0.5} + 0.79 \tag{5.76}$$

To construct interpolation formulas, we can use a more accurate method, which consists of a more complete use of asymptotic information about the mean number Sh. Assume that the asymptotic of a number Pe for small numbers defines the expression

$$\frac{\mathrm{Sh}}{\mathrm{Sh}_0} = 1 + A\mathrm{Pe}^n \tag{5.77}$$

Here, Sh_0 is the average number Sh, corresponding to the mass exchange of a fixed particle with the surrounding medium at $\mathrm{Pe} = 0$. The approximate dependence of the mean number Sh can be sought in the form

$$\frac{\mathrm{Sh}}{\mathrm{Sh}_0} = 1 + A\mathrm{Pe}^n\Phi(\mathrm{Pe}) \tag{5.78}$$

Using the asymptotic conditions, we can obtain the interpolation formula [41]

$$\frac{\mathrm{Sh}}{\mathrm{Sh}_0} = 1 + \frac{A\mathrm{Pe}^n}{1 + \left(\frac{A}{B}\right)\mathrm{Pe}^{n-m}} \tag{5.79}$$

In particular, for mass transfer in a drop and a bubble of moderate viscosity for which the limiting cases $\mathrm{Sh} = 1 + 0.5\mathrm{Pe}$, $\mathrm{Pe} \to 0$; $\mathrm{Sh} = \left(\frac{2\mathrm{Pe}}{3\pi}\right)^{0.5}$, $\mathrm{Pe} \to \infty$ the expression (5.79) is represented in the form

$$\frac{\mathrm{Sh}}{\mathrm{Sh}_0} = 1 + \frac{0.5\mathrm{Pe}}{1 + 1.1\mathrm{Pe}^{0.5}} \tag{5.80}$$

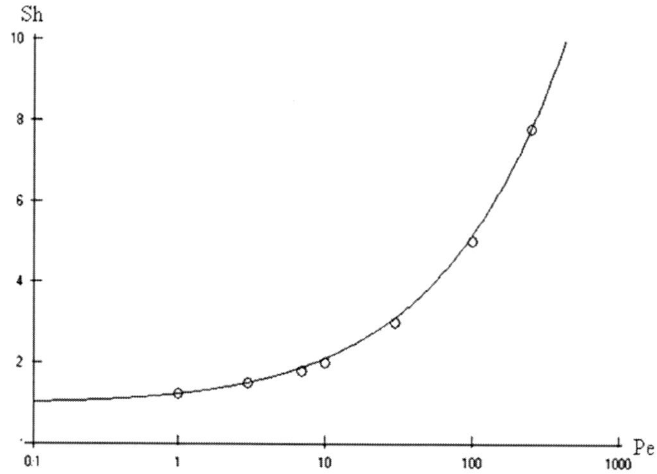

FIGURE 5.4 The dependence of the Sherwood number on the Peclet number (points are experiment).

In Figure 5.4 shows a comparison of equation (5.80) with the experiment.

It should be noted that this method satisfactorily describes the experimental data for single particles, droplets, and bubbles if the derivative $\partial \mathrm{Sh}/\partial \mathrm{Pe}$ is a continuous function in the entire interval of the number change Pe. The calculation of the number Sh by the method of model equations in more complicated cases of flow past a particle surface is given in the paper [41].

5.5.2 The Reynolds Analogy

An analogy between mass transfer, heat transfer, and momentum transfer is used to determine the mass-transfer coefficient, based on the similarity of the equations describing the corresponding flows. Consider the following relations

$$\mathrm{Sh} = \frac{\beta L}{D}, \quad \beta \Delta C = D \nabla C_w \tag{5.81}$$

From these relations one can write

$$D\frac{\nabla C_w}{\Delta C} = D\frac{\mathrm{Sh}}{L} \tag{5.82}$$

For the coefficient of resistance, we can write

$$\xi_T = \frac{\tau_w}{\frac{1}{2}\rho_c V_\infty^2} = \frac{\eta_c \nabla V|_w}{\frac{1}{2}\rho_c V_\infty^2}, \quad \nu_c \frac{\nabla V|_w}{\Delta V} = \nu_c \frac{V_\infty \xi_T}{2\nu_c} \tag{5.83}$$

Using these relations, we finally obtain

$$\frac{D\dfrac{\nabla C_w}{\Delta C}}{\nu_c \dfrac{\nabla V_w}{\Delta V}} = 1 = \frac{D\dfrac{\mathrm{Sh}}{L}}{\nu_c \dfrac{V_\infty \xi_T}{2\nu_c}} = \frac{\mathrm{Sh}}{\dfrac{\xi_T}{2}\mathrm{Re}\,\mathrm{Sc}} \tag{5.84}$$

or we have

$$\mathrm{Sh} = \left(\frac{\xi_T}{2}\right)\mathrm{Re}\,\mathrm{Sc} \tag{5.85}$$

which is an analogy of Reynolds. In this way, the simplest form of such relations is the well-known Reynolds analogy between heat transfer and mass transfer

$$\mathrm{St} = \frac{\mathrm{Nu}}{\mathrm{RePr}} = \frac{\mathrm{Sh}}{\mathrm{ReSc}} = \frac{\xi_T}{2} = \phi(\mathrm{Re}) \tag{5.86}$$

where ξ_T is the coefficient of friction, defined as $\xi_T = {}^{\tau_w}\!/_{\rho V^2}$, τ_w is the tangential stress on the wall. This expression can be written in another form

$$\frac{\lambda}{\rho C_P V} = \frac{\beta}{V} = \frac{\xi_T}{2} = \varphi(\mathrm{Re}) \tag{5.87}$$

When flowing in pipes, the friction coefficient for the region $2.10^3 \le \mathrm{Re} \le 10^5$ is equal $\xi_T \approx {}^{0.3164}\!/_{\mathrm{Re}^{1/4}}$, substituting this in (5.72) for the mass-transfer coefficient we have

$$\mathrm{Sh} = 0.1582\mathrm{Re}^{3/4}\mathrm{Sc} \tag{5.88}$$

It should be noted that the Reynolds analogy is the simplest analogy, and it can be used for laminar flow at $\mathrm{Pr} = \mathrm{Sc} = 1$.

5.5.3 The Chilton–Colborns Analogy

For a laminar boundary layer on a plane surface, one can write

$$\mathrm{Sh} = \mathrm{Nu} = 0.33\mathrm{Re}^{1/2}\,\mathrm{Pr}^{1/3} = 0.33\mathrm{Re}^{1/2}\mathrm{Sc}^{1/3},\ \xi_T = \frac{0.66}{\mathrm{Re}^{1/2}} \tag{5.89}$$

Then we have

$$\mathrm{Sh} = \mathrm{Nu} = \frac{\xi_T}{2}\mathrm{Re}^{1/2}\,\mathrm{Pr}^{1/3} = \frac{\xi_T}{2}\mathrm{Re}^{1/2}\,\mathrm{Sc}^{1/3},\quad \mathrm{St} = \frac{\mathrm{Nu}}{\mathrm{RePr}} = \frac{\mathrm{Sh}}{\mathrm{ReSc}} = \frac{\xi_T}{2\,\mathrm{Pr}^{2/3}} = \frac{\xi_T}{2\mathrm{Sc}^{2/3}} \tag{5.90}$$

The Chilton–Coburn analogy is one of the most useful and definitely the simplest among the set of expressions connecting the transfer of mass, heat, and momentum and is expressed by the equation

$$\mathrm{St_T}\mathrm{Pr}^{2/3} = \mathrm{St_D}\mathrm{Sc}^{2/3} = \frac{\xi_T}{2} = \phi(\mathrm{Re}) \tag{5.91}$$

This expression can also be written in another form

$$\frac{\mathrm{Nu}}{\mathrm{RePr}^{1/3}} = \frac{\mathrm{Sh}}{\mathrm{ReSc}^{1/3}} = \frac{\xi_T}{2} = \phi(\mathrm{Re}) \tag{5.92}$$

It is experimentally established that the relation (5.92) is valid for turbulent flows of liquids and gases within $0.6 < \mathrm{Sc} < 3000$ and above.

Assuming that for straight and smooth tubes for $\mathrm{Re} > 10^4$ and $C_D = 0.046\,\mathrm{Re}^{-0,2}$, substituting in (5.92), we obtain an expression for calculating the mass-transfer coefficient

$$\mathrm{Sh} = 0.023\,\mathrm{Re}^{0.8}\,\mathrm{Sc}^{1/3} \tag{5.93}$$

For a turbulent flow, putting $\xi_T = 0.3164/\mathrm{Re}^{1/4}$, we obtain an expression for calculating the mass-transfer coefficient

$$\mathrm{Sh} = 0.1582\,\mathrm{Re}^{3/4}\,\mathrm{Sc}^{1/3} \tag{5.94}$$

It was established experimentally that the relation (5.94) is valid for turbulent flows of liquids and gases in the range Sc from 0.6 to 3000 and higher. The validity and accuracy of the formulas obtained for calculating the mass-transfer coefficients depends on the accuracy of the expression for the coefficient of friction. Essential for these analogies is that they take into account the coefficient of friction. Therefore, in the case of poorly streamlined bodies, when the shape resistance prevails, these relations are inapplicable and give large deviations from the true values $\xi_T = \xi_T(\mathrm{Re})$. The methods and analogies previously described make it possible to determine the mass-transfer coefficients in moving continuous phases only in isolated cases, albeit numerous, on surfaces with a simple configuration. Usually, the specific form of the dependence of these coefficients on the operating parameters of the process is established experimentally for each type of mass-transfer apparatus.

5.5.4 The Karman Analogy

More complex analogies between mass transfer and the amount of motion in a turbulent flow are established by Karman, improved on the basis of the Prandtl–Taylor theory and taking into account the fact that the resistance to transfer of heat and mass in the turbulent boundary layer consists of three parts corresponding to the laminar sublayer $y_+ < 5$, transition zone $5 < y_+ < 30$, and turbulent core flow. As a result, in order to determine the number during flow in the pipes, the following expression is proposed [18]

$$\mathrm{Sh} = \frac{\frac{\xi_T}{2}\,\mathrm{Re}\,\mathrm{Sc}}{1 + 5\sqrt{\frac{\xi_T}{2}}\left(\mathrm{Sc}\text{-}1\right)} \tag{5.95}$$

If we assume that in the pipes the friction coefficient for the region $2.10^3 \le \mathrm{Re} \le 10^5$ is determined by the formula, $\xi_T = 0.3164/\mathrm{Re}^{1/4}$, then we get

$$\mathrm{Sh} = \frac{0.1582\,\mathrm{Re}^{3/4}\,\mathrm{Sc}}{1 + 2\,\mathrm{Re}^{-1/8}\left(\mathrm{Sc}\text{-}1\right)} \tag{5.96}$$

The advantage of these methods of analogy is the relationship between the mass-transfer coefficient and the coefficient of resistance.

5.5.5 Methods of Empirical Modeling

Theoretical solutions of mass transfer cannot cover a wide class of polydisperse granular materials with irregularly shaped particles. Approximate solutions of mass-transfer problems using experimental data allow us to obtain empirical or semi-empirical dependencies for mass-transfer processes with sufficient accuracy. The essence of the method of empirical modeling of mass-transfer processes is to choose such a formula with the smallest number of unknown coefficients, in order to satisfy the

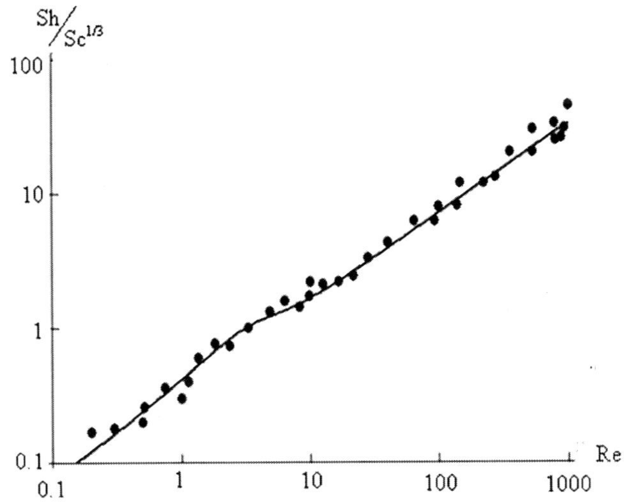

FIGURE 5.5 Evaporation (sublimation) of elements of a granular layer from naphthalene to gas (points—experiment). (From Soo, S.L. and Ihrig, H.K., *Transactions ASME*, 82, 609–621, 1960.)

smallest deviation of the experimental and calculated information. Thus, for complex processes of convective mass transfer, it can be shown that the determination of the number for large regions of the number change can be achieved by an empirical approach using the experimental data in the form

$$\frac{\mathrm{Sh}}{\mathrm{Sc}^{1/3}} = f(\kappa, \mathrm{Re}) \tag{5.97}$$

where $f(\mathrm{Re})$ is an arbitrary function that can be represented both as a series and have any empirical structure satisfying experimental data, κ a lot of unknown coefficients, the choice of which ensures the adequacy of the model. The choice of coefficient estimates κ can be constructed on the basis of minimizing the experimental and calculated values of the number or

$$\underset{min\,\kappa}{I(\kappa)} = \sum_{n=1}^{N} \left[f_n(\kappa, \mathrm{Re}) - f_{ne}(\kappa, \mathrm{Re}) \right]^2 \tag{5.98}$$

Figure 5.5 shows a comparison of the empirical dependence of the number Sh on Re for the process of evaporation of naphthalene

$$\frac{\mathrm{Sh}}{\mathrm{Sc}^{1/3}} = \frac{1}{3}\mathrm{Re}^{2/3}\left(1 + \frac{1.2\mathrm{Re}^{5/6}}{5 + 0.1\mathrm{Re}^3}\right), \quad 0.2 < \mathrm{Re} < 1000 \tag{5.99}$$

obtained on the basis of experimental data in the following form, although the same data can be described for individual regions of the change in the number Re

$$\begin{aligned} &\mathrm{Re} < 2, \quad \frac{\mathrm{Sh}}{\mathrm{Sc}^{1/3}} = 0.515\,\mathrm{Re}^{0.85}; \\ &2 < \mathrm{Re} < 30, \quad \frac{\mathrm{Sh}}{\mathrm{Sc}^{1/3}} = 0.725\,\mathrm{Re}^{0.47}; \\ &\mathrm{Re} > 30, \quad \frac{\mathrm{Sh}}{\mathrm{Sc}^{1/3}} = 0.395\,\mathrm{Re}^{0.64} \end{aligned} \tag{5.100}$$

As appears from pilot studies and Figure 5.5 at a mass transfer in a granular layer, three models are observed: $\mathrm{Re} < 2$ – laminar flow of particles; $2 < \mathrm{Re} < 30$ – transitional mode; $\mathrm{Re} > 30$ – turbulent mode. Existence of these modes of flow of a particle gives a curvature of a curve or change of a derivative in transitional area.

For process of evaporation of moisture, the following empirical equation is offered

$$\frac{\mathrm{Sh}}{\mathrm{Sc}^{1/3}} = \frac{1}{4}\mathrm{Re}^{2/3}\left(1 + \frac{0.25\,\mathrm{Re}^{5/4}}{18 + 0.015\,\mathrm{Re}^2}\right), \quad 16 < \mathrm{Re} < 3000 \tag{5.101}$$

A comparison of settlement and experimental data for calculation of coefficient of a mass transfer at evaporation of moisture from a surface of spherical particles is given in Figure 5.6.

The similar situation can be observed at a mass transfer in the mixing devices. For the mixing devices, specific dissipation of energy can be counted as $\varepsilon_R = \frac{kn^3D^3}{\upsilon_0}$, where n is the mixer velocity, D is the diameter of the mixer, and υ_0 is the mixer volume.

A change of coefficient of a mass transfer depending on the velocity of the mixing device is given in Figure 5.7 [52]. Having put $\varepsilon_R^{1/2} \sim n^{3/2}$, the description of the experimental values given in work [52], can highly be submitted in the expression

$$\beta_\upsilon \approx 5.0 + 1.85 \times 10^{-3} n^{3/2} \exp\left(-1.8 \times 10^{-4} n^{3/2}\right) \tag{5.102}$$

with the correlation coefficient equal $r^2 = 0.963$.

The analysis of experimental data (Figures 5.4 through 5.6) at a mass exchange in granular layers shows that this derivative at a certain value of numbers Pe or Re have a weak gap (Figure 5.8) that is possible will smooth a certain choice of structure of function $f(\kappa, \mathrm{Re})$.

It, perhaps, is connected with the fact that the majority of the given equations doesn't consider influence of character and the mode of a current of a layer of a particle or flow of a single particle on a mass transfer, or only laminar flow of particles is considered. Experimental values for a mass transfer in a layer of particles are given in Figure 5.9 [52].

It should be noted that the method of empirical modeling of process with the set of regional conditions demands from the researcher a big intuition and practical experience. It, first of all, is connected with the choice of the corresponding function, corresponding on character to a type of an

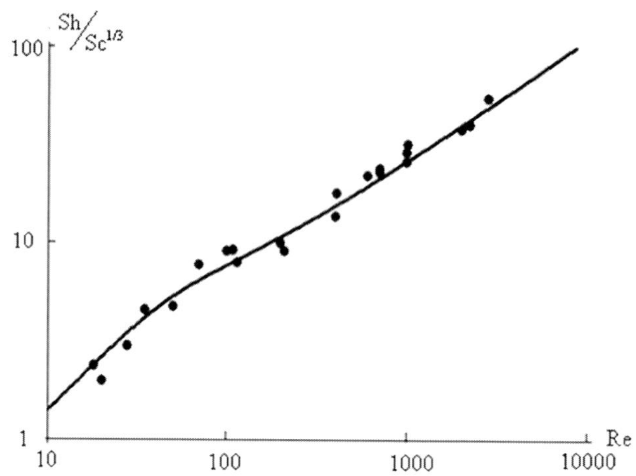

FIGURE 5.6 Evaporation of moisture from a surface of spherical particles of zeolite (points—an experiment). (From Soo, S.L. and Ihrig, H.K., *Transactions ASME*, 82, 609–621, 1960.)

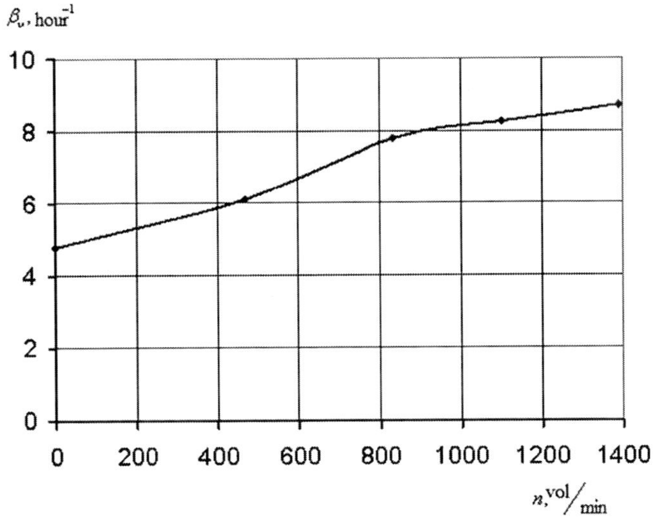

FIGURE 5.7 Dependence of the mass-transfer coefficient on the velocity of the mixer.

experimental curve with observance of all boundary conditions and with successful assessment of the unknown coefficients entering this equation.

In literature [19, 55], there is very large amount of dependences for definition of the number of Sherwood or coefficients of a mass transfer for various cases of a current (Table 5.1).

Here, $\mathrm{Gr} = a^3 g v^2 \frac{\Delta\rho}{\rho}$ is a Grashof's number, and η_w is a value of dynamic viscosity on a surface. In work [19], formulas for calculation of coefficient of a mass transfer are given in a fluidized layer of spherical particles

$$\begin{aligned} \mathrm{Sh} &= \frac{2.06}{\varepsilon}\mathrm{Re}^{0.425}\,\mathrm{Sc}^{1/3}, \quad \mathrm{Re} > 1900 \\ \mathrm{Sh} &= \frac{0.357}{\varepsilon}\mathrm{Re}^{0.641}\,\mathrm{Sc}^{1/3}, \quad 3 < \mathrm{Re} < 900 \end{aligned} \tag{5.103}$$

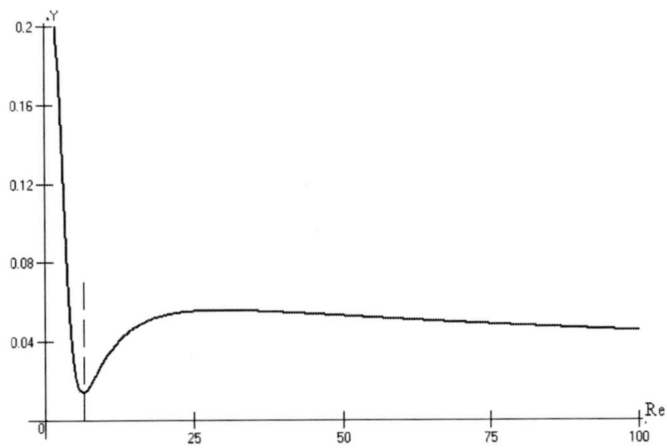

FIGURE 5.8 Change derivative $Y = \left(\partial \mathrm{Sh} / \partial \mathrm{Sc}^{1/3}\right) / \partial \mathrm{Re}$ of number agrees to equation (5.101).

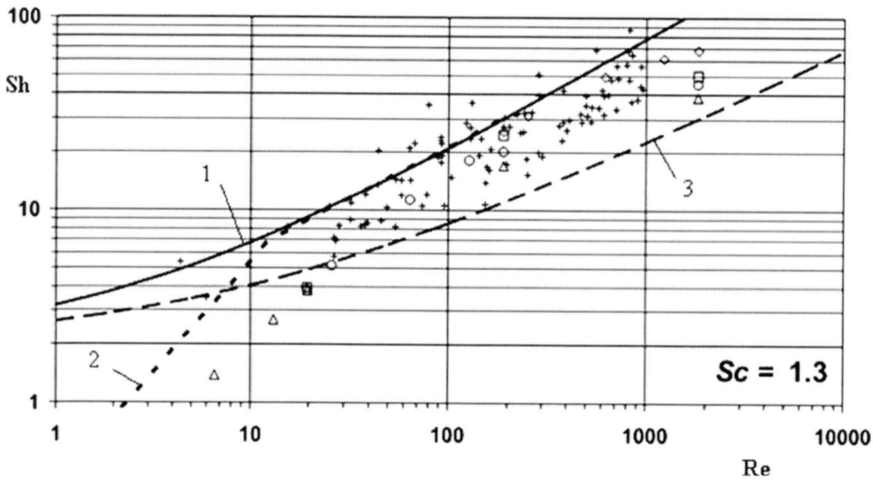

FIGURE 5.9 Dependence of number of Sherwood on Reynolds number Re for a layer of particles: 1 – [52]; 2 – [53]; 3 – [54] (the points in experiment taken from various references).

TABLE 5.1
Formulas for Determination of Coefficient of a Mass Transfer

	Expressions	The Condition
1.	$Sh = 0.332Re^{1/2}Sc^{1/3}$	Flat surface, laminar current, $0.6 < Sc < 50$
2.	$Sh = 0.0296Re^{4/5}Sc^{1/3}$	Turbulent flow, $Re < 10^8$, $0.5 < Sc < 50$
3.	$Sh = 2 + 0.6Re^{1/2}Sc^{1/3}$	Spherical firm particle, $Re < 0.2$
	$Sh = 2 + \left(0.4Re^{1/2} + 0.06Re^{2/3}\right)Sc^{0.4}\dfrac{\eta}{\eta_w}$	$3.5 < Re < 7.6 \times 10^4$, $0.71 < Sc < 380$, $1 < \eta/\eta_w < 3.2$
	$Sh = 2 + 0.59\left(GrSc\right)^{1/2}$	free convection $2 \times 10^8 < GrSc < 1.5 \times 10^{10}$
4.	$Sh = 0.61Re^{1/2}Sc^{1/3}$	Cylindrical particle, $Re < 1$
	$Sh = 0.3 + \left[0.62Re^{1/2}Sc^{1/3}\left(1 + 0.4/Sc^{2/3}\right)^{-1/4}\right] \times \left[1 + \left(\dfrac{Re}{282000}\right)^{5/3}\right]^{4/5}$	$Pe > 0.2$
5.	$Sh = 25Re^{0.45}Sc^{1/2}$	Current of liquid in a motionless layer
	$Sh = 1.2\left(1-\varepsilon\right)^{0.36}Re^{0.64}Sc^{1/3}$	Current of gas in a motionless porous layer
6.	$Sh = 0.31Gr^{1/3}Sc^{1/3}$	Bubbles in liquid in mixers
	$Sh = 0.42Gr^{1/3}Sc^{1/2}$	Big drops in mixers
	$Sh = 1.13Re^{0.8}$	Small drops in mixers

Except for a laminar current in pipes and channels and flows of particles of the simplest form (a flat surface, the sphere, the cylinder), the given formulas have semi-empirical or empirical character and are constructed on the basis of the simplest formulas and pilot studies by interpolation and extrapolation. As it appears from these experimental data of a mass transfer in a layer of particles for big area of change of values of number Re, on curve dependence $\mathrm{Sh} = f(\mathrm{Re})$ the break is also observed that is characteristic for a mass transfer in a layer.

Despite a large number of works and a variety of approaches in the field of a convective mass exchange of disperse particles, for the present there is no satisfactory quantitative theory connecting number Sh with their parameters (numbers Pe, Re, geometry of a form of single particles, geometry of structure of a granular layer, etc.) on the basis of model and mechanism of a mass transfer. The offered models of a convective mass exchange, as we know, aren't physical laws, and represent the empirical or semi-empirical approximations describing experimental data. The analysis of literature shows that a variety of models of a mass exchange reflects fundamental differences of views of the current or flow, which are observed in various fluid structures. The results of comparing mass transfer models with an experiment allow us to conclude that, with approximately the same accuracy, the same physicochemical nature of the system can be described by the same equation, and the same dispersed system can be described by equations of a fundamentally different structure. Finally, it is possible to choose the appropriate mass transfer equation for describing any experiment, without considering the real mechanism of mass transfer, which is quite acceptable for engineering applications; however, it cannot be considered normal from the basic science point of view. The variety of models of a convective mass transfer presents researchers with a choice of the most adequate equation, best approximating experimental data, disregarding the mechanism of transfer of weight.

5.6 APPROXIMATE SOLUTIONS OF THE CONVECTIVE TRANSPORT EQUATION

Approximate solutions of the equations of convective transfer of weight are valuable in the fact that they allow us to receive certain simple ratios suitable for further research of the process. The equations of stationary convective transfer of weight at small numbers $\mathrm{Pe} \ll 1$ will be presented in the dimensionless form

$$\frac{\partial^2 X}{\partial \xi^2} + \alpha f(\xi) \frac{\partial X}{\partial \xi} = 0 \tag{5.104}$$

where α is the dimensionless parameter, ξ is the function depending on dimensionless coordinate, and $X = C/C_0$, $f(\xi)$ is the function, depending on distribution of velocity. Equation (5.104) describes problems of flow of a cylindrical body whose axis is focused across a stream $\left(\alpha = \mathrm{Sc},\ \xi = \left(U/DR\right)^{1/2} \frac{y}{x^{1/3}},\ \xi = \left(4U/\nu d\right)^{1/2} \frac{y}{2}\right)$, diffusion in a laminar stream on a pipe surface $\left(f(\xi) = \xi^2,\ \alpha = 2/3,\ \xi = \left(U/DR\right)^{1/2} \frac{y}{x^{1/3}}\right)$, transfer of weight in a forward critical point of a spherical drop, streamline by a Stokes stream ($\alpha = \mathrm{Pe}/(\beta+1)$, $f(\xi) = \xi$, β is the parameter), and many other tasks whose description can be consolidated to equation (5.104).

We will consider the simplest problem of the convective transfer of weight described by the equation

$$\frac{d^2 X}{d\xi^2} + \mathrm{Pe}\,\xi^n \frac{dX}{d\xi} = 0 \tag{5.105}$$

$$\xi = 0,\ X = 0;\ \xi \to \infty,\ X = X_0 \tag{5.106}$$

where $X = {}^{c}\!/_{c_0}$.

Integrating this equation twice, we have

$$X(\xi) = X_0 \frac{\int_0^{\xi} \exp\left(-\mathrm{Pe}\frac{\xi^{n+1}}{n+1}\right) d\xi}{\int_0^{\infty} \exp\left(-\mathrm{Pe}\frac{\xi^{n+1}}{n+1}\right) d\xi} \tag{5.107}$$

Having put that

$$\int_0^{\infty} \exp\left(-\mathrm{Pe}\frac{\xi^{n+1}}{n+1}\right) d\xi = \frac{1}{n+1}\left(\frac{n+1}{\mathrm{Pe}}\right)^{1/n+1} \Gamma\left(\frac{1}{n+1}\right) \tag{5.108}$$

we will define value of this integral for various values n:

$$n = 1, \int_0^{\infty} \exp\left(-\mathrm{Pe}\frac{\xi^2}{2}\right) d\xi = \frac{1}{2}\sqrt{\frac{2}{\mathrm{Pe}}}\,\Gamma\left(\frac{1}{2}\right) \approx 1.25\mathrm{Pe}^{-1/2} \tag{5.109}$$

$$n = 2, \int_0^{\infty} \exp\left(-\mathrm{Pe}\frac{\xi^3}{3}\right) d\xi = \frac{1}{3}\sqrt[3]{\frac{3}{\mathrm{Pe}}}\,\Gamma\left(\frac{1}{3}\right) \approx 1.287\mathrm{Pe}^{-1/3} \tag{5.110}$$

$$n = 3, \int_0^{\infty} \exp\left(-\mathrm{Pe}\frac{\xi^4}{4}\right) d\xi = \frac{1}{4}\sqrt[4]{\frac{4}{\mathrm{Pe}}}\,\Gamma\left(\frac{1}{4}\right) \approx 1.28\mathrm{Pe}^{-1/4} \tag{5.111}$$

We will receive the common decision (5.105) in a look

$$X(\xi) = X_0(n+1)\frac{\int_0^{\xi} \exp\left(-\mathrm{Pe}\frac{\xi^{n+1}}{n+1}\right) d\xi}{\left(\frac{n+1}{\mathrm{Pe}}\right)^{\frac{1}{n+1}}\left(\frac{1}{n+1}\right)} \tag{5.112}$$

The weight stream from the surface unit for a unit of time will be defined as

$$j = -\left(\frac{dX}{d\xi}\right)\Big|_{\xi=0} = \frac{X_0(n+1)}{\left(\frac{1}{n+1}\right)\left(\frac{n+1}{\mathrm{Pe}}\right)^{\frac{1}{n+1}}} \tag{5.113}$$

The dimensionless number of Sherwood will be presented as

$$\mathrm{Sh} = \frac{(n+1)^{\frac{n}{n+1}}}{\Gamma\left(\frac{1}{n+1}\right)} \mathrm{Pe}^{\frac{1}{n+1}} \tag{5.114}$$

In particular, at $n = 1$ we have

$$\mathrm{Sh} = \sqrt{\frac{2}{\pi}} \mathrm{Pe}^{1/2} \tag{5.115}$$

In practical calculations, this equation, taking into account a laminar current, is also usually $\mathrm{Pe} = \mathrm{ReSc}$ presented in the form

$$\mathrm{Sh} = A \mathrm{Re}^n \mathrm{Sc}^m \tag{5.116}$$

where A, m and $n = \frac{1}{3}$ are the coefficients defined experimentally or from the solution of the corresponding problem of a mass transfer.

5.6.1 Approximate Methods in the Solution of the Equations Are Mass Transfer

We will give two methods of the solution of the equations of a convective mass transfer by method of the smallest squares and a method of approximations below. In both cases, the solution of the equations of transfer turns out in the form of some approach with unknown coefficients, which are defined from regional statements of the problem.

1. **Method of the smallest squares.** For use of this method, we will consider the simplest description of processes of transfer in a look

$$\frac{d^2X}{d\xi^2} + \mathrm{Pe}\frac{dX}{d\xi} + k_0 X = 0 \tag{5.117}$$

with boundary conditions

$$\xi = 0,\ X = 1;\ \xi = 1,\ X = 0 \tag{5.117a}$$

Here, X is dimensionless concentration, and ξ is dimensionless coordinate.

The analytical solution of this regional task in the absence of chemical reaction can be presented in the form

$$X(\xi) = \frac{\exp(-\mathrm{Pe}\xi) - \exp(-\mathrm{Pe})}{1 - \exp(-\mathrm{Pe})} \tag{5.118}$$

If the decision to present in the form

$$X(\xi) = a_0 + a_1\xi + a_2\xi^2 \tag{5.119}$$

the essence of a method of the smallest squares comes down to minimization of a square discrepancy of the right and left members of equation (5.119)

$$I = \int \Delta^2(\xi, a)\, d\xi = \int_0^1 \left(\frac{d^2X}{d\xi^2} + \mathrm{Pe}\frac{dX}{d\xi} \right)^2 d\xi \tag{5.120}$$

From regional statements of the problem (5.117a) and decisions (5.119), we will define coefficients

$$a_0 = 1,\ a_1 = -(1 + a_2) \tag{5.121}$$

substituting which in (5.120) taking into account derivatives, we will receive

$$I = \int_0^1 \left[a_1\mathrm{Pe} + 2a_2(1 + \xi\mathrm{Pe}) \right]^2 d\xi = a_1^2\mathrm{Pe}^2 + 4a_1a_2\mathrm{Pe} + 2a_1a_2\mathrm{Pe}^2 + 4a_2^2\left(1 + \mathrm{Pe} + \frac{\mathrm{Pe}^2}{3} \right) \tag{5.122}$$

From a condition of a minimum (5.122)

$$\frac{\partial I}{\partial a_2} = a_1\mathrm{Pe}(2 + \mathrm{Pe}) + 4a_2\left(1 + \mathrm{Pe} + \frac{\mathrm{Pe}^2}{3} \right) = 0 \tag{5.123}$$

we have

$$a_2 = -\frac{\mathrm{Pe}(2 + \mathrm{Pe})}{4\left(1 + \mathrm{Pe} + \frac{\mathrm{Pe}^2}{3} \right)} a_1 \tag{5.124}$$

Having put that $a_1 = -(1 + a_2)$, we will receive

$$a_1 = \frac{4\left(1 + \mathrm{Pe} + \frac{\mathrm{Pe}^2}{3} \right)}{4 + 2\mathrm{Pe} + \frac{\mathrm{Pe}^2}{3}},\ a_2 = \frac{\mathrm{Pe}(2 + \mathrm{Pe})}{4 + 2\mathrm{Pe} + \frac{\mathrm{Pe}^2}{3}} \tag{5.125}$$

Finally, the approximate solution of equation (5.122) will be presented in the form

$$X(\xi) = 1 - \frac{4\left(1 + \mathrm{Pe} + \frac{\mathrm{Pe}^2}{3} \right)}{4 + 2\mathrm{Pe} + \frac{\mathrm{Pe}^2}{3}}\xi + \frac{\mathrm{Pe}(2 + \mathrm{Pe})}{4 + 2\mathrm{Pe} + \frac{\mathrm{Pe}^2}{3}}\xi^2 \tag{5.126}$$

A comparison of results of the analytical decision (5.118) with the approximate decision (5.126) is given in Table 5.2.

TABLE 5.2
Comparison of the Analytical and Approximate Solution of Equation (5.117)

	Pe = 0.1		Pe = 0.2		Pe = 0.5		Pe = 0.8	
ξ	(5.62)	(5.66)	(5.62)	(5.66)	(5.62)	(5.66)	(5.62)	(5.66)
0.1	0.895	0.895	0.893	0.891	0.877	0.871	0.859	0.865
0.2	0.792	0.792	0.786	0.784	0.759	0.760	0.730	0.738
0.4	0.588	0.588	0.580	0.576	0.540	0.541	0.501	0.507
0.6	0.388	0.388	0.380	0.376	0.343	0.341	0.306	0.307
0,8	0.192	0.192	0.188	0.184	0.163	0.160	0.140	0.138
1.0	0	0	0	0	0	0	0	0

	Pe = 1.0		Pe = 1.2		Pe = 4.0	
ξ	(5.62)	(5.66)	(5.62)	(5.66)	(5.62)	(5.66)
0.1	0.849	0.850	0.837	0.849	0.960	0.725
0.2	0.713	0.724	0.694	0.710	0.439	0.578
0.4	0.478	0.486	0.457	0.466	0.187	0.286
0.6	0.286	0.286	0.265	0.266	0.0737	0.068
0.8	0.138	0.129	0.1172	0.1107	0.0228	0.021

As appears from these comparisons, the approximate decision (5.126) with in admissible accuracy well describes only at $\mathrm{Pe} \le 1.2$. For a more exact decision, it is necessary to increase the number of members in equation (5.126). At $\mathrm{Pe} \le 0.1$ the approximate decision, it is possible to present in the form $X(\xi) \approx 1-\xi$.

2. **Method of consecutive approximations.** Submission of the decision (5.117) in the form of consecutive ranks is the basis for a method of consecutive approximations

$$X(\xi) = \sum_{n=0}^{\infty} a_n \xi^n \tag{5.127}$$

We will define the corresponding derivatives entering equation (5.117)

$$\frac{dX}{d\xi} = \sum_{n=0}^{\infty} a_n n \xi^{n-1}, \quad \frac{d^2X}{d\xi^2} = \sum_{n=0}^{\infty} a_n n(n-1)\xi^{n-2} \tag{5.128}$$

substituting which in (5.117), we will receive

$$\sum_{n=0}^{\infty} a_n n(n-1)\xi^{n-2} + \mathrm{Pe}\sum_{n=0}^{\infty} a_n n \xi^{n-1} + k_0 \sum_{n=0}^{\infty} a_n \xi^n = 0 \tag{5.129}$$

In this equality, grouping identical degrees and equating zero, finally we will define a recurrent formula for calculation of coefficients

$$a_n = -\frac{(n-1)\mathrm{Pe}\, a_{n-1} + k_0 a_{n-1}}{n(n-1)} \tag{5.130}$$

If the chemical reaction is reaction $k_0 = 0$, then we will receive $a_n = -\frac{\mathrm{Pe}}{n} a_{n-1}$ or it is possible to write down: $a_0 = 1$, $a_1 = -1$, $a_2 = -\frac{\mathrm{Pe}}{2} a_1 = \frac{\mathrm{Pe}}{2}$, $a_3 = -\frac{\mathrm{Pe}}{3} a_2 = -\frac{\mathrm{Pe}^2}{6}$,.... Then the common decision of equation (5.72) is possible and will present in the form

TABLE 5.3
Comparison of the Analytical Decision (5.118) with Method of Consecutive Approximations (5.131)

	Pe = 0.1		Pe = 0.2		Pe = 0.8		Pe = 0.01	
ξ	(5.118)	(5.131)	(5.118)	(5.131)	(5.118)	(5.131)	(5.118)	(5.131)
0.1	0.895	0.900	0.893	0.900	0.859	0.900	0.900	0.900
0.2	0.792	0.802	0.786	0.804	0.730	0.815	0.800	0.800
0.4	0.588	0.608	0.580	0.615	0.500	0.650	0.600	0.600
0.6	0.388	0.400	0.380	0.430	0.306	0.520	0.400	0.402
0.8	0.192	0.203	0.188	0.260	0.140	0.400	0.200	0.203
1.0	0.0	0.048	0.0	0.093	0.0	0.290	0.0	0.03

$$X(\xi) = 1 - \xi + \frac{\mathrm{Pe}}{2}\xi^2 - \frac{\mathrm{Pe}^2}{6}\xi^3 + \ldots \tag{5.131}$$

A comparison of the analytical decision (5.118) with method of consecutive approximations is given (5.131) in Table 5.3.

At small values of number $\mathrm{Pe} \le 0.01$ the decision becomes simpler to a look

$$X(\xi) \approx 1 - \xi \tag{5.132}$$

With an increase of the value of a number, the error of the numerical decision increases, which demands an increase in the number of members in a row (5.131).

5.7 ANALOGY AND SIMILARITY OF PROCESSES OF MASS, HEAT TRANSFER, AND MOMENTUM

Processes of transfer of weight, heat, and impulse are described by the interconnected difficult nonlinear equations with variable physical and chemical properties in private derivatives with nonlinear regional conditions. The complexity and difficulty of the solution of these equations have forced us to look for necessary decisions experimentally, using the theory of similarity and the analysis of dimensions. The main objectives of the theory of similarity are an establishment of similarity of various processes by means of criteria of similarity and definition of a possibility of generalization of results of the solution of the corresponding task in the absence of ways of finding their full analytical decisions. It is important to note that at physical similarity in space and time, fields of the corresponding physical parameters of two various processes are similar, for example, at hydrodynamic similarity (the systems of operating forces or force fields [forces of inertia, gravitation, viscosity, pressure, etc.]), at similarity of thermal processes (the respective fields of temperatures and thermal streams), and at similarity of mass-exchanged processes (streams of substances and fields of concentration, etc.). Theories of similarity are the cornerstone three theorems that are formulated as follows:

1st theorem. If physical processes are similar to each other, then the criteria of the same name similarity of these processes have the identical size.

2nd theorem. The equations describing physical processes can be presented in the form of functional communication between criteria of similarity.

3rd theorem. In order that physical processes are similar to each other, it is also necessary also enough that these processes are qualitatively identical, that is, the equations of their description coincide, except the parameters that are contained in them and they of the same name defining criteria are equal.

The main criteria of similarity characterizing transfer of weight, heat, and an impulse have been previously stated, and their equations are very similar in structure

$$V_x \frac{\partial V_x}{\partial x} + V_y \frac{\partial V_x}{\partial y} = -\frac{1}{\rho}\frac{\partial P}{\partial x} + \nu\left(\frac{\partial^2 V_x}{\partial x^2} + \frac{\partial^2 V_x}{\partial y^2}\right)$$

$$V_x \frac{\partial T}{\partial x} + V_y \frac{\partial T}{\partial y} = a_T\left(\frac{\partial^2 T}{\partial x^2} + \frac{\partial^2 T}{\partial y^2}\right) \quad (5.133)$$

$$V_x \frac{\partial C}{\partial x} + V_y \frac{\partial C}{\partial y} = D\left(\frac{\partial^2 C}{\partial x^2} + \frac{\partial^2 C}{\partial y^2}\right)$$

Also, specific streams of an impulse (superficial friction), heat, and weight are similar

$$\tau = -(\nu + \nu_T)\frac{d\rho V_x}{dy}$$

$$q = -(a_m + a_T)\frac{dT}{dy} \quad (5.134)$$

$$j = -(D + D_T)\frac{dC}{dy}$$

where V_x, T, C are the velocity, temperature, and concentration; ν, a_m, D are the coefficients of kinematic viscosity, molecular thermal diffusivity, and diffusion; ν_T, a_T, D_T are the coefficients of turbulent viscosity of thermal diffusivity and diffusion; and τ, q, j are the superficial friction, specific amounts of transferable heat, and mass.

As appears from these equations, the transfer of weight, heat, and an impulse are similar, though differ in the variables and coefficients entering these equations. It is also important to note that dimensions of all coefficients of transfer—viscosity, thermal diffusivities and diffusions—are identical and equal m^2/s. Moreover, the criteria of similarity for processes of transfer of heat and weight are given in Table 5.4.

Here, $\mathrm{St}_T, \mathrm{St}_D$ are the thermal and diffusive numbers of Stanton defined in a look

$$\mathrm{St}_T = \frac{\alpha}{V\rho C_P}, \mathrm{St}_D = \frac{\beta}{V} \quad (5.135)$$

Combining these criteria of similarity variously, it is possible to receive the new numbers of similarity Sc more accurately reflecting features of course of processes of transfer. The most important characteristic for processes of transfer of weight is the number Sc consisting of physical parameters. In the physical sense, Sc is a measure of similarity of fields of concentration and velocity, that is, at $\mathrm{Sc} = 1$ or $\nu_c = D$ fields of concentration and velocities are similar. Therefore, at the compelled convective current, Schmidt's number reflects the relation of thickness of diffusive and dynamic boundary layers: $\delta_D/\delta \sim \mathrm{Sc}^{-1/2}$.

In liquids, Schmidt's number is more than a unit; therefore, the thickness of a dynamic boundary layer is more than diffusive. In various oils, the value of numbers of Schmidt reaches several tens

TABLE 5.4
Basis Criteria of Similarity of Heat Transfer and Mass Transfer

Criteria of Similarity Heat Exchange	Formula	Criteria of Similarity Mass Exchange	Formula
Reynolds number characterizes the power ratio of inertia to forces of molecular friction	$\mathrm{Re} = VL/\nu$	Reynolds number	$\mathrm{Re} = VL/\nu$
Prandtl's number characterizes similarity of velocity and thermal fields	$\mathrm{Pr} = \nu/a_m$	Schmidt's number characterizes similarity of high-velocity and mass fields	$\mathrm{Sc} = \nu/D$
Nusselt's number characterizes heat exchange on border of liquid and a surface	$\mathrm{Nu} = \alpha L/\lambda$	Sherwood number characterizes a mass exchange on border of liquid and a surface	$\mathrm{Sh} = \beta L/D$
Peclet's number characterizes the convective transfer ratio to diffusive	$\mathrm{Pe} = VL/a_m$	Peclet's number diffusion characterizes the relation of convective transfer to diffusive	$\mathrm{Pe}_D = VL/D$
Stanton's number expresses the intensity ratio of a heat transfer to the specific heat content of the stream	$\mathrm{St_T} = \mathrm{Nu}/\mathrm{RePr}$	Stanton's number	$\mathrm{St_D} = \mathrm{Sh}/\mathrm{ReSc}$
Grashof's number characterizes the lift ratio, due to the difference in density to forces of molecular friction	$\mathrm{Gr} = \dfrac{gL\beta_l \Delta T}{\nu^2}$	Lewis's number characterizes similarity of mass and thermal fields in a liquid stream	$\mathrm{Le} = \dfrac{D}{a_T}$

of thousands; therefore, the concentration layer is actually located in a viscous underlayer. It should be noted that from the physical parameters that are among Schmidt, only the viscosity of liquids strongly depends on temperature, so the nature of the change of number Sc with a temperature is similar to the change of viscosity.

Due to such an analogy for determination of coefficients of heat exchange and a mass exchange, use the criteria equations of the type

$$\mathrm{Nu} = A_1 \mathrm{Re}^m \mathrm{Pr}^n$$
$$\mathrm{Sh} = A_2 \mathrm{Re}^m \mathrm{Sc}^n \tag{5.136}$$

In these equations, coefficients A_1, A_2, m, n are defined experimentally or receive as a result of transformation to a dimensionless look of the theoretical decision for a certain area of change of numbers Sc and Pr.

Example 5.2

The spherical particle of naphthalene with a diameter of 10 mm is flowed round by an air stream with a velocity $V = 5\,{}^{M}\!/_{sec}$ and with a temperature $T = 20°C$. In the course of sublimation, naphthalene evaporates in air. Having put air density equal $\rho_c = 1.29\,{}^{KT}\!/_{M^3}$, its viscosity $\eta_c = 1.8\times10^{-5}\,{}^{kq}\!/_{ms}$ and coefficient of diffusion of vapors of naphthalene in air $D = 4.5\times10^{-6}\,{}^{m^2}\!/_{s}$, we will define coefficient of mass transfer, using the criteria equation $\mathrm{Sh}=0.023\mathrm{Re}^{0.8}\mathrm{Sc}^{1/3}$.

First of all, we will calculate numbers

$$\mathrm{Re} = \frac{Vd\rho_c}{\eta_c} = \frac{5\times0.01\times1/29}{1.8.10^{-5}} = 3583$$

$$\mathrm{Sc} = \frac{\eta}{\rho D} = \frac{1.8\times10^{-5}}{1.29\times4.5\times10^{-6}} = 3.1$$

Using these values, we will define Sherwood number

$$\mathrm{Sh} = 0.023\mathrm{Re}^{0.8}\,\mathrm{Sc}^{1/3} = 0.023 \times 3583^{0.8} \times 3.1^{1/3} = 23.38$$

On the other hand, having put that $\mathrm{Sh} = {}^{\beta_L d}\!/_{D}$ we will define value of coefficient of a mass exchange as

$$\beta_L = \frac{\mathrm{Sh}\,D}{d} = \frac{23.38 \times 4.5 \times 10^{-6}}{0.01} = 10.52 \times 10^{-3}\,{}^{\mathrm{m}}\!/_{\mathrm{s}}$$

5.7.1 Method of the Analysis of Dimensions

The main method of the theory of similarity is this analysis of dimensions of the physical quantities characterizing a condition of the studied process and parameters, which define this state. The dimension of a physical quantity is understood as the expression corresponding to the ratio between the measured physical quantities that underlie the system of units of measure. The basis of dimensional analysis is the rule that the basic equations expressing the relationship between variables and process parameters should be valid for any choice of units of measurement. From this rule followed that all members of each equation must have the same size.

At the solution of the practical problems connected with determination of coefficient of a mass transfer, the method of the analysis of dimensions is of particular importance. When studying the vast majority of the phenomena of chemical technology, it is enough to enter independent main units of measure—lengths (*m*), the mass (*kq*), time (*sec*), and temperature (°C). Usually the dimension is written down symbolically in the form of a formula in which it is accepted to designate a symbol of unit of length L, units of mass M, units of time T, and temperature unit θ.

The main variables and their corresponding dimensions are given in Table 5.5. Other variables can be expressed, using these dimensions.

We will express as an example 1 C, 1 Vt, and 1 kal in units m, kq, sec, and °C . We can write down

$$1C = 1\,m = \frac{\mathrm{kg}\cdot\mathrm{m}^2}{\mathrm{sec}^2},\ 1Vt = 1\frac{C}{\mathrm{sec}} = \frac{\mathrm{kg}\cdot\mathrm{m}^2}{\mathrm{sec}^3},\ 1\ \mathrm{kal} = 4.186C = 4.186\frac{\mathrm{kg}\cdot\mathrm{m}^2}{\mathrm{sec}^3}$$

TABLE 5.5
Variables and Their Dimensions

Type	Variable	Symbols	Dimension
The basis parameters	Mass	m	$[M]$
	Length	l	$[L]$
	Time	t	$[T]$
	Temperature	T	$[\theta]$
Mechanical	Velocity	V	$[LT^{-1}]$
	Acceleration	g	$[LT^{-2}]$
	Density	ρ	$[ML^{-3}]$
	Dynamic viscosity	η	$[ML^{-1}T^{-1}]$
	Force	F	$[MLT^{-2}]$
Thermal	Heat conductivity	λ	$[MLT^{-3}\theta^{-1}]$
	Thermal capacity	c	$[LT^{-2}\theta^{-1}]$
	Coefficient of mass transfer	α	$[MT^{-3}\theta^{-1}]$
Diffusive	Concentration	C	$[ML^{-3}]$
	Diffusion coefficient	D	$[LT^{-2}]$
	Coefficient of mass transfer	β	$[LT^{-1}]$

It is similarly possible to express coefficients of heat conductivity, a heat transfer, and thermal capacity

$$\lambda = 1\frac{Vt}{\text{m}\,°\text{C}} = \frac{\text{kg}\cdot\text{m}}{\sec^3\ \text{C}}, \alpha = 1\frac{Vt}{\text{m}^2°\text{C}} = \frac{\text{kg}}{\sec^3 °\text{C}}, \text{C} = 1\frac{\text{kal}}{\text{kg}°\text{C}} = \frac{\text{m}^2}{\sec^2 °\text{C}}$$

The main contents of the theory of dimension are made by the π theorem, which is formulated as follows. Let some functional dependence between various sizes be $f(X_1, X_2, \ldots X_n) = 0$. Let the maximum number of these dimensional sizes with independent dimensions equal m.

Then the initial communication between n dimensional sizes expressing some physical law can be represented as a ratio between dimensionless sizes $(n-m)$, each of which has an appearance of a sedate monomial. The number of the main units of measurements by means of which all these variables are measured makes:

$$\Delta P = \left[\text{H}/\text{m}^2\right] = \left[\text{kg}/\text{s}^2\,\text{m}\right] = ML^{-1}T^{-2}; V = \left[\text{m}/\text{s}\right] = LT^{-1};$$
$$\rho = \left[\text{kg}/\text{m}^3\right] = ML^{-3}; \nu = \left[\text{m}^2/\text{s}\right] = L^2T^{-1} \tag{5.137}$$

So, let us assume that some parameter N of the process is connected with other parameters A, B, C, D of the process by dependence

$$N = f(A, B, C, D) = kA^aB^bC^cD^d \tag{5.138}$$

where k, f, b, c, d are unknown coefficients defined on the basis of pilot studies. We will assume that process parameters N, A, B, C, D depend on physical properties of a stream (viscosity, density, velocity, temperature), that in symbols of dimension will be presented in the form

$$\begin{aligned}
N &= L^{\alpha_0}T^{m_0}\theta^{k_0}M^{n_0} \\
A &= L^{\alpha_1}T^{m_1}\theta^{k_1}M^{n_1} \\
B &= L^{\alpha_2}T^{m_2}\theta^{k_2}M^{n_2} \\
C &= L^{\alpha_3}T^{m_3}\theta^{k_3}M^{n_3} \\
D &= L^{\alpha_4}T^{m_4}\theta^{k_4}M^{n_4}
\end{aligned} \tag{5.139}$$

Then equation (5.138) can be written down in a look

$$L^{\alpha_0}T^{m_0}\theta^{k_0}M^{n_0} = k\left(L^{\alpha_1}T^{m_1}\theta^{k_1}M^{n_1}\right)^a \left(L^{\alpha_2}T^{m_2}\theta^{k_2}M^{n_2}\right)^b \left(L^{\alpha_3}T^{m_3}\theta^{k_3}M^{n_3}\right)^c \left(L^{\alpha_4}T^{m_4}\theta^{k_4}M^{n_4}\right)^d \tag{5.140}$$

Comparing degrees at identical dimensions, we will receive

$$\begin{aligned}
\alpha_0 &= a\alpha_1 + b\alpha_2 + c\alpha_3 + d\alpha_4 \\
m_0 &= am_1 + bm_2 + cm_2 + dm_2 \\
k_0 &= ak_1 + bk_2 + ck_3 + dk_4 \\
n_0 &= an_1 + bn_2 + cn_3 + dn_4
\end{aligned} \tag{5.141}$$

In this equation, the number of unknown coefficients (k,a,b,c,d) is more than number of the equations, and the value of coefficients $(\alpha_i, m_i, k_i, n_i)$ are known according to the dimension of the corresponding parameters. For the solution of this system of the equations, relatively *a*, *b*, *c*, *d* and *k* having accepted *(n-m)* (*n* number of unknown coefficients, *m* number of the equations) key coefficients, we will express other coefficients through these. Key coefficients can be defined experimentally.

1. **Coefficient of mass transfer**. Using a method of dimensions, we will define empirical expression for coefficient of mass exchange $\beta_L[LT^{-1}]$, having put that the last is function of the characteristic size of a body $r[L]$, stream velocity $V[LT^{-1}]$, density of a stream $\rho[ML^{-3}]$, viscosity $\eta[MT^{-1}L^{-1}]$, and coefficient of diffusion $D[L^2T^{-1}]$. We will define coefficient of mass transfer in a look

$$\beta_L = kr^a V^b \rho^c \eta^d D^e \tag{5.142}$$

Having substituted values of dimension, we will receive

$$LT^{-1} = kL^a \left(LT^{-1}\right)^b \left(ML^{-3}\right)^c \left(MT^{-1}L^{-1}\right)^d \left(L^2T^{-1}\right)^e \tag{5.143}$$

Comparing degrees for the corresponding dimensions, we will receive

$$\begin{aligned} L &:.1 = a + b - 3c - d + 2e \\ T &:.-1 = -b - d - e \\ M &:.0 = c + d \end{aligned} \tag{5.144}$$

In this equation, the number of unknown coefficients equals $n = 5$, and the number of the equations $-m = 3$. The number of key parameters equals $n - m = 2$. As key parameters, we will accept *b*, *d* and we will express other coefficients through these coefficients. From the last equation (5.144), we have $c = -d$, and from the second equation—$e = 1 - b - d$. Then from the first equation, we will receive $a = b - 1$. Considering these values, it is possible to write down

$$\beta_L = kr^{b-1} V^b \rho^{-d} \eta^d D^{1-b-d} \tag{5.145}$$

We will write down this equation in a look

$$\frac{\beta_L r}{D} = k \frac{r^b V^b}{D^b} \frac{\eta^d}{\rho^d D^d} \tag{5.146}$$

Having entered criteria $\mathrm{Sh} = \frac{\beta_L r}{D}$, $\mathrm{Pe} = \frac{Vr}{D}$, and $\mathrm{Sc} = \frac{\eta}{\rho D}$, we will present the equation for calculation of coefficient of a mass transfer in the form

$$\mathrm{Sh} = k\mathrm{Pe}^b \mathrm{Sc}^d \tag{5.147}$$

If Pe = ReSc, then it is possible to write down

$$\mathrm{Sh} = k\,\mathrm{Re}^b\,\mathrm{Sc}^{b+d} \tag{5.148}$$

Coefficients (k,b,d) for any process of a mass exchange can be defined with use of experimental data.

2. **Coefficient of heat transfer**. Using a method of dimensions, we will define an empirical formula for calculation of the coefficient of a heat transfer $\alpha = {}^{\mathrm{Vt}}\!/_{\mathrm{m}^2} = {}^{\mathrm{kg}}\!/_{\mathrm{sec}^3} = \left[MT^{-3}\theta^{-1}\right]$ depending on the size of a body $r[L]$, stream velocity $V\left[LT^{-1}\right]$, on environment density $\rho\left[ML^{-3}\right]$, on viscosity of the environment $\eta\left[MT^{-1}L^{-1}\right]$, heat conductivity of the environment, and thermal capacity $C_P\left[L^2T^{-2}\theta^{-1}\right]$ of a stream, that is

$$\alpha = kr^aV^b\rho^c\eta^d\lambda^eC_P^f \tag{5.149}$$

Considering dimensions of the corresponding sizes, it is possible to write down

$$MT^{-3}\theta^{-1} = kL^a\left(LT^{-1}\right)^b\left(ML^{-3}\right)^c\left(MT^{-1}L^{-1}\right)^d\left(MLT^{-3}\theta^{-1}\right)^e\left(L^2T^{-2}\theta^{-1}\right)^f \tag{5.150}$$

Comparing degrees of the identical sizes, we will receive

$$\begin{aligned} L&:.0 = a + b - 3c + d + e + 2f \\ T&:.-3 = -b - d - 3e - 2f \\ M&:.1 = c + d + e \\ \theta&:.-1 = -e - f \end{aligned} \tag{5.151}$$

In this system, there exist $n = 6$ of unknown coefficients and $m = 4$ equations. Taking for key coefficients c and f, we will express other coefficients through them. From the last equation, we have $e = 1 - f$, from the third equation $d = f - c$, from the second equation $b = c$, then from the first equation we have $a = c - 1$. Substituting these values of coefficients in (5.131) and grouping the corresponding variables, we will receive

$$\frac{\alpha.r}{\lambda} = k\left(\frac{rV\rho}{\eta}\right)^c\left(\frac{\eta C_P}{\lambda}\right)^f \tag{5.152}$$

Having put $\mathrm{Nu} = \frac{\alpha.r}{\lambda}$, $\mathrm{Re} = \frac{Vr\rho}{\eta} = \frac{Vr}{\nu}$, $\mathrm{Pr} = \frac{\eta C_P}{\lambda}$, then we can write down

$$\mathrm{Nu} = k\mathrm{Re}^c\mathrm{Pr}^f \tag{5.153}$$

In this expression, coefficients k, c, f are defined, proceeding from experimental values of the measured sizes.

3. **Drag coefficient of resistance of firm spherical particles**. The drag force of solid spherical particles $F_D[MLT^{-2}]$ depends on: diameter of particles $a[L]$, velocity of a stream $V[LT^{-1}]$, density $\rho[ML^{-3}]$, and dynamic viscosity $\eta[ML^{-1}T^{-1}]$. We can write down the general expression for force as

$$F_D = Kd_P^aV^b\rho^c\eta^d \tag{5.154}$$

Passing to dimensional values, we will receive

$$MLT^{-2} = KL^a\left(LT^{-1}\right)^b\left(ML^{-3}\right)^c\left(ML^{-1}T^{-1}\right) \tag{5.155}$$

Comparing identical degrees, we have

$$M:.1 = c + d$$

$$L:.1 = a + b - 3c - d \tag{5.156}$$

$$T:.-2 = -b - d$$

Having accepted coefficient d key, we will express other coefficients as

$$a = 2 - d \quad b = 2 - d \quad c = 2 - d \tag{5.157}$$

Then we can write down

$$F_D = K d_P^{2-d} V^{2-d} \rho^{1-d} \eta^d \tag{5.158}$$

Having entered dimensionless number $\mathrm{Re}_d = \dfrac{V a \rho}{\eta}$, we will receive

$$F_D = K d_P^2 V^2 \mathrm{Re}_d^{-d} \tag{5.159}$$

Having expressed resistance force as $F_D = C_D S \dfrac{\rho V^2}{2}$, $S = \dfrac{\pi\ d_P^2}{4}$ we will finally receive

$$C_D = \frac{8}{\pi}\left(\frac{F_D}{d_P^2 V^2}\right) = \frac{8K}{\pi}\mathrm{Re}_d^{-d} = A\mathrm{Re}_d^{-d} = f\left(\mathrm{Re}_d\right) \tag{5.160}$$

where A is some experimentally defined coefficient. As it has been noted above, for small values $\mathrm{Re}_d << 1$, $A = 24$ and $d = 1$, i.e., $C_D = {}^{24}\!/_{\mathrm{Re}_d}$. Similarly, we will determine drag force for a particle in non-Newtonian liquid, having put $F_d = f\left(\rho, d_p, k, n, V\right)$, k is consistence coefficient, and n is an exponent. Except given above, we conduct the following dimensions: $k = \left[ML^{-1}T^{n-2}\right]$, $n = \left[M^0 L^0 T^0\right]$. Similarly, to the previously stated calculations, for force of resistance or coefficient of resistance, we will receive

$$C_D = \frac{F_d}{\rho V^2 d_p^2} = f\left(\frac{\rho V^{2-n} d_p^n}{k}, n\right) = f\left(\mathrm{Re}_t, n\right) \tag{5.161}$$

The theory of similarity and dimensions is the strong tool in the analysis of processes of transfer, if it isn't possible to receive analytical solutions of the differential equations with regional conditions. However, it should be noted that in the analysis of dimensions, the right choice of parameters on which transfer coefficients depend is important.

5.8 INFLUENCE OF THE MARANGONI EFFECT ON CONVECTIVE MASS TRANSFER AND CAPILLARY FLOW

The liquid drop applied on the surfaces of other liquids, depending on the properties of the liquids of the two drops, can remain in the form of a drop or spread on a surface of the other drop (oil–water). In both cases, the system accepts a state with the minimum energy of Gibbs. If conditions of the spreading of one liquid on the surface another are met, then the process of spreading happens rather quickly. Thanks to aspiration of the lower liquid to the reduction of superficial energy, as a result of molecular–kinetic motion, molecules of the spreading substance extend to surfaces, forming a monomolecular layer. The spreading of liquid with a smaller coefficient of a superficial tension on the surface of liquid with a big superficial tension makes a basis of effect of Marangoni. Heterogeneity on a superficial tension is caused by heterogeneity of structure and temperature and concentration in different points of an interphase surface. In moving streams of disperse systems, such heterogeneity depends also on the distribution of velocity of a stream at the flow of drops and bubbles. The current comes from a small area toward a big superficial tension, owing to the reduction of energy of Gibbs. Marangoni's effect is characteristic for systems—liquid–liquid and liquid–gas—and is the result of emergence of a gradient of concentration in the phenomena of a mass transfer and a temperature gradient in processes of interphase transfer of heat.

According to Marangoni's effect, at the contact of two liquid phases with superficial tension σ_1 and σ_2, the liquid with a smaller superficial tension σ_2 spreads on the surface of the first liquid $(\sigma_2 < \sigma_1)$. In this case, boundary conditions between the adjoining liquids has to consider a superficial tension taking into account, which will register in a look

$$\left[P_1 - P_2 - \sigma\left(\frac{1}{R_1} + \frac{1}{R_2}\right)\right] = \left(\tau_{W1} - \tau_{w2}\right) + \frac{\partial \sigma}{\partial x} \tag{5.162}$$

where P_1 and P_2 are the pressure in both liquids, and τ_{W1} and τ_{W2} are the shift tension on the surface of both liquids. It should be noted that the last member reflects change of a superficial tension in expression (5.133) or changes of temperature and expresses itself the tangential force operating per unit area surfaces from capillary forces, that is, $\tau_\sigma = {}^{\partial\sigma}/_{\partial x}$ and is friction tension on a surface, caused by change of a superficial tension. If with growth of concentration of the surface the substance decreases σ, then forces on the surfaces of liquid directed lengthways seek to increase the area of area with the increased concentration. Superficial forces cause the movement of near-surface layers of liquid, which because of viscosity, in turn, carry away deeper layers; therefore, there is a convective movement, passing in certain cases during a turbulent flow. Now there are two competing mechanisms of free convection: (1) the gravitational convection of Rayleigh arising spontaneously in the field of gravity, owing to dependence of density on temperature; and (2) the capillary convection of Marangoni arising, owing to dependence of a superficial tension on temperature and from concentration of surface substance (capillary). The fundamental characteristic of process is the existence of a threshold above which there is an organized movement.

For spherical drops, the presence of the Marangoni effect substantially changes the view of the internal flow. At large values of this change, the dependence of surface tension from concentration increases. Characteristic pictures of an internal current in drops with and without Marangoni's effect taken into account are given in Figure 5.10.

As it appears from Figure 5.10a with Marangoni's effect, a current in a drop becomes casual, that is, explained by uneven distribution of coefficient of a superficial tension or concentration to surfaces.

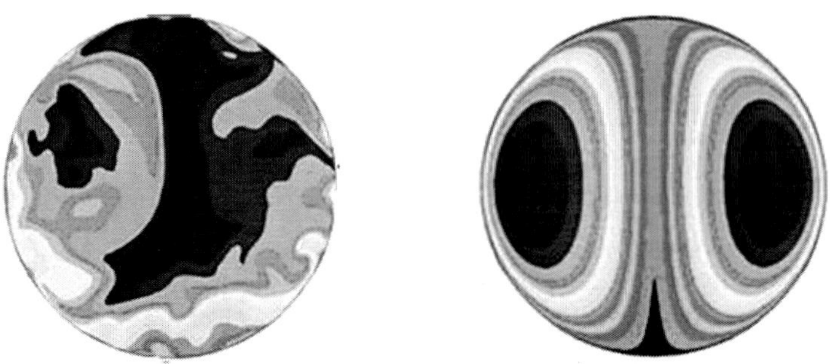

(a) (b)

FIGURE 5.10 A current in a drop: (a) with Marangoni's effect and (b) without Marangoni's effect (usual flow of a drop).

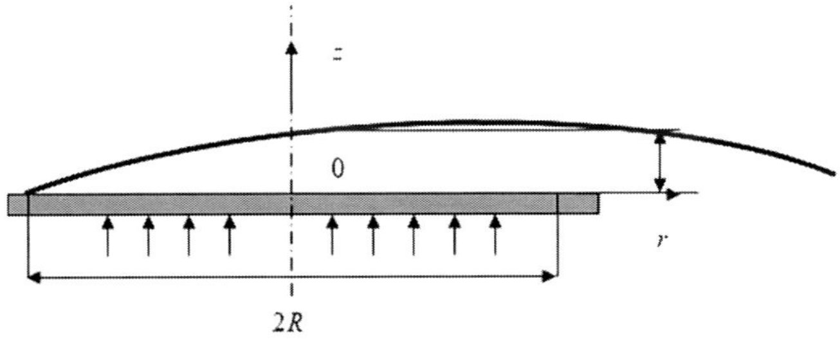

FIGURE 5.11 Spreading of a drop on a flat surface.

5.8.1 Capillary Flow of Liquid on a Flat Surface

We will consider the simplest flat problem of spreading of a drop of one liquid on a flat surface or on the surface of another liquid (Figure 5.11).

On the free surface of a flat spreadable liquid at its insignificant thickness, the condition (5.133) becomes simpler to simpler look

$$\eta \frac{dV}{dx} = \frac{d\sigma}{dx} \tag{5.163}$$

We will assume that the spreadable layer of liquid on the surface of the other liquid forms a spot of round section. Obviously, such assumption is not absolutely exact, as at thermocapillary effect, the instability and fluctuation of a surface of liquid promotes existence of a convective turbulent flow and different types of transfer. At the constant density and viscosity of the spreadable liquid, we will determine the liquid layer thickness, having put $V_z << V_r$ and $\partial V_z/\partial r << \partial V_r/\partial r$, that the first liquid is in rest, a current in a film laminar and in view of a subtlety of a spreadable layer $V_z << V_r$ and $\partial V_z/\partial r << \partial V_r/\partial r$, the equations of the movement in cylindrical coordinates take a form

$$\eta \frac{\partial^2 V_r}{\partial z^2} = \frac{\partial P}{\partial r}, \ \frac{\partial P}{\partial z} = 0 \tag{5.164}$$

$$\frac{1}{r}\frac{\partial (rV_r)}{\partial r} + \frac{\partial V_z}{\partial z} = 0 \tag{5.165}$$

with boundary conditions

$$z = 0, V_r = V_z = 0 \tag{5.166}$$

$$z = h, \ \eta \frac{dV_r}{dz} = \frac{\partial \sigma}{\partial r}, \ P = P_0 \tag{5.167}$$

Here, it is supposed that z is the layer thickness coordinate, r is the coordinate in the direction of the movement, $P = P_0 + \rho g(h - z)$ is the pressure in liquid, P_0 is the external pressure, h is the variable thickness of spreadable liquid, and ρ is the liquid density. Then equation (5.166), taking into account $\partial P/\partial r = \rho g\, \partial h/\partial r$, will be presented as

$$\eta \frac{d^2 V_r}{dz^2} = \rho g \frac{\partial h}{\partial r} \tag{5.168}$$

Integrating equation (5.168) on the condition of (5.167), we have

$$\eta \frac{\partial V_r}{\partial z} = \rho g \frac{\partial h}{\partial r}(z-h) + \frac{\partial \sigma}{\partial r} \tag{5.169}$$

Repeatedly integrating this equation under the second boundary condition, we will determine the velocity of a current of a layer in a look

$$\eta V_r = \rho g z \frac{dh}{dr}\left(\frac{z}{2} - h\right) + z\frac{\partial \sigma}{\partial r} \tag{5.170}$$

This equation demonstrates that stream velocity in each section is the variable, thanks to a change of thickness of a layer and depends on change of a superficial tension. From equation (5.165) we have

$$V_z = -\frac{1}{r}\frac{\partial}{\partial r}\int_0^h rV_r dz \tag{5.171}$$

Suppose that $V_z = \partial h/\partial t$, we will receive

$$\frac{\partial h}{\partial t} = -\frac{1}{r}\frac{\partial}{\partial r}\left[r\left(-\frac{\rho g h^3}{3\eta}\frac{\partial h}{\partial r} + \frac{h^2}{2\eta}\frac{\partial \sigma}{\partial r}\right)\right] \tag{5.172}$$

This equation determines the change of thickness of a layer by time, and in the radial direction is the nonlinear equation. At stationary spreading of a layer $\partial h/\partial t = 0$ and considering (5.172), we have

$$-\frac{dh^2}{dr} + \frac{3}{\rho g}\frac{\partial \sigma}{\partial r} = 0 \tag{5.173}$$

The equation represents the change of thickness of spreadable liquid at change of a superficial tension and weight of the drop. The decision (5.173) at $h = h_0$ under an entry and at $\partial\sigma/\partial r = \frac{\partial\sigma}{\partial c}\frac{\partial c}{\partial r}$ can also be presented in the form

$$h/h_0 = \sqrt{1 + \frac{3}{\rho g h_0^2}\frac{\partial \sigma}{\partial c}\Delta c} \tag{5.174}$$

where $\Delta c = c - c_0 < 0$. At dependence of density of a layer and superficial tension on temperature, in work [5] the following equation is offered

$$gh^2 = 3\rho^{-3/4}\left[\int \rho^{-1/4} d\sigma + A\right] \tag{5.175}$$

where A is the some constant. This task is similar to a problem of running off from a vertical surface under the influence of layer weight, with honors of the fact that in this case spreading of liquid results from change of a superficial tension in communication on a horizontal surface in that the common decision of equation (5.168) can have wave character. Substituting (5.172) in equation (5.167), we will receive distribution of velocity of spreadable liquid on a section

$$V_r = \frac{z}{2\eta}\frac{\partial \sigma}{\partial r}\left(\frac{3}{2}\frac{z}{h} - 1\right) \tag{5.176}$$

On a layer surface at $z = h$, the velocity is equal

$$V_r = \frac{h}{4\eta}\frac{\partial\sigma}{\partial c}\frac{dc}{dr} \tag{5.177}$$

From this equation, it follows that the velocity of the surface current of the substance is higher than the surface activity of the dissolved substance.

Let's say that on the motionless surface of liquid the drop with a radius R_0 falls and spreads on a surface. Believing that the volume of a spreadable drop V_k remains constant, it is possible to write

$$V_r = \frac{4}{3}\pi R_0^3 = \pi . r_S^2 h \tag{5.178}$$

From this equation, we will define the radius of the runoff spot

$$r_S \approx 1{,}15 R_0^{3/2} h^{-1/2} \tag{5.179}$$

Considering (5.174), we will finally receive

$$r_S \approx 1{,}15 R_0^{3/2}\left(h_0^2 + \frac{3}{\rho g}\frac{\partial\sigma}{\partial c}\Delta c\right)^{-1/4} \tag{5.180}$$

Entering equation (5.180), distribution of concentration of the spreadable liquid is defined by laws of convective diffusion with mobility of a surface, phase transformations (evaporation about surfaces, dissolution in an interphase layer, etc.), and a surface configuration. For a flat, motionless surface distribution of concentration of spreadable liquid in a symmetric round section, it is defined in a look

$$\frac{D_S}{r}\frac{\partial}{\partial r}\left(r\frac{\partial c}{\partial r}\right) = 0, \; r = r_0, \; c = c_0 \tag{5.181}$$

where D_S is the coefficient of superficial diffusion of spreadable liquid, and r_0 is the radius of an initial spot of a layer. The solution of this equation under the set boundary conditions and the assumption that, on border of spreadable liquid $r \geq r_S$ concentration is equal to zero, will be defined as

$$c = c_0 \frac{\ln {}^{r}/_{r_S}}{\ln {}^{r_0}/_{r_S}} \tag{5.182}$$

As $r \geq r_0$ follows from this expression

$$\Delta c = c - c_0 = c_0\left(1 - \frac{\ln r/r_S}{\ln r_0/r_S}\right) < 0 \tag{5.183}$$

When the spreading distribution of concentration can be determined by a surface of a drop of the other liquid from the solution of the equation [6], then

$$\frac{1}{a\sin\theta}\frac{\partial}{\partial\theta}\left(cV_\theta \sin^2\theta\right) = \frac{D}{a^2}\frac{1}{\sin\theta}\frac{\partial}{\partial\theta}\left(\sin\theta\frac{\partial c}{\partial\theta}\right) \tag{5.184}$$

where θ is the corner counted from the equator of a drop on surfaces, and V_θ is the stream velocity on the equator of a drop.

5.8.2 Influence of the Marangoni Effect on Mass Transfer

Marangoni's effect plays an essential role in processes of a mass transfer (rectification, absorption, liquid extraction) and also influences coalescence and crushing of drops in a stream. It is shown in the change of coefficients of a mass transfer, thanks to emergence of interphase convection and in change of a surface of phase contact. Besides, Marangoni's effect influences the stability of interphase films, and depending on the conditions and properties of the adjoining liquids, the stability of films can increase or decrease. In this regard, it should be noted that in the presence of a set of forces, it makes sense to differentiate areas on the dominating forces: inertia, capillary, gravitational. In the case of a current of liquid at small values of numbers, We and Vo capillary forces are dominating, and the effects of gravitation can be neglected. At small values of number Vonda, Weber's number defines a crucial role of capillary and gravitational forces. As it is previously noted, Marangoni's effect is shown at the contact of two phases with different coefficients of a superficial tension. At the same time, there is an apparent motion of the environment with smaller coefficient of a superficial tension by analogy with a convective current of liquid. Such movement in an interphase layer (in certain cases a turbulent flow) where the main phenomena of transfer proceed, it has significant effect on transfer coefficients and also changes the size of a contact surface between the adjoining phases.

The set of works from which it is important to distinguish, Sterling and Scriven [56], Marra and Huethorst [57], Semkov and Kolev [58], and Li et al. [59], are devoted to a research of influence of the Marangoni effect on a mass exchange. For accounting of this effect in processes of transfer, Marangoni's number is defined as is usually entered

$$\mathrm{Ma} = \frac{\left(-\partial\sigma/\partial C\right)C_0 L}{\eta D} \tag{5.185}$$

where D is the diffusion coefficient, C_0 is the characteristic concentration, η is the dynamic viscosity, L is the characteristic length, and $\partial\sigma/\partial C$ is the dependence of a superficial tension on concentration change. In literature, there are also other expressions of Marangoni's number. Brian [60] shows this number is defined in a look

$$\mathrm{Ma} = \frac{(\sigma_i - \sigma_k)}{\eta K_L} \tag{5.186}$$

(where K_L is the coefficient of mass transfer), Van Kloster and Drinkenberg [61] in a look

$$\mathrm{Ma} = \frac{(\sigma_i - \sigma_k)H}{\eta D} \tag{5.187}$$

Crumzin [62] in a look

$$\mathrm{Ma} = \frac{\left(-\partial\sigma/\partial C\right)\left(\frac{C_k - C_i}{d}\right)H^2}{\eta D} \tag{5.188}$$

Here, H is the thickness of a liquid film, and d is the penetration depth.

It is important to note that change of a superficial tension is possible as a result of the existence of different points of a surface and various values of temperature and concentration. The current of liquid in an interphase layer is possible also at various densities of liquid at different points of a surface. In this case, it is useful to use the Rayleigh number presented as

$$\mathrm{Ra} = \frac{\left(-\partial\rho/\partial C\right)C_0 g H^3}{\eta D} \tag{5.189}$$

For assessment of influence of the effect of Marangoni, various parameters defining change of coefficients of a mass transfer and surface of an interphase layer are entered. In particular, a factor of improvement is entered of coefficient of a mass transfer in a look $\phi_m = {}^{K_L}\!/_{K_L^*}$ and assessment ϕ_m to carry out a formula $\varphi_m \approx 1 + \alpha \mathrm{Ma}$ (α is the coefficient defined experimentally).

5.8.3 Mass Transfer during Thermocapillary Flow

Once again, we will note that the change of a superficial tension is caused by change of temperature (a thermocapillary current) when heating from below of liquid with a free surface and concentration of surface substance (a capillary current). Researches of free convection has begun at the end of the last century with works of Benar, who, heating from below on a flat surface a thin layer of liquid, observed the steady structures reminiscent of bee honeycombs. Benar explained the emergence of hexagonal cellular structures with the influence of viscosity and a superficial tension of liquid. The considered stationary cells of Benar (Figure 5.12) and Prigozhin [63] called dissipative structures, as unlike equilibrium structures, are formed and remain, thanks to the exchange of energy and substance with the external environment in nonequilibrium conditions. The fundamental characteristic of process is the existence of a threshold above which there is an organized movement. Later, pilot studies have revealed differences in structures of currents in cases of a free surface of liquid with prevalence of the thermocapillary mechanism and a surface, a limited firm plate where convection is caused by gravitational forces (Figure 5.12). As experiments have shown, with a temperature increase, the size the of cells at first decreases, then begin to grow. The temperature gradient also influences the direction of the movement of liquid in the cells of Rayleigh–Benard.

The convective structure of a heat transfer is completely defined by a geometrical configuration of walls. If liquid is limited to two planes, the lower of which is warmed up, then the structure of convective streams consists of the direct rollers parallel to the short party (Figure 5.13). Convection

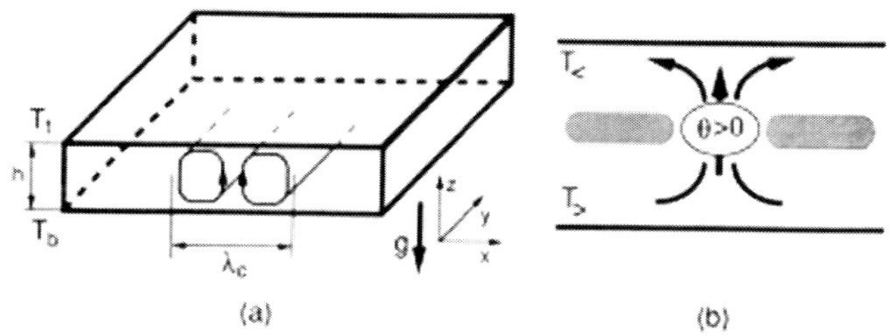

FIGURE 5.12 Geometry of thermoconvective flows of Rayleigh–Bernard: (a) a type of a surface and circulation in cells and (b) the scheme of convective cells with the ascending stream in the center.

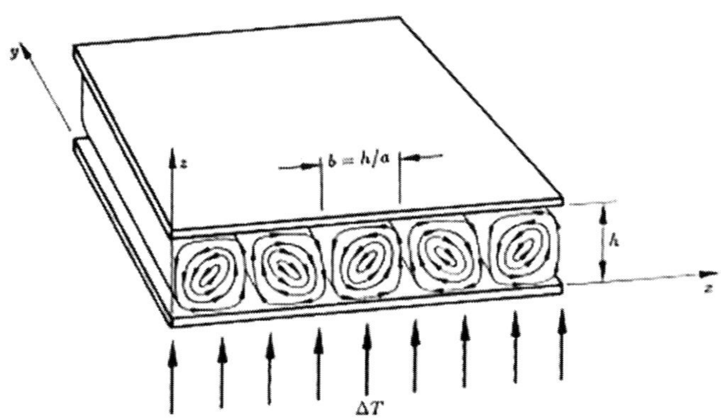

FIGURE 5.13 Geometry of convective streams between two surfaces.

in the flat horizontal layer of liquid, which is warmed up from below, has the features characteristic of many phenomena of hydrodynamic stability.

At a thermocapillary current, the specified mechanisms can set the liquid environment to motion, if the operating force exceeds dissipation force $F_d = \rho v a_T$. In this case, operating forces can be classified as follows: (1) the buoyancy force $F_{tp} = \rho\beta g\text{Sh}^4$ generated by volume expansion of liquid (β is the coefficient of volume expansion of liquid, S is the cross-sectional area, and h is the liquid layer thickness) and characterized by Grashof's number; and (2) the thermocapillary force $F_{t\sigma} = \sigma(T)\text{Sh}^2$ generated by effect of a capillarity. And action of these forces is characterized by Rayleigh and Marangoni's dimensionless numbers

$$\text{Ra} = \frac{\beta g\text{Sh}^4}{v\, a_T},\ \text{Ma} = \frac{\sigma(T)\text{Sh}^2}{\rho v a_T} \tag{5.190}$$

In some works [64] Marangoni's number is defined as

$$\text{Ma} = \frac{\left(-\partial\sigma/\partial T\right)h\Delta T}{\eta a_T} \tag{5.191}$$

These numbers show in how many times force, the exciting movement, surpasses the dissipation force. At $\text{Ra} > 660$, $\text{Ma} > 80$ achievement, the thermoconvective current, even if action of other exciting forces is absent, arises to at least one of the numbers of critical value.

We will consider the movement of liquid of very small h thickness on a flat, firm surface at uniform heating from below. At the same time, we will adopt the law of change of a superficial tension along the surface of liquid as

$$\text{grad}\sigma = \frac{\partial\sigma}{\partial T}\text{grad}T \tag{5.192}$$

For all liquids, the temperature coefficient of a superficial tension $(\partial\sigma/\partial T) < 0$, that is, a superficial tension, decreases with the growth of temperature. The general equations of transfer of weight and an impulse in a laminar stream for a flat current will be presented in the form

$$\frac{\partial P}{\partial x} = \eta\left(\frac{\partial^2 V_x}{\partial x^2} + \frac{\partial^2 V}{\partial y^2}\right)$$

$$\frac{\partial V_x}{\partial x} + \frac{\partial V_y}{\partial y} = 0 \tag{5.193}$$

$$V_x\frac{\partial C}{\partial x} + V_y\frac{\partial C}{\partial y} = D\left(\frac{\partial^2 C}{\partial x^2} + \frac{\partial^2 C}{\partial y^2}\right)$$

with regional conditions

$$y = 0, V_x = 0, C = 0 \tag{5.193a}$$

$$y = h, \eta\frac{\partial V_x}{\partial y} = \frac{\partial\sigma}{\partial x} = \frac{\partial\sigma}{\partial T}\frac{\partial T}{\partial x}, C = C_0$$

As a very small thickness of a layer is accepted, some members of equation (5.193) can be neglected

$$\frac{\partial P}{\partial x}=\eta\left(\frac{\partial^2 V}{\partial y^2}\right),\int_0^h V_x dy=0,V_x\frac{\partial C}{\partial x}=D\left(\frac{\partial^2 C}{\partial y^2}\right) \tag{5.194}$$

The solution of a hydrodynamic task, that is, the first two equations (5.193) with boundary conditions (5.193a), does not present great difficulties and is provided in work Levich [6] in a look

$$V_x=\frac{1}{4\eta h}\frac{\partial\sigma}{\partial T}\left(3y^2-4hy+h^2\right)\frac{\partial T}{\partial x} \tag{5.195}$$

The maximum velocity of liquid on a surface at $y=0$ is equal

$$V_{xm}=\frac{h}{4\eta}\frac{\partial\sigma}{\partial T}\frac{\partial T}{\partial x} \tag{5.196}$$

The solution of a problem of diffusion provided in work [6] has shown that the full transfer of weight is equal

$$I\approx 0.6\frac{C_0 SD^{2/3}}{L^{1/3}}\left(\frac{1}{\eta}\frac{\partial\sigma}{\partial x}\right)^{1/3} \tag{5.197}$$

where L is the surface length, and $S=hL$ is the surface area. Having entered Sherwood $\mathrm{Sh}=\frac{IL}{SDC_0}$ and Schmidt's $\mathrm{Sc}=\frac{\nu}{D}$ dimensionless numbers, we will rewrite this equation in a look

$$\mathrm{Sh}\approx 0.6\mathrm{Ma}^{1/3}\mathrm{Sc}^{1/3} \tag{5.198}$$

where $\mathrm{Ma}=\frac{\rho_c L^2}{\eta_c^2}\frac{\partial\sigma}{\partial T}\frac{\partial T}{\partial x}$ is Marangoni's number. This equation is fair at values of Reynolds number $\mathrm{Re}=\frac{h^2}{\nu_c}\left(\frac{1}{\eta_c}\frac{\partial\sigma}{\partial x}\right)<<1$, that is, for very thin layers of liquid at small changes of a superficial tension. Obviously, at great values of Reynolds number, the current on a surface gains turbulent character.

5.8.4 Mass Transfer with Capillary Flow

As pilot studies have shown, the existence of the Marangoni effect increases the coefficient of a mass transfer and influences the size of an interphase surface. The general equations of transfer of an impulse and weight will be presented as

$$\begin{gathered}\frac{\partial}{\partial t}\left(\rho\vec{V}\right)+\nabla\rho\vec{V}\vec{V}=\eta\nabla^2\vec{V}+\rho\vec{g}\\ \nabla\vec{V}=0\\ \frac{\partial C_A}{\partial t}+\vec{V}\nabla C_A=D_A\nabla^2 C_A\end{gathered} \tag{5.199}$$

Boundary conditions on a surface of spherical drops and bubbles (5.189) in spherical coordinates will be presented as:

θ in the direction

$$\eta_d\left(\frac{\partial V_\theta}{\partial r}-\frac{V_\theta}{r}\right)_d=\eta_c\left(\frac{\partial V_\theta}{\partial r}-\frac{V_\theta}{r}\right)_c+\frac{1}{R\sin\varphi}\frac{\partial\sigma}{\partial C}\frac{\partial C}{\partial\theta} \tag{5.199a}$$

and φ in the direction

$$\eta_d\left(\frac{\partial V_\varphi}{\partial r}-\frac{V_\varphi}{r}\right)_d=\eta_c\left(\frac{\partial V_\varphi}{\partial r}-\frac{V_\varphi}{r}\right)_c+\frac{1}{R}\frac{\partial\sigma}{\partial C}\frac{\partial C}{\partial\varphi} \tag{5.199b}$$

For processes of transfer of weight on a surface of a spherical drop, the following boundary conditions of the fourth sort can be satisfied

$$D_{Ad}\left(\frac{\partial C_{Ad}}{\partial r}\right)_{r=R}=D_{Ac}\left(\frac{\partial C_{Ac}}{\partial r}\right)_{r=R} \tag{5.199c}$$

In these superficial conditions, indexes (c,d) characterize a continuous and disperse phase. Relative velocity V of emersion of bubbles or subsidence of drops can be determined, proceeding from the balance of forces operating on a particle

$$m\frac{\partial V}{\partial t}=F_g+F_A+F_C \tag{5.200}$$

where m is the mass of a particle, $F_g=\rho_d v_d g$ is the gravitational force, v_d is the particle volume, $F_A=\rho_C v_d g$ the pushing-out Archimedes force, $F_C=C_D\rho_C\left(\pi/4\right)a^2\left(v^2/2\right)$ is the resistance force, and a is the diameter of a particle.

The solution of the specified task in a general view is possible only in the numerical way, though at certain assumptions (in particular, about a subtlety of an interphase layer, about weak dependence of a superficial tension on temperature and concentration, etc.) it is possible to receive also theoretical decisions.

For a flat surface, the solution of a problem of a mass exchange at a capillary current will be similar (5.173) and (5.174), where the capillary number of Marangoni will be defined in a look $\mathrm{Ma}=\frac{\rho_c L^2}{\eta_c^2}\frac{\partial\sigma}{\partial C}\frac{\partial C}{\partial x}$, that is, the number of transferable weight is proportional to Marangoni and Schmidt's number in degree 1/3.

On the basis of pilot studies in literature, a series of the semi-empirical formulas defining the influence of the Marangoni effect on the coefficient of a mass transfer is received

$$K_M=K_D\left[1+\alpha\left(\Delta\sigma-\Delta\sigma_{kp}\right)\right] \tag{5.201}$$

$$K_M^2=K_D^2+\frac{\alpha K_M}{\eta_c}\frac{\partial\sigma}{\partial C}\Delta C \tag{5.202}$$

In these formulas, K_M is the coefficient of a mass transfer taking into account Marangoni's effect, and α is the parameter that is defined on the basis of an experiment. Using the theory of isotropic turbulence, the coefficient of a mass transfer is defined in a look

$$K_M=K_D\left[1+\left(\frac{K_\sigma}{K_D}\right)^4\right]^{1/4} \tag{5.203}$$

where K_σ is the empirical parameter of isotropic turbulence.

5.8.5 Use of a Method of Dimensions

Using a method of the analysis of dimensions, we will estimate influence of effect of Marangoni on a mass transfer. For assessment of influence of the Marangoni effect, various parameters defining change of coefficients of a mass transfer and surface are entered. We will assume that the coefficient of a mass transfer $\beta_L[\mathrm{m/s}]$ depends on the characteristic size $H[m]$, density $\rho\left[\mathrm{kg/m^3}\right]$, dynamic viscosity $\eta\left[\mathrm{kg/ms}\right]$, coefficient of diffusion $D\left[\mathrm{m^2/s}\right]$, and coefficient of a superficial tension $\sigma\left[\mathrm{kg/s^2}\right]$. Then we can write down

$$\beta_L = kH^a\rho^b\eta^d D^e\sigma^m \tag{5.204}$$

This equation is representable in symbols of dimensions

$$LT^{-1} = kL^a\left(ML^{-3}\right)^b\left(MT^{-1}L^{-1}\right)^d\left(L^2T^{-1}\right)^e\left(MT^{-2}\right)^m \tag{5.205}$$

For degrees of identical dimensions, we will write down

$$\begin{aligned} &\text{for } L: 1 = a - 3b - d + 2e \\ &\text{for } T: -1 = -d - e - 2m \\ &\text{for } M: 0 = b + d + m \end{aligned} \tag{5.206}$$

Having accepted as key coefficients b and d, from the third equation we have

$$m = -(b + d) \tag{5.207}$$

substituting which in the second equation, we will receive

$$e = 1 + d + 2b \tag{5.208}$$

Then from the first equation we have

$$a = -b - d - 1 \tag{5.209}$$

Having substituted these values of coefficients in equation (5.204), we will receive

$$\beta_L = kH^{-(b+d)-1}\rho^b\eta^d D^{1+d+2b}\sigma^{-(b+d)} \tag{5.210}$$

Grouping dimensionless parameters according to the following criteria: Sherwood number $\mathrm{Sh} = \beta_L H/D$, Marangoni's number and Schmidt's number $\mathrm{Sc} = \eta/\rho D$, multiplying and dividing on ΔC, in a limit case we will receive

$$\mathrm{Sh} = k\mathrm{Ma}^{-(b+d)}\mathrm{Sc}^{-b} \tag{5.211}$$

The expression represents the equation for calculation of a mass transfer at variable superficial tension in an interphase layer. Values of coefficients b and d can be also defined on the basis of pilot studies. Comparing with the results of convective transfer for a flat surface, it is possible that those coefficients are equal in equation (5.194), respectively: $b = -1/3,\ d = 0$.

In the work [64], more complicated formulas are proposed for determining the Sherwood number as a function of the Marangoni number. For the liquid–gas system in this work, the following formula is offered

$$\mathrm{Sh} = \left[\left(\frac{1}{4}\frac{y^2}{b^2\mathrm{Ma}} + y\right)^{1/2} - \frac{y}{2b\sqrt{\mathrm{Ma}}}\mathrm{Sc}^{1/4}\right]^2 \tag{5.212}$$

where $y = \frac{m\beta_L h}{D}$, m is the coefficient of distribution of weight, h is the characteristic length of transfer of weight, b is the empirical parameter characterizing extent of updating of a surface, and D is the diffusion coefficient. In a limit case $y/\mathrm{Ma} \rightarrow \infty$, this equation becomes simpler to the look similar to (5.211)

$$\mathrm{Sh} = b^2 \mathrm{Ma}\mathrm{Sc}^{-1/2} \tag{5.213}$$

The equations presented in this section are very complex, and numerical solutions must be used to solve them. However, when solving practical problems of mass transfer, it is sufficient to use equations in the form of criteria that allow us to estimate the mass-transfer coefficients within the limits of accuracy.

6 Analytical Solutions of the Problem of Mass Transfer with Different Peclet Numbers

In this chapter, we will look at private stationary and non-stationary solutions of the equations of interphase mass transfer for various boundary conditions in a fluid at rest or at various Peclet numbers [4]. Moreover, the following assumptions are accepted: (1) the flow is laminar, with the exception of the last section, where mass transfer is considered in an isotropic turbulent flow; and (2) the physical and chemical properties of particles and the environment are constant. These solutions are valuable in that they allow one to investigate the processes of mass transfer without the influence of hydrodynamics on the flow of the process and, thus, to estimate certain parameters of mass transfer (mass-transfer coefficient, diffusion coefficient, flow of the transferred mass, etc.). Below we will consider the solution of various problems of mass transfer for a plane, cylindrical, and spherical surface, as well as in extraction, absorption, evaporation, etc. The main task of these solutions is to determine the mass flux from the surface of the particle, which is important for determining the mass-transfer coefficients or the Sherwood number $\text{Sh} = \frac{JL}{D\Delta CS}$, defined in the form where J is the total mass flow of matter to the particle surface, L is the characteristic particle size, D is the diffusion coefficient, ΔC is the mass-transfer force, and S is surface area of a particles. Certain problems represent the problems associated with the transfer of matter with the presence of a chemical reaction. To estimate the degree of influence of mass transfer or the rate of chemical reaction, it is necessary to introduce a dimensionless number $\text{K}_\text{D} = \frac{kL^2}{D}$, where k is the reaction rate constant. The square root of this number $\varphi_0 = \sqrt{\text{K}_\text{D}} = \sqrt{\frac{k}{D}}L$ is the Thiele modulus, a parameter characterizing the degree of influence of mass transfer and chemical reaction on the course of the process. In this section, we will consider particular solutions of the problems of mass transfer between a drop or a gas bubble and an isotropic turbulent flow.

6.1 MASS TRANSFER FOR SMALL PECLET NUMBERS

At low Peclet numbers, diffusion mass transfer is mainly observed, as a result of which, the general equations of mass transfer are substantially simplified. In this section, various analytical solutions of stationary mass-transfer equations are proposed with allowance for chemical reactions and the mass flux per unit surface is determined, which is an important parameter for calculating the transport coefficients.

6.1.1 Stationary Solutions of the Mass Transfer Problem for a Plane Surface

1. We consider one-dimensional solutions of problems of mass transfer of matter for a plane surface (Figure 6.1), described by an equation of the type

$$D_A \frac{d^2C_A}{dx^2} = 0 \tag{6.1}$$

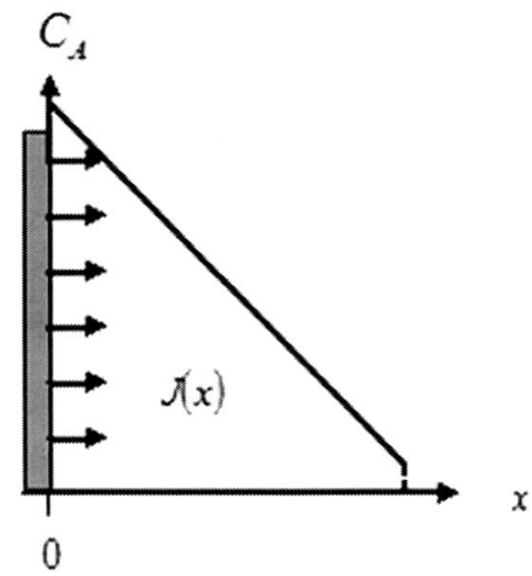

FIGURE 6.1 Concentration distribution scheme and flow for a flat surface.

with boundary conditions

$$x = 0,\ C_A = C_{A0};\ x = x_L,\ C_A = C_{AL} \tag{6.2}$$

where C_{A0}, C_{AL} are the concentration of the substance on the surface and in the volume of flow. The solution of this simplest problem is obtained by double integration of equation (6.1)

$$C_A = \left(C_{AL} - C_{A0}\right)\frac{x}{L} + C_{A0} \tag{6.3}$$

The quantity of the mass to be transferred to the volume of the flow is defined in the form

$$j(x) = -D_A \frac{dC_A}{dx} = \frac{D_A}{L}\left(C_{AL} - C_{A0}\right) \tag{6.4}$$

2. Suppose that a chemical reaction of the first order occurs on a flat surface $A \rightarrow B$. The boundary conditions for this problem will be

$$x = 0,\ -D_A \frac{dC_A}{dx} = kC_A \tag{6.5}$$

$$x = L,\ C_A = C_{AL}$$

The solution of equation (6.1) with these boundary conditions is represented in the form

$$C_A(x) = -\frac{C_{AL}}{\left(1 - \frac{kL}{D_A}\right)}\left(1 - \frac{kx}{D_A}\right) \tag{6.6}$$

The amount of mass carried from surface to volume is

$$j(x) = \frac{kC_{Al}}{1 - \frac{kL}{D_A}} \tag{6.7}$$

Taking into account equation (6.6), we determine the concentration of the substance on the surface

$$C_A\big|_{x=0} = C_{A0} = \frac{C_{AL}}{1-\dfrac{kL}{D_A}} \tag{6.8}$$

Then the solution of (6.6) can be written in the form

$$C_A(x) = C_{A0} = \left(1-\frac{k}{D_A}x\right) \tag{6.9}$$

If the substance is transferred from volume to surface, the concentration profile and the amount of the substance to be transferred are determined as

$$C_A(x) = \frac{C_{AL}}{1+\dfrac{kL}{D_A}}\left(1+\frac{kx}{D_A}\right)$$
$$j(x) = \frac{kC_{AL}}{1+\dfrac{kL}{D_A}} \tag{6.10}$$

3. If a chemical reaction of the first order occurs in the volume of a flat wall of equal thickness $2l$ and the transfer from the surface carried out is convective, the mass-transfer equation and the boundary conditions of the problem will be represented as

$$D_A\frac{d^2C_A}{dx^2} - kC_A = 0,\ x = 0,\ \frac{dC_A}{dx} = 0,\ x = l,\ D_A\frac{dC_A}{dx} = \beta_L\left(C_A - C_{AL}\right) \tag{6.11}$$

The solution of this problem is presented in the form

$$C_A(x) = \frac{\beta_L C_{AL}\,\mathrm{ch}\left(\sqrt{\dfrac{k}{D_A}}x\right)}{\beta_L\mathrm{ch}\left(\sqrt{\dfrac{k}{D}}l\right) - \sqrt{kD_A}\,\mathrm{sh}\left(\sqrt{\dfrac{k}{D_A}}l\right)} \tag{6.12}$$

$$j(x) = -D_A\frac{\partial C_A}{\partial x}\Big|_{x=l} = \beta_L\sqrt{kD_A}\,\frac{C_{AL}\,\mathrm{th}\left(\sqrt{\dfrac{k}{D_A}}l\right)}{\beta_L - \sqrt{kD_A}\,\mathrm{th}\left(\sqrt{\dfrac{k}{D_A}}l\right)} \tag{6.12a}$$

where $\mathrm{sh}(x), \mathrm{ch}(x), \mathrm{th}(x)$ are the hyperbolic sine, cosine, and tangent; and β is the mass-transfer coefficient. If $\beta^2 >> kD_A$, those mass transfer by convective way is much more than diffusion; mass flow is equal to

$$j(x) \approx C_{AL}\sqrt{kD_A}\,\mathrm{th}\left(\sqrt{\frac{kl}{D_A}}\right) \tag{6.12b}$$

The limiting stage is the chemical reaction and the diffusion transfer of the mass inside the particle.

4. Consider the case in which an nth-order reaction occurs in the volume of a solid particle $(n > 1)$

$$D_A \frac{d^2C_A}{dx^2} = kC_A^n$$
$$x = 0, \quad \frac{dC_A}{dx} = 0, \quad x \to \infty, \quad C_A = 0 \tag{6.13}$$

Introducing a new variable $q = {dC_A}/{dx}$ and defining the derivative as

$$\frac{d^2C_A}{dx^2} = \frac{dq}{dx} = \frac{dq}{dC_A}\frac{dC_A}{dx} = q\frac{dq}{dC_A} \tag{6.14}$$

we transform (6.13) to the form

$$q\frac{dq}{dC_A} - \frac{k}{D_A}C_A^n = 0 \tag{6.15}$$

The solution of this equation is defined as

$$\frac{q^2}{2} - \frac{1}{n+1}\frac{k}{D_A}C_A^{n+1} = B \tag{6.16}$$

The integration constant is determined from the boundary conditions as. Then

$$q^2 = \left(\frac{dC_A}{dx}\right)^2 = \frac{2}{n+1}\frac{k}{D_A}C_A^{n+1}, \quad \frac{dC_A}{dx} = -\sqrt{\frac{2}{n+1}}\sqrt{\frac{k}{D_A}}C_A^{n+1/2} \tag{6.17}$$

The mass flux per unit surface is defined as

$$j = -D_A\frac{dC_A}{dx} = \sqrt{\frac{2kD_A}{n+1}}\,C_A^{\frac{n+1}{2}} \tag{6.18}$$

Further, integrating (6.17), we obtain

$$\frac{2}{1-n}C_A^{\frac{1-n}{2}} = -\sqrt{\frac{2}{n+1}}\sqrt{\frac{k}{D_A}}x + B_0 \tag{6.19}$$

Having determined the integration constant B_0 from the boundary conditions $B_0 = \frac{2}{1-n}C_{A0}^{\frac{1-n}{2}}$, we find the concentration profile for the thickness of a planar particle

$$C_A^{\frac{1-n}{2}} - C_{A0}^{\frac{1-n}{2}} = (n-1)\sqrt{\frac{k}{2(n+1)D_A}}x \tag{6.20}$$

Consider the solution of equation (6.13) for $n = 1$ with the following boundary conditions

$$x = 0, \quad \frac{dC_A}{dx} = 0, \quad x = L, \ C_A = C_{AL} \tag{6.21}$$

The solution of this problem can be represented in the form

$$C_A = C_{AL} \frac{\exp\left(\sqrt{\frac{k}{D_A}}x\right) - \exp\left(-\sqrt{\frac{k}{D_A}}x\right)}{\exp\left(\sqrt{\frac{k}{D_A}}L\right) + \exp\left(-\sqrt{\frac{k}{D_A}}L\right)} \tag{6.22}$$

The flow of mass of matter per unit time will be determined as

$$j = -D_A \frac{dC_A}{dx}\Big|_{x=L} = C_{AL}\sqrt{kD_A} \frac{\exp\left(\sqrt{\frac{k}{D_A}}L\right) - \exp\left(-\sqrt{\frac{k}{D_A}}L\right)}{\exp\left(\sqrt{\frac{k}{D_A}}L\right) + \exp\left(-\sqrt{\frac{k}{D_A}}L\right)} = C_{AL}\sqrt{kD_A}\,\mathrm{th}\left(\sqrt{\frac{k}{D_A}}L\right) \tag{6.23}$$

Introducing the Thiele module $\varphi_0 = \sqrt{\frac{k}{D}}L$ and the reaction rate in the specific volume, defined as the ratio of the volume of the particle to its surface $w = kC_A L$, we find the efficiency factor in the form

$$\eta = \frac{j}{w} = \frac{1}{\varphi_0}\mathrm{th}\varphi_0 \tag{6.24}$$

It should be noted that if $\varphi_0 < 0.3$, then $\mathrm{th}\varphi_0 \to \varphi_0$ and $\eta \to 1$. In this region, the effect of mass transfer of matter is insignificant, and the process is determined by the chemical reaction. If $\varphi_0 > 3$, then $\mathrm{th}\varphi_0 \to 1$ or $\eta \to \varphi_0^{-1}$ in this area and the mass transfer dominates. And, finally, if $0.3 < \varphi_0 < 3$, then the process is equally influenced by mass transfer and chemical reaction.

6.1.2 Stationary Solutions of Mass Transfer Problems for Cylindrical Particles

Cylindrical surfaces in chemical technology are often found during flow in pipes (tubular apparatus, transport tubes) and in the form of cylindrical particles (Rashig rings) and others.

1. Consider a one-dimensional solution of mass transfer for a cylindrical surface in the case when the length of a cylindrical particle is much larger than its diameter. The transport equation in cylindrical coordinates is represented as

$$\frac{D_A}{r}\frac{d}{dr}\left(r\frac{dC_A}{dr}\right) = 0 \tag{6.25}$$

with boundary conditions

$$r = R,\ C_A = C_{A0};\ r = L,\ C_A = C_{AL} \tag{6.26}$$

The solution of this problem will be presented as

$$C_A(x) = C_{A0} + \frac{C_{A0} - C_{AL}}{\ln R/L} \ln \frac{r}{R} \tag{6.27}$$

and the amount of mass flux per unit time as

$$j = -D_A \frac{dC_A}{dr}\bigg|_{r=R} = -\frac{D_A}{R} \frac{C_{A0} - C_{Al}}{\ln R/L} \tag{6.28}$$

The total flux of matter to the surface of a cylindrical particle of radius R and length L will be determined as

$$J = \frac{2\pi L D_A}{\ln R/L} \Delta C \tag{6.29}$$

2. If a first-order reaction occurs on a cylindrical surface, then the boundary conditions are represented as

$$r = R,\ -D_A \frac{dC_A}{dr} = kC_A;\ r = L,\ C_A = C_{AL} \tag{6.30}$$

In this case, the concentration profile and the flux of matter are determined in the form

$$C_A(r) = \frac{C_{AL}}{\ln L/R - \frac{D_A}{kR^2}} \left(\ln \frac{r}{R} - \frac{D_A}{kR^2} \right) \tag{6.31}$$

$$j(r) = -\frac{D_A}{R} \frac{C_{AL}}{\ln L/R - \frac{D_A}{kR^2}} \tag{6.32}$$

The total flux through the lateral surface of a cylindrical particle is determined as

$$J = -\frac{2\pi L D_A C_{AL}}{\ln L/R - \frac{D_A}{kR^2}} \tag{6.33}$$

3. Suppose that a chemical reaction of the first order $A \to B$ occurs in the volume of a cylindrical particle. Then the transport equations and the boundary conditions of the problem are the following:

$$\frac{D_A}{r} \frac{d}{dr}\left(r \frac{dC_A}{dr} \right) + kC_A = 0 \tag{6.34}$$

$$r = 0,\ \frac{dC_A}{dr} = 0;\ r = R,\ D_A \frac{dC_A}{dr} = \beta_L (C_A - C_{AL})$$

The solution of this problem is presented in the form

$$C_A(r) = \beta_L C_{AL} \frac{\mathrm{J}_0\left(\sqrt{\frac{k}{D_A}}r\right)}{\sqrt{kD_A}\mathrm{J}_1\left(\sqrt{\frac{k}{D_A}}R\right) + \beta_L \mathrm{J}_0\left(\sqrt{\frac{k}{D_A}}R\right)} \tag{6.35}$$

where $J_0(r)$, $J_1(r)$ are the Bessel functions of zero and the first order.
The mass flow per unit time is

$$j = -D_A \frac{\partial C_A}{\partial r}\bigg|_{r=R} = \beta_L \sqrt{kD_A} \frac{C_{AL}\mathrm{J}_1\left(\sqrt{\frac{k}{D_A}}R\right)}{\beta_L J_0\left(\sqrt{\frac{k}{D_A}}R\right) + \sqrt{kD_A}\mathrm{J}_1\left(\sqrt{\frac{k}{D_A}}R\right)} \tag{6.36}$$

If mass transfer by convective means is much more than diffusion $\beta_L^2 >> kD_A$, mass flow can be defined as

$$j = C_{AL}\beta_L\sqrt{kD_A} \frac{\mathrm{J}_1\left(\sqrt{\frac{k}{D_A}}R\right)}{\mathrm{J}_0\left(\sqrt{\frac{k}{D_A}}R\right)} = C_{AL}\beta_L\sqrt{kD_A}\frac{\mathrm{J}_1\left(\sqrt{\varphi_0}\right)}{\mathrm{J}_0\left(\sqrt{\varphi_0}\right)} \tag{6.36a}$$

where $\varphi_0 = {kR^2}/{D_A}$ is the module Thiele. For small values, that is, the limiting stage is a chemical reaction, using decomposition $J(x) \approx \frac{x^n}{2^n \Gamma(n+1)}$, the mass flow of matter can be determined in the form

$$j = C_{AL}\beta_L\sqrt{k\varphi_0 D_A} \tag{6.36b}$$

If mass transfer by convective means is much less than diffusion $\beta_L^2 << kD_A$, then given $j \approx \beta_L C_{AL}$.
Consider the solution of equation (6.34) with the following boundary conditions

$$r = 0, \quad \frac{\partial C_A}{\partial r} = 0, \quad r = R, \quad C_A = C_{AL} \tag{6.37}$$

The general solution of equation (6.34) is represented as

$$C_A = B_1\mathrm{J}_0\left(\sqrt{\frac{k}{D_A}}r\right) + B_2 \tag{6.38}$$

Using the boundary conditions, we define the integration coefficients and mass flow in the form

$$j = D_A \frac{\partial C_A}{\partial r} = C_{AL}\sqrt{kD_A}\,\mathrm{J}_1\left(\sqrt{\frac{k}{D}}R\right), \qquad C_A = C_{AL}\frac{\mathrm{J}_0\left(\sqrt{\frac{k}{D_A}}r\right)}{\mathrm{J}_1\left(\sqrt{\frac{k}{D_A}}R\right)} \tag{6.39}$$

The total mass flux to the surface of the particle is

$$J = 2\pi RLC_{AL}\sqrt{kD_A}\,\mathrm{J}_1\left(\sqrt{\frac{k}{D_A}}R\right) \tag{6.40}$$

Introducing the Thiele module $\varphi_0 = \sqrt{\frac{k}{D_A}}R$, both for the reaction in the volume of a particle $w = \pi R^2 kC_A$, we determine the efficiency factor in the form

$$\eta = \frac{j}{w} = \frac{2}{\varphi_0}\mathrm{J}_1(\varphi_0) \tag{6.41}$$

4. We consider the problem of mass transfer in a cylindrical tube with a weak flow for the case with uniform diffusion of matter, both along the radius and along the length, when the initial concentration of matter at the inlet is known. The stationary equation of mass transfer is represented in the form

$$\frac{1}{r}\frac{\partial}{\partial r}\left(r\frac{\partial C}{\partial r}\right) + \frac{\partial^2 C}{\partial z^2} = 0 \tag{6.42}$$

$$C(r,L) = C(R,z) = 0,\ C(r,0) = C_0,\ 0 \le r \le R,\ 0 \le z \le L$$

The solution of this boundary-value problem by the method of separation of variables is represented in the form

$$C(r,z) = 2C_0\sum_{k=1}^{\infty}\frac{\mathrm{ch}\left(\frac{\mu_k z}{R}\right) - \mathrm{cth}\left(\frac{\mu_k L}{R}\right)\mathrm{sh}\left(\frac{\mu_k z}{R}\right)}{\mu_k \mathrm{J}_1(\mu_k)}\mathrm{J}_0\left(\frac{\mu_k r}{R}\right) \tag{6.42a}$$

Here, μ_k is the positive roots of the equation $\mathrm{J}_0(\mu) = 0$ $\left(\mathrm{J}(\mu_1) = 0.5191,\ \mu_1 = 2.4048\right)$. The mass flow per unit internal surface of the pipe per unit time is

$$j = -D\frac{\partial C}{\partial r}\bigg|_{r=R} = \frac{2C_0 D}{R}\sum_{k=1}^{\infty}\left(\mathrm{ch}\left(\frac{\mu_k z}{R}\right) - \mathrm{cth}\left(\frac{\mu_k L}{R}\right)\mathrm{sh}\left(\frac{\mu_k z}{R}\right)\right) \tag{6.43}$$

Mass flow to the entire inner surface of the pipe is equal to

$$J = \int_{(s)} jdS = 4\pi C_0 DR\sum_{k=1}^{\infty}\frac{1}{\mu_k \mathrm{sh}(\mu_k L/R)} \tag{6.44}$$

Let us consider the solution of the same problem for the porous wall of a cylindrical body in the case when the distribution of the concentration of matter along the length of the body is given in the form $C(R,z) = Az(L-z)$, and the boundary conditions are given by $C(r,0) = C(r,L) = 0$ (here, L, R are the length and radius of the cylindrical body, coefficient). The solution of this boundary-value problem is presented as

$$C(r,z)=\frac{8AL^2}{\pi^2}\sum_{n=0}^{\infty}\frac{\mathrm{I}_0\left[\frac{(2n+1)\pi r}{L}\right]}{(2n+1)^2\,\mathrm{I}_0\left[\frac{(2n+1)\pi R}{L}\right]}\sin\frac{(2n+1)\pi z}{L} \tag{6.45}$$

The flow of matter from the unit surface of the pipe to the flow volume per unit time is

$$C(r,z)=\frac{8AL^2}{\pi^2}\sum_{n=0}^{\infty}\frac{\mathrm{I}_0\left[\frac{(2n+1)\pi r}{L}\right]}{(2n+1)^2\,\mathrm{I}_0\left[\frac{(2n+1)\pi R}{L}\right]}\sin\frac{(2n+1)\pi z}{L}. \tag{6.46}$$

The flow of material through the entire surface of the pipe is

$$j=-D\frac{\partial C}{\partial r}\Big|_{r=R}=\frac{8ADL}{\pi^2}\sum_{n=0}^{\infty}\frac{\mathrm{I}_1\left[\frac{(2n+1)\pi R}{L}\right]}{(2n+1)\mathrm{I}_0\left[\frac{(2n+1)\pi R}{L}\right]}\sin\left[\frac{(2n+1)\pi z}{L}\right] \tag{6.46a}$$

where $\mathrm{I}_0(x)$, $\mathrm{I}(x)$ are the Bessel functions of the imaginary argument.

6.1.3 Stationary Solutions of Mass Transfer Problems for Spherical Particles

In a number of processes of chemical technology, problems of mass transfer to the surface of spherical particles are often encountered. Such particles are droplets, bubbles, and spherical solid particles (adsorbent, catalyst), as well as other types of particles having a spherical shape. In convective transfer, the hydrodynamics of deformable particles (droplets, bubbles) at moderate and large numbers completely differ from mass transfer to a solid surface. First of all, this difference is explained by the presence of internal flow in droplets and bubbles, as well as deformation and change in shape, which is absent in solid particles. Hydrodynamics of very small particles, that is, $\mathrm{Re} \ll 1$, for drops and solid particles is close in many respects. For a weak laminar flow or for small values of the numbers Re and Pe, we can consider stationary and non-stationary solutions without taking into account convective transfer.

1. For a spherical surface, the one-dimensional diffusion transport of mass is described by equation

$$\frac{D_A}{r^2}\frac{d}{dr}\left(r^2\frac{dC_A}{dr}\right)=0 \tag{6.47}$$

$$r=R\,,\ C_A=C_{A0};\ r=L\ M\,,\ C_A=C_{AL}$$

where L is the distance is far from the particle. The solution of equation (6.47) with boundary conditions is represented as

$$C_A(r)=C_{A0}+\frac{C_{A0}-C_{AL}}{1-R/L}\left(\frac{R}{r}-1\right) \tag{6.47a}$$

The mass flux per unit time per unit surface and the total surface is equal to

$$j = -D_A \frac{dC_A}{dr}\Big|_{r=R} = \frac{D_A}{R}\frac{C_{A0} - C_{AL}}{1 - R/L} \tag{6.47b}$$

$$J = 4\pi R D_A \frac{C_{A0} - C_{AL}}{1 - R/L} \tag{6.47c}$$

2. If a chemical reaction of the first order occurs on the surface of a spherical particle, then the boundary conditions are represented as

$$r = R,\ -D_A \frac{dC_A}{dr}\Big|_{r=R} = kC_A;\ r = L,\ C_A = C_{AL} \tag{6.48}$$

The concentration profile is determined by solving equation (6.47) under the boundary conditions (6.48)

$$C_A = \frac{C_{AL}}{1 - \frac{R}{L} - \frac{D_A}{kR}}\left(1 - \frac{R}{r} - \frac{D_A}{kR^2}\right) \tag{6.49}$$

Then the mass flux per unit surface and on the total surface is

$$j = -\frac{D_A}{R}\frac{C_{AL}}{1 - \frac{R}{L} - \frac{D_A}{kR^2}} \tag{6.50}$$

$$J = -4\pi R D_A \frac{C_{AL}}{1 - \frac{R}{L} - \frac{D_A}{kR^2}} \tag{6.51}$$

3. If a first-order chemical reaction occurs in the volume of the spherical particle with an external mass transfer from the surface of the particle, then the mass-transfer equation is determined in the form

$$\frac{D_A}{r^2}\frac{d}{dr}\left(r^2 \frac{dC_a}{dr}\right) - kC_A = 0 \tag{6.52}$$

with boundary conditions

$$r = 0,\quad \frac{dC_A}{dr} = 0;\quad r = R,\ \ -D_A \frac{dC_A}{dr} = \beta_L\left(C_A - C_{AL}\right) \tag{6.52a}$$

To solve equation (6.52), we introduce an additional function $\theta = rC_A$, as a result of which we have

$$\frac{d^2\theta}{dr^2} - m^2\theta = 0,\ m^2 = k/D_A \tag{6.53}$$

The solution of equation (6.53) is represented in the form

$$\theta(r) = B_1\mathrm{ch}(rm) + B_2\mathrm{sh}(rm) \tag{6.53a}$$

The values of the integration coefficients u are determined using the boundary conditions (6.52a). Then the solution of (6.52) is defined in the form

$$C_A = \frac{R}{r}\frac{\beta_L}{D_A}\frac{C_{AL}\mathrm{ch}\left(\sqrt{\frac{k}{D_A}}r\right)}{\sqrt{\frac{k}{D_A}}\mathrm{sh}\left(\sqrt{\frac{k}{D_A}}R\right)-\left(\frac{1}{R}+\frac{\beta_L}{D_A}\right)\mathrm{ch}\left(\sqrt{\frac{k}{D_A}}R\right)} \tag{6.54}$$

The mass flow per unit surface is

$$j = \beta_L C_{AL}\frac{\mathrm{th}\left(\sqrt{\frac{kR^2}{D_A}}\right)-\sqrt{\frac{D_A}{kR^2}}}{\mathrm{th}\left(\sqrt{\frac{kR^2}{D_A}}\right)-\sqrt{\frac{D_A}{kR^2}}-\frac{\beta_L}{\sqrt{kD_A}}} \tag{6.54a}$$

Let us consider special cases:

1. If $\beta_L << D_A/R$, then the convective transport is insignificant, and the mass flow is determined by a simpler equation; $j \approx \beta_L C_{AL}$
2. If $\beta_L >> D_A/R$ and taking into account that $\mathrm{th}(\infty) \to 1$, then the mass flow is equal to

$$j \approx \frac{C_{AL}}{\frac{1}{\beta_L}-\frac{1}{\sqrt{kD_A}}} \tag{6.54b}$$

3. Consider mass transfer in a spherical particle with a chemical reaction, when the external convective mass transfer can be neglected. This task is important for evaluating the efficiency of a single grain of the catalyst and allows us to assess the degree of influence of mass transfer on the chemical reaction, that is, equation (6.52) with the following boundary conditions:

$$r = 0, \quad \frac{dC_A}{dr} = 0, \quad r = R, \quad C_A = C_{A0} \tag{6.55}$$

We represent equation (6.14) in the form

$$\frac{d^2C_A}{dr^2} + \frac{2}{r}\frac{dC_A}{dr} - \lambda^2 C_A = 0, \quad \lambda^2 = \frac{k}{D_A} \tag{6.56}$$

Introducing a new variable $\psi = rC_A$, we transform equation (6.56) to the form

$$r^2\frac{d^2\left(\psi/r\right)}{dr^2} + 2r\frac{d\left(\psi/r\right)}{dr} - \lambda^2 r\psi = 0 \tag{6.57}$$

Having determined the corresponding derivatives

$$\frac{d\left(\psi/r\right)}{dr}=\frac{1}{r}\frac{d\psi}{dr}-\frac{1}{r^2}\psi,\qquad \frac{d^2\left(\psi/r\right)}{dr^2}=\frac{1}{r}\frac{d^2\psi}{dr^2}-\frac{2}{r}\frac{d\psi}{dr}+\frac{2}{r^3}\psi \tag{6.58}$$

we transform equation (6.57) to the form

$$\frac{d^2\psi}{dr^2}=\lambda^2\psi \tag{6.59}$$

The solution of this equation is represented as

$$\psi=B_1\exp(\lambda r)+B_2\exp(-\lambda r),\qquad C_A=\frac{B_1}{r}\exp(\lambda r)+\frac{B_2}{r}\exp(-\lambda r) \tag{6.60}$$

Define the derivative

$$\frac{dC_A}{dr}=B_1\left[\frac{\lambda}{r}\exp(\lambda r)-\frac{1}{r^2}\exp(\lambda r)\right]+B_2\left[-\frac{\lambda}{r}\exp(-\lambda r)-\frac{1}{r^2}\exp(-\lambda r)\right] \tag{6.61}$$

Using the first boundary condition, it is not difficult to obtain from this equation that $B_1=-B_2$. Using the second boundary condition, we define

$$\frac{C_A}{C_{A0}}=\frac{R}{r}\frac{\exp(\lambda r)-\exp(-\lambda r)}{\exp(\lambda R)-\exp(\lambda R)}=\frac{R}{r}\frac{\mathrm{sh}(\lambda r)}{\mathrm{sh}(\lambda R)} \tag{6.62}$$

The mass flux through the surface of a spherical particle can be determined in the form

$$J=-4\pi R^2 D_A\left.\frac{\partial C_A}{\partial r}\right|_{r=R}=4\pi RD_AC_{A0}\left[\lambda R\,\mathrm{cth}(\lambda R)-1\right] \tag{6.63}$$

Let us determine the reaction rate in the volume of the particle in the form $w=\frac{4}{3}\pi R^3kC_A$. Then the efficiency factor is defined as

$$\eta=\frac{J}{w}=\frac{4\pi RD_AC_{A0}\left[\lambda R\,\mathrm{cth}(\lambda R)-1\right]}{\frac{4}{3}\pi R^3kC_{A0}}=\frac{3D_A}{kR^2}\left[\lambda R\,\mathrm{cth}(\lambda R)-1\right]$$
$$=\frac{3}{R^2\lambda^2}\left[\lambda R\,\mathrm{cth}(\lambda R)-1\right] \tag{6.64}$$

If we assume that $\varphi_0=\lambda R=R\sqrt{k/D_A}$ is a module Tile, then for a spherical particle the efficiency factor is defined in the form

$$\eta=\frac{3}{\varphi_0}\mathrm{cth}\varphi_0-\frac{3}{\varphi_0^2}=\frac{3}{\varphi_0}\left(\mathrm{cth}\varphi_0-\frac{1}{\varphi_0}\right) \tag{6.65}$$

As follows from this equation, with an increase in the Thiele modulus, the efficiency factor decreases $\lim_{\varphi_0\to\infty}\eta\to 0$, and with the reduction of the Thiele modulus, the condition is satisfied $\lim_{\varphi_0\to 0}\eta\to 1$ (Figure 6.2).

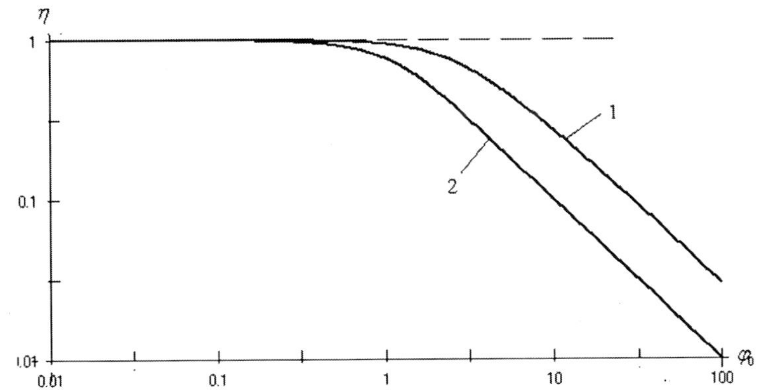

FIGURE 6.2 Dependence of the efficiency coefficient on the Thiele module: (1) for a spherical particle and (2) for a planar particle.

Example 6.1

Evaluate the Thiele module and the efficiency factor for catalyst particles with a diameter of 5 mm if the rate constant of the reaction $k = 5\times10^{-3}\ \mathrm{s}^{-1}$ is equal and the diffusion coefficient in the pores is equal $D_A = 5\times10^{-9}\ \mathrm{m^2/s}$.

Solution

Define $\lambda = \sqrt{\frac{k}{D_A}} = \sqrt{\frac{5\times10^{-3}}{5\times10^{-9}}} = 10^3\ \mathrm{m}^{-1}$. Then the Thiele module is equal to $\varphi_0 = \lambda R = 10^3 \times 2.5 \times 10^{-3} = 2.5$. The efficiency factor is determined from equation (6.65): $\eta = \frac{3}{2.5}\left(\mathrm{cth}2.5 - \frac{1}{2.5}\right) = 0.736$.

4. We consider the general problem of mass transfer to the spherical surface of a fixed drop (external problem) and inside the drop (internal problem). These problems usually characterize processes such as absorption, liquid extraction, and dissolution of gases in a liquid. The mass transfer is described by equation

$$\frac{1}{r^2}\frac{\partial}{\partial r}\left(r^2\frac{\partial C}{\partial r}\right) + \frac{1}{r^2\sin\theta}\frac{\partial}{\partial\theta}\left(\sin\theta\frac{\partial C}{\partial\theta}\right) + \frac{1}{r^2\sin^2\theta}\frac{\partial^2 C}{\partial\varphi^2} = 0 \tag{6.66}$$

with boundary conditions

$$0 \le r \le R,\ 0 \le \theta \le \pi,\ \ 0 \le \phi \le 2\pi,\ C(R,\theta,\phi) = f(\theta,\phi) \tag{6.67}$$

The solution of the boundary-value problem (6.18) for the case of mass transfer inside a drop is represented as

$$C(r,\theta,\varphi) = \sum_{n=0}^{\infty}\sum_{k=0}^{\infty}\left(\frac{r}{R}\right)^n \left(A_{nk}\cos k\varphi + B_{nk}\sin k\varphi\right)\mathrm{P}_n^k(\cos\theta) \tag{6.68}$$

Here,

$$A_{nk} = \frac{(2n+1)(n-k)!}{2\pi\alpha_m(n+k)!}\int_0^{\pi}\int_0^{2\pi} f(\theta,\varphi)\cos k\varphi\,\mathrm{P}_n^k(\cos\theta)\sin\theta\, d\theta\, d\varphi \tag{6.69}$$

$$B_{nk} = \frac{(2n+1)(n-k)!}{2\pi\alpha_m (n+k)!} \int_0^{\pi} \int_0^{2\pi} f(\theta, \varphi) \sin k\varphi \, \mathrm{P}_n^k(\cos\theta) \sin\theta \, d\theta d\varphi \tag{6.70}$$

$\alpha_m = 2$, if $k = 0$ and $\alpha_m = 1$, if k not equal to zero, P_k^n are the Legendre polynomials. For an external problem $R < r < \infty$ the solution of this boundary-value problem has the form

$$C(r,\theta,\varphi) = \sum_{n=0}^{\infty} \sum_{k=0}^{\infty} \left(\frac{R}{r}\right)^{n+1} (A_{nk} \cos k\varphi + B_{nk} \sin k\varphi) \mathrm{P}_n^k(\cos\theta) \tag{6.71}$$

If the concentration of the substance, both inside and outside the droplet, does not depend on φ, then solutions (6.69) and (6.71) will be simplified to the following form

$$0 \le r \le R, \; C(r,\theta) = \sum_{n=0}^{\infty} \frac{(2n+1)}{2} \left[\int_0^{\pi} f(\xi) \mathrm{P}_n(\cos\xi) \sin\xi d\xi \right] \left(\frac{r}{R}\right)^n \mathrm{P}_n(\cos\theta) \tag{6.72}$$

$$R \le r \le \infty, \; C(r,\theta) = \sum_{n=0}^{\infty} \frac{(2n+1)}{2} \left[\int_0^{\pi} f(\xi) \mathrm{P}_n(\cos\xi) \sin\xi d\xi \right] \left(\frac{R}{r}\right)^{n+1} \mathrm{P}_n(\cos\theta) \tag{6.73}$$

If the concentration distribution on the drop surface is constant, that is, $C(R,\theta,\varphi) = C_0$, then we have the following solutions

$$0 \le r \le R, \; C(r,\theta) = C_0 \sum_{n=0}^{\infty} \frac{(2n+1)}{2} \left[\int_0^{\pi} \mathrm{P}_n(\cos\xi) \sin\xi d\xi \right] \left(\frac{r}{R}\right)^n \mathrm{P}_n(\cos\theta) \tag{6.74}$$

$$R \le r \le \infty, \; C(r,\theta) = C_0 \sum_{n=0}^{\infty} \frac{(2n+1)}{2} \left[\int_0^{\pi} \mathrm{P}_n(\cos\xi) \sin\xi d\xi \right] \left(\frac{R}{r}\right)^{n+1} \mathrm{P}_n(\cos\theta) \tag{6.75}$$

The mass flux per unit surface of the drop per unit time is

$$j = -D \frac{\partial C}{\partial r}\Big|_{r=R} = \frac{DC_0}{R} \sum_{n=0}^{\infty} (n+1)\left(n + \frac{1}{2}\right) \left[\int_0^{\pi} \mathrm{P}_n(\cos\xi) \sin\xi d\xi \right] \mathrm{P}_n(\cos\theta) \tag{6.76}$$

6.2 MASS TRANSFER IN HETEROGENEOUS CHEMICAL REACTIONS

Most chemical processes with the presence of chemical reactions occur under heterogeneous conditions, that is, with the presence of several immiscible phases. Since the reagents are under certain conditions in the gaseous, liquid, and solid state, depending on their phase state and the combination of phases, the reactions can proceed in one phase, at the interface, or in several phases. Evidently, the diffusion and dynamic boundary layers play an important role in the course of chemical reactions at the interface between phases. Depending on the nature of the phases, there are: (1) homophase reactions that occur within a single phase, since all components and the catalyst are in the same phase. Such reactions are divided into the gas phase, when all components are in the gaseous state and liquid phase, when all the components and reaction products are in the liquid state; and (2) heterophase reactions, when the components are in two or more phases. In this case, systems such as gas–liquid, two immiscible liquid–liquid systems, gas–solid, and liquid–solid are distinguished heterogeneous catalytic systems. The processes occurring in chemical reactors are determined by the phase state of the initial reactants, the type of catalyst (solid or liquid), the nature and order of the chemical reactions complicated by the transfer of momentum, heat and mass, liquid-phase reactions by the Marangoni effect, boundary-layer thickness, etc. The driving force of diffusion transfer of matter from one phase to the other is the difference in concentration in both phases and the nature of the flow (laminar or turbulent), and this transfer is completed when the equilibrium state of the system is reached. Heterogeneous catalytic processes are carried out by the following consecutive stages: (1) convective and diffusive transport of matter to the outer surface of catalyst particles, (2) diffusion of the components to the internal surface of the porous particle, (3) adsorption of reagents on the surface, (4) chemical reaction; (5) desorption of the reaction products from the surface of the catalyst, and (6) convective diffusion and transfer of reaction products from the inner and outer surfaces of the catalyst to the volume of the stream. If either of these stages turns out to be the slowest or the most restrictive, then it determines the velocity of the process as a whole. Thus, depending on the rate of chemical reactions and the rate of diffusion transport of matter to the reaction surface, the kinetic, transition, and diffusion regions of the course of heterogeneous catalytic processes are distinguished. At velocities of mass transfer considerably exceeding the velocity of chemical reactions, the process proceeds in the kinetic region, where the rate of reactions is determined by their kinetics. Obviously, this region is characterized by small values of the number and the reaction rate constant. In the transition region, the reaction rate depends both on the laws of chemical kinetics and on the rate and intensity of mass transfer. The diffusion reaction region corresponds to the conditions under which the reaction rate is much higher than the rate of convective mass transfer. Under these conditions, the concentration of components on the surface does not depend on the kinetics of the chemical reaction and, entirely, are determined by the conditions of thermodynamic equilibrium. We have considered previously the solutions of the mass-transfer equations in the presence of chemical reactions for various flow surfaces and boundary conditions, which made it possible to determine the Thiele module and the efficiency factor of the surface-dependent use of the surface $k/_D$. For a stationary transport process on the surface of a particle, the rates of chemical reaction and convective mass transfer must equal

$$\beta_L\left(C_\infty - C_R\right) = kC_R^n \tag{6.77}$$

where C_∞, C_R are the concentration of the substance far from the surface in the volume of the flow, the order of the reaction. Such a method of determining the concentration on the surface with a mass-transfer coefficient that is independent of distance, according to [36], is called the "equipartition surface method" or "quasi-stationary method." We transform this equation to the form

$$1 - \frac{C_R}{C_\infty} = \frac{k}{\beta_L}\left(\frac{C_R}{C_\infty}\right)^n C_\infty^{n-1} \tag{6.78}$$

or in dimensionless form

$$\xi = z\left(1 - \xi\right)^n \tag{6.79}$$

In the case of a finite surface reaction rate for which the concentration of the reacting substance at local equilibrium on the wall is zero, along with the defining numbers Re and Sc, characteristic for any problem of convective mass transfer, additional $z = {}^{k_0 C_\infty}/_{\beta_L C_\infty} = {}^{q_{kin}}/_{q_{dif}}$, $k_0 = {}^{k}/_{C_\infty^{1-n}}$, $\xi = {}^{C_R}/_{C_\infty}$, q_{kin} is mass flow in the kinetic region, mass flow in the diffusion region appear. The parameter ξ equal to the ratio of the true reaction rate to the diffusion rate of a substance $\xi = {}^{kC_R}/_{\beta_L C_L}$ is called the efficiency coefficient of surface reactions. The dependences for a transversely streamlined plate are shown in Figure 6.3.

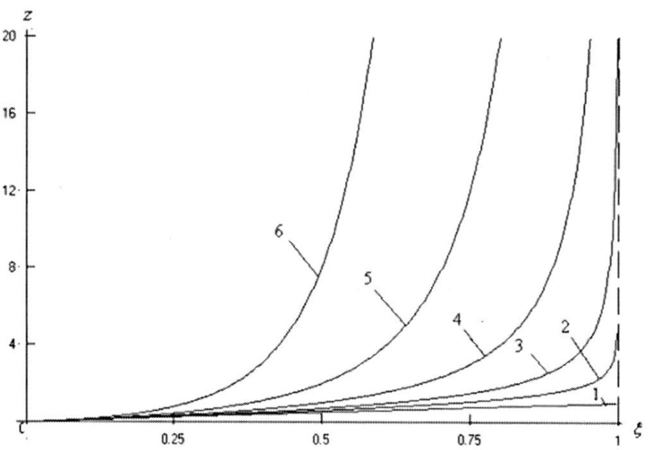

FIGURE 6.3 Dependence ξ on z for n, equal to: (1) $n = 0$; (2) $n = 1/4$; (3) $n = 1/2$; (4) $n = 1$; (5) $n = 2$; and (6) $n = 4$.

6.3 NON-STATIONARY MASS TRANSFER

The equations of non-stationary mass transfer by diffusion describe many mass-transfer processes and are suitable for studying the rate of mass transfer at low flow velocities and convective transport for small Peclet and Reynolds numbers. Below we will consider general and particular solutions of the mass-transfer equation for surfaces with different geometries under different boundary and initial conditions. It should be noted that with the change in the boundary conditions, depending on the nature of the problem being solved, the final solutions turn out to be different.

6.3.1 Non-Stationary Mass Transfer on a Flat Surface

Consider a planar surface in an infinitely large volume of a liquid that has an initial concentration distribution $C(x,0)=\varphi(x)$. The non-stationary mass transfer by diffusion is described by equation

$$\frac{\partial C}{\partial t}=D\left(\frac{\partial^2 C}{\partial x^2}+\frac{\partial^2 C}{\partial y^2}+\frac{\partial^2 C}{\partial z^2}\right) \tag{6.80}$$

We consider the simplest problem of one-dimensional mass transfer under constant initial conditions

$$\frac{\partial C}{\partial t}=D\frac{\partial^2 C}{\partial x^2} \tag{6.81}$$

$$C(0,t)=C_L,\ C(\infty,t)=C_0,\ C(x,0)=C_0 \tag{6.81a}$$

where C_L is the concentration of the transferred substance on the surface, the concentration of the substance in the volume of the flow. To solve this problem, we introduce a new variable $\eta = x/\sqrt{4Dt}$, taking into account that we define the derivatives entering into this equation

$$\frac{\partial^2 C}{\partial x^2}=\frac{1}{4Dt}\frac{\partial^2 C}{\partial \eta^2},\ \frac{\partial C}{\partial t}=-\frac{x^2}{8Dt^2\eta}\frac{\partial C}{\partial \eta} \tag{6.82}$$

Substituting these derivatives into equation (6.81), we obtain

$$\frac{d^2C}{d\eta^2} = -2\eta\frac{dC}{d\eta} \tag{6.83}$$

which is an ordinary differential equation. Introducing the variable $Z = {dC}/{d\eta}$, we obtain the equation

$$\frac{dZ}{d\eta} = -2\eta Z \tag{6.84}$$

the solution of which is presented in the form

$$Z = B_1 \exp\left(-\eta^2\right) = \frac{dC}{d\eta} \tag{6.85}$$

Reintegrating this equation, we have

$$C = B_1 \int_0^{\eta} \exp\left(-\eta^2\right) d\eta + B_2 \tag{6.86}$$

We find the integration coefficients from the boundary conditions

$$t = 0,\ \eta \to \infty,\ C = C_0;\ \ x = 0,\ \eta = 0,\ C = C_L \tag{6.87}$$

The integration coefficients in equation (6.86), with allowance for the expression $\int_0^{\infty} \exp\left(-\eta^2\right) d\eta = \frac{\sqrt{\pi}}{2}$, are respectively equal to

$$B_1 = \left(C_0 - C_L\right)\frac{2}{\sqrt{\pi}},\ B_2 = C_L \tag{6.88}$$

Then the general solution of equation (6.81) with the boundary conditions (6.81a) can be represented in the form

$$\frac{C - C_0}{C_L - C_0} = 1 - \operatorname{erf}\left({x}/{\sqrt{4Dt}} \right) \tag{6.89}$$

where $\operatorname{erf}\left({x}/{\sqrt{4Dt}} \right) = \int_0^{x/\sqrt{4Dt}} \exp\left(-\eta^2\right) d\eta$ is the error function. Having determined the derivative of equation (6.89) as

$$\left.\frac{\partial C}{\partial x}\right|_{x=0} = -\left(C_L - C_0\right)\frac{2}{\sqrt{\pi}}\frac{1}{\sqrt{4Dt}} = -\left(C_L - C_0\right)\frac{1}{\sqrt{\pi Dt}} \tag{6.90}$$

calculate the amount of mass carried through unit of area per unit time in the form

$$j = -D\left.\frac{\partial C}{\partial x}\right|_{x=0} = \left(C_L - C_0\right)\sqrt{\frac{D}{\pi . t}} \tag{6.91}$$

The mass-transfer coefficient is defined in the form

$$\beta_L = \frac{j}{\Delta C} = \left(\frac{D}{\pi\, t}\right)^{1/2} \tag{6.91a}$$

As follows from these equations, the amount of mass transferred and the mass-transfer coefficient are inversely proportional to the phase contact time.

For the total time t_P, the amount of mass carried is

$$J = \int_0^{t_P} j dt = 2\left(C_L - C_0\right)\sqrt{\frac{Dt_P}{\pi}} \tag{6.92}$$

Using the surface update model with the update probability equal to $P(t,\tau)$

$$P(t,\tau) = \frac{\exp\left(-t/\tau\right)}{\tau} \tag{6.93}$$

(where τ is the surface renewal time) and equation (6.91), we determine the total mass flux to the surface

$$\mathrm{J} = \int_0^{\infty} P(t,\tau) j dt = \int_0^{\infty} \frac{\exp\left(-t/\tau\right)}{\tau}\sqrt{\frac{D}{\pi t}}\Delta C dt = \sqrt{\frac{D}{\pi}}\frac{\Delta C}{\tau}\int_0^{\infty} t^{1/2}\exp\left(-t/\tau\right) dt = \sqrt{\frac{D}{\tau}}\Delta C \tag{6.94}$$

The mass-transfer coefficient is defined as: $\beta_L = \left(D/\tau\right)^{1/2}$. From this equation, it follows that both for the penetration model and for the surface renewal model, the mass-transfer coefficient is proportional to the square root of the diffusion coefficient.

For different boundary conditions, these solutions will change accordingly, which will be considered below.

1. We consider the boundary conditions for a plane semi-infinite surface in the form

$$C(0,t) = 0,\ C(x,0) = \varphi(x),\ \ 0 < x < \infty \tag{6.95}$$

The one-dimensional solution of the mass-transfer problem (6.81) can be represented in the form

$$C(x,t) = \frac{1}{2\sqrt{D\pi.t}}\int_0^{\infty}\left[\exp\left(-\frac{(x-\xi)^2}{4Dt}\right) - \exp\left(-\frac{(x+\xi)^2}{4Dt}\right)\right]\varphi(\xi)d\xi \tag{6.96}$$

If we assume that $\varphi(x) = C_0$, then, dividing the integral into two terms $\alpha = \frac{\xi - x}{\sqrt{4Dt}}$ and $\alpha_1 = \frac{\xi + x}{\sqrt{4Dt}}$ introducing the variables u, the solution (6.96) is transformed to the form analogous to (6.89)

$$C(x,t) = C_0 \operatorname{erf}\left(x/\sqrt{4Dt}\right) \tag{6.96a}$$

The mass flow is defined as

$$j = C_0\sqrt{D/\pi t},\ J = 2C_0\sqrt{Dt_P/\pi} \tag{6.96b}$$

For finite dimensions of a flat surface under boundary conditions

$$C(0,t) = 0,\ C(L,t) = 0,\ C(x,0) = \varphi(x) = Ax \tag{6.97}$$

the solution of the one-dimensional mass-transfer equation is represented as

$$C(x,t) = \frac{2LA}{\pi}\sum_{n=1}^{\infty}\frac{(-1)^{n+1}}{n}\exp\left[-\left(\pi^2 n^2 \mathrm{Fo}\right)\right]\sin\frac{\pi n}{L} \tag{6.98}$$

The mass flux per unit surface per unit time is determined in the form

$$j = D\frac{\partial C}{\partial x}\Big|_{x=0} = AD\sum_{n=1}^{\infty}(-1)^{n+1}\exp\left(-\pi^2 n^2 \mathrm{Fo}\right) \tag{6.99}$$

where $\mathrm{Fo} = Dt/L^2$ is the Fourier number.

2. If a chemical reaction in the bulk of a planar particle is of the first order $A \to B$, then for a semi-infinite surface in the isothermal case, the mass-transfer equation can be written in the form

$$\frac{\partial !}{\partial t} = D\frac{\partial^2 C}{\partial x^2} + kC \tag{6.100}$$

$$C(0,t) = 0,\ C(x,0) = \varphi(x),\ 0 < x < \infty$$

where k is the rate constant of the reaction. The solution of this problem will be presented as

$$C(x.t) = \frac{\exp(-kt)}{\sqrt{4\pi Dt}}\int_0^{\infty}\left[\exp\left(-\frac{(x-\xi)^2}{4Dt}\right) - \exp\left(-\frac{(x+\xi)^2}{4Dt}\right)\right]\varphi(\xi)d\xi \tag{6.100a}$$

If the initial distribution is constant $\varphi(x) = C_0$, then the solution is simplified to the form

$$C(x.t) = C_0\exp(-kt)\operatorname{erf}\left(x/4Dt\right) \tag{6.101}$$

For finite dimensions of the surface, the solution of equation (6.100), with boundary conditions $C(0,t) = 0,\ C(L,t) = 0,\ C(t,0) = \varphi(x)$

$$C(x,t) = \frac{2}{L}\sum_{n=1}^{\infty}B_n\exp\left[-\left(\pi^2 n^2 \mathrm{Fo} + kt\right)\right]\sin\frac{\pi n}{L}x \tag{6.102}$$

where $B_n = \int_0^L \varphi(\xi)\sin\left(\frac{\pi n}{L}\xi\right)d\xi$, and k is the reaction rate constant. The mass flux per unit surface per unit time is

$$j = \frac{2\pi D}{L^2} \exp(-kt) \sum_{n=1}^{\infty} B_n n \exp\left(-\pi^2 n^2 \text{Fo}\right) \tag{6.103}$$

3. For an arbitrary volumetric chemical reaction, the mass-transfer equation with boundary conditions is written in the form

$$\frac{\partial C}{\partial t} = D \frac{\partial^2 C}{\partial x^2} + f(x) \tag{6.104}$$

$$C(0,t) = 0,\ C(x,0) = 0,\ 0 < x < \infty$$

The solution of this problem is presented in the form

$$C(x,t) = \frac{1}{\sqrt{4Dt}} \int_0^t \int_0^\infty \frac{1}{\sqrt{t-\tau}} \left[\exp\left(-\frac{(x-\xi)^2}{4D(t-\tau)} \right) - \exp\left(-\frac{(x+\xi)^2}{4D(t-\tau)} \right) \right] f(\xi) d\xi d\tau \tag{6.104a}$$

4. For mass propagation in three-dimensional space with the same diffusion coefficient, the mass-transfer equation is represented in the form

$$\frac{\partial C}{\partial t} = D\left(\frac{\partial^2 C}{\partial x^2} + \frac{\partial^2 C}{\partial y^2} + \frac{\partial^2 C}{\partial z^2} \right)$$
$$C(x, y, 0, t) = 0\ ,\ C(x, y, z, 0) = \varphi(x, y, z) \tag{6.105}$$

$$-\infty < x, y < \infty,\ 0 < x < \infty$$

The solution of this problem in general form will be presented as

$$C(x,y,z,t) = \frac{1}{(4D\pi t)^{3/2}} \int_{-\infty}^{\infty} \int_{-\infty}^{\infty} \exp\left[-\frac{(x-\xi)^2 + (y-\eta)^2}{4Dt} \right] \cdot$$
$$\times \int_0^\infty \left[\exp\left(-\frac{(z-\zeta)^2}{4Dt} \right) - \exp\left(-\frac{(z+\xi)^2}{4Dt} \right) \right] \varphi(\xi,\eta,\zeta) d\xi d\eta d\zeta \tag{6.106}$$

This problem is typical for the propagation of a mass in infinite space $0 \le x, y, z \le \infty$. For the one-dimensional case, equation (6.30) in spherical coordinates is written as

$$\frac{\partial C}{\partial t} = \frac{D}{r^n} \frac{\partial}{\partial r} \left(r^n \frac{\partial C}{\partial r} \right) \tag{6.107}$$

The solution of this equation is represented as

$$C(r,t) = \frac{m_i \exp\left(-\frac{r^2}{4Dt} \right)}{(4\pi Dt)^{\frac{1+n}{2}}} \tag{6.107a}$$

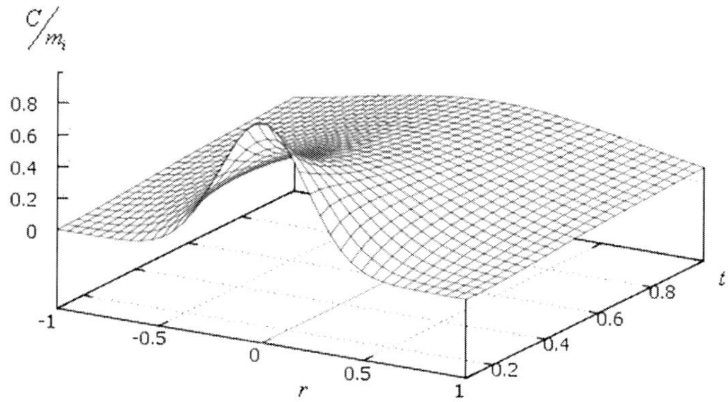

FIGURE 6.4 Distribution of concentration along the propagation radius mass.

Here, $n = 0$ is a flat surface, $m_i = \int_{-\infty}^{\infty} C(r,t)\,dr$; $n = 1$ is a cylindrical surface $m_i = \int_0^{\infty} C(r,t)\,2\pi r\,dr$; $n = 2$ is a spherical surface, and $m_i = \int_0^{\infty} C(r,t)\,4\pi r^2 dr$. The value of the maximum concentration is

$$C(0,t) = \frac{m_i}{(4\pi Dt)^{\frac{1+n}{2}}} \tag{6.108}$$

The distribution of the concentration of matter in space for different values of time is given in Figure 6.4. As follows from this figure, the maximum value of the concentration depends on the diffusion coefficient and the propagation time of the mass. Far from the source of mass, the concentration of matter decreases monotonically and tends asymptotically to zero.

5. In a three-dimensional bounded space with a first-order volume chemical reaction, where $x \le L$, $y \le L$, and $z \le L$ (z is vertical coordinate), when the substance concentration at the plate ends is zero, the mass-transfer equation and the boundary conditions are represented as

$$\frac{\partial C}{\partial t} = D\left(\frac{\partial^2 C}{\partial x^2} + \frac{\partial^2 C}{\partial y^2} + \frac{\partial^2 C}{\partial z^2}\right) - kC$$

$$C(0,y,z,t) = C(L,y,z,t) = 0,\ C(x,0,z,t) = C(x,L,z,t) = 0 \tag{6.109}$$

$$C(x,y,0,t) = C(x,y,L,t) = 0,\ C(x,y,z,0) = C_0$$

The solution of this boundary-value problem by the method of separation of variables is represented in the form

$$C(x,y,z,t) = \frac{64C_0}{\pi^3}\exp(-kt)\sum_{k=0}^{\infty}\sum_{n=0}^{\infty}\sum_{m=0}^{\infty} A_{kmn}\exp\left(-\pi^2\lambda_{kmn}\mathrm{Fo}\right)\sin\frac{(2k+1)\pi x}{L}$$
$$\times\sin\frac{(2m+1)\pi y}{L}\sin\frac{(2n+1)\pi z}{L} \tag{6.110}$$

Here, $A_{kmn} = \left[(2k+1)(2m+1)(2n+1)\right]^{-1}$, $\lambda_{kmn} = (2k+1)^2 + (2m+1)^2 + (2n+1)^2$.

For small values of the number $\mathrm{Fo} = {Dt}/{L^2} < 0,3$ it suffices to restrict ourselves to the first term of the series

$$C(x,y,z,t) \approx \frac{64C_0}{\pi^3}\exp\left(-kt - \pi^2\mathrm{Fo}\right)\sin\frac{\pi x}{L}\sin\frac{\pi y}{L}\sin\frac{\pi z}{L} \tag{6.110a}$$

The mass flow of matter per unit surface per unit time in the direction of the coordinate is defined as

$$\frac{\partial C}{\partial z} = \frac{64C_0}{\pi^2 L}\exp\left(-kt - \pi^2\mathrm{Fo}\right)\sin\frac{\pi x}{L}\sin\frac{\pi y}{L} \tag{6.111}$$

$$j = D\frac{\partial C}{\partial z}\Big|_{z=L} = \frac{64DC_0}{\pi^2 L}\exp\left(-kt - \pi^2\mathrm{Fo}\right)\sin\frac{\pi x}{L}\sin\frac{\pi y}{L} \tag{6.112}$$

The mass flux to the entire wall surface per unit time, taking into account $\int_0^L\int_0^L \sin\frac{\pi.x}{L}\sin\frac{\pi.y}{L}\,dxdy = \frac{4L^2}{\pi^2}$, is equal to

$$j = \int_{(s)} j ds = \frac{256}{\pi^4}DC_0L\exp\left(-kt - \pi^2\mathrm{Fo}\right) \tag{6.113}$$

The total mass flow over time will be determined as

$$J = \int_0^{t_P} J dt = 0,657\frac{DC_0L^3}{kL^2 - \pi^2 D}\left(1 - \exp\left(-kt_P - \pi^2\mathrm{Fo}_P\right)\right) \tag{6.114}$$

6.3.2 Non-Stationary Mass Transfer on a Cylindrical Surface

Consider the transfer of the mass to the cylindrical surface of a particle or pipe in a stationary liquid. The solution of the problems of mass transfer on a cylindrical surface will also be realized under different boundary conditions.

1. Assume that the mass transfer is described by equation

$$\frac{\partial C}{\partial t} = D\left[\frac{1}{r}\frac{\partial}{\partial r}\left(r\frac{\partial C}{\partial r}\right) + \frac{\partial^2 C}{\partial z^2}\right] \tag{6.115}$$

with boundary conditions

$$C(r,0,t) = C(r,L,t) = C(R,z,t) = C_0,\ C(r,z,0) = 0 \tag{6.116}$$

The solution of this boundary-value problem by the method of separation of variables is represented in the form

$$\frac{C(r,z,t)}{C_0} = 1 + \frac{4}{\pi}\sum_{k=0}^{\infty}\sum_{n=0}^{\infty}\frac{1}{(2k+1)J_1(\mu_n)}\exp\left[-(2k+1)^2\pi^2 \mathrm{Fo}_L\right] \times\exp\left(-\mu_n^2 \mathrm{Fo}_R\right)J_0\left(\frac{\mu_n r}{R}\right)\sin\frac{(2k+1)\pi z}{L} \tag{6.117}$$

where $\mathrm{Fo}_L = {Dt}/{L^2}$, $\mathrm{Fo}_R = {Dt}/{R^2}$ μ_n are eigenvalues, which are the positive roots of the equation $J_0(\mu) = 0$. Restricting ourselves to the first term in (6.117) for small values of Fourier numbers, we can write

$$\frac{C(r,z,t)}{C_0} = 1 + \frac{4J_0\left(\frac{\mu_1 r}{R}\right)}{\pi J_1(\mu_1)}\exp\left(-\pi^2 \mathrm{Fo_L} - \mu_1^2 \mathrm{Fo_R}\right)\sin\frac{\pi z}{L} \tag{6.118}$$

The amount of mass carried per unit time per unit cylindrical surface is

$$\left.\frac{\partial C}{\partial r}\right|_{r=R} = A_0\frac{C_0}{R}\sin\frac{\pi . z}{L}\exp\left(-\pi^2 \mathrm{Fo_L} - \mu_1^2 \mathrm{Fo_R}\right) \tag{6.119}$$

$$j = A_0\frac{DC_0}{R}\sin\frac{\pi . z}{L}\exp\left(-\pi^2 \mathrm{Fo_L} - \mu_1^2 \mathrm{Fo_R}\right) \tag{6.120}$$

where $A_0 = \frac{4\mu_1}{\pi J_1(\mu_1)} = 5{,}9$, $\mu_1 = 2{,}4048$, $J_1(\mu_1) = 0{,}5191$. Mass flow to the surface of a cylindrical particle in length L per unit time, provided that $S = 2\pi R\int_0^L \sin\frac{\pi z}{L}dz$

$$J = \int_{(s)} j\,dS = 23{,}6DC_0L\exp\left(-\pi^2 \mathrm{Fo_L} - \mu_1^2 \mathrm{Fo_R}\right) \tag{6.121}$$

The total mass flow in time is

$$J = \int_0^{t_P} J dt = \frac{23{,}6L^3R^2C_0}{\pi^2R^2 + \mu_1^2L^2}\left[1 - \exp\left(-\pi^2 \mathrm{Fo_{PL}} - \mu_1^2 \mathrm{Fo_{PR}}\right)\right] \tag{6.122}$$

where $\mathrm{Fo}_{PL} = {Dt_P}/{L^2}$, $\mathrm{Fo}_{PR} = {Dt_P}/{R^2}$

6.3.3 Non-Stationary Mass Transfer on a Spherical Surface

These problems relate to the case of an inhomogeneous distribution of the concentration over the surface of a spherical particle. The general equation of non-stationary mass transfer on the surface of a spherical particle for an asymmetric case is described in the form

$$\frac{\partial C}{\partial t} = D\left(\frac{1}{r^2}\frac{\partial}{\partial r}\left(r^2\frac{\partial C}{\partial r}\right) + \frac{1}{r^2\sin\psi}\frac{\partial}{\partial\psi}\left(\sin\psi\frac{\partial C}{\partial\psi}\right) + \frac{1}{r^2\sin^2\psi}\frac{\partial^2 C}{\partial\varphi^2}\right) + w(C) \tag{6.123}$$

with the corresponding boundary conditions (where ψ, φ are polar angles). The solution of equation (6.123) for various inhomogeneous boundary conditions presents certain difficulties.

Suppose that in the absence of an internal mass source $w(C)=0$, the concentration on the surface of the sphere is constant $C(R,\psi,\phi,t)=0$ (a boundary condition of the first kind), and the initial concentration of the surface is an arbitrary function, $C(r,\psi,\phi,0)=f(r,\psi,\phi)$ and $0\le r\le R$, $0\le\psi\le\pi$, $0\le\phi\le 2\pi$. The general solution of this boundary-value problem in the absence of a chemical reaction is represented in the form

$$C(r,\psi,\phi,t)=\sum_{n=0}^{\infty}\sum_{m=1}^{\infty}\sum_{k=0}^{n}\exp\left(-\mu_{nm}^{2}\text{Fo}\right)\frac{1}{\sqrt{r}}J_{n+1/2}\left(\frac{\mu_{n,m}r}{R}\right)\times \left(A_{mnk}\cos k\phi+B_{mnk}\sin k\phi\right)P_{n}^{k}\left(\cos\psi\right) \tag{6.124}$$

where $\mu_{n,m}$ is the positive roots of the equation, $J_{n+1/2}(\mu)=0$

$$A_{mnk}=\frac{(2n+1)(n-k)!}{\pi\alpha_k(n+k)!\left[J_{n+1/2}'(\mu_{n,m})\right]^2R^2}\int_0^R\int_0^\pi\int_0^{2\pi}r^{3/2}f(r,\psi,\phi)J_{n+1/2}\left(\frac{\mu_{n,m}r}{R}\right)\times \cos k\phi\, P_n^k(\cos\psi)\sin\psi\, dr\, d\psi\, d\phi$$

$$B_{mnk}=\frac{(2n+1)(n-k)!}{\pi\alpha_k(n+k)!\left[J_{n+1/2}'(\mu_{n,m})\right]^2R^2}\int_0^R\int_0^\pi\int_0^{2\pi}r^{3/2}f(r,\psi,\phi)J_{n+1/2}\left(\frac{\mu_{n,m}r}{R}\right)\times \sin k\phi\, P_n^k(\cos\psi)\sin\psi\, dr\, d\psi\, d\phi \tag{6.125}$$

$\alpha_k=2$, if $k=0$ and $\alpha_k=1$, if $k\neq 0$; $P_n^k(\cos\psi)$ is the ad joint Legendre functions, some of which are described in the corresponding reference books. If we assume that the concentration distribution does not depend on the angle, then the mass-transfer equation is represented as

$$\frac{\partial C}{\partial t}=D\left[\frac{1}{r^2}\frac{\partial}{\partial r}\left(r^2\frac{\partial C}{\partial r}\right)+\frac{1}{r^2\sin\phi}\frac{\partial}{\partial\phi}\left(\sin\phi\frac{\partial C}{\partial\phi}\right)\right] \tag{6.126}$$

$$C(r,\phi,t)=0\,,\ C(r,\phi,0)=f(r,\phi),\quad 0\le r\le R,\ 0\le\phi\le\pi$$

The solution of the boundary-value problem is represented in the form

$$C(r,\phi,t)=\sum_{n=0}^{\infty}\sum_{m=0}^{\infty}\frac{a_{nm}}{\sqrt{r}}\exp\left(-\mu_{nm}^{2}\text{Fo}\right)P_n(\cos\phi)J_{n+1/2}\left(\frac{\mu_{nm}r}{R}\right) \tag{6.126a}$$

Here, $a_{nm}=\dfrac{2n+1}{R^2\left[J'_{n+1/2}(\mu_{nm})\right]^2}\displaystyle\int_0^R\int_0^\pi r^{3/2}f(r,\psi)J_{n+1/2}\left(\frac{\mu_{nm}r}{R}\right)P_n(\cos\phi)\sin\phi\, dr\, d\phi$, μ_{nm} is the positive root of the equation $J_{n+1/2}(\mu)=0$.

If the concentration of matter in the volume is a constant $C(r,\phi,0)=C_\infty$, then the coefficient in equation (6.126a) is transformed to the form

$$a_{nm} = C_\infty \frac{2n+1}{R^2\left[J'_{n+1/2}(\mu_{nm})\right]^2}\int_0^R\int_0^\pi r^{3/2} J_{n+1/2}\left(\frac{\mu_{nm} r}{R}\right) P_n(\cos\phi)\sin\phi\, dr\, d\phi \tag{6.127}$$

Taking this equation into account, we define the mass flow to a spherical surface in the form

$$j = \frac{DC_\infty}{R}\sum_{n=0}^{\infty}\sum_{m=1}^{\infty} b_{nm}\exp(-\mu_{nm}\mathrm{Fo})P_n(\cos\phi) \tag{6.128}$$

Here, $b_{nm} = \frac{2n+1}{R^{5/2}}\frac{\mu_{nm}}{J'_{n+1/2}}\int_0^R\int_0^\pi r^{3/2}J_{n+1/2}\left(\frac{\mu_{nm}r}{R}\right)P_n(\cos\phi)\sin\phi\, dr\, d\phi$. If we confine ourselves to the first two terms of the indicated series $(n=0;1)$ for a number Fo < 0.3, setting $P_0(\cos\phi)=1,\ P_1(\cos\phi)=\sin\phi$ and $\int_0^1 r^{3/2}J_{1/2}\left(\frac{\mu_0 r}{R}\right)dr = \frac{R^{5/2}}{\mu_0}J_{3/2}\left(\frac{\mu_0 r}{R}\right)$, we obtain

$$b_0 = \frac{2C_\infty R^{1/2}}{\mu_0}\frac{J_{3/2}(\mu_0)}{\left[J'_{1/2}(\mu_0)\right]^2} \tag{6.129}$$

$$C(r,t) = 2C_\infty\left(\frac{R}{r}\right)^{1/2}\sin\phi\frac{J_{3/2}(\mu_0)}{\mu_0\left[J'_{1/2}(\mu_0)\right]^2}J_{1/2}\left(\frac{\mu_0 r}{R}\right)\exp\left(-\mu_0^2\mathrm{Fo}\right) \tag{6.130}$$

The mass flow per unit surface is

$$\left.\frac{\partial C}{\partial r}\right|_{r=R} = \frac{2C_\infty}{R}\sin\phi\exp\left(-\mu_0^2\mathrm{Fo}\right)\frac{J_{3/2}(\mu_0)}{J'_{1/2}(\mu_0)} \tag{6.131}$$

$$j = -\frac{2C_\infty D}{R}\sin\phi\exp\left(-\mu_0^2\mathrm{Fo}\right)\frac{J_{3/2}(\mu_0)}{J'_{1/2}(\mu_0)}$$

Assuming that, $\mu_0 \approx 3{,}5$, $J_{3/2}(\mu_0)\approx 0.48$ and $J'_{1/2}(\mu_0)\approx J_{-1/2}(\mu_0)\approx 0.45$, then (6.131) is represented in the form

$$j \approx -\frac{2C_\infty D}{R}\exp\left(-\mu_0^2\mathrm{Fo}\right)\sin\phi \tag{6.132}$$

1. We consider the non-stationary transport of matter in a spherical particle with a convective transport of matter on the surface, described by the equation

$$\frac{\partial C}{\partial t} = \frac{D}{r^2}\frac{\partial}{\partial r}\left(r^2\frac{\partial C}{\partial r}\right)$$
$$\left.\frac{\partial C}{\partial r}\right|_{r=R} = -\frac{\beta_L}{D}(C - C_p);\quad r=0,\quad C(r,t)=C_0 \tag{6.133}$$

The solution of the boundary-value problem (6.133) can be represented in the form

$$C(r,t)=C_p+\frac{2R}{r}\frac{\beta_L}{D}(C_0-C_p)\sum_{n=1}^{\infty}A_n\exp\left(-\mu_n^2Dt\right)\sin\left(\mu_nr\right)$$

$$A_n=\frac{\cos\left(\mu_nR\right)}{\mu_n\left(1-\frac{R\beta_L}{D}-\cos^2\left(\mu_nR\right)\right)} \tag{6.133a}$$

where μ_n are the positive roots of the equation $\mathrm{tg}\left(\mu R\right)=\frac{\mu R}{1-\mu R}$. The flow of matter from a surface unit is defined as

$$j=-D\frac{\partial C}{\partial r}\bigg|_{r=R}=\frac{2\beta_L\left(C_0-C_p\right)}{R}\sum_{n=0}^{\infty}\tilde{A}_n\exp\left(-\mu_n^2\mathrm{R}^2\mathrm{Fo}\right)\left[\mathrm{tg}\left(\mu_nR\right)-\mu_nR\right]$$

$$\tilde{A}_n=1\Big/\mu_n\left(1-{}^{R\beta_L}\!/_D-\cos^2\left(\mu_nR\right)\right) \tag{6.133b}$$

where $\mathrm{Fo}=\frac{Dt}{R^2}$ is the Fourier diffusion number.

2. Consider the transport of matter in a spherical particle with an arbitrary distribution of the initial concentration along the radius and the external convective transport of matter

$$\frac{\partial C}{\partial t}=\frac{D}{r^2}\frac{\partial}{\partial r}\left(r^2\frac{\partial C}{\partial r}\right)$$

$$\frac{\partial C}{\partial r}\bigg|_{r=R}=-\frac{\beta_L}{D}\left(C-C_p\right);\quad t=0,\quad C\left(r,t\right)=f\left(r\right) \tag{6.134}$$

The solution of this boundary-value problem by the method of separation of variables is represented in the form

$$C(r,t)=C_p+\frac{2\left(1-{}^{R\beta_L}\!/_D\right)}{Rr}\sum_{n=1}^{\infty}A_n\exp\left(-\mu_n^2Dt\right)\sin\left(\mu_nr\right)$$

$$A_n=\frac{1}{1-{}^{R\beta_L}\!/_D-\cos^2\left(\mu_n^2R\right)}\left[\int_0^R rf\left(r\right)\sin\left(\mu_nr\right)dr\right] \tag{6.134a}$$

The flux of matter per unit surface per unit time is

$$j = -D\left.\frac{\partial C}{\partial r}\right|_{r=R} = \frac{2(D-\beta_L R)}{R^3}\sum_{n=1}^{\infty} A_n \exp\left(-\mu_n^2 \mathrm{R}^2 \mathrm{Fo}\right)\left[\sin(\mu_n R) - \mu_n R\cos(\mu_n R)\right] \tag{6.134b}$$

3. Consider the process of gas absorption in a spherical drop of an absorbent described by the equation (penetration model)

$$\frac{\partial C}{\partial t} = \frac{D}{r^2}\frac{\partial}{\partial r}\left(r^2 \frac{\partial C}{\partial r}\right)$$
$$t = 0,\ C = C_0;\quad r = R,\ \ C = C_s \tag{6.135}$$

where C_s is the equilibrium concentration. The solution of this equation with given boundary conditions, in analogy with (6.134), can be represented in the form

$$\theta = \frac{C - C_0}{C_s - C_0} = 1 + \frac{2R}{\pi r}\sum_{n=1}^{\infty}\frac{(-1)^n}{n}\sin\left(\frac{n\pi r}{R}\right)\exp\left(-\frac{D\pi^2 t}{R^2}n^2\right) \tag{6.136}$$

The average concentration of the absorbed gas is determined as

$$\theta_s = \int_0^R \frac{4\pi r^2 \theta}{4\pi R^3/3}dr = 1 - \frac{6}{\pi^2}\sum_{n=1}^{\infty}\frac{1}{n^2}\exp\left(-4n^2\pi^2 \mathrm{Fo}\right) \tag{6.137}$$

Here, $\mathrm{Fo} = \frac{Dt}{a^2}$ is the Fourier number. The total mass flux to the surface of the drop in time will be determined in the form

$$J = \left(\frac{1}{4\pi R^2 t_c}\right)\left[\frac{4\pi R^3 (C_s - C_0)}{3}\right] = \left(\frac{D}{6t_c}\right)(C_s - C_0)\left[1 - \frac{6}{\pi^2}\sum_{n=1}^{\infty}\frac{1}{n^2}\exp\left(-4n^2\pi^2 \mathrm{Fo}\right)\right] \tag{6.138}$$

The Sherwood number is defined as

$$\mathrm{Sh} = \frac{Ja}{D\Delta C} = \frac{1}{6\mathrm{Fo}}\left[1 - \frac{6}{\pi^2}\sum_{n=1}^{\infty}\frac{1}{n^2}\exp\left(-4n^2\pi^2 \mathrm{Fo}\right)\right] \tag{6.139}$$

The connection between the Sherwood number and the Fourier number according to the penetration model (6.139) is shown in Figure 6.5.

These solutions allow us to determine the mass-transfer coefficient between a gas stream and a single drop of a liquid absorbent. Below we will consider mass transfer in absorption processes using a film model.

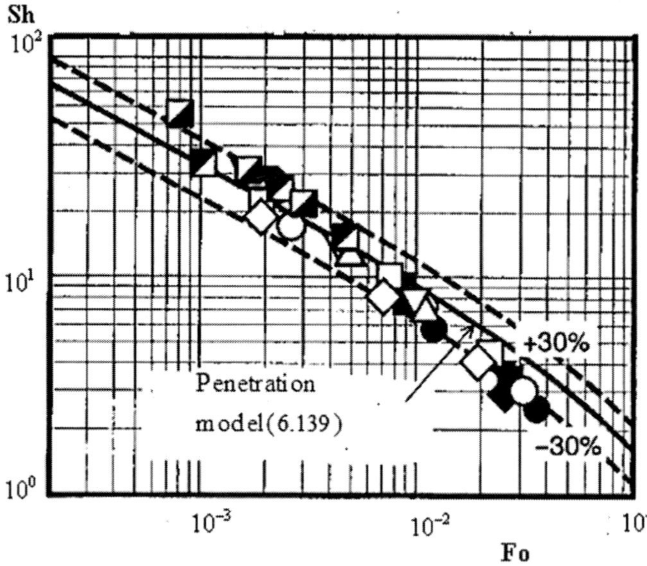

FIGURE 6.5 Dependence Sh on Fo (points-the experiment on the absorption of carbon dioxide in water). (From Koisi Asano Mass Transfer, *From Fundamentals to modern Industrial Application*, WILEY–VSH Verlag GmbH and Co, Weinheim, 2006.)

6.4 APPLIED PROBLEMS OF MASS TRANSFER

This section will cover application tasks associated with the transfer of mass in various processes of chemical technology. As a result of solving the mass-transfer equations for a particular process, the mass flux to the surface of the particle, the Sherwood number, and the mass-transfer coefficient will be determined.

6.4.1 Mass Transfer in the Processes of Gas–Liquid Absorption

In absorption processes, the absorption of a component from the gas phase is realized as a result of the transfer of the mass of matter from the gas phase to the liquid through the interfacial film (film model), that is, the solvent selectively absorbs this or that gas from the gas mixture (Figure 6.6). In the boundary layer, the equilibrium between the gas and liquid phases is described by Henry's law $y_P = mx_P$ (where $m = {}^{H(T)}\!/_{P}$, P is total pressure, and $H(P)$ is Henry's coefficient).

The transport equations for small values of the number for the gas and liquid phases are described by equations

$$D_g \frac{\partial^2 y}{\partial z_1^2} = \frac{\partial y}{\partial t} \tag{6.140}$$

$$D_S \frac{\partial^2 x}{\partial z_2^2} = \frac{\partial x}{\partial t} \tag{6.140a}$$

Here, x is the concentration of the absorbed substance in the liquid phase, y is the concentration of the absorbed substance in the gas phase, and D_g, D_S are the diffusion coefficients of the substance in the gas and liquid phases. We formulate the boundary conditions for the solution of equations (6.140)

$$z_1 = z_2 = 0,\ y_P = mx_P,\ y = y_P,\ x = x_P \tag{6.141}$$

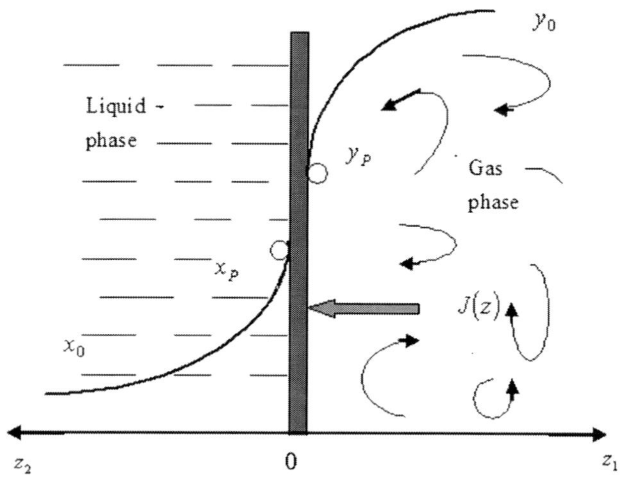

FIGURE 6.6 Mass-transfer scheme during absorption.

$$D_g \frac{\partial y}{\partial z_1} = -D_S \frac{\partial x}{\partial z_2} \tag{6.141a}$$

$$z_1 = z_2 = \infty, \frac{\partial y}{\partial z_1} = \frac{\partial x}{\partial z_2} = 0 \tag{6.141b}$$

$$t = 0, \ y = y_0, \ x = x_0 \tag{6.141c}$$

To solve this boundary-value problem, we introduce new variables

$$\eta_1 = \frac{z_1}{\sqrt{4D_g t}}, \eta_2 = \frac{z_2}{\sqrt{4D_S t}} \tag{6.142}$$

Using these variables, the solution of equations (6.140) will be represented as

$$y = (y_0 - mx_P)\operatorname{erf}(\eta_1) + mx_P, \ x = (x_0 - x_P)\operatorname{erf}(\eta_2) + x_P \tag{6.143}$$

Taking into account the values $y_P = mx_P$ of the derivatives

$$\left.\frac{\partial y}{\partial z_1}\right|_{\substack{z_1=0\\ \eta_1=0}} = \frac{2}{\sqrt{\pi}}(y_0 - mx_P)\frac{1}{\sqrt{4D_g t}} \tag{6.144}$$

$$\left.\frac{\partial x}{\partial z_2}\right|_{\substack{z_2=0\\ \eta_2=0}} = \frac{2}{\sqrt{\pi}}(x_0 - x_P)\frac{1}{\sqrt{4D_S t}}$$

we obtain a relation between the equilibrium values in the form

$$x_P = \frac{\frac{1}{\sqrt{D_S}} y_0 + \frac{1}{\sqrt{D_g}} x_0}{\frac{1}{\sqrt{D_g}} + m\frac{1}{\sqrt{D_S}}} \tag{6.145}$$

As a result, solutions (6.143) are represented in the form

$$y = \frac{\frac{1}{\sqrt{D_g}}\left(y_0 - mx_0\right)}{\frac{1}{\sqrt{D_g}} + \frac{m}{\sqrt{D_S}}} \operatorname{erf}\left(\frac{z_1}{\sqrt{4D_g t}}\right) + \frac{\frac{my_0}{\sqrt{D_S}} + \frac{mx_0}{\sqrt{D_g}}}{\frac{1}{\sqrt{D_g}} + \frac{m}{\sqrt{D_S}}} \tag{6.146}$$

$$x = \frac{\frac{1}{\sqrt{D_S}}\left(mx_0 - y_0\right)}{\frac{1}{\sqrt{D_g}} + \frac{m}{\sqrt{D_S}}} \operatorname{erf}\left(\frac{z_2}{\sqrt{4D_S t}}\right) + \frac{\frac{y_0}{\sqrt{D_S}} + \frac{x_0}{\sqrt{D_g}}}{\frac{1}{\sqrt{D_g}} + \frac{m}{\sqrt{D_S}}}$$

These equations determine the profile of the substance to be absorbed in the gas and liquid phases. The amount of the substance transferred from the gas phase to the liquid phase will be determined as

$$j = D_g \frac{\partial y}{\partial z_1}\Big|_{z_1=0} = D_g \frac{\frac{1}{\sqrt{D_g}}\left(y_0 - mx_0\right)}{\frac{1}{\sqrt{D_g}} + \frac{m}{\sqrt{D_S}}} \exp\left(\eta_1^2\right)\Big|_{\eta_1=0} \frac{2/\sqrt{\pi}}{\sqrt{4D_g t}}$$

$$= \frac{y_0 - mx_0}{\frac{1}{\sqrt{D_g}} + \frac{m}{\sqrt{D_S}}} \frac{1}{\sqrt{\pi . t}} \frac{P}{RT} \tag{6.147}$$

The amount of the transferred substance over time is defined as

$$J = \int_0^{t_P} j dt = \frac{y_0 - mx_0}{\frac{1}{\sqrt{D_g}} + \frac{m}{\sqrt{D_S}}} \left(\frac{4t_P}{\pi}\right)^{1/2} \frac{P}{RT} \tag{6.148}$$

For the gas phase, the absorption coefficient is determined from equation (6.148)

$$K_g = \frac{P}{RT} \frac{\left(\frac{4t_P}{\pi}\right)^{1/2}}{\frac{1}{\sqrt{D_g}} + \frac{m}{\sqrt{D_S}}} \tag{6.149}$$

If the concentration of the absorbed substance in the solvent is zero $x_0 = 0$, the flow of matter is determined in the form

$$J = \left(\frac{4t_P}{\pi}\right)^{1/2} \frac{y_0 P}{RT} \frac{1}{\frac{1}{\sqrt{D_g}} + \frac{m}{\sqrt{D_S}}} \tag{6.150}$$

As the temperature decreases, the absorption coefficient (6.149) increases, which leads to an increase in the solubility of gases. The increase in the mass-transfer coefficients can occur under the influence of convection caused by local gradients of surface tension, especially with the

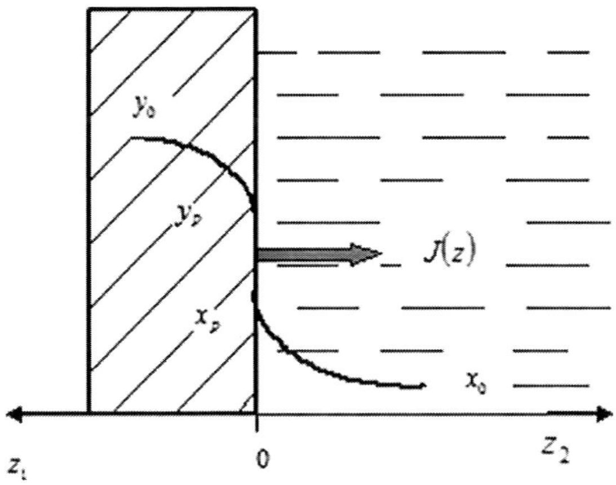

FIGURE 6.7 Flow pattern during the extraction of matter.

simultaneous occurrence of chemical reactions. Absorption is often carried out in the form of an absorption-desorption cycle; however, the desorption stage may be absent if a finished product is obtained as a result of absorption or regeneration of the absorber is impossible. The height of the nozzle or the height of the column is calculated on the basis of determining the number of units of transfer.

6.4.2 Mass Transfer in the Extraction Processes in the Liquid–Solid Phase

Extraction of a substance from the solid of a substance by a liquid solvent is realized as a result of diffusion transfer of the absorbed substance from the solid to the liquid phase. In principle, the mechanism of matter transport inside the solid phase is diffusive, and from the surface to the volume of the liquid phase it can be both diffusive and convective (Figure 6.7).

The equations of mass transfer in the solid and liquid phases are presented in the form

$$D_K \frac{\partial^2 y}{\partial z_1^2} = \frac{\partial y}{\partial t} \tag{6.151}$$

$$D_S \frac{\partial^2 x}{\partial z_2^2} = \frac{\partial x}{\partial t} \tag{6.151a}$$

Here, y, x are the concentration of the substances absorbed in the solid and liquid phases; D_K, D_S are the diffusion coefficients of the substance in the solid and liquid phases; y_0, x_0 are the initial concentration of the substance in the solid and liquid phases; and y_K, x_K are the equilibrium concentrations of the substance at the contact surface of the solid and liquid phases, which obeys Henry's law. Assuming that the substance is transported from the surface to the volume of the liquid by convection, we formulate the boundary conditions of the problem in the following way

$$z_1 = z_2 = 0,\ y = y_P,\ x = x_P,\ y_P = m x_P \tag{6.152}$$

$$D_K \frac{\partial y}{\partial z_1} = \beta\left(x - x_0\right) \tag{6.152a}$$

$$t = 0,\ y = y_0,\ x = x_0 \tag{6.152b}$$

where β is the mass-transfer coefficient. Introducing the new variables $\eta_1 = {}^{z_1}\!/\!\sqrt{4D_K t}$ and $\eta_2 = {}^{z_2}\!/\!\sqrt{4D_S t}$, with the boundary conditions (6.152) are represented in the form

$$y = (y_0 - mx_P)\operatorname{erf}(\eta_1) + mx_P \tag{6.153}$$

$$x = (x_0 - x_P)\operatorname{erf}(\eta_2) + x_P \tag{6.153a}$$

Then, using condition (6.152a), we can write

$$D_K(y_0 - mx_P)\frac{2}{\sqrt{\pi}}\exp\left(-\eta_1^2\right)\Big|_{\eta_1=0}\frac{1}{\sqrt{4D_K t}} = \beta\left[(x_0 - x_P)erf(\eta_1)| + x_P - x_0\right]_{\eta_1=0} \tag{6.154}$$

From this expression we have

$$x_P = \frac{y_0\sqrt{\frac{D_K}{\pi t}} + \beta x_0}{m\sqrt{\frac{D_K}{\pi t}} + \beta} = \frac{y_0 + \gamma x_0}{m + \gamma} \tag{6.155}$$

Then the solutions of (6.153) are written in the form

$$\begin{aligned} y &= \frac{m}{m+\gamma}\left[(y_0 - mx_0)\operatorname{erf}(\eta_1) + y_0 + \gamma x_0\right] \\ x &= \frac{1}{m+\gamma}\left[(mx_0 - y_0)\operatorname{erf}(\eta_2) + y_0 + \gamma x_0\right] \end{aligned} \tag{6.156}$$

The amount of matter transported per unit surface per unit time is equal to

$$j = D_K\frac{\partial y}{\partial z_1}\Big|_{z_1=0} = \frac{m}{m+\gamma}\sqrt{\frac{D_K}{\pi . t}}(y_0 - mx_0)\frac{P}{RT} \tag{6.157}$$

The amount of substance carried over time in time is equal to

$$J = \int_0^{t_K} j dt = \frac{m}{m+\gamma}\sqrt{\frac{4D_K t}{\pi}}(y_0 - mx_0)\frac{P}{RT} \tag{6.158}$$

The proposed solutions are simplified if at the initial instant of time the solvent is considered to be pure, that is, $x_0 = 0$

$$y = \frac{my_0}{m+\gamma}\left(1 + \operatorname{erf}(\eta_1)\right),\ x = \frac{y_0}{m+\gamma}\left(1 - \operatorname{erf}(\eta_2)\right) \tag{6.159}$$

$$j = \frac{m}{m+\gamma}\sqrt{\frac{D_K}{\pi t}}\frac{y_0 P}{RT}$$

6.4.3 Non-Stationary Dissolution of a Flat Particle

Dissolution of solids is a process of heterophasic physicochemical interaction of a solid and a liquid, accompanied by the transition of the solid phase into a solution. In contrast to the extraction process, in which the solids contacted with the solution consist of two or more soluble and inert solid phases, the dissolution is selective and is mainly carried out by molecular diffusion.

Let us consider the diffusion one-dimensional dissolution of a planar solid particle in a liquid flow with an initial solute concentration equal to zero and described by the mass-transfer equation in the form

$$\frac{\partial C}{\partial t} = D\frac{\partial^2 C}{\partial x^2} \tag{6.160}$$

$$x = 0,\ C = C_0,\ t = 0,\ C = 0$$

where C_0 is the concentration of soluble substance on the surface.

Introducing a new variable $\eta = \frac{x}{\sqrt{4Dt}}$, the solution of this boundary-value problem is represented in the form

$$C(x,t) = C_0 - \frac{2}{\sqrt{\pi}}C_0\int_0^{\eta}\exp(-\eta^2)d\eta = C_0\left(1 - \text{erf}\left(\frac{x}{\sqrt{4Dt}}\right)\right) \tag{6.161}$$

The mass flux per unit surface is determined in the form

$$j = -D\frac{\partial C}{\partial x}\Big|_{x=0} = D\frac{2C_0}{\sqrt{\pi}}\exp(-\eta^2)\Big|_{\eta=0}\frac{1}{\sqrt{4Dt}} = C_0\sqrt{\frac{D}{\pi . t}} \tag{6.162}$$

6.4.4 Non-Stationary Evaporation of a Fixed Unit Drop

Evaporation of a liquid is widely used in mass-exchange processes of chemical and oil refining technology. This is the evaporation of a liquid during the separation of liquid mixtures in distillation columns, in the drying of a porous material, in the processes of liquid spraying and combustion, etc.

The problem of evaporation of a single spherical drop in a gaseous medium with given parameters can be regarded as classical. In the simplest case, it was solved by Maxwell, using for the spherical symmetric geometry the diffusion equation (Fick's first law) for the flux of evaporated particles from the droplet

$$J = -4\pi DR^2\frac{\partial C}{\partial r}\Big|_{r=R} \tag{6.163}$$

Integrating this equation, we can determine the distribution of the vapor concentration outside the drop

$$C = \frac{J}{4\pi RD} + C_\infty \tag{6.164}$$

where C_∞ is the concentration of steam is at a great distance from the surface of the drop. Assuming the equality of the concentration above the droplet surface and the saturation concentration, we obtain

$$C_S = \frac{J}{4\pi RD} + C_\infty,\ J = 4\pi RD(C_S - C_\infty) \tag{6.165}$$

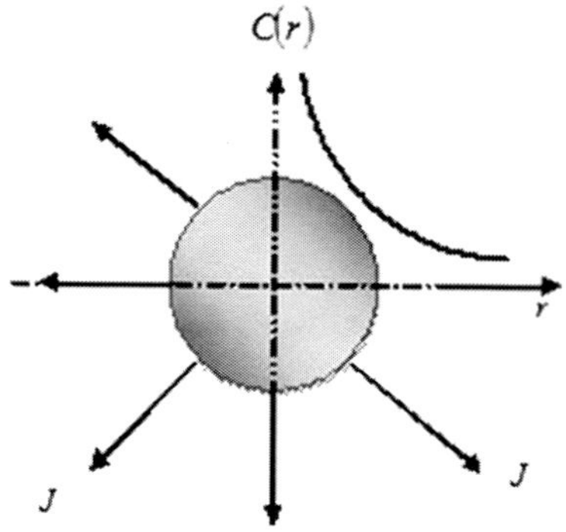

FIGURE 6.8 Uniform intensity of evaporation of a spherical drop.

Passing to partial pressures, Maxwell's equation will be presented as

$$J = \frac{4\pi RDM}{RT}\left(p_S - p_\infty\right) \tag{6.166}$$

Assuming that $J = {}^{dm}\!/_{dt}$ and $dm = \frac{\pi}{2}d^2\rho da$ (here, m is the mass of the droplet) can determine the time for which the drop to change its size to a_0

$$t = \frac{\rho RT\left(a^2 - a_0^2\right)}{8DM\left(p_S - p_\infty\right)} \tag{6.167}$$

At first glance, the evaporation process seems to be very simple, concealing in itself complex phenomena related to the thermodynamics of the surface and the diffusion drift of particles from the surface of the droplet into the volume. It should be noted that the vapor at the surface of the drop will be saturated if the rate of arrival from the drop surface is sufficiently large in comparison with the rate of their diffusion entrainment. In addition, evaporation is accompanied by the removal of energy from the surface of the drop, which leads to a decrease in temperature with respect to the ambient temperature. This is fraught with the reverse phenomenon of vapor condensation directly on the surface of the drop and a decrease in the rate of evaporation. The formation of the diffusion flux of vapor molecules occurs at a distance of the order of the mean free path of molecules λ from the surface of the liquid phase. The nature of the flow of vapor near a spherical droplet should depend on the Knudsen number $\mathrm{Kn} = {}^{\lambda}\!/_{R}$. Calculation of the flow of steam is greatly simplified when the number $\mathrm{Kn} \ll 1$.

Thus, the problem of describing the evaporation process leads to a significant complication of the equations used. In the general case, the exact solution of the process of evaporation of a drop with all the indicated phenomena can be realized only numerically. However, the numerical results obtained are devoid of visibility characteristics of analytic expressions.

Let us consider the process of non-stationary evaporation of a stationary unit drop, for the case of complete diffusion transfer of vapor from a surface to a volume, the absence of its condensation on the surface of a drop, at a constant temperature of the surface and the medium for $\mathrm{Pe} \ll 1$ (Figure 6.8). In the simplest case, the process of evaporation of a stationary single spherical droplet is described by the equation following from equation (6.147) in the case of uniform evaporation from the surface

$$\frac{\partial C}{\partial t} = \frac{D}{r^2}\frac{\partial}{\partial r}\left(r^2\frac{\partial C}{\partial r}\right) = \frac{D}{r}\frac{\partial^2 (rC)}{\partial r^2} \tag{6.168}$$

$$t=0,\ C(r,0)=C_0,\ r=R,\ C(R,t)=C_R,\ r\to\infty,\ C(r,\infty)=C_0$$

Introducing a new variable $\theta = C - C_0$, equation (6.168) can be written in the form

$$\frac{\partial \theta}{\partial t}=\frac{D}{r}\frac{\partial^2 (r\theta)}{\partial r^2} \tag{6.169}$$

$$t=0,\ \theta(r,0)=0;\ r=R,\ \theta(R,t)=C_R-C_0;\ r\to\infty,\ \theta(r,\infty)=0$$

Introducing a new variable $v = r(C - C_0) = r\theta$, we obtain

$$\frac{\partial v}{\partial t}=D\frac{\partial^2 v}{\partial r^2} \tag{6.170}$$

$$t=0,\ v(r,0)=0;\ r=R,\ v(r,t)=R(C_R-C_0);\ r\to\infty,\ v(r,\infty)=0$$

For the solution of (6.169) we introduce a new variable $\eta = r/\sqrt{4Dt}$, as a result of which this equation is transformed into an ordinary differential equation

$$\frac{d^2 v}{d\eta^2}+2\eta\frac{dv}{d\eta}=0 \tag{6.171}$$

The solution of this equation for given boundary conditions is represented as

$$v(r,t)=R(C_R-C_0)\frac{1-\operatorname{erf}\left(r/\sqrt{4Dt}\right)}{1-\operatorname{erf}\left(R/\sqrt{4Dt}\right)} \tag{6.172}$$

Passing to the concentrations, we obtain

$$C(r,t)=C_0+\frac{R}{r}(C_R-C_0)\frac{1-\operatorname{erf}\left(r/\sqrt{4Dt}\right)}{1-\operatorname{erf}\left(R/\sqrt{4Dt}\right)} \tag{6.173}$$

This equation can be used to describe the process of dissolution of solid spherical particles. The flux of vapor from the surface of the droplet is determined as $\left(\int_0^\infty \exp(-\eta^2)d\eta=\sqrt{\pi}/2\right)$

$$\left.\frac{\partial C}{\partial r}\right|_{r=R}=-(C_R-C_0)\left[\frac{1}{R}+\frac{1}{\sqrt{\pi Dt}}\frac{\exp\left(-R^2/4Dt\right)}{1-\operatorname{erf}\left(R/\sqrt{4Dt}\right)}\right] \tag{6.174}$$

$$j=(C_R-C_0)\left[\frac{D}{R}+\sqrt{\frac{D}{\pi t}}\frac{\exp\left(-R^2/4Dt\right)}{1-\operatorname{erf}\left(R/\sqrt{4Dt}\right)}\right] \tag{6.175}$$

Here, C_R is the concentration of steam on the surface of the drop, and C_0 is the concentration of steam in the volume of the stream. Introducing the Fourier number $\text{Fo} = {}^{Dt}/_{R^2}$, equation (6.175) can be written as

$$j = (C_R - C_0)\left[\frac{D}{R} + \sqrt{\frac{D}{\pi t}}\frac{\exp\left(-{}^{1}/_{4\text{Fo}}\right)}{1-\text{erf}\left({}^{1}/_{2\sqrt{\text{Fo}}}\right)}\right] \tag{6.176}$$

For large values of the number Fo >> 1, expression (6.176) can be written in the form

$$j = (C_R - C_0)\left[\frac{D}{R} + \sqrt{\frac{D}{\pi .t}}\right] \tag{6.177}$$

The total mass flux from the surface of the drop is defined as

$$J = 4\pi R^2 j = 4R(C_R - C_0)D\left[\pi + \sqrt{\frac{\pi}{\text{Fo}}}\right] \tag{6.178}$$

The Sherwood number is defined as

$$\text{Sh} = \frac{J\,2R}{DS\Delta C} = 2\left(1 + \frac{1}{\sqrt{\pi \text{Fo}}}\right) \tag{6.179}$$

This expression coincides with the mass-transfer equation given in the works [1,8,21].

For the region of variation of the number $0.1 < \text{Pe} < 1000$ for a solid spherical particle moving in a steady flow, a semi-empirical formula [21]

$$\text{Sh}_T = 2 + \left[0.333\text{Pe}^{0.840}\left(1 + 0.331\text{Pe}^{0.507}\right)\right] \tag{6.179a}$$

for the gas bubble

$$\text{Sh}_g = 2 + \left[0.651\text{Pe}^{1.72}\left(1 + 0.331\text{Pe}^{1.22}\right)\right] \tag{6.179b}$$

In the paper [17] for a spherical particle, using the Chilton–Colborn analogy, the following formula is proposed

$$\text{Sh} = 2 + 0.6\text{Re}_d^{1/2}\text{Sc}^{1/3} \tag{6.179c}$$

Then for the drop the Sherwood number for Re < 1 and 1 < Pe < 1000 is equal to

$$\text{Sh} = \frac{\text{Sh}_g + \gamma \text{Sh}_T}{1 + \gamma} \tag{6.179d}$$

The mass transfer by free convection for a spherical particle is determined in the form

$$\text{Sh} = 2 + 0.59(\text{GrSc})^{1/4}, \quad 2\times10^8 < \text{GrSc} < 1.5\times10^{10} \tag{6.180}$$

The asymptotic value of the Sherwood test for $Fo \to \infty$, as follows from equation (6.179), is equal to two. As a result of evaporation, the mass of the fixed spherical drop varies according to equation

$$\frac{dm}{dt} = -J \tag{6.181}$$

Or, for a spherical droplet $m = \frac{4}{3}\pi R^3 \rho_d$, this equation, taking (6.176) into account, is represented as

$$\frac{dR}{dt} = -\frac{C_R - C_0}{\rho_d}\left[\frac{D}{R} + \sqrt{\frac{D}{\pi t}}\right], \ R(t) = R_0 \tag{6.182}$$

where C_R, C_0 are the concentration of vapor on the surface of the droplet and away from it.
Consider the solution of this equation under various conditions:

1. If $D/R >> \sqrt{\frac{D}{\pi . t}}$ or Fo >> $1/\pi \approx 0,3$, then equation (6.182) is represented as

$$R\frac{dR}{dt} = -\frac{C_R - C_0}{\rho_d} D, \ R(t)\big|_{t=0} = R_0 \tag{6.183}$$

   The solution of this equation is represented in the form

$$R = \sqrt{R_0^2 - \frac{C_R - C_0}{\rho_d} Dt} = R_0\sqrt{1 - t/t_i} \tag{6.184}$$

   From this equation, we determine the time for complete evaporation $t = t_i$, $R = 0$

$$t_i = \frac{R_0^2 \rho_d}{(C_R - C_0)D} \tag{6.185}$$

2. If $Fo << 0,3$, then equation (6.182) is represented as

$$\frac{dR}{dt} = -\frac{C_R - C_0}{\rho_d}\sqrt{\frac{D}{\pi . t}} \tag{6.186}$$

   The solution of this equation is defined as

$$R = R_0 - 2\frac{C_R - C_0}{\rho_d}\sqrt{\frac{Dt}{\pi}} \tag{6.187}$$

   For an evaporating droplet in a moving medium, the vapor flux per unit surface per unit time will be determined as

$$j = \frac{\text{Sh}\, D\, \Delta C}{2R} \tag{6.188}$$

Example 6.2

A drop of oil with a diameter of 1 mm evaporates at a temperature of 350 K in a stream of air having a velocity of 2.88 m/s. Determine the rate of evaporation and the rate of decrease in the droplet size if the mass fraction of the vapor on the surface is $x_m = 0.278$, the droplet density is $\rho_d = 677 \,{}^{kg}\!/_{m^3}$, the vapor density is $\rho_p = 1.46 \,{}^{kg}\!/_{m^3}$, the viscosity of the vapor is $\nu_p = 2.06\times10^{-5}\,Pa.s$, the viscosity of environment is $\nu_c = 1.68\times10^{-5}\,Pa.s$ and the diffusion coefficient of the vapor is $D = 7.55\times10^{-6}\,{}^{m^2}\!/_{s}$.

Solution

Define the Schmidt and Reynolds numbers

$$\text{Sc} = \frac{\nu_c}{\rho_p D} = \frac{1.68\times10^{-5}}{1.46\left(7.55\times10^{-6}\right)} = 1.53$$

$$\text{Re}_d = \frac{aV}{\nu_c} = \frac{10^{-3}\times2.88}{2.06\times10^{-5}} = 141$$

The Sherwood number is

$$\text{Sh} = 2 + 0.6\,\text{Re}_d^{1/2}\,\text{Sc}^{1/3} = 2 = 0.6\times141^{1/2}\times1.53^{1/3} = 10.2$$

The mass flux from the drop surface is determined as

$$j = \frac{\text{Sh} D \rho_p \Delta C}{a} = \frac{10.2\times\left(7.55\times10^{-6}\right)\times1.46\times0.278}{10^{-3}} = 3.125\times10^{-2}\,\frac{\text{kg}}{\text{m}^2\text{s}}$$

The rate of evaporation of a drop is

$$V_s = j\pi a^2 = 3.125\times10^{-2}\times3.14\times10^{-6} = 9.81\times10^{-8}\,{}^{kg}\!/_{s}$$

The rate of reduction of the droplet size is defined as

$$\frac{da}{dt} = -\frac{2j}{\rho_d} = \frac{2\times3.125\times10^{-2}}{677} = -9.23\times10^{-2}\,{}^{mm}\!/_{s}$$

6.5 MASS TRANSFER FOR DRYING POROUS MATERIALS

Drying porous materials is widely used in the processes of chemical, food technology, and in an agrarian economy. Experimental studies of the drying processes of porous materials, in particular for food products, have been carried out in many works [64–66], where on the basis of which various empirical dependences of the distribution of moisture concentration over time and the estimation of the coefficient of molecular diffusion were obtained. The theoretical studies related to the solution of certain issues of mass exchange under laminar flow are devoted to the work. In the drying process, the transfer of the mass of moisture in the dried material to the surface from the body volume is mainly carried out by molecular and Knudsen diffusion, and from the surface to the airflow by convective diffusion [67,68]. This is explained by the fact that in porous moist bodies the convective component of mass transfer is very small in comparison with the diffusion one. Moreover, with $(\text{GrPr}) < 10^3$ (Gr is the Grashof number, Pr is the Prandtl number), the total mass-transfer coefficient in disperse media is approximately equal to the molecular diffusion coefficient. The process of mass transfer in porous media is also determined by the nature and geometry of the pores (tortuosity, blockage of pore channels, etc.) and their dimensions, which can be taken into account in determining Knudsen diffusion. The next important step in the drying process is the convective transfer of moisture from the surface of the body to the volume of moving air, which can be especially related to the nature of the flow—laminar or turbulent. For large values of the Reynolds number, moisture transfer from the surface is effected by turbulent diffusion [7,8], which depends on the main characteristics of the turbulent flow (specific dissipation energy, turbulence scale, etc.).

Mass transfer in a porous body is carried out by molecular diffusion (D) and Knudsen diffusion (D_{Kn}), as a result of which the effective diffusion coefficient is defined as

$$\frac{1}{D_{ef}}=\frac{1}{D}+\frac{1}{D_{Kn}}. \tag{6.189}$$

From the surface of the porous body, mass transfer is realized by molecular D and turbulent diffusion (D_T), as a result of which the effective diffusion coefficient is defined as $D_{ef}=D+D_T$. It is important to note that for large Reynolds numbers (Re), that is, in the region of developed turbulence, we can take $D=D_T$.

6.5.1 Mass Transfer in the Processes of Drying Porous Bodies at a Constant Diffusion Coefficient

At a constant temperature and the condition that the moisture flow in the porous body is laminar, the transfer of moisture mass in a flat layer is described by the convective transfer equation (Figure 6.7) [69]:

$$\frac{\partial M}{\partial t}+V_x\frac{\partial M}{\partial x}+V_y\frac{\partial M}{\partial y}+V_z\frac{\partial M}{\partial z}=D\left(\frac{\partial^2 M}{\partial x^2}+\frac{\partial^2 M}{\partial y^2}+\frac{\partial^2 M}{\partial z^2}\right), \tag{6.190}$$

where M is the moisture concentration, and V_x and V_y are the flow velocity components in the x and y directions. For small values of the Peclet number $Pe=\frac{VL}{D}<<1$ (L is the size of the material layer) and the insignificance of the translations along the x, y, and z axes, equation (6.190) simplifies to the form (i.e., moisture transfer occurs in the normal direction to the surface) (Figure 6.9):

$$\frac{\partial M}{\partial t}=D\frac{\partial^2 M}{\partial y^2} \tag{6.191}$$

with boundary conditions

$$\begin{aligned} &t=\ 0, M=M_0,\\ &t>0,\ y=\ 0, \frac{\partial M}{\partial y}=0,\\ &t>0,\ y=L,\ M=M_p \end{aligned} \tag{6.192}$$

where M_0 is the initial moisture content, M_P is the equilibrium concentration with the flow, and L is the thickness of the porous material.

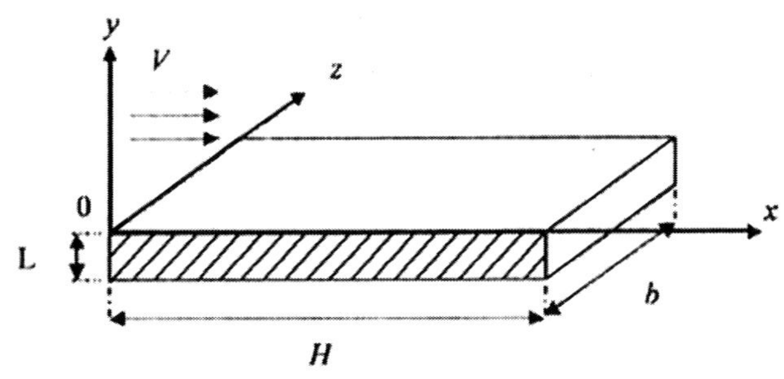

FIGURE 6.9 Orientation of streams during drying.

The solution of equation (6.191) with the boundary conditions (6.192) can be represented in the form:

$$M_R = \frac{M - M_0}{M_p - M_0} = \frac{8}{\pi^2}\sum_{n=0}^{\infty}\frac{1}{(2n+1)}\exp\left(-\frac{(2n+1)^2\pi^2}{4}Fo\right) \tag{6.193}$$

where $Fo = \frac{Dt}{L^2}$ is the Fourier number. It is important to note that for $Fo > 0{,}3$ this series converges, and it suffices to take into account only the first term of expression (6.193)

$$M_R = \frac{8}{\pi^2}\exp\left(-\frac{\pi^2}{4}Fo\right). \tag{6.194}$$

Using the experimental data [66] and equation (6.194), we can estimate the diffusion coefficient in the form

$$D = D_0 e^{-Q/T} \tag{6.195}$$

where $D_0 = 2.11 \times 10^{-5}$ m²/s, $Q = 3135.2$, activation energy $E = Q.R$ (R = 0.008314 kJ/(mol K^{-1}) determined according to the experimental data [66].

The study and calculations showed that the value of the diffusion coefficient in a porous medium varies in the range of $D \approx 9.5 \times 10^{-10}$–$1.3 \times 10^{-9}$ m²/s.

The effect of the air flow rate on the drying process starts from the surface of the dried material and extends over the entire volume of the flow. If evaporation of moisture occurs on the surface of the body, a movement of moisture into the evaporation zone is observed in the volume of the porous bodies, which is due to the action of various causes (changes in pressure in the pores, changes in the meniscus of the surface, etc.).

Assuming that mass transfer on the surface of the material to be dried is carried out convective, we change the boundary conditions (6.192) to the form

$$\begin{aligned} &t = 0, M = M_0, \\ &t > 0, y = L, -\frac{\partial M}{\partial \xi} = \text{Bi}\,(M - M_p),\ \xi = y/L, \\ &t > 0,\ y = 0, \frac{\partial M}{\partial y} = 0, \end{aligned} \tag{6.196}$$

where $Bi = \frac{\alpha L}{D}$ is the Bio number, and α is the mass-transfer coefficient. The second condition (6.196) determines the equality of the diffusion and convective flows on the surface of the layer. The solution of equation (6.191), taking into account the boundary conditions (6.196), can be represented in the form:

$$M_R = \sum_{n=1}^{\infty} C_n \cos(\mu_n \frac{y}{L})\cdot\exp(-\mu_n^2 Fo), \tag{6.197}$$

where $C_n = \dfrac{2\sin\mu_n}{\mu_n + \sin\mu_n\cos\mu_n}$ are the coefficients of the series, and μ_n is the positive root of equation

$$\frac{\mu}{Bi} = ctg\,\mu. \tag{6.198}$$

Theoretical studies of equations (6.197) and (6.198) for different values of the numbers Bi have shown that for

1. $Bi \to \infty$ the solution (6.197) can be represented in the form

$$M_R \approx \frac{4}{\pi}\sum_{n=1}^{\infty}\frac{(-1)^n}{2n-1}\cos\left[(2n-1)\frac{\pi}{2}\frac{y}{L}\right]\exp\left[-(2Bi-1)\frac{\pi^2}{4}Fo\right], \tag{6.199}$$

2. $Bi \to 0,\ C_1 = \lim\limits_{\mu_1 \to 0}\dfrac{2\sin\mu_1}{\mu_1 + \sin\mu_1\cos\mu_1} \to 1,\ C_2 = C_3 = \ \dots\ = C_n = 0.$

In this case, it follows from the solution (6.199) that $M_R|_{Bi\to 0} \to 1$, or the concentration of moisture on the surface of the body, is equal to its equilibrium value $(M \to M_p)$. If $0 < \mathrm{Bi} \ll 1$ and putting tg $(\mu_1) \approx \mu_1$, the solution (6.199) is transformed to the form

$$M_R = \cos\left(\frac{y}{L}\sqrt{Bi}\right)\exp(-Bi\,Fo). \tag{6.200}$$

As a result of the analysis of the experimental studies and the preceding equations (6.197)–(6.200) for the porous layer, the following dependence was established for determining the number Bi:

$$Bi = 0.35(T/T_0)^{1.4}\,\mathrm{Re}^{0.15}, \tag{6.201}$$

where T_0 is the initial temperature of the porous layer, $\mathrm{Re} = \frac{VL}{\nu_B}$ is the Reynolds number, and ν_B is the air viscosity ($\nu_B \approx 0.153 \times 10^{-4}$ m²/s).

From the surface of the dried material, the vapor is carried away to the flow volume by convective diffusion. In this case, the nature of the hydrodynamic flow can have a significant effect on the mass-transfer coefficient. According to Figure 6.9, for a thin horizontal layer of the dried material, consider convective moisture transfer in air, assuming that in equal

$$\frac{\partial^2 M}{\partial z^2} \ll \frac{\partial^2 M}{\partial y^2}\cdot\frac{\partial^2 M}{\partial x^2} \ll \frac{\partial^2 M}{\partial y^2}, \tag{6.202}$$

Then the convective transfer equation (6.190) for the stationary case is written in the form

$$V_x\frac{\partial M_R}{\partial x} + V_y\frac{\partial M_R}{\partial y} = D_p\frac{\partial^2 M_R}{\partial y^2}, \tag{6.203}$$

$$y = 0,\ M_R = 0,\ y \to \infty,\ M_R = 1,$$

where $M_R = \frac{M-M_0}{M_p-M_0}$, and D_p is the diffusion coefficient of moisture in the air. According to [6], the solution (6.203) can be represented in the form

$$M_R(x,y) = \frac{\int_0^{\eta}\exp\left(-\frac{2}{9}Sc\eta^3\right)d\eta}{\int_0^{\infty}\exp\left(-\frac{2}{9}Sc\eta^3\right)d\eta}, \tag{6.204}$$

where $\eta = \frac{1}{2}\sqrt{\frac{Vy^2}{\nu x}}$, $Sc = \nu/D_p$ is the Schmidt number. Assuming that

$$\int_0^\infty \exp\left(-\frac{2}{9}Sc\eta^3\right)d\eta = \frac{(1/3)}{\sqrt[3]{6}\,Sc^{1/3}}, \tag{6.205}$$

we rewrite equation (6.204) in the form

$$M_R(x,y) = \frac{\sqrt[3]{6}}{(1/3)}Sc^{1/3}\int_0^\eta \exp\left(-\frac{2}{9}Sc\eta^3 d\eta\right) \tag{6.206}$$

where Γ (1/3) = 2.678 is the gamma function. From equation (6.206), the distribution of the moisture concentration is determined in the form

$$M_R(x,y) = \frac{\sqrt[3]{6}}{(1/3)}Sc^{1/3}\eta\sum_{n=0}^{\infty}\frac{(-1)^n}{n!(3n+1)}\left(\frac{2Sc}{9}\right)^n \eta^{3n} \tag{6.207}$$

If $(L/H)^2 < 30$, then this equation becomes simpler to the form

$$M_R(x,y) \approx \frac{\sqrt[3]{6}}{(1/3)}Sc^{1/3}\eta\left(1 - \frac{1}{18}Sc\eta^3\right) \tag{6.208}$$

The amount of moisture transferred per unit time from a unit surface is defined as

$$j = D_p\left(\frac{\partial M_R}{\partial y}\right)_{y=0} = \frac{\sqrt[3]{6}}{2(1/3)}D_p\Delta M\sqrt{\frac{V}{\nu x}}Sc^{1/3}, \tag{6.209}$$

where $\Delta M = M_0 - M_p$. The amount of moisture transported through the entire outer surface of the porous layer is determined in the form

$$I = \int_0^H\int_0^b j\,dxdz = \frac{\sqrt[3]{6}}{2(1/3)}bD\Delta M\,Sc^{1/3}\,\mathrm{Re}^{1/2}. \tag{6.210}$$

The Sherwood number, using (6.210), is defined as

$$Sh = \frac{HI}{D_p F\Delta M} = 0.6785\,\mathrm{Re}^{1/2}\,Sc^{1/3}, \tag{6.211}$$

where F = H.b is the surface area of the layer (Figure 6.7). Equation (6.211) can be used to estimate the mass-transfer coefficient $\beta = \frac{D\,\mathrm{Sh}}{H}$. It should be noted that the coefficients before the expression (6.211) and the exponent of the number Re can be variable and depend on the temperature and nature of the flow. For a turbulent flow, the expression for determining the Sherwood number is given in the form

$$Sh = 0.0296\,\mathrm{Re}^{4/5}\,Sc^{1/3} \tag{6.212}$$

The difference in formulas (6.211) and (6.212) is explained by the nature of the flow (laminar and turbulent) and drying conditions. In the general case, taking into account the nature of the air flow, the mass-transfer coefficient can be determined from the criterion equation

$$Sh = A(T)\,\mathrm{Re}^{n}\cdot Sc^{1/3} \tag{6.213}$$

And the determining size in the calculation of the criterion Sh and Re is the length of the evaporation surface H in the direction of motion of the hot air. When intensive drying with high air temperatures, the mass-transfer coefficient can be higher than the calculated one. Under these conditions, the material can undergo deformation of the shape and the formation of cracks of a bulk character, on which the diffusion coefficient depends. This condition for some materials requires moderate drying conditions. In this study, we consider the solution of mass-transfer problems for three drying cases: (1) molecular vapor transfer in a porous material; (2) on the surface of the material to be dried; and (3) convective transfer of steam from the surface to the volume of the air flow.

Using experimental studies [66] for temperatures of 40°C $\leq T \leq$ 50°C and a number of $1.8\times10^4 \leq \mathrm{Re} \leq 4\times10^4$, equation (6.212) is represented in the form

$$Sh = A(T)\,\mathrm{Re}^{3/8}\cdot Sc^{1/3} \tag{6.214}$$

where $A(T) = 0.0148(T/T_0)^{1.25}$, $\mathrm{Re} = \frac{VH}{v_B}$ is the Reynolds number. It is important to note that the expression (6.214) does not take into account the horizontal surface moisture transfer by natural convection. To estimate the effect of natural convection, we give the following equation:

$$\frac{I}{I_g} \approx 0.754 Sc^{1/12}\,\mathrm{Re}^{1/2}\,Gr^{-1/4}, \tag{6.215}$$

where $Gr = \frac{gH^3\alpha_0}{4v^2}$ is the Grashof number, $\alpha_0 = \frac{M_0}{\rho M_p}\left(\frac{\partial\rho}{\partial M}\right)\Big|_{M=M_p}$, I_g is the mass transfer due to natural convection. Using the experimental data and equation (6.215) ($\mathrm{Gr} \approx 6.8\times10^5$), we obtain $I = 4.86 I_g$, that is, the convective mass transfer is about five times a more natural convection.

Thus, theoretical and experimental studies of drying processes allowed us: (1) to estimate the mass transfer in a porous layer and to obtain, on the basis of experimental data, the proportionality of the number Bi ~ $\mathrm{Re}^{0.15}$; (2) to analyze the convective transfer of the mass of moisture from the surface to the air, and on the basis of experimental data, estimate the number of Sh as a function of the numbers Re and Sc; and (c) to evaluate the contribution of natural convection to the drying process for horizontal surfaces. In general, the conducted studies allow analyzing of the mass transfer in the drying processes, which is important for the calculation and design of the drying unit.

6.5.2 Mass Transfer in the Drying Process with Variable Diffusion Coefficient

Drying processes in porous media are characterized by very complex physical phenomena (the movement of moisture in capillaries of irregular shape, phase transformations, etc.), as a result of which many parameters of the porous body depend on the moisture and the degree of porosity. Drying processes are accompanied by a change in volume (decrease), geometry, and shape of the body, associated with deformation and compaction of the porous layer under the influence of internal and external deforming stresses and proceed with microstructural changes in the porous structure. These phenomena lead to a change in the porosity, density, and effective diffusion coefficient from time. Thus, the relationship between the density and porosity of the material with relative humidity was experimentally investigated and a correlation dependence of the form [68]

$$\varepsilon = 0.038 + 0.03\ \exp\left(-{}^{x}\!/_{x_0}\right) \tag{6.216}$$

The main internal parameters of the porous medium during its drying are the density and porosity, which are connected by a single equation $\varepsilon = 1 - {}^{\rho}\!/_{\rho_d}$, and also the effective diffusion coefficient, which depends on the porosity of the medium. These parameters mainly depend on drying methods and are determined by its velocity, because at high velocity in the material due to internal deforming stresses depending on temperature, the material is deformed and compressed, as a result of which cracks and fractures appear in it. Experimental studies on the effect of moisture content on the parameters of the process showed that in the early stages of drying the material density and the diffusion coefficient increase, reaching a certain maximum, after which they decrease monotonically. Thus, the drying of porous deformable bodies is characterized by varying values of density, porosity, and the coefficient of effective diffusion. At constant temperature and laminar flow in an isotropic porous body, the mass transfer in a flat layer (Figure 6.7) is described by equation

$$\frac{\partial M}{\partial t} + V_x \frac{\partial M}{\partial x} + V_y \frac{\partial M}{\partial y} = D_*(\varepsilon, t)\left(\frac{\partial^2 M}{\partial x^2} + \frac{\partial M}{\partial y^2}\right) \tag{6.217}$$

As a result of the densification of the layer under the action of internal deforming stresses, the change in porosity of the layer is described by equation

$$\varepsilon = \varepsilon_0 \exp\left(-\frac{3}{4}\frac{\int_0^t \sigma_D dt}{\xi_s}\right) \tag{6.218}$$

Here, $D_*(\varepsilon, t)$ is the effective diffusion coefficient, which depends on the porosity of the layer and time; $\varepsilon, \varepsilon_0$ is the current and initial porosity of the layer; and ξ_s is the internal deforming stress, the shear viscosity of the skeleton of the porous medium. The effective diffusion coefficient depends on the porosity of the medium in the form (4.67) [69]. The coefficient of effective diffusion can be calculated from expression, $\varepsilon < 0.55$

$$D_*\!/_{D_0} \approx 0.62\,\varepsilon\,\eta_d \tag{6.219}$$

where η_d is the coefficient of tortuosity pore. It should be noted that the coefficient of effective diffusion in an inhomogeneous anisotropic medium also depends on the spatial coordinate and time. We consider the one-dimensional solution of (6.217) for small numbers Pe << 1 for an insignificant deformation of the skeleton of a porous medium in the form

$$\frac{\partial M}{\partial t} = D_*(\varepsilon, t)\frac{\partial^2 M}{\partial y^2}, \quad t = 0,\ M(y, t) = M_0,\ t > 0,\ y = L,\ M(t, y) = M_p \tag{6.220}$$

Here, M_0, M_p are the initial and equilibrium moisture contents. In the case of a change in time and porosity, we define it in the form [70]

$$D_*(\varepsilon, t) = D_0 \psi_*(t) \tag{6.221}$$

where $\psi_*(t) = A\varepsilon_0 \exp\left(-\frac{3}{4}\frac{\int_0^t \sigma_D dt}{\xi_s}\right)$, $A = 0.62\eta_p$. Introducing the new variables,

$$\varsigma = \varsigma(l_0) + \int_{l_0}^{l} \frac{dy}{D_* s_0}, \quad f(\varsigma) = \int_{l_0}^{l} \frac{dy}{\sqrt{D_*}} \tag{6.222}$$

after simple transformations, we rewrite equation (6.220) in the form

$$\frac{\partial^2 M}{\partial \varsigma^2} = \left(\frac{\partial f}{\partial \varsigma}\right)^2 \frac{\partial M}{\partial t} \tag{6.223}$$

where l is the distance along the streamline, s_0 is the cross-sectional area of the pore channel. The independent variable ς that performs the role of the spatial coordinate is numerically equal to the stationary transport potential realized in the current tube under consideration $\partial M/\partial t = 0$. For the solution of (6.223), it is more expedient to approximate the function $f(\varsigma)$ as follows:

$$f(\varsigma) = b\varsigma^{\alpha} \tag{6.224}$$

Then equation (6.223) with regard to (6.224) and the Laplace transform takes the form

$$\frac{d^2 \vartheta}{d\varsigma^2} - \left(\frac{\partial f}{\partial \varsigma}\right)^2 \lambda \vartheta = 0$$
$$\vartheta(l,\lambda) = \phi\left[F(l,t)\right] = \int_t^{\infty} \exp(-\lambda t) F(l,t)\,dt, \quad F(l,t) = M(l,t) - M(l,0) \tag{6.225}$$

Substituting (6.224) into (6.225), we obtain the modified Bessel equation and, expressing in terms of expression (6.224) in the form, the solution (6.225) can be represented in the form

$$\vartheta(f,\lambda,\alpha) = \left(\frac{f}{b}\right)^{1/2\alpha} \left[A_1 K_{1/2\alpha}\left(f\sqrt{\lambda}\right) + A_2 I_{1/2\alpha}\left(f\sqrt{\lambda}\right)\right] \tag{6.226}$$

where $I_{1/2\alpha}\left(f\sqrt{\lambda}\right)$, $K_{1/2\alpha}\left(f\sqrt{\lambda}\right)$ are the modified Bessel functions of the imaginary argument of the first and second kind, and A_1, A_2 are the coefficients. We shall seek particular solutions satisfying the boundary conditions

$$\lim_{f\to 0} \vartheta(f,\lambda,\alpha) = \vartheta(0,\lambda,\alpha), \quad \lim_{f\to\infty} \vartheta(f,\lambda,\alpha) = 0 \tag{6.227}$$

The second condition (6.227) corresponds to the first term (6.226)

$$\vartheta(f,\lambda,\alpha) = \left(\frac{f}{b}\right)^{1/2\alpha} \left[A_1 K_{1/2\alpha}\left(f\sqrt{\lambda}\right)\right] \tag{6.228}$$

Taking into account the asymptotic equality for small $K_n(z) = 2^{n-1}\Gamma(n) z^{-n}$, using the first condition (6.227), we define

$$A_1 = \frac{2^{1-\frac{1}{2\alpha}}}{\Gamma\left(\frac{1}{2\alpha}\right)} \vartheta(0,\lambda,\alpha)\left(b\sqrt{\lambda}\right)^{\frac{1}{2\alpha}} \tag{6.229}$$

and we represent the solution of (6.228) in the form

$$\vartheta(f,\lambda,\alpha) = \lambda\vartheta(0,\lambda,\alpha)\omega(f,\lambda,\alpha), \quad \omega(f,\lambda,\alpha) = 2^{1-n} f^n \lambda^{\frac{n}{2}-1} \mathrm{K}\left(\mathrm{f}\sqrt{\lambda}\right) \tag{6.230}$$

The incomplete gamma function serves as the original $\omega(f,\lambda,\alpha)$

$$\Gamma\left(\frac{f^2}{4t}, n\right) = \phi^{-1}\left[\omega(f,\lambda,n)\right] = \int_\delta^\infty \exp(-y)\, y^{n-1} dy, \quad \delta = \frac{f^2}{4t} \tag{6.231}$$

Hence, $n = \frac{1}{2\alpha}$, by the inverse transformation theorem, assuming that the original (6.230) can be represented in the form

$$F(l,t) = F(f,t,\alpha) = \frac{\partial}{\partial t}\int_0^t F(0,t-\tau,\alpha)\psi\left(\frac{f^2}{4t},\alpha\right)d\tau,$$
$$\psi\left(\frac{f^2}{4t},\alpha\right) = \frac{1}{\Gamma\left(\frac{1}{2\alpha}\right)}\int_\delta^\infty \exp(-y) y^{\frac{1}{2\alpha}-1} dy \tag{6.232}$$

Using in the last equation (6.232) $x = y^{\frac{1}{2\alpha}}$ and the relation for the gamma function $n\Gamma(n) = \Gamma(1+n)$, we obtain

$$\psi\left(\frac{f^2}{4t},\alpha\right) = \frac{1}{\Gamma\left(1+\frac{1}{2\alpha}\right)}\int_x^\infty \exp\left(-x^{2\alpha}\right)dx, \quad x = \left(\frac{f^2}{4t}\right)^{\frac{1}{2\alpha}} \tag{6.233}$$

Producing in (6.232) differentiation with respect to the second equation, we obtain

$$F(f,t,\alpha) = \frac{1}{\Gamma\left(1+\frac{1}{2\alpha}\right)}\int_0^\infty F(0,t-\tau,\alpha)\left(\frac{f^2}{4\tau}\right)^{\frac{1}{2\alpha}} \exp\left(-\frac{f^2}{4\tau}\right)\frac{d\tau}{\tau} \tag{6.234}$$

It is easy to verify that the solution (6.234) satisfies equation (6.223) and the boundary conditions (6.228). Then the solution (6.220), taking into account the boundary conditions, is represented in the form

$$\Delta M(f,t,\alpha) = \frac{1}{\Gamma\left(\frac{1}{2\alpha}\right)}\int_0^\infty \Delta M(0,t-\tau,\alpha)\left(\frac{f^2}{4\tau}\right)^{\frac{1}{2\alpha}} \exp\left(-\frac{f^2}{4\tau}\right)\frac{d\tau}{\tau} \tag{6.235}$$

where $\Delta M(f,t,\alpha) = M(l,t) - M(l,0)$, $M(l,0) = M_0$. Equation (6.235), which in general characterizes the transfer of moisture in a porous body, is very complicated and its solution for an arbitrary specification of a function is possible only by numerical means or using exponential series. However, if we assume that when the moisture concentration in the porous medium is fully formed $l = L$ and equal $M(L,t,\alpha) = M_p$, then the solution (6.235) can be represented in the form

$$M_s(f,t,\alpha)=\frac{M(f,t,\alpha)-M_0}{M_p-M_0}=1-\psi\left(\frac{f^2}{4\tau},\alpha\right) \tag{6.236}$$

If we assume that $\alpha=1$, that is, to take a linear relationship $\left(\Gamma\left(\frac{1}{2}\right)=\frac{\pi}{2}\right)$, then the expression on the right side becomes the probability integral and the solution (6.236) can be written as

$$M_s(y,t)=\frac{2}{\pi}\int_0^x \exp\left(-\lambda^2\right)d\lambda=\operatorname{erf}\left(\frac{y}{2\sqrt{D_* t}}\right)=\operatorname{erf}\left(\frac{y}{2L\sqrt{Fo}}\right)=\Phi(\eta) \tag{6.237}$$

where $Fo = {}^{D_*t}/_{L^2}$ is the Fourier number, $\eta = {}^{y}/_{2\sqrt{D_*t}}$ and the effective diffusion coefficient is determined according to the expression (6.219). It is obvious that $\alpha \neq 1$, when the moisture distribution differs from (6.237), in view of the nonlinearity of the solution (6.235). It should be noted that if the effective diffusion coefficient depends only on time, then, using expression (6.221), expression (6.220) is represented as

$$\frac{\partial M}{\partial t_*}=\frac{\partial^2 M}{\partial y^2} \tag{6.238}$$

where $t_* = D_0\int_0^t \psi_+(\tau)d\tau$. Introducing the variable $\xi = {}^{y}/_{2\sqrt{t_*}}$, we transform (6.238) into an ordinary differential equation with a solution $M_s(y)=\operatorname{erf}(\xi)$ from which we determine the mass flux of vapor from the surface of the layer in the form

$$J=-SD_*\left.\frac{\partial M}{\partial y}\right|_{y=L}=\frac{bH}{L}\frac{D_*\left(M_p-M_0\right)}{\sqrt{\pi Fo}} \tag{6.239}$$

Then the dimensionless mass flux or the Sherwood number is defined as

$$\operatorname{Sh}(\tau)=\frac{H/L}{\sqrt{\pi Fo}},\qquad \overline{\operatorname{Sh}}=\frac{1}{t}\int_0^t \operatorname{Sh}(\tau)d\tau \tag{6.240}$$

Naturally, this expression, unlike (6.214), does not take into account the convective transfer of the mass of moisture from the surface to the gas stream, although it takes into account the variable nature of the effective diffusion coefficient. The results of experimental studies of the drying of porous material carried out by us for various temperatures T = 40°C, 45°C, and 50°C and air velocity V = 1.0 m/s are shown in Figure 6.10.

An estimate of the dependence of the change in the diffusion coefficient on time is given in Figure 6.11.

6.5.3 The Solution of the Inverse Ill-Posed Problem of Mass Transfer

The problem of determining the value of the effective diffusion coefficient or the general variable η from the measured and calculated values $M_s(\eta)$ is the inverse of the incorrectly posed problem. The inverse problem of estimating the effective diffusion coefficient is determined from (6.237) by the expression

$$\eta=\Phi^{-1}(M_s) \tag{6.241}$$

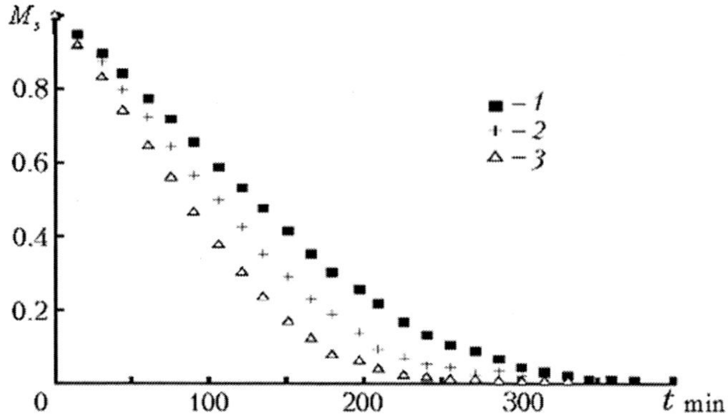

FIGURE 6.10 Dependence of the relative concentration of moisture on the drying time at a flow rate $V = 1\ ^m/_s$ and temperature: (1) 40°C; (2) 45°C; and (3) 50°C.

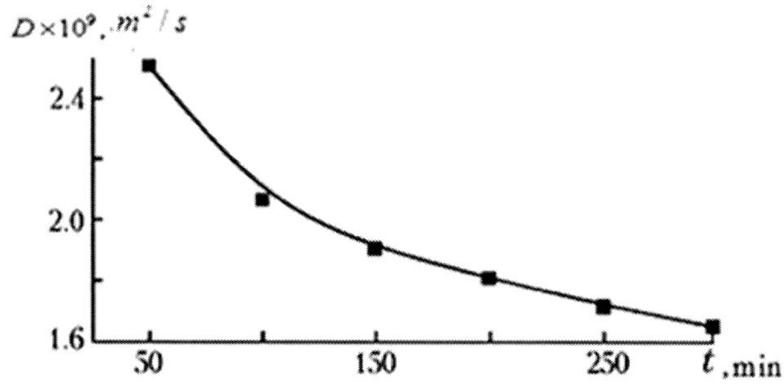

FIGURE 6.11 Dependence of the diffusion coefficient on time.

and is an incorrectly posed problem, because very small errors in experimental measurements and calculations of moisture concentration values can lead to large errors in the estimation or variable diffusion coefficient. At the same time, there is a need to determine the region of regularity of the solution of the inverse problem; determine the sensitivity of solutions to errors in measurements and moisture concentrations. Differentiating (6.241) with respect to the moisture concentration, we have

$$\frac{\partial \eta}{\partial M} = \frac{\partial \Phi^{-1}(M)}{\partial M} = \frac{\partial M_s}{\partial M}\left(\frac{\partial \Phi^{-1}}{\partial \eta}\right) \tag{6.242}$$

Considering the small variations in the concentration errors, and having determined the derivative from (6.237) $\partial\Phi/_{\partial\eta} = 1/_{\sqrt{2\pi}}\exp(-\eta^2)$, after simple transformations from (6.242) we obtain

$$\delta(\eta) = \frac{\Delta\eta}{\eta} = K(\eta)\frac{\delta M_s}{M_0 - M} \tag{6.243}$$

where $K(\eta) = \frac{\sqrt{2\pi}}{\eta}\exp(\eta^2)$, δM_s is the absolute error. Expression (6.243) defines the region of regularity of expression (6.241) under the condition that the relative error is sufficiently small $\delta(\eta)$. Starting from (6.243), setting $\eta = {}^{y}/_{\sqrt{2D*t}}$, we estimate the error for the effective diffusion coefficient in the form

$$\frac{\Delta D_*}{D_*} = K^{-2}(\eta)\left(\frac{\Delta M_s}{M_0 - M_p}\right)^2 \tag{6.244}$$

and the value $K^{-1}(\eta)$ varies from 0 to 1. As follows from (6.244), the error in determining the diffusion coefficient depends on the square of the relative error in the measurement of the moisture concentration. Since the relative error $\delta(M) = {}^{\Delta M_s}\!/\!_{(M_0 - M_p)}$, $K(\eta) \le 1$, then for $\delta(D_*) = {}^{\Delta D_*}\!/\!_{D_*} \to 0$, at $\delta(M) \to 0$, that is, infinitesimal variations correspond to infinitesimal increments $\delta(D_*)$. Thus, the inverse problem of estimating the effective diffusion coefficient under these conditions can be considered arbitrarily correct.

6.6 MASS TRANSFER BETWEEN A DROP OR A GAS BUBBLE AND ISOTROPIC TURBULENT FLOW

Most of the processes of chemical technology (absorption, distillation, extraction from the liquid phase, dissolution, gas–liquid reactors) are characterized by a mass transfer between a drop or gas bubbles and the surrounding turbulent medium. The study of the problem of mass transfer is associated, first of all, with the characteristics of turbulence (energy dissipation, turbulence scale) and with the properties of the medium [71–75].

The problems associated with mass exchange between particles (a drop, gas bubbles) are considered in many works. Despite the empirical nature of most of the formulas obtained for calculating the mass-transfer coefficient, most of them are associated with energy dissipation in turbulent flow. So, in work [72], the coefficient of mass transfer is proportional to the dissipation of energy $\beta_L \sim \varepsilon_R^{1/4}$, in work [74,75] it is $-\beta_L \sim \varepsilon_R^{0.4}$. In the works $[76-80]$ based on the analysis of dimensions, based on the Higby penetration model for a single gas bubble, the following model is proposed for calculating the mass-transfer coefficient at low turbulence intensity in the following form

$$\beta_L = C_{12} D^{1/2}\left(\frac{\varepsilon_R \rho_c}{k}\right)^{1/(2(1+n))} \tag{6.245}$$

where D is the coefficient of molecular diffusion, k is the viscosity index for non-Newtonian fluids, or $k = \nu_c$ is the viscosity of the medium for Newtonian liquids at $n = 1$. In this model, C_{12} is a coefficient determined experimentally and having different values: for Newtonian liquids $C_{12} = 1.13$ [78], in mixing devices [77], $C_{12} = 0.13$ [72], $C_{12} = 0.4$ [79], $C_{12} = 0.523$ [80], $C_{12} = 0.301$ [81,82], etc. With such a spread of the coefficient, it is difficult to judge the value of the mass-transfer coefficient for certain conditions of the process. In the paper [72], in order to determine the mass-transfer coefficient for various particles and bubbles on the basis of experimental studies, the following correlation is

$$\beta_L \mathrm{Sc}^{2/3} = 0.13\left(\frac{\varepsilon_v \eta_c}{\rho_c^2}\right)^{1/4} \tag{6.246}$$

where $\varepsilon_v = \varepsilon_R \rho_c$ is the energy dissipation per unit volume.

It is important to note that the Higby model does not explicitly take into account the convective mass transfer and reflects the more qualitative side of the transfer. The presented analysis [81,82] shows that the determination of the mass-transfer coefficient entirely depends on the characteristics of the isotropic turbulent flow.

It should be noted that, other things being equal, the diffusion flux per drop is greater than the hard sphere, since this is due to the more favorable flow conditions caused by the circulation of the internal liquid, the mobility of the droplet surface, and the small angle of flow separation from the surface. In this paper [6], the region of flow separation from the surface of spherical droplets and bubbles is estimated in the form $\theta_1 \approx \mathrm{Re}_d^{-1/2}$. The paper [83] gives the following empirical correlation of the hydrodynamic trace for a solid spherical particle

$$\theta_1 \approx 180 - 42.5\left[\ln\left(\frac{\mathrm{Re}_d}{20}\right)\right]^{0.483} \tag{6.247}$$

where θ_1 is the angle of flow separation from the surface of droplets, which is accounted for from the aft part, $\theta = 180 - \theta_1$. Using the experimental data [83,84], for the region $\theta \le 180^0$ it is possible to obtain a simpler formula with the correlation coefficient $r^2 = 0.995$ $\left(\theta_1 = \frac{218}{\mathrm{Re}_d^{1/7}}\right)$. As follows from these formulas for large values of the number Re_d, the angle of separation of the flow from the bubble surface is negligible, of the order of $\theta_1 < 1^0$, which has an insignificant effect on the mass-transfer coefficient. In this connection, below, when deriving the appropriate formulas for calculating the mass-transfer coefficient, we assume that the flow around bubbles and droplets is continuous.

6.6.1 Mass Transfer in an Isotropic Flow at Low Peclet Numbers

Most of the processes of chemical technology take place in the region of large numbers Re, that is, in a developed turbulent flow. Let us consider the mass transfer to the surface of a single spherical droplet at small numbers $\mathrm{Pe} < 1$ and $C = \tilde{C} + C'$ (where C' is the pulsation component of the concentration of matter) described by a non-stationary equation of the form

$$\frac{\partial \tilde{C}}{\partial t} = \frac{1}{r^2}\frac{\partial}{\partial r}\left(r^2 D_T \frac{\partial \tilde{C}}{\partial r}\right) \tag{6.248}$$

with boundary conditions

$$\begin{aligned} t = 0, \quad & r > R, \quad \tilde{C} = \tilde{C}_0 \\ t > 0, \quad & r = R, \quad \tilde{C} = \tilde{C}_P \\ & r \to \infty, \quad \tilde{C} = \tilde{C}_0 \end{aligned} \tag{6.248a}$$

Here D_T is the coefficient of turbulent diffusion, and $\tilde{C}_0$, $\tilde{C}_P$ is the concentration of matter in the flow and on the surface. According to [6], the coefficient of turbulent diffusion in an isotropic turbulent flow is determined by the following expressions

$$\lambda > \lambda_0, \qquad D_T \approx \alpha\left(\varepsilon_R \lambda\right)^{1/3}\lambda \tag{6.249}$$

$$\lambda > \lambda_0, \qquad D_T \approx \alpha\left(\varepsilon_R \lambda\right)^{1/3}\lambda \tag{6.249a}$$

where $\lambda_0 = \left(\frac{v_c^3}{\varepsilon_R}\right)^{1/4}$ is the Kolmogorov scale of turbulence. In the region $\lambda = \lambda_k$, the coefficient of turbulent diffusion decreases and at a certain value becomes equal to the coefficient of molecular diffusion $D_T = D$. Thus, using (6.249a) in the limiting case, we can write

$$\alpha = \frac{D}{\varepsilon_R^{1/3}\lambda_k^{4/3}} \tag{6.250}$$

If we assume that $\lambda_k \le \lambda_0 = \left(\frac{\nu_c^3}{\varepsilon_M}\right)^{1/4}$, then from equation (6.250) we have

$$\alpha \approx \frac{D}{\nu_c} = \mathrm{Sc}^{-1} \tag{6.251}$$

where $\mathrm{Sc} = \frac{\nu_c}{D}$ is the Schmidt number. An analogous estimate (6.251) can be obtained using equation (6.249a). It is important to note that for gases $\mathrm{Sc} \approx 1$, and for liquids of the order $\mathrm{Sc} \approx 10^3$. Taking into account (6.250) and (6.251), the mass-transfer equation can be represented in the form

$$\frac{\partial \tilde{C}}{\partial t} = \frac{\varepsilon_R^{1/3}}{r^2 \mathrm{Sc}} \frac{\partial}{\partial r}\left(r^{10/3} \frac{\partial \tilde{C}}{\partial r}\right) \tag{6.252}$$

The solution of this equation by the method of separation of variables with boundary conditions (6.248a) can be represented in the form [85–87]

$$\tilde{C}(r,t) \approx \sum_{n=1}^{\infty} A_n \mathrm{J}_2\left[\mu_n \left(\frac{r}{R}\right)^{1/3}\right] \exp\left(-\mu_n^2 t\right) + \tilde{C}_P \tag{6.253}$$

where μ_n is the eigenvalues defined in the form $\mu_n = q_n \left(\frac{\varepsilon_R^{1/3}}{3\mathrm{Sc}R^{2/3}}\right)^{1/2}$, q_n are the roots of the equation $J_2(q_n) = 0$ $(q_1 = 5.05;\ q_2 = 8.45;\ q_3 = 11.8, \ldots)$, $A_n = \frac{2}{R^2} \frac{\int_0^R \tilde{C}_0 J_2\left[\mu_n \left(\frac{r}{R}\right)^{1/3}\right] r\,dr}{\mathrm{J}_1^2(\mu_n)}$. The series (6.253) converges sufficiently rapidly, and therefore we can confine ourselves to the first term with the following coefficients

$$\mu_1 = 5.05\left(\frac{\varepsilon_R^{1/3}}{3R^{2/3}\mathrm{Sc}}\right)^{1/2}, \qquad A_1 = \frac{2\tilde{C}_0 \mathrm{J}_3(q_1)}{q_1 \mathrm{J}_2^2(q_1)} \tag{6.253a}$$

Thus, an approximate solution of equation (6.253) can be represented in the form

$$\frac{\tilde{C}(r,t) - \tilde{C}_P}{\tilde{C}_0} \approx 1.2 \mathrm{J}_2\left[5.05\left(\frac{r}{R}\right)^{1/3}\right] \exp\left[-8.45\alpha\left(\frac{\varepsilon_R}{R^2}\right)^{1/3} t\right] \tag{6.253b}$$

The mass flow to the surface of the drop is defined in the form

$$J = D_T \frac{\partial \tilde{C}}{\partial r} \approx 0.67 R\alpha \left(\frac{\varepsilon_R}{R^2}\right)^{1/3} \Delta\tilde{C} \exp\left[-8.45\alpha\left(\frac{\varepsilon_R}{R^2}\right)^{1/3} t\right] \tag{6.254}$$

where $\Delta\tilde{C} = \tilde{C}_0 - \tilde{C}_P$. As follows from equation (6.254), the mass flow of matter to the surface of the drop depends essentially on the energy dissipation $J \sim \left(\frac{\varepsilon_R}{R^2}\right)^{1/3}$. Comparing with the equation of convective mass transfer $J = \beta_L \Delta\tilde{C}$, the mass-transfer coefficient is defined in the form

$$\beta_L \approx 0.67 R\alpha \left(\frac{\varepsilon_R}{R^2}\right)^{1/3} \exp\left[-8.45\alpha\left(\frac{\varepsilon_R}{R^2}\right)^{1/3} t\right] \tag{6.255}$$

For the case $\lambda < \lambda_0$, that is, for a low turbulence intensity, the solution of equation (6.248), taking into account (6.249a), is represented in the form

$$\frac{\tilde{C}(r,t)-\tilde{C}_P}{\tilde{C}_0}=\left[1-\left(\frac{R}{r}\right)^3\right]\exp\left[-\left(\frac{\varepsilon_R}{v_c \mathrm{Sc}}\right)^{1/2} t\right] \tag{6.256}$$

The mass flux to the surface of the drop is expressed in the form

$$J = 3R\alpha\Delta\tilde{C}_0\left(\frac{\varepsilon_R}{v_c}\right)^{1/2}\exp\left[-\alpha\left(\frac{\varepsilon_R}{v_c}\right)^{1/2} t\right] \tag{6.257}$$

For a small turbulence intensity $\lambda < \lambda_0$, the mass flow to the droplet surface depends on the energy dissipation and the viscosity of the medium $J \sim \left(\frac{\varepsilon_R}{v_c}\right)^{1/2}$. To calculate the mass-transfer coefficient from equation (6.257), we can obtain the following formula [71]

$$\beta = 3R\alpha\left(\frac{\varepsilon_R}{v_c}\right)^{1/2}\exp\left[-\alpha\left(\frac{\varepsilon_R}{v_c}\right)^{1/2} t\right] \tag{6.258}$$

The proposed expressions show that the mass-transfer coefficient for small values of the number Pe depends on the dissipation of energy per unit mass for $\lambda < \lambda_0$ of the viscous flow regime $\sim \varepsilon_R^{1/2}$ and $\sim \varepsilon_R^{1/3}$ for the developed turbulence $\lambda > \lambda_0$.

6.6.2 Mass Transfer between a Drop and an Isotropic Flow at Large Peclet Numbers

For large values of the number $\mathrm{Pe} > 10^3$, the transfer process with sufficient accuracy for practical purposes can be considered steady and can be considered in the diffusion boundary layer approximation. The distribution of the concentration of matter in the boundary-layer approximation is described by an equation of the form

$$\tilde{V}_r\frac{\partial \tilde{C}}{\partial r}+\frac{\tilde{V}_\theta}{r}\frac{\partial \tilde{C}}{\partial \theta}=\frac{\partial}{\partial r}\left[D_T(r)\frac{\partial \tilde{C}}{\partial r}\right] \tag{6.259}$$

The boundary conditions are

$$\begin{aligned} r \to \infty, &\quad \tilde{C}=\tilde{C}_0 \\ r = R, &\quad \tilde{C}=\tilde{C}_P \end{aligned} \tag{6.260}$$

where $\tilde{C}_0$ and $\tilde{C}_P$ are the concentration of matter in the volume and on the surface of the drop. We introduce the stream function associated with the velocities V_r and V_θ relations

$$V_\theta = -\frac{1}{r\sin\theta}\frac{\partial \psi}{\partial r}, \qquad V_r = \frac{1}{r^2\sin\theta}\frac{\partial \psi}{\partial \theta} \tag{6.261}$$

Using (6.249) for the case $\lambda > \lambda_0$, we define the diffusion coefficient in the form $D_T = \alpha\varepsilon_R^{1/3} r^{4/3}$. Passing to the new variables θ and ψ, after simple transformations for the right-hand side of (6.259), we obtain

$$\frac{\partial}{\partial r}\left[D_T(r)\frac{\partial \tilde{C}}{\partial r}\right]=\alpha\varepsilon_R^{1/3}R^{7/3}\frac{\partial}{\partial \psi}\left[V_\theta \sin^2\theta\frac{\partial \tilde{C}}{\partial \psi}\right] \tag{6.262}$$

Taking into account (6.259) and $V_\theta \approx V_0 \sin\theta$ on the surface of the drop $r = R$, we rewrite equation (6.262) in the form

$$\frac{\partial \tilde{C}}{\partial \theta}=\alpha\varepsilon_R^{1/3}R^{13/3}V_0 \sin^3\theta\frac{\partial^2 \tilde{C}}{\partial \psi^2} \tag{6.263}$$

Having neglected the details of the solution of this type of equation, given in $[6,88, 89]$, finally we define the diffusion flux in the form

$$J = D_T\frac{\partial \tilde{C}}{\partial y}\Big|_{y=0}=\frac{\alpha\,\varepsilon_R^{1/3}R^{7/3}V_0\Delta\tilde{C}\sin^2\theta}{\sqrt{\alpha\,\pi\,\varepsilon_R^{1/3}R^{13/3}V_0\left(\frac{2}{3}-\cos\theta+\frac{\cos^3\theta}{3}\right)}}$$
$$=\sqrt{\frac{3\alpha}{\pi}}\left(\varepsilon_R R\right)^{1/6}V_0^{1/2}\Delta\tilde{C}\frac{1+\cos\theta}{\sqrt{2+\cos\theta}} \tag{6.264}$$

where $y = r - R$ is the new variable representing the distance to the surface of the drop, $y << R$.

Comparing this equation with the expression for the mass transfer represented in the form $J = \beta_L\Delta\tilde{C}$ and expressing the velocity on the surface of the drop through the velocity of its motion $V_0 = \frac{U}{2(1+\gamma)}$, we determine the mass-transfer coefficient in the form

$$\beta_L=\sqrt{\frac{3\alpha}{2\pi}}\left(\varepsilon_R R\right)^{1/6}\left(\frac{U}{1+\gamma}\right)^{1/2}\frac{1+\cos\theta}{\sqrt{2+\cos\theta}} \tag{6.265}$$

Similarly, for the weak intensity of the turbulent flow, using expression (6.249), the diffusion flux to the surface of the drop is defined in the form

$$J=\sqrt{\frac{3\alpha}{\pi}}\left(\frac{\varepsilon_R}{\nu_c}\right)^{1/4}\left(V_0R\right)^{1/2}\Delta\tilde{C}\frac{1+\cos\theta}{\sqrt{2+\cos\theta}} \tag{6.266}$$

The mass-transfer coefficient is defined as

$$\beta_L=\sqrt{\frac{3\alpha}{\pi}}\left(\frac{\varepsilon_R}{\nu_c}\right)^{1/4}\left(\frac{UR}{1+\gamma}\right)^{1/2}\frac{1+\cos\theta}{\sqrt{2+\cos\theta}} \tag{6.267}$$

Thus, using (6.266) and (6.267), the mass-transfer coefficient between the turbulent flow and the drop is defined as

$$\lambda > \lambda_0,\qquad \beta_L = C_{11}\left(\varepsilon_R R\right)^{1/6}U^{1/2}\left(\frac{\alpha}{1+\gamma}\right)^{1/2}$$
$$\lambda < \lambda_0,\qquad \beta_L = C_{12}\left(\frac{\varepsilon_R}{\nu_c}\right)^{1/4}\left(UR\right)^{1/2}\left(\frac{\alpha}{1+\gamma}\right)^{1/2} \tag{6.268}$$

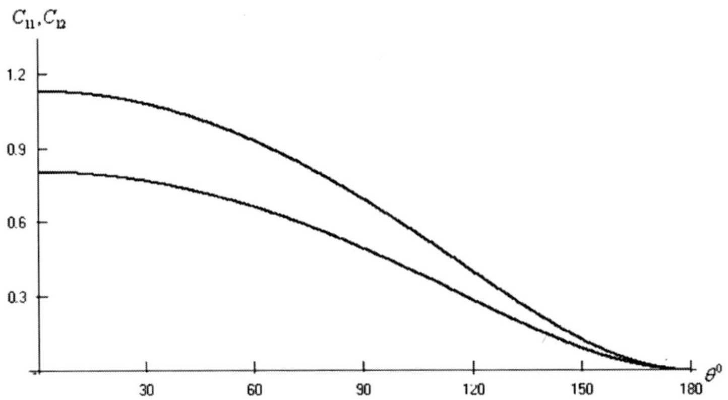

FIGURE 6.12 Change of coefficients depending on the angle θ.

Here, C_{11} and C_{12} are the coefficients defined as follows

$$\lambda > \lambda_0, \qquad C_{11} = \sqrt{\frac{3}{2\pi}} \frac{1+\cos\theta}{\sqrt{2+\cos\theta}}$$
$$\lambda < \lambda_0, \qquad C_{12} = \sqrt{\frac{3}{4\pi}} \frac{1+\cos\theta}{\sqrt{2+\cos\theta}} \tag{6.269}$$

As the angle increases θ, the mass flux density gradually decreases and reaches a value equal to zero in the aft part of the droplet $\theta = \pi$ with and correspondingly zero mass-transfer coefficients for any values λ with an insignificant separation of the flow from the droplet surface (Figure 6.12). Similar to this picture, the mass flow to the surface of the bubble will change, although the separation of the flow from the surface is observed for solid particles (Figure 6.13).

It is important to note that this result is characteristic only for the continuous flow of droplets and bubbles, although for solid particles this result may be somewhat different (Figure 6.13). It follows from Figure 6.12, the coefficients C_{11} and C_{12} can take different values depending on θ. At the point where the stream $\theta = 0$ hits the surface of the drop, we have

$$\lambda > \lambda_0, \qquad C_{11} = \sqrt{\frac{2}{\pi}}$$
$$\lambda < \lambda_0, \qquad C_{12} = \frac{2}{\sqrt{\pi}} \tag{6.270}$$

In the literature $[76-82]$, different values are given for the coefficient $C_{12} = 0.13-1.13$ for low turbulence intensity $(\lambda < \lambda_0)$. It follows from Figure 6.12, the coefficient C_{12} determined from the expression (6.269) varies from 1.1286 $(\lambda > \lambda_0)$ and 0.798 $(\lambda < \lambda_0)$ at the bubble flow point $(\alpha = 1, \gamma = 0)$ to zero in its aft part.

Taking into account expression (6.251), we define the turbulent Schmidt number in the form

$$\lambda > \lambda_0, \qquad \mathrm{Sc_T} = \frac{\nu_c}{D_T} = \mathrm{Sc}\left(\frac{\nu_c^3}{\varepsilon_R R^4}\right)^{1/3}$$
$$\lambda < \lambda_0, \qquad \mathrm{Sc_T} = \mathrm{Sc}\left(\frac{\nu_c^3}{\varepsilon_R R^4}\right)^{1/2} \tag{6.271}$$

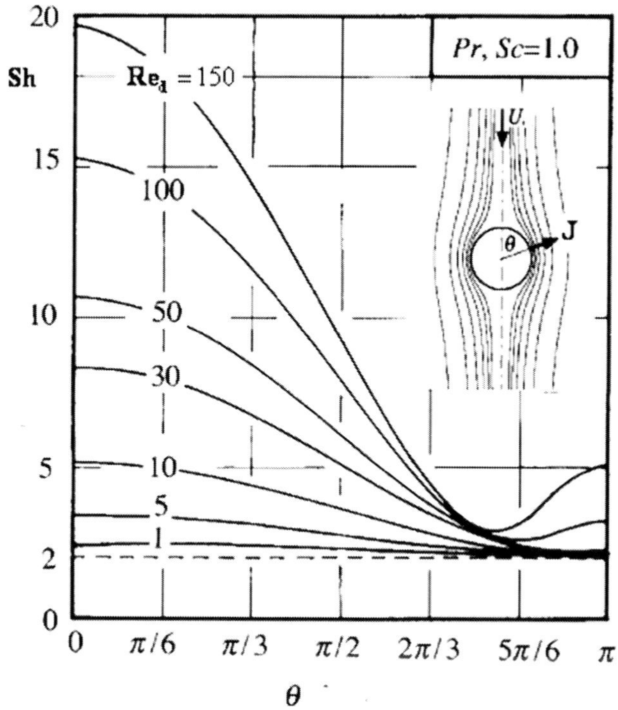

FIGURE 6.13 The change in mass flow on the surface of solid spherical particles.

We define the turbulent Sherwood number $(\lambda < \lambda_0)$, using expression (6.268) in the form

$$\mathrm{Sh} = \frac{\beta_L R}{D_T} = C_{12}\mathrm{Sc}^{1/2}\left(\frac{v_c^3}{\varepsilon_R R^4}\right)^{1/4}\left(\frac{\mathrm{Re}_d}{1+\gamma}\right)^{1/2} = C_{12}\mathrm{Sc}_{\mathrm{T}}^{1/2}\left(\frac{\mathrm{Re}_d}{1+\gamma}\right)^{1/2} = C_{12}\mathrm{Pe}_{\mathrm{d}}^{1/2}\left(1+\gamma\right)^{-1/2} \quad (6.272)$$

Similarly, for $\lambda > \lambda_0$, using the expressions (6.249) and (6.268), we define a number in the form

$$\mathrm{Sh} = C_{11}\mathrm{Sc}^{1/2}\left(\frac{v_c^3}{\varepsilon_R R^4}\right)^{1/6}\left(\frac{\mathrm{Re}_d}{1+\gamma}\right)^{1/2} \quad (6.273)$$

Here, $\mathrm{Pe}_{\mathrm{d}} = \mathrm{Sc}_{\mathrm{T}}\,\mathrm{Re}_d$. Experimental studies of the mass exchange between bubbles and liquid in vertical and horizontal pipes show that equations (6.272) and (6.273) also depend on the ratio of the particle size to the channel size.

6.6.3 Energy Dissipation in a Turbulent Flow and Its Influence on Mass Transfer

In Section 2.6, we considered the definition of the dissipative function, the dissipation of energy in different flows, and its effect on the heat transfer. In the same chapter, let us consider the effect of dissipation on the determination of the mass-transfer coefficients in an isotropic turbulent flow. Energy dissipation, characterizing all dissipative systems, is one of the important characteristics of the turbulent flow and forms the basis for calculating the parameters (turbulent diffusion, turbulent thermal conductivity, turbulent viscosity, etc.) of all transport phenomena occurring in the processes of chemical technology. At present, there is no complete quantitative evaluation of the parameters of the developed turbulence, although under certain assumptions of the isotropy of the turbulent flow, a number of important qualitative results can be obtained for the estimation of mass transfer.

The dissipative function in spherical coordinates for small numbers Re is defined as

$$\Phi_D = 2\eta_c\left[\left(\frac{\partial V_r}{\partial r}\right)^2+\left(\frac{1}{r}\frac{\partial V_\theta}{\partial \theta}+\frac{V_r}{r}\right)^2+\left(\frac{V_\theta}{r}ctg\theta+\frac{V_r}{r}\right)^2\right]+\eta_c\left[r\frac{\partial}{\partial r}\left(\frac{\partial V_\theta}{\partial r}\right)+\frac{1}{r}\frac{\partial V_r}{\partial \theta}\right]^2 \tag{6.274}$$

Setting for the external flow past the drop, the following expressions for the normal and tangential velocity components [88]

$$V_{r1} = U\left[1-\frac{2+3\gamma}{2(1+\gamma)}\frac{R}{r}+\frac{\gamma}{2(1+\gamma)}\frac{R^3}{r^3}\right]\cos\theta \tag{6.275}$$

$$V_{\theta 2} = -U\left[1-\frac{2+3\gamma}{2(1+\gamma)}\frac{R}{r}-\frac{\gamma}{4(1+\gamma)}\frac{R^3}{r^3}\right]\sin\theta \tag{6.276}$$

and, having determined the corresponding derivatives and omitting simple transformations, from (6.274) we have

$$\Phi_D = \frac{3\eta_c}{4}\frac{U^2R^2}{r^4}\left[\cos^2\theta\left(\psi_1^2-6\psi_1\psi_2\frac{R^2}{r^2}+6\psi_2^2\frac{R^4}{r^4}\right)+3\frac{R^4}{r^4}\psi_2^2\right] \tag{6.277}$$

Here, $\psi_1 = \frac{2+3\gamma}{1+\gamma}$, $\psi_2 = \frac{\gamma}{1+\gamma}$. Let us determine the total dissipation energy in the form

$$E_D = -\frac{dE}{dt} = \int_0^\pi\int_{(S)}\Phi(r)2\pi r^2 dr\big|_{r=R}\sin^2 d\theta \tag{6.278}$$

Here, E is the kinetic energy of internal friction. Using (6.277), the expression (6.278) is finally representable in the form

$$E_D = -\frac{dE}{dt} = -\pi\,\eta_c U^2 Rf(\psi_1,\psi_2) \tag{6.279}$$

Here, $f(\psi_1,\psi_2) = \psi_1^2 - 2\psi_1\psi_2 + 3\psi_2^2$. The resistance force for a spherical particle in a flow is defined as

$$F_S = -\frac{1}{2}\frac{\partial}{\partial U}\left(\frac{dE}{dt}\right) = \pi\,\eta_c U\,R\,f(\psi_1,\psi_2) \tag{6.280}$$

Considering U the change in the mean velocity over the scale of pulsations λ, the dissipation energy per unit mass $m_\lambda = \frac{\pi}{6}\rho_c\lambda^3$ for a turbulent flow is defined as the energy dissipation per unit mass in the form

$$\varepsilon_R = \frac{E_D}{m_\lambda} = 6\frac{\nu_T V'^2}{\lambda^2}f(\psi_1,\psi_2) \tag{6.281}$$

Expressing the pulsation velocity through the total flow velocity $V' = \sqrt{\frac{\xi}{8}}U$, we obtain for the maximum turbulence scale in the pipe (the minimum turbulence scale value is $\lambda = \frac{d_T}{\mathrm{Re}^{11/12}}$ [90])

$$\varepsilon_R = \frac{3}{2}\xi(\mathrm{Re})\frac{U^3}{d_T} f(\psi_1, \psi_2) \tag{6.282}$$

This expression coincides with the equations for the dissipation of energy per unit mass for isotropic turbulence, obtained from the analysis of the dimensionality of variables [6,78,79,90], for the difference between the coefficient and the function that takes into account the influence of the droplet viscosity and the medium. In particular, for $4\times10^3 < \mathrm{Re} < 10^5$, using the well-known Blass's formula $\xi(\mathrm{Re}) = \frac{0.3164}{\mathrm{Re}^{1/4}}$, from equation (6.282), by simple transformations, we obtain

$$\varepsilon_R = 0.2373 f(\psi_1, \psi_2)\frac{\nu_c^3}{d_T^4}\mathrm{Re}^{11/4} \tag{6.283}$$

In particular, for bubbles $f(\psi_1, \psi_2) = 4$ this expression takes the form

$$\varepsilon_R \approx \frac{\nu_c^3}{d_T^4}\mathrm{Re}^{11/4} \tag{6.284}$$

Since the mass-transfer coefficients are practically expressed in terms of the main flow parameters, we determine the relationship between the numbers Re and Re_d, based on the equality of the inertial forces in the turbulent flow and the resistance force of the particle. The inertial force acting on the drop from the side of the turbulent flow, taking into account the turbulent acceleration $a_\lambda = \frac{\varepsilon_R^{2/3}}{\lambda^{1/3}}$ and equation (6.284), is defined in the form

$$F_i = m_\lambda a_\lambda = \frac{\pi}{6}\frac{\eta_c^2}{\rho_c}\left(\frac{d}{d_T}\right)^{8/3} f^{2/3}(\psi_1, \psi_2)\mathrm{Re}^{11/6} \tag{6.285}$$

Equating this force to the drag force $F_i = F_S = \frac{3\pi\eta_c^2 \mathrm{Re}_d}{\rho_c}$ for small numbers $\mathrm{Re}_d \le 1$, we obtain

$$\mathrm{Re}_d = \frac{1}{18}\left(\frac{d}{d_T}\right)^{8/3} f^{2/3}(\psi_1, \psi_2)\mathrm{Re}^{11/6} \tag{6.286}$$

The resulting equation (6.286), which relates the number for a drop Re_d with the number Re for a turbulent flow and depends on the ratio of the particle size and the channel, should be used in estimating the mass-transfer coefficients. It is important to note that this expression is characteristic for droplets in the Stokes flow regime and for large droplet sizes can be a very complex expression. For numbers $2 < \mathrm{Re}_d < 100$ and small values $\mathrm{Mo} < 9\times10^{-7}$, the drag coefficient of gas bubbles can be determined as $C_D = \frac{14.9}{\mathrm{Re}_d^{0.78}}$ [92–94] and the drag force, respectively, as $F_S = 14.9\pi\eta_c^2 \frac{\mathrm{Re}_d^{1.22}}{8\rho_c}$. Comparing with (6.285), we obtain

$$\mathrm{Re}_d = 0.1383\left(\frac{d}{d_T}\right)^{2.185} f^{2/3}(\psi_1, \psi_2)\mathrm{Re}^{1.5} \tag{6.287}$$

As the droplet size increases, the effect of the channel size becomes more significant. For deformable droplets, the determination of the drag force is related to the degree of deformation, the properties of the droplet and the medium, and for bubbles, deformation of the shape begins at $\mathrm{Re}_d > 100$ and small numbers Mo. In this paper [94], the condition of deformation of gas bubbles is given in the form of an expression $\mathrm{Re}_d\,\mathrm{Mo}^{1/6} > 7$, where the Morton number is $\mathrm{Mo} = \frac{g\eta_c^4}{\rho_c\sigma^3}\frac{\Delta\rho}{\rho_c} = \frac{4}{3}C_D\mathrm{We}^3\,\mathrm{Re}_d^{-4}$.

In the expressions for the mass-transfer coefficient, we can set $\mathrm{Sh} \sim \mathrm{Re}_d^{1/2} \sim f^{1/3}\left(\frac{d}{d_T}\right)^{4/3}\mathrm{Re}^{11/12}$ for small drops and $\mathrm{Sh} \sim \mathrm{Re}_d^{1/2} \sim \mathrm{Re}^{0.75}\left(\frac{d}{d_T}\right)^{1.09}$ for large drops (6.287) for the region $\lambda < \lambda_0$.

The preceding formulas (6.255), (6.258), (6.265), and (6.267) for calculating the mass-transfer coefficients indicate their different dependence on the energy dissipation of an isotropic turbulent flow as a function of the scale of the turbulence. Experimental studies of the mass exchange between small bubbles and liquid under turbulent flow in horizontal and vertical pipes and in mixing devices are given in the paper [90,94,95].

Theoretical results show that for any value Pe with a low turbulence intensity for very small bubble sizes $(\lambda < \lambda_0)$, the mass-transfer coefficient is directly proportional to the value $\beta_L \sim \left(\frac{\varepsilon_R}{\nu_c}\right)^{1/4} = \tau_k^{-1/2}$. Here, $\tau_k = \left(\frac{\varepsilon_R}{\nu_c}\right)^{-1/2} = \frac{R}{U}$ is the Kolmogorov time scale of turbulence or phase contact time, defined for individual drops and bubbles, as the time of passage of a particle of distance equal to its radius. In the particular case, if we assume in the mixing systems that the diffusion coefficient $D_T = UR$, that is, considered as a scale of pulsations $\lambda \approx R$, and U regarded as a change in the mean velocity over the scale of pulsations $(\gamma << 1, \alpha \approx 1)$ for gas bubbles, then equation (6.255) takes the form

$$\beta_L = C_{12}D_T^{1/2}\left(\frac{\varepsilon_R}{\nu_c}\right)^{1/2} = C_{12}D_T^{1/2}\left(\frac{U}{R}\right)^{1/2} \tag{6.288}$$

and, in a particular case, coincides with the semi-empirical expression (6.245), for the mass-transfer coefficient in mixing systems. Figure 6.14 compares this equation with the experimental mass transfer data in mixing devices [90,94]. The wide spread of the experimental values of the mass-transfer coefficient does not allow us to make an adequate assumption about the inverse proportionality of the square of the size of the bubbles. For these experimental data, a simple correlation gives a simple correlation: $\beta_L = 7.926 - 0.776d$.

As follows from equation (6.288), the theory leads to a relationship in which the mass-transfer coefficient is determined by a quantity proportional to the square root of the diffusion coefficient and inversely proportional to the size of the gas bubble for small values of the surface renewal time τ_k.

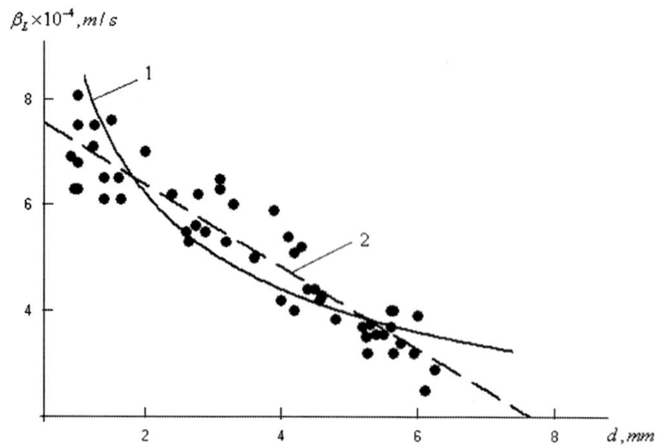

FIGURE 6.14 Dependence of the mass-transfer coefficient on the particle size: (1) equation (6.288) and (2) $\beta_L = 7.926 - 0.776d$.

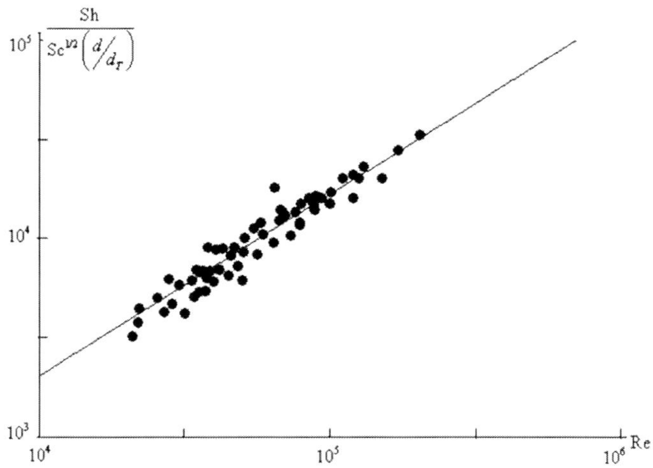

FIGURE 6.15 Comparison of calculated values and experimental data mass-transfer coefficient in horizontal pipes. (From Kress, T.S. Mass transfer between small bubbles and liquids in co-current turbulent pipeline flow. Dissertation of University of Tennessee (UK) for degree doctor of philosophy, 167 p, 1972.)

The coefficients of mass transfer between the turbulent flow and the droplet change somewhat during the flow of the dispersed system in the tubes for $\mathrm{Re} \geq 10^4$.

Using (6.271), (6.287), and experimental data for the mass-transfer coefficient, Figure 6.15 compares the experimental and calculated values, $f = 4$ and for gas air bubbles of dimensions $d = 0.38 - 0.5$ mm in a solution of glycerol with water in a horizontal tube (*sm*).

As the calculation formula, we used expression (6.273) with allowance for (6.286) of the following form

$$\frac{\mathrm{Sh}}{\mathrm{Sc}^{1/2}\left(\dfrac{d}{d_T}\right)} = a_0\,\mathrm{Re}^{11/12} \tag{6.289}$$

As follows from Figure 6.15, this expression $a_0 = 0.436$ with and the correlation coefficient $r^2 = 0.968$, the component, gives a satisfactory correspondence with the experimental data. Using (6.271) and (6.273), we obtain

$$\frac{\mathrm{Sh}}{\mathrm{Sc_T}} = \frac{\mathrm{Sh}}{\mathrm{Sc}^{1/3}} = C_{11}\left[\left(\frac{\nu_c^3}{\varepsilon_R R^4}\right)^{1/6} : \left(\frac{\nu_c^3}{\varepsilon_R R^4}\right)^{1/2}\right]\left(\frac{\mathrm{Re}_d}{1+\gamma}\right)^{1/2} = C_{11}\left(\frac{\varepsilon_R R^4}{\nu_c^3}\right)^{1/3}\left(\frac{\mathrm{Re}_d}{1+\gamma}\right)^{1/2} \tag{6.290}$$

However, taking into account the influence of various parameters (gravity, deformation, etc.), we rewrite this equation in the form

$$\frac{\mathrm{Sh}}{\mathrm{Sc}^{1/3}} = C_{13}\left[\left(\frac{\varepsilon_R R^4}{\nu_c^3}\right)^{1/3}\right]^n \tag{6.291}$$

where $C_{13} = \frac{C_{11}}{\mathrm{Sc}^{1/6}}\frac{\mathrm{Re}^{11/12}}{(1+\gamma)^{1/2}}$. Using the experimental data [90] for mass transfer between a gas bubble and a solution of glycerin and water in mixing devices ($d = 0.38 - 0.76$ mm, $\mathrm{Sc} = 419 - 2013$), it can be shown that the value n varies within $0.5 \leq n \leq 1.5$. In particular $n = 0.5$, using the first equation (6.271), we can show that (Figure 6.16).

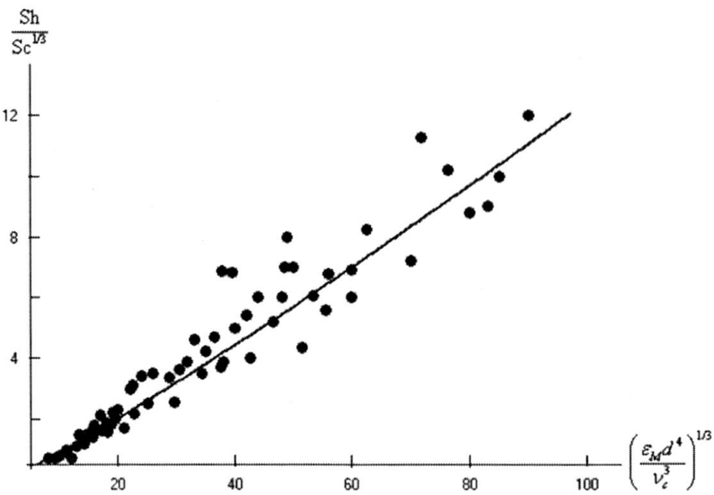

FIGURE 6.16 Comparison of the mass-transfer coefficient with the experimental data for different values of the number Sc. (From Kress, T.S. Mass transfer between small bubbles and liquids in co-current turbulent pipeline flow. Dissertation of University of Tennessee (UK) for degree doctor of philosophy, 167 p, 1972.)

A comparison of the equation for determining the mass-transfer coefficient between the bubble and turbulent flow in the Stokes mode for different values of the Schmidt number Sc = 400 − 2000 for different concentrations of glycerol in an aqueous solution (from 12.5 to 37.5%) with experimental data

$$\frac{\mathrm{Sh}}{\mathrm{Sc}^{1/3}} = 0.0686\left[\left(\frac{\varepsilon_R d^4}{\nu_c^3}\right)^{1/3}\right]^{1.13} \tag{6.292}$$

Figure 6.17 shows the change in the mass-transfer coefficient as a function of the number Re and dissipation of energy in the mixing devices and a comparison with the experimental data of the equation, which is represented in the form

$$\frac{\mathrm{Sh}}{\mathrm{Sc}^{1/3}} = 0.985\times10^{-5}\left[\left(\frac{\varepsilon_R d^4}{\nu_c^3}\right)^{1/3}\right]^{0.7}\mathrm{Re}^{11/12} \tag{6.293}$$

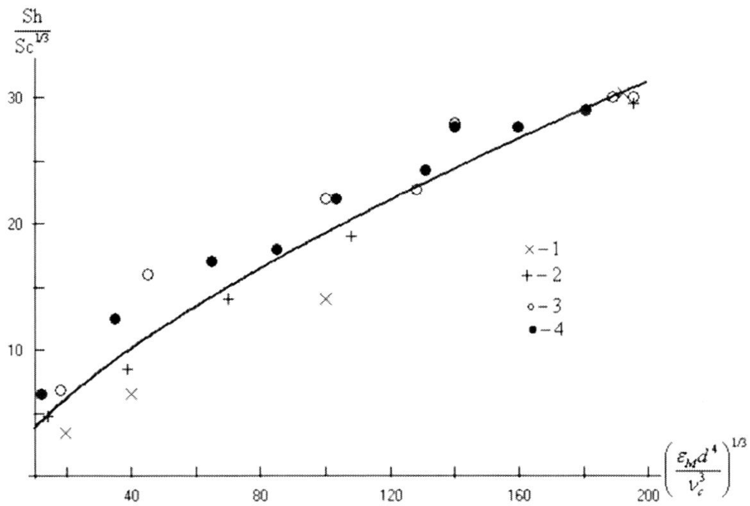

FIGURE 6.17 Comparison of experimental and calculated values of mass-transfer coefficients in mixing devices for different values of the number Re: (1) 12205; (2) 15272; (3) 20342; and (4) 22374.

As follows from Figures 6.13 through 6.15, the proposed equations for calculating the mass-transfer coefficient are in satisfactory agreement with the experimental data.

Thus, the results and formulas for determining the mass-transfer coefficient show that mass transfer in an isotropic turbulent flow is determined by the main parameters of turbulence (dissipation of specific energy, turbulence scale, turbulent diffusion) and flow properties (viscosity, density). It should be noted that the formulas for calculating the Sherwood number are mainly related to the dimensionless complex $\left(\frac{\varepsilon_R d^4}{v_c^3}\right)$, the number Re for the flow to a power close to 0.9–1.2.

In conclusion, we note that in the literature [1,10,11,17,83,89,96] one can find a lot of empirical and semi-empirical formulas for calculating mass-transfer coefficients or numbers Sh. It is regrettable that most of the formulas given do not specify the limits of their use for the range of the numbers Re and Sc; most of these formulas are applicable only for laminar flow. The effectiveness of these formulas is determined by the degree of dispersion of the experimental measurements of the mass-transfer coefficients and taking into account the necessary process parameters.

7 Hydrodynamics and Mass Transfer in Non-Newtonian Fluids

In the chemical and other industries, there is a wide class of liquids for which Newton's law of viscous friction does not hold. Such liquids include highly concentrated suspensions and emulsions, polymer solutions and melts, composite materials, heavy oil fractions, etc. Some questions related to the transfer of momentum, mass, and heat in non-Newtonian fluids are considered in the papers. This chapter will deal with the formation of structures in concentrated dispersive media, the general concepts of the rheology of non-Newtonian fluids and the rheological models of heavy oils, mass transfer in non-Newtonian liquids, the rheology of oil emulsions, and many other issues.

7.1 GENERAL INFORMATION ABOUT NON-NEWTONIAN FLUIDS

Non-Newtonian fluids, first of all, are characterized by the fact that the nature and regularities of their flow are predetermined by the special effect of the velocity gradient on the shear resistance. A general equation that describes a rheological curve for non-Newtonian fluids:

$$\tau = \tau_0 + \eta_p \left(\frac{dV}{dy} \right)^n = \tau_0 + \eta_p \dot{\gamma}^n \tag{7.1}$$

Here, τ_0 is the yield strength, η_p is the effective viscosity, and $\dot{\gamma} = {}^{dV}\!/_{dy}$ is the velocity gradient, the exponent. In particular, this equation describes the flow of Herschel–Bulkley liquids and is often used when approximating a small section of the flow curve at high shear rates.

At $\tau_0 = 0$, the flow of pseudoplastic and dilatant fluids is considered, whose flow curve is described by the following equation:

$$\tau = \eta_p \left(\frac{dV}{dy} \right)^n \tag{7.2}$$

Here, η_p is the plastic viscosity. It should be noted when this expression $n < 1$ becomes the equation for a Newtonian fluid. Curves describing the rheological properties of non-Newtonian fluids are given in Figure 7.1. Let us consider the characteristic features of non-Newtonian liquids: (1) pseudoplastic liquids, for which $n = 1$. Examples of such liquids are polymer solutions, suspensions and emulsions, including oil emulsions and sludges and many petroleum products; (2) dilatant fluids for which, $n > 1$, $\tau_0 = 0$. Examples of such liquids are concentrated suspensions, pastes, etc.; and (3) Bingham liquids, for which, $n = 1, \tau_0 > 0$ they are characterized in that they can flow only at a shear stress greater than the yield strength.

If, over time, the flow curves change their character, then such liquids are called non-stationary, for which the dependence of the effective viscosity on the velocity gradient and on the duration of the shear stress is characteristic. Such liquids are divided into thixotropic liquids, for which the liquid structure is destroyed with increasing duration of the shear stress, and it becomes more fluid, for which the fluidity decreases with increasing duration of the shear stress. It is believed that after the complete destruction of the structure of thixotropic liquids, a Newtonian flow arises, although

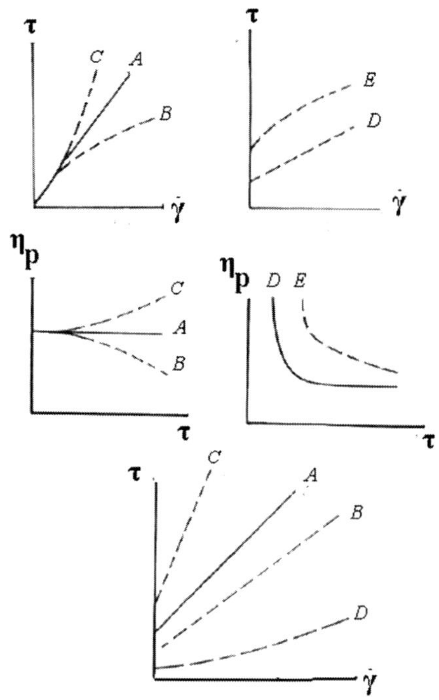

FIGURE 7.1 Characteristic curves for various types of non-Newtonian liquids: (A) Newtonian fluid; (B) pseudoplastic fluid; (C) dilatant fluid; (D) Bingham liquid; and (E) plastic fluid.

the latter is difficult to fix. For the described flow case, the concepts of the greatest viscosity of a practically undisturbed structure and the smallest viscosity of an ultimately destroyed structure are introduced.

Thixotropic properties are most characteristic for structured disperse systems, accompanied by aggregative instability accompanied by aggregation, coagulation, coalescence, and fragmentation of dispersed particles, that is, the state and the property of the disperse system will change with time. This is due to the fact that the effective viscosity of a disperse system depends significantly on the concentration of particles per unit volume and on the particle size or aggregates, as noted in Section 1.7.

The formation or destruction of aggregates is characterized in due course by some equilibrium state, meaning that the processes of destruction and recovery are mutually compensated and have some equilibrium distribution of aggregates in size. It should be noted that thixotropic properties are characterized by oil and oil products, which are usually described by the Herschel–Bulkley equation and at high shear rates by the Bingham equation (Table 7.1), which remains the basic equation for describing abnormal oils, such as heavy oils.

In general, all researchers involved in the study of oil properties agree that non-Newtonian properties of oils, oil emulsions and sludges, heavy oil fractions, and lubricating oils are associated with the presence of a structured disperse system. The equation for the thixotropic behavior of such systems can be represented by different dependencies of the shear stress on time. In particular, for systems, a more suitable differential equation may be an expression of the type

$$\frac{d\tau}{dt} = -k_0\left(\tau - \tau_\infty\right), \quad \tau\left(t\right)\Big|_{t=0} = \tau_0 \tag{7.3}$$

TABLE 7.1
Rheological Models for Non-Newtonian Fluids

№	Equation	Formula
1	Bingham	$\tau = \tau_0 + \eta\dot{\gamma}, \ \tau > \tau_0$
		$\dot{\gamma} = 0, \quad \tau < \tau_0$
2	Casson	$\tau^{1/2} = \tau_0^{1/2} + \eta\dot{\gamma}^{1/2}, \ \tau > \tau_0$
		$\dot{\gamma} = 0, \quad \tau < \tau_0$
3	Herschel–Bulkley	$\tau = \operatorname{sign}\tau_0 + k\lvert\dot{\gamma}\rvert^{n-1}\dot{\gamma}^n, \ \tau > \tau_0$
		$\dot{\gamma} = 0, \quad \tau < \tau_0$
4	Shulman	$\tau^{1/n} = \tau_0^{1/n} + (\eta\dot{\gamma})^{1/n}, \quad \tau > \tau_0$
		$\dot{\gamma} = 0, \quad \tau < \tau_0$
5	Oswald–de-Ville	$\tau = k\lvert\dot{\gamma}\rvert^{n-1}\dot{\gamma}^n$
6	Bingham	$\tau = \tau_0 \operatorname{sign} \dot{\gamma} + \eta\dot{\gamma}$

where k_0 is a certain coefficient, and τ_∞ is a finite equilibrium value of the shear stress. The solution of this equation can be represented in the form

$$\tau(t) = \tau_\infty + (\tau_0 - \tau_\infty)\exp\left(-\frac{t}{t_p}\right) \tag{7.4}$$

where $k_0 = 1/t_p$ is the time constant, and t_p is the relaxation time.

Highly paraffinic oils, heavy oils, oils with high solids content (clay, sand, and paraffin particles, etc.), and water droplets always have ultimate shear stress τ_0. One of the main indicators of thixotropic liquids is the thixotropy period, that is, the recovery time of fluidity, which is different for different types of liquids. Examples of such structures may be various types of paints, finely divided clay, and bentonite suspensions widely used in drilling oil wells. The following group of unsteady liquids include viscoelastic or Maxwell fluids that flow under the action of shear stress, but after the stress is removed, they regain their former shape. Examples of viscoelastic liquids are resins, asphaltenes, and substances of a dough-like structure.

Sometimes the consequence of such unsteadiness is the change in the properties of the fluid over time. Some liquids, having a yield strength and being Bingham, exhibit pseudoplastic properties when the shear stress is increased, and with a further increase in stress, they can behave like dilatant liquids. Thus, in practice, there are liquids of a variety of non-permanent properties, each of which finds its practical application.

To date, many concepts and models have been put forward to describe the shear flow of disperse systems, resulting in a wide variety of dependencies of effective viscosity on shear stress τ and shear rate $\dot{\gamma}$. Characteristic flow patterns for highly concentrated liquids are presented in Table 7.1 [19,21–23].

The main drawbacks of the Herschel–Bulkley, Ostwald, and Schulman equations are the absence of a theoretical justification for the microrheological model linking the rheological coefficients with the structural-rheological characteristics of the system. The following classification of non-Newtonion flow types has now been adopted.

If flow is observed at extremely low values of the shear rate, but the effective viscosity decreases with increasing shear rate, the fluid is called pseudoplastic. A fluid whose equation contains the ultimate shear stress is called plastic or nonlinearly plastic. A non-Newtonian fluid with a constant differential viscosity is called an ideal plastic fluid or a Bingham fluid.

7.1.1 The Flow of a Non-Newtonian Fluid on a Surface

Non-Newtonian fluids are characterized by the fact that the nature and regularities of their flow are determined by the different effects of the velocity gradient on the shear resistance (Table 7.1). Let us consider the steady-state flow of non-Newtonian power fluids on a flat and cylindrical surface, described by a general equation of the form

$$\frac{k}{\rho_c r^m}\frac{\partial}{\partial r}\left[r^m\left(\frac{\partial V}{\partial r}\right)^n\right]-\frac{g}{\rho_c}\frac{\partial P}{\partial x}=0$$

$$r=0,\ \frac{dV}{dr}=0,\ r=R,\ V=0 \tag{7.5}$$

Here, $m=0$ is a flat surface, $m=1$ is a cylindrical surface, and $m=2$ is a spherical surface. We rewrite this equation in the form

$$\frac{\partial}{\partial r}\left[r^m\left(\frac{\partial V}{\partial r}\right)^n\right]-\frac{gr^m}{k}\frac{\partial P}{\partial x}=0 \tag{7.6}$$

Integrating this equation with the first condition in mind, we obtain

$$\frac{\partial V}{\partial r}=\left[\frac{r}{(m+1)k}\right]^{1/n}\left(g\frac{\partial P}{\partial x}\right)^{1/n} \tag{7.7}$$

Reintegrating the given equation with allowance for the second boundary condition, we have

$$V=\frac{n}{n+1}\left[\frac{g}{(m+1)k}\frac{\partial P}{\partial r}\right]^{1/n}\left(R^{\frac{n+1}{n}}-r^{\frac{n+1}{n}}\right)=V_{\max}\left[1-\left(\frac{r}{R}\right)^{\frac{n+1}{n}}\right]$$

$$V_{\max}=\frac{n}{n+1}\left[\frac{g}{(m+1)k}\frac{\partial P}{\partial r}\right]^{1/n}R^{\frac{n+1}{n}} \tag{7.8}$$

From these expressions, we can obtain the relationship between the velocity at any point of the flow and its maximum value

$$V/V_{\max}=1-\left(\frac{r}{R}\right)^{\frac{n+1}{n}} \tag{7.9}$$

Expression (7.8) allows us to determine the velocity distribution of a non-Newtonian fluid along the radius or along the transverse coordinate for a plane $(m=0)$, cylindrical $(m=0)$, and spherical surface $(m=2)$.

Using the flow velocity distribution, it is possible to determine the relationship between the volumetric flow rate of a non-Newtonian fluid, the driving force of the flow process, and the plastic viscosity of the fluid.

7.2 STRUCTURING OF DISPERSE SYSTEMS

The spatial mutual arrangement of the constituent parts under the structure of the body is usually understood: atoms, molecules, small particles, etc. It is necessary to distinguish between ordered and disordered structures. In ordered structures, the distance between bound particles of the same size is the same (crystal lattice). In disordered structures, the bonds and the distances between the particles are not equal, and the formation of the structure is characterized by some randomness. In disperse systems, the concepts of structure and structure formation are usually associated with coagulation. In the coagulation process, a spatial structural grid is formed from the particles of the dispersed phase, which dramatically increases the strength of the system.

7.2.1 Structure Formation in Disperse Systems

The state of disperse systems is characterized by aggregative and sedimentary instability, accompanied by the formation of coagulation structures, aggregates of particles, cluster aggregates, and ultimately, the framework.

Coagulation structures are formed due to intermolecular bonds between the particles, and if liquid interlayers remain between the particles, the thickness of this interlayer significantly affects the strength of the coagulation structure. Aggregative unstable oil systems are characterized by the volatility of the state of the medium, due to the continuous structure formation and changes in the physical properties of the particles, that is, change in the volume and size of asphaltene particles as a result of their interaction, collision, coagulation, and crushing at a certain concentration in a closed volume. The relationship between the structure and viscosity of oil-dispersed systems, as well as the features of their non-Newtonian flow, is explained by a change in structure as a result of the formation and destruction of aggregates of asphaltene particles. Oil-structured systems containing crystals of high-molecular-weight paraffin, resins, and asphaltene particles and at very low laminar flow velocities or in the absence of flow form a chain or, in the extreme case, a continuous grid (skeleton) between themselves and the structure of the porous medium (Figure 7.2). Sequential coagulation or agglomeration of individual asphaltene nanoparticles into nanoaggregates and clusters of nanoaggregates eventually form a viscoelastic framework that imparts certain rheological properties to oil characterizing non-Newtonian fluids [2–4].

For aggregate-stable disperse systems, the coordinates and moment of the particles are independent, that is, the position, direction, and velocity of each particle do not depend on the position and

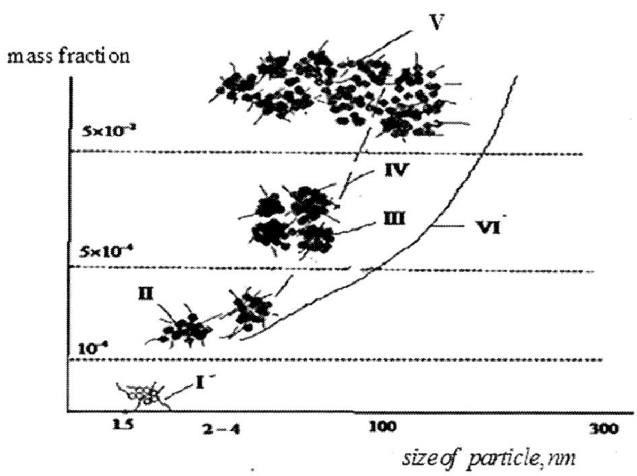

FIGURE 7.2 Aggregation of particles of asphaltenes in oils: (I) individual molecules and particles; (II) nanoaggregates; (III) clusters of nanoaggregates; (IV) an unstable suspension; (V) the viscoelastic skeleton; and (VI) stable emulsion with the participation of toluene.

velocity of the other particles. Such a state basically characterizes dilute disperse systems, where there is no probability of particle interaction, their collisions, and their coagulation. The distance between the particles, corresponding to the absence of their collision and depending on the concentration of particles in the volume, can be determined approximately by the formula [6]: $l \approx 80a\sqrt[3]{\rho_d/C_m}$ where ρ_d is the particle density, g/sm^3, and C_m is the mass concentration of the particles, g/m^3. As follows from this formula, as the concentration of particles in a closed volume increases, the distance between particles decreases and the probability of their collision increases, which leads to the formation of coagulation structures. Further formation of aggregates, clusters of aggregates, and carcass leads to the formation of a continuous loose network of interconnected particles, which substantially increase the viscosity of the system, accompanied by a decrease in its mobility and flow velocity. The destruction of bonds between particles and the destruction of the structure can be caused by mechanical external influences, for example, an increase in the shear rate due to the growth of a velocity gradient or pressure, etc.

Under the action of hydrodynamic forces, all bonds in the aggregate are stretched to a critical value, after which the aggregate breaks down into aggregates of smaller size, and a further increase in force leads to complete destruction. In a state of rest, the broken bonds, and with them the structural state of the system, can be completely restored, which is characterized by the thixotropic properties of the system. It should be noted that a change in temperature in some cases can also affect the thixotropic properties of the system. In particular, this refers to the appearance of physical phenomena (melting, evaporation) at which the phase state of a particle changes. In oil, this applies primarily to paraffin particles, which can melt at high temperatures, changing their phase state, leading to the destruction of certain structures. With a drop in temperature, crystallization processes appear with the formation of solid paraffin particles capable of creating certain structures. Similar processes occur in oil-containing particles of asphaltenes, if it contains aromatic hydrocarbons (toluene, benzene), which dissolve the asphaltenes contained in the oil and destroy coagulation structures and aggregates. A significant influence on the interaction of the part is due to the turbulence of the flow of the disperse system, which increases the collision frequency by several times compared with the laminar flow. However, the presence of turbulence not only contributes to the creation of coagulation structures but also to the destruction of the formed aggregates and the framework when the equilibrium between the phenomena of coagulation and crushing is disturbed. Due to turbulence of the flow, aggregates and clusters deform and undergo compression, stretching, and breaking of bonds between particles, which leads to their destruction at high values of turbulent pulsations. The particle size also affects the frequency of their collision and, accordingly, the creation of coagulation structures. In some cases, the collision of droplets and bubbles contributes to their fusion and enlargement, which contributes to the disruption of the aggregative and sedimentation (kinetic) stability of the medium, which results in a qualitative change in the structure of the dispersed medium. The potential for the formation of coagulation structures in the collision of solid particles is also associated with many other factors: adhesion and cohesion, surface properties of particles, propensity to adhere to them, their size and shape, etc. If dry particles stick together to form a structure, then such a state is possible for each particle size up to a certain critical flow rate, after which they are destroyed. Coarse-grained solid particles in most cases do not form coagulation structures and at high concentrations form only a dense packing of a disordered structure characterized by a minimum porosity and a sufficiently large shear or bulk viscosity. To some extent, this fact is explained by the presence of elastic collisions during their interaction. In the presence of any substance on the surface of the particles of a liquid interlayer, which increases the adhesion properties of the surface, it is possible to form aggregates of particles characterized by a loose structure. Such physical phenomena as sintering at high temperatures, crystallization from the liquid phase of particles create the possibility of forming aggregates of sufficiently high strength and hardness, and the physical processes of melting and dissolution of particles are capable of destroying the structure of any aggregates and framework. The formation of coagulation structures and aggregates gives dispersed systems the character of non-Newtonian fluids with inherent rheological properties. The change in the mass of non-deformable nanoaggregates is defined as

$$\frac{dm}{dt} = (m_\infty - m)\omega \tag{7.10}$$

The solution of this equation is represented in the form

$$m = m_\infty\left[1 - \exp(-\omega t)\right] \tag{7.11}$$

where m_∞ is the limiting mass of the aggregate, which depends on its stability.

Putting the spherical shape of the nanoaggregates and bearing in mind that $m = \frac{\pi}{6}a^3\rho$, the size of the nanoaggregates, taking into account (7.11), we define in the form

$$a_g = a_{g\infty}\left[1 - \exp\left(-C_0\varphi_0\left(\frac{\varepsilon_R}{\nu_c}\right)^{1/2} t\right)\right]^{1/3} \tag{7.12}$$

For laminar flow, the formation of aggregates describes an equation of the form

$$a_g = a_{g\infty}\left[1 - \exp(-8\pi DN_0 a_0 t)\right] \tag{7.13}$$

As follows from Figure 7.3, the dimensions of the nanoaggregates vary within limits $a_{g\infty} = 8 - 10\,\text{nm}$, and the maximum size of the frame is limited by the presence of pore walls or pipes.

As the volume fraction of asphaltene particles increases, the frequency of collisions between them increases. The relaxation time for turbulent flow is determined by the expression $\tau_R = (\nu_C/\varepsilon_R)^{1/2}$ for the laminar flow $\tau_R = 3\nu_C/(8kTN_0)$, which facilitates the rapid achievement of the final size of the aggregate. With increasing viscosity of oil, both for laminar and turbulent flow, the frequency of collisions of asphaltene particles decreases, which slows the formation rates of nanoaggregates.

7.2.2 The Main Stages of Structure Formation

The flow and viscosity of non-Newtonian fluids depend on the external action (shear stress) and acquire the ability to flow only after the destruction of this structure at $\tau >> \tau_0$ (where τ_0 is yield strength), and small external stresses produce an elastic deformation of the mesh or skeleton.

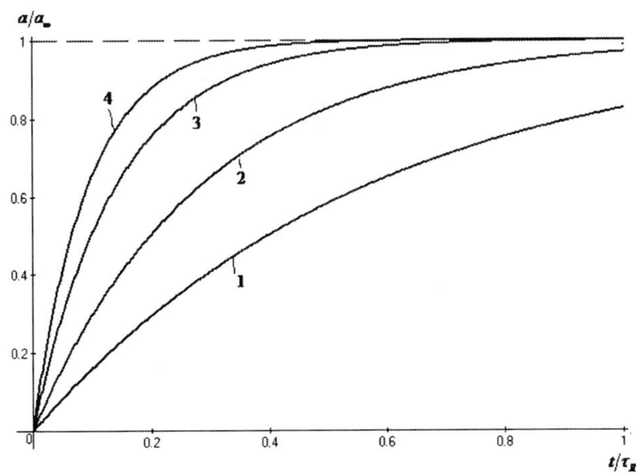

FIGURE 7.3 Change in the size of the nanoaggregates in time, depending on the content of particles of asphaltenes in oil: (1) $\varphi = 0.05$; (2) $\varphi = 0.1$; (3) $\varphi = 0.2$; and (4) $\varphi = 0.3$.

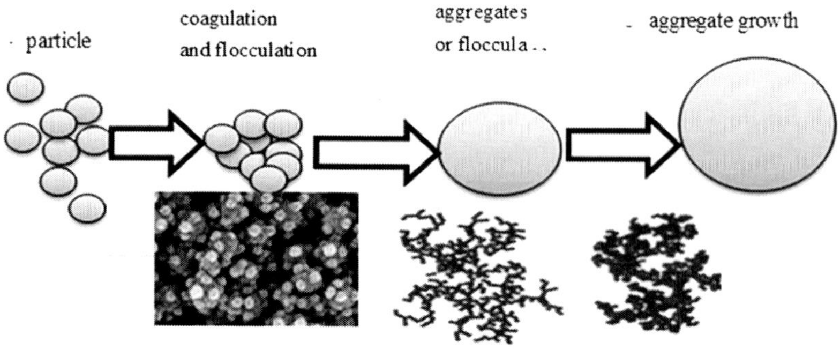

FIGURE 7.4 Scheme of coagulation and formation of aggregates of particles.

The formation of structures in a dispersed stream is associated with the physical processes of coagulation and crushing and proceeds in several stages: the creation of coagulation structures, the formation of aggregates, aggregate clusters, and finally the skeleton corresponding to the maximum viscosity of the dispersed medium and the minimum flow velocity (Figure 7.4). These structures break up into separate particles as a result of the destruction of aggregates under the influence of shear flow; moreover, the equilibrium shifts toward the formation of individual particles as the shear rate increases.

Further, with increasing a shear stress τ, gradual destruction of temporary contacts between the elements of the structure and the formation of others begins. As a result, a dynamic equilibrium arises, the rate of deformation (flow) increases sharply, and for many plastic systems, the rheological curve goes to a level corresponding to the plastic viscosity (η^*) of the system.

Extrapolation of the curve of the dependence of the flow velocity on the shear stress gives a value that quantitatively characterizes the shear strength of the structure and is the limiting dynamic shear stress corresponding to the ultimate stress of structural failure. Thus, the structuring of disperse systems can be broken down into the following stages: (1) the first state is characterized by the condition $0 < \tau < \tau_0$ (τ is shear stress, τ_0 is yield strength). In this state $\tau > \tau_0$ the flow is absent, and the external action cannot break the strength of the system; (2) with a further increase in voltage, when, the system begins to flow. The velocity of movement in this case is insignificant and the bonds between the particles have time to recover again after their destruction. The structure does not collapse, only the movement of particles relative to each other is observed. A similar movement is called creep or flowing current. The viscosity of the system under creep conditions will be greatest; in practice it will correspond to the viscosity of an undeveloped structure; (3) the third state of the dispersed system is characterized by the process of structural failure at a stress equal to the ultimate strength $\tau = \tau_\gamma$. Irreversible destruction of the structure begins at the boundary of the second and third states, and ends at the boundary of the third and fourth states. In this state, the disperse system of communication between the particles is not restored, the viscosity decreases, and the velocity of the system's motion increases; and (4) the fourth state of the structure is completely destroyed (or individual aggregates of particles oriented in the flow are formed). The viscosity in this state becomes constant, and its value is minimal (η_{min}). The velocity of a system with a disrupted structure increases in proportion to the external action. The stress characterizing the loss of strength and complete destruction of the structure is usually denoted by τ_γ.

7.2.3 Deformation and Structural Failure, Thixotropy

Aggregates formed as a result of coagulation and aggregation contain a multitude of tiny interconnected particles and have a loose structure, although their density toward the center rises. Due to the presence of voids in the aggregates, they cannot be strong, and as a result of an increase in the

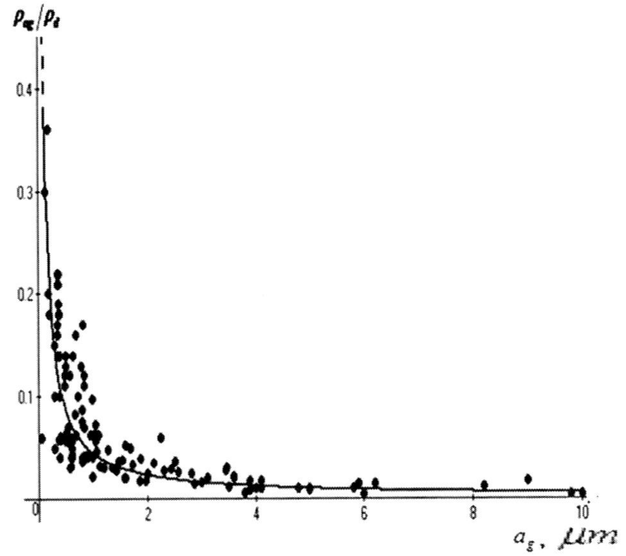

FIGURE 7.5 Dependence of the density of the structure of the carcass on its dimensions (points are experiment).

external load in the hydrodynamic field they are broken up to a single particle. The density of aggregates is much less than the density of asphaltene particles, although they behave as a single particle with corresponding dimensions. With an increase in the cluster size of aggregates or a viscoelastic framework, their state becomes less stable as a result of a decrease in density (Figure 7.5). Using the experimental data, we obtain

$$\frac{\rho_{ag}}{\rho_d} = 0.011 + \frac{0.433}{a_g} \tag{7.14}$$

To improve the rheological properties of structured oil-dispersed systems, it is necessary to mechanically destroy the viscoelastic framework of asphaltenes under the influence of external forces (velocity gradient or pressure). The main types of deformation of the framework or clusters of nanoaggregates, leading to a change in its size and shape are: compression, stretching, bending, and torsion. In the process of elastic deformation, particles of asphaltenes in the framework of the particle slightly shift relative to each other, and the more the distance between the particles changes, the more the forces of particle interaction become. Upon removing the external load on the carcass under the effect of these forces, the particle returns to its original position and its resulting distortion of the carcass is paused. In plastic deformation, the particles in the structure are displaced over long distances, the skeleton is stretched, and the bonds between clusters and particles are destroyed with large velocity gradients or pressure gradients. This displacement becomes irreversible with increasing load, which leads to the destruction of the framework. Elastoplastic deformation upon reaching high stresses can result in the dispersion of a system of coupled particles into individual particles. As a result, the flow of the structured disperse system becomes the usual flow of the Newtonian fluid. The destruction of the carcass leads, first of all, to a significant decrease in viscosity and an increase in the mobility of the disperse system. The formation of aggregates of free-bound particles or a framework of aggregates is also characterized by the reverse process—their destruction due to the low strength of bonds under the action of external stresses, acquiring the ability to flow. In structured systems, the particles of a dispersed phase tend to interact strongly with the formation of a single spatial grid, the skeleton that facilitates the loss of mobility of the medium. The viscosity of structured fluids is usually high and

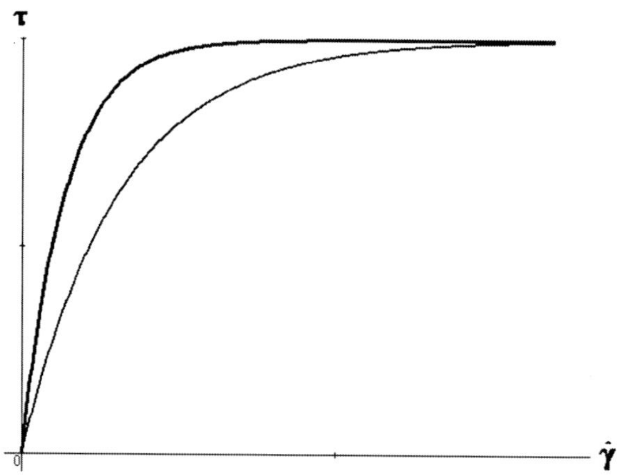

FIGURE 7.6 Thixotropic properties of structured disperse systems.

rapidly increases, even with a small increase in concentration. The viscosity coefficient of structured free-disperse systems is not constant and depends on the applied voltage. The mechanical properties of structured systems are determined not so much by the properties of the particles that form the structure, as by the nature and characteristics of the inter particle bonds and inter layers of the medium

Oil-dispersed systems are characterized by thixotropic properties. With a decrease in voltage or pressure, the reverse process occurs, that is, the dispersed system due to the aggregation of particles is structured with the formation of nanoaggregates and a framework. With increasing voltage, the structure skeleton, clusters of nanoaggregates, and aggregates themselves are destroyed, and the structured dispersed system becomes a normal liquid with dispersed inclusions. Thus, the rheological lines of thixotropic liquids form a characteristic hysteresis loop (Figure 7.6).

7.2.4 Features of Rheology of Structured Systems

The flow of structured disperse systems is described by rheological models for non-Newtonian liquids, reflecting the main stages of structure formation. Non-Newtonian fluids, first of all, are characterized by the fact that the nature and regularities of their flow are predetermined by the special effect of the velocity gradient on the shear resistance. A general equation that describes a rheological curve for non-Newtonian fluids:

$$\tau = \tau_0 + \eta_p \left(\frac{dV}{dy} \right)^n = \tau_0 + \eta_p \dot{\gamma}^n \quad (7.15)$$

Here, τ_0 is the yield strength, η_p is the effective viscosity, $\dot{\gamma} = {}^{dV}\!/_{dy}$ is the velocity gradient, and n is the exponent. In particular, this equation describes the flow of Bingham liquids and is often used when approximating a small section of the flow curve at high shear rates.

At $\tau_0 = 0$ the flow of pseudoplastic and dilatant fluids is considered, whose flow curve is described by the following equation:

$$\tau = \eta_p \left(\frac{dV}{dy} \right)^n \quad (7.16)$$

Here, η_p is the plastic viscosity. It should be noted that when this expression becomes the equation for a Newtonian fluid. Let us consider the characteristic features of non-Newtonian liquids:

(1) pseudoplastic liquids, for which $n < 1$, $\tau_0 = 0$. Examples of such liquids are polymer solutions, suspensions, and emulsions, including oil emulsions and sludges and many petroleum products; (2) dilatant fluids for which $n > 1$, $\tau_0 = 0$. Examples of such liquids are concentrated suspensions, pastes, etc.; and (3) Bingham liquids, for which $n = 1, \tau_0 > 0$,and they are characterized in that they can flow only at a shear stress greater than the yield strength.

If, over time, the flow curves change their character, then such liquids are called non-stationary, for which the dependence of the effective viscosity on the velocity gradient $\dot{\gamma}$ and on the duration of the shear stress is characteristic. Such liquids are divided into thixotropic liquids, for which the liquid structure is destroyed with increasing duration of the shear stress, and it becomes more fluid, for which the fluidity decreases with increasing duration of the shear stress. It is believed that after the complete destruction of the structure of thixotropic liquids, a Newtonian flow arises, although the latter is difficult to fix. For the described flow case, the concepts of the greatest viscosity of a practically undisturbed structure and the smallest viscosity of an ultimately destroyed structure are introduced.

Thixotropic properties are most characteristic for structured disperse systems, accompanied by aggregative instability accompanied by aggregation, coagulation, coalescence, and fragmentation of dispersed particles, that is, the state and the property of the disperse system will change with time. This is due to the fact that the effective viscosity of a disperse system depends essentially on the concentration of particles per unit volume and on the particle or aggregate sizes. The formation or destruction of aggregates over time is characterized by a certain equilibrium state, meaning that the processes of destruction and restoration are mutually compensated and have some equilibrium distribution of aggregates in size. It should be noted that thixotropic properties are characterized by oil and oil products, which are usually described by the Herschel–Bulkley equation, and at high shear rates by the Bingham equation (Table 7.3), which remains the basic equation for describing abnormal oils, such as heavy oils.

In general, all researchers who study the properties of oils agree that non-Newtonian properties of oils, oil emulsions and sludges, heavy oil fractions, and lubricating oils are associated with the presence of a structured disperse system. The equation for the thixotropic behavior of such systems can be represented by different dependencies of the shear stress on time. In particular, for systems, a more suitable differential equation may be an expression of the type

$$\frac{d\tau}{dt} = -k_0\left(\tau - \tau_\infty\right), \quad \tau\left(t\right)\Big|_{t=0} = \tau_0 \tag{7.17}$$

where k_0 is a certain coefficient, and τ_∞ is a finite equilibrium value of the shear stress. The solution of this equation can be represented in the form

$$\tau\left(t\right) = \tau_\infty + \left(\tau_0 - \tau_\infty\right)\exp\left(-\frac{t}{t_p}\right) \tag{7.18}$$

where $k_0 = \frac{1}{t_p}$ is the time constant, and t_p is the relaxation time.

Highly paraffinic oils, heavy oils, oils with high solids content (clay, sand and paraffin particles etc.), and water droplets always have ultimate shear stress τ_0. One of the main indicators of thixotropic liquids is the thixotropy period, that is, the recovery time of fluidity, which is different for different types of liquids. Examples of such structures may be various types of paints, finely divided clay, and bentonite suspensions widely used in drilling oil wells. The following group of unsteady liquids are viscoelastic or Maxwellian fluids that flow under the influence of shear stress, but after the stress is removed they restore their former shape. Examples of viscoelastic liquids are resins, asphaltenes, and substances of a dough-like structure.

Sometimes the consequence of such non-stationary is the change in the properties of the fluid over time. Some liquids, having yield strength and being Bingham, exhibit pseudoplastic properties when the shear stress is increased, and with a further increase in stress, they can behave like dilatant liquids. Thus, in practice, there are liquids of a variety of non-permanent properties, each of which finds its practical application. To date, many concepts and models have been put forward to describe the shear flow of disperse systems, resulting in a wide variety of dependencies of effective viscosity on shear stress τ and shear rate $\dot{\gamma}$.

7.2.5 The Non-Stationary Equation of Structure Formation

The following assumptions can be based on the rheological model of the flow of oil-dispersed systems:

- In a structured oil system, there are nanoaggregates that have arisen as a result of collision, coagulation, and aggregation of asphaltene particles due to diffusion under laminar and turbulent shear flow and sedimentation (gravitational coagulation); the formed aggregates of asphaltenes can be deposited on the surface, forming a sufficiently thick layer of deposits on the walls of the collector pores and the discharge lines. And the pressure drop, depending on the temperature, can lead to a repeated dissolution or tearing off of the particles of precipitated asphaltenes with intensive mixing or turbulent flow;
- Nanoaggregates move as independent flow units prior to collision with other similar aggregates or asphaltene particles. Nanoaggregates collide with each other into clusters of nanoaggregates and then create a viscoelastic framework with maximum viscosity and loose coagulation structure. The maximum size of the frame of the nanoaggregates is determined by the size of the channels (pores, pipes) through which the flow flows;
- Nanoaggregates are able to rotate in a gradient field and burst under the action of tensile hydrodynamic forces, depending on the pressure gradient or flow velocity;
- The linear dimensions of the nanoaggregates are in the interval of the size of a single particle of asphaltenes up to the maximum size of the cluster or skeleton;
- In the limiting case of infinite velocity $\lim_{\tau\to\infty}(\tau_0/\tau)\to 0$, all the aggregates under the condition are destroyed up to individual particles, as a result of which the flow of the disperse system approaches Newtonian; and
- In the presence of aromatic hydrocarbons, asphaltenes dissolve well, thereby preventing the formation of a structure, that is, formation of clusters and viscoelastic framework. In addition, the solubility of asphaltenes is influenced by the presence of other compounds contained in oil, such as resins.

When $\tau >> \tau_0$, this expression goes into the usual Darcy equation for unstructured oil-dispersed systems. The rheological equation of a viscoelastic Maxwell fluid in substantial derivatives is written in the form

$$\lambda\left(\frac{\partial \tau}{\partial t}+U\frac{\partial \tau}{\partial y}\right)+\tau=\eta_c\dot{\gamma} \tag{7.19}$$

$$t=0,\ \tau=\tau_0,\ \dot{\gamma}=0$$

A particular form of equation (7.19) is represented in the form

$$\lambda\left(\frac{\partial \tau}{\partial t}+U\frac{\partial \tau}{\partial y}\right)+\tau=0 \tag{7.20}$$

The solution of equation (7.20) can be represented in the form

$$\tau = C_1 f\left(y - Ut\right)\exp\left(-\tau/\lambda\right) \tag{7.21}$$

Substituting this solution in (7.20), we obtain the identity. Here, $\lambda = \eta_c/G$ is the Maxwell relaxation time, $f\left(y - Ut\right)$ is the deformation front velocity, U is the function determining the deformation moving front in the framework, y is the coordinate, G is the shear elastic modulus, $\dot{\gamma} = d\gamma/dt$ is shear rate, γ is the shear gradient, and τ_0 is the shear stress or yield strength. If $\tau \le \tau_0$, then $\dot{\gamma} = 0$. The complete solution of equation (7.20) is presented in the form

$$\tau = C_1 f\left(y - Ut\right)\exp\left(-t/\lambda\right)\tau_0 \tag{7.22}$$

or this equation can be represented in logarithmic form

$$\ln\tau = \ln\tau_0 - t/\lambda + \ln\left(C_1 f\left(y - Ut\right)\right),\ \ \tau_0 = \eta_c\dot{\gamma} \tag{7.23}$$

It is obvious that the quantity t/λ in equation (7.23) characterizes the deformation of the viscoelastic framework in time and depends on the velocity or pressure gradient, which in the approximation this dependence can be represented as $t/\lambda = t\dot{\gamma}/\mathrm{We} = f\left[\mathrm{grad}P/\left(\mathrm{grad}\ P\right)_0\right]$ (where $\mathrm{We} = \lambda\dot{\gamma}$ is the Weissenberg number).

It should be noted that in some rheological models, instead of the shear stress, a pressure gradient is used. Consider the equation of fluid flow in the form

$$\frac{1}{\rho}\frac{\partial P}{\partial x} = \nu\frac{\partial^2 V}{\partial y^2} = \frac{\eta}{\rho}\frac{\partial \tau}{\partial y} \tag{7.24}$$

Integrating this equation, we obtain an expression that establishes a relationship between the shear stress and the pressure gradient in the form

$$\tau = y\frac{\partial P}{\partial x} = y\ \mathrm{grad}\ P \tag{7.25}$$

Omitting the last term in (7.23), the next section gives rheological models for the filtration of non-Newtonian oils, which made it possible to obtain a semi-empirical rheological equation for the viscoelastic framework in the form

$$\ln\tau = \ln\tau_0 - \alpha\left[\frac{\mathrm{grad}\,P}{\left(\mathrm{grad}\,P\right)_0}\right]^{\mathrm{n}} \tag{7.26}$$

Obviously, the exponent is determined depending on the degree of destruction of the carcass, temperature, reservoir properties, and other parameters of various oil fields.

It should be noted that the oil of different deposits can exhibit properties corresponding to different rheological models. So, the work presents experimental studies on the rheology of Azerbaijani oil with different water content. The results of these studies have shown that these oils, depending on the content of water droplets, exhibit the properties of power dilatant non-Newtonian fluids. In particular, using experimental data for conventional oils, a number of empirical expressions were obtained that describe the rheological properties of this oil.

The dependence of the tangential stress on the gradient of the shear rate is proposed in the form (Figure 7.7)

$$\tau = 12.5\dot{\gamma}^{1/2} \tag{7.27}$$

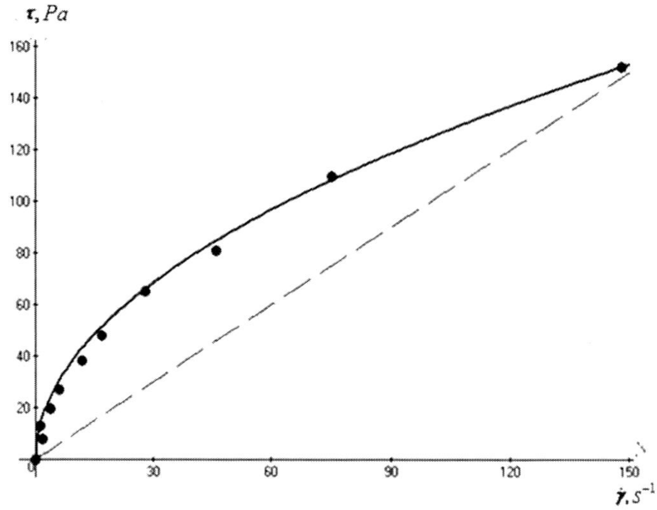

FIGURE 7.7 The character of the rheological curve for the Azeri oil.

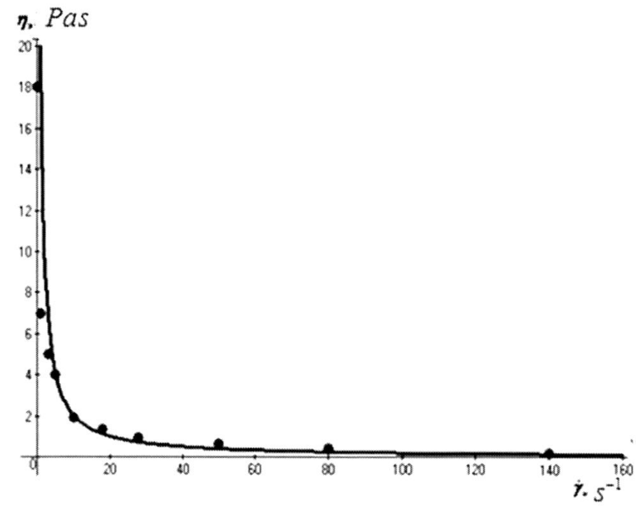

FIGURE 7.8 Dependence of viscosity on the gradient of shear rate.

Using the experimental data, the dependence of viscosity on the gradient of the shear rate was obtained (Figure 7.8)

$$\eta = \eta_0 \frac{\dot{\gamma}_0}{\dot{\gamma}} = \frac{20}{\dot{\gamma}} \tag{7.28}$$

and the dependence of the viscosity on the water content (Figure 7.8)

$$\eta = (-11.6 + 0.6\varphi)\exp\left(\frac{2.8\varphi}{(\varphi - 100)^2}\right) - 7\exp\left(-0.01(\varphi - 60)^2\right) \tag{7.29}$$

where φ is the water content of oil in percent.

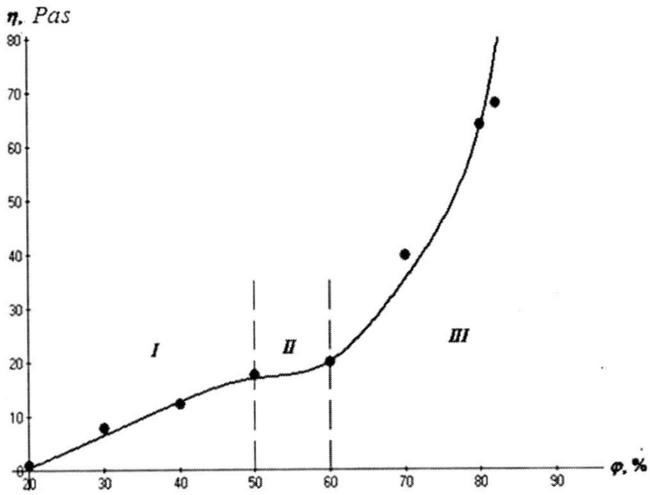

FIGURE 7.9 Dependence of viscosity on water content: I is the flow area without the formation of structures; II is the area of formation of structures; and III is the scope of completion of structure formation.

In Figure 7.9, the first region corresponds to the usual increase in the effective viscosity with increasing particle concentration, the second region corresponds to the beginning of the formation of aggregates of particles, and the third region corresponds to the complete structure formation of the system with the maximum viscosity increase and a decrease in the flow velocity to almost zero. The curve in Figure 7.9 is a characteristic rheological curve with inherent refraction in the field of structure formation.

Figure 7.10 shows the decrease in oil mobility with increasing water content with a permeability coefficient equal to $k = 2.61$ μm^2.

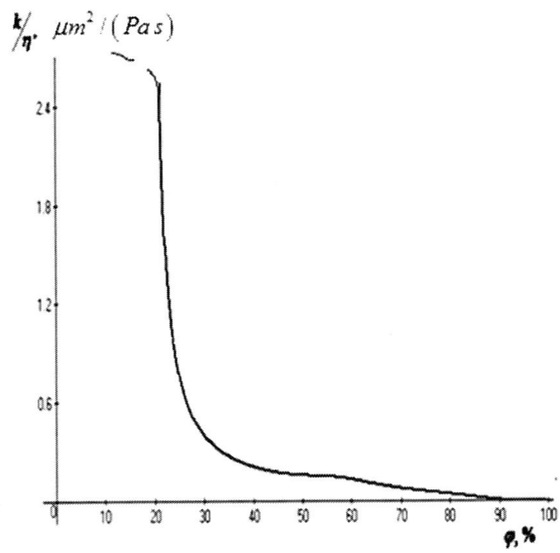

FIGURE 7.10 Dependence of fluid mobility with the increase in the amount of water in the volume.

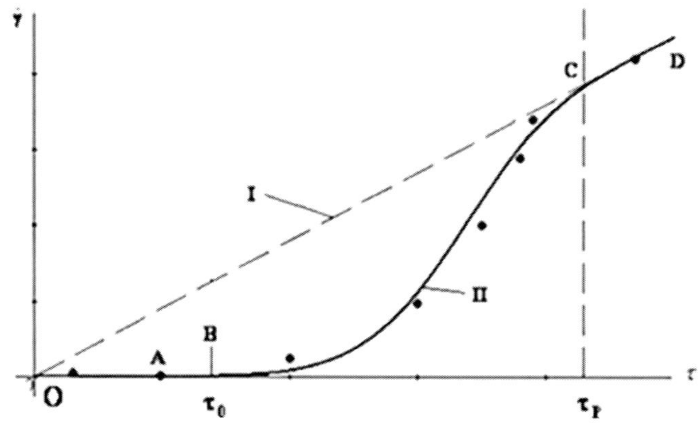

FIGURE 7.11 Characteristic curve of the flow of structured oil: I corresponds to the flow of Newtonian fluid; point A is the condition $\tau \geq \tau_0$; point B is the beginning of the destruction of the structure, $\tau \geq \tau_0$; section BC (II) is the structure destruction $\tau \geq \tau_0$; and CD is the area of complete destruction of the structure.

7.3 RHEOLOGY OF THE COURSE OF NON-NEWTONIAN OILS

Abnormal oils differ in their properties from ordinary oils, and their description obeys the laws of flow of non-Newtonian fluids.

The characteristic curves for the flow of structured oil are shown in Figure 7.11. The flow curve has two asymptotic states: the Newtonian rectilinear dependence for $\dot{\gamma} \to 0$ and the dependence described by the Bingham equation for $\dot{\gamma} \to \infty$, in connection with which the limiting values of the viscosity η_0 and appear η_∞. In Figure 7.11, line I corresponds to the behavior of the Newtonian fluid, curve II corresponds to the flow of structured (non-Newtonian) oil. On this curve II, the section OA corresponds to low-flow velocities (the phenomenon of creep), in which the system is insignificantly damaged, since the disruptions associated with the current can be thixotropically restored, that is, the flow of the system takes place without destroying its structure. This, the so-called phenomenon of creep in the flow of oil, occurs at maximum viscosity of the system η_∞.

Further, with increasing stress τ, a gradual destruction of temporary contacts between the elements of the structure and the formation of others begins. As a result, there is a dynamic equilibrium, the rate of deformation (flow) increases sharply, and for many plastic systems the rheological curve goes to the section of curve BC-II corresponding to the plastic viscosity (η^*) of the system. Extrapolation of the CS curve to the axis gives a value that quantitatively characterizes the shear strength of the structure and is the limiting dynamic shear stress corresponding to the ultimate stress of structural failure. The section of the plastic flow of the Sun is described by the Bingham equation

$$\tau = \tau_0 + \eta^* \dot{\gamma} \tag{7.30}$$

For anomalous oils, the Darcy filtering law deviates from the classical form and can be written in a nonlinear form

$$V = -\frac{k}{\eta(T, \mathrm{grad}\, P)} \frac{\partial P}{\partial x} \tag{7.31}$$

From Newton's equation ($\eta = \frac{\tau}{\dot{\gamma}}$), taking (7.30) into account, we obtain

$$\eta = \eta^* + \frac{\tau_0}{\dot{\gamma}} = \eta^* \left(\frac{\tau_0}{\eta^* \dot{\gamma}} + 1 \right), \tag{7.32}$$

Having determined from (7.30)

$$\eta^{*}\dot{\gamma} = \tau - \tau_0 \tag{7.33}$$

Finally, we obtain an expression for the effective viscosity in the form

$$\eta = \eta^{*} + \frac{\tau_0}{\dot{\gamma}} = \eta^{*}\frac{\tau}{\tau - \tau_0}, \tag{7.34}$$

from which it follows that with increasing τ, the value of $\tau >> \tau_0$ and $\eta \to \eta^{*}$, corresponding to a system with a completely destroyed structure (a section of CD) also decreases in the limit for $\tau_0/\tau \to 0$. Thus, the viscosity of a structured system in the course of flow under the influence of an η_0, increasing shear stress η_∞ varies from that corresponding to an undamaged structure to that characteristic of a completely destroyed structure. Substituting (7.34) into expression (7.31), we obtain the Darcy equation for a structured oil system

$$V = -\frac{k}{\eta^{*}}\left(1 - \frac{\tau_0}{\tau}\right)\frac{\partial P}{\partial x} \tag{7.35}$$

When $\tau >> \tau_0$, this expression goes into the usual Darcy equation for unstructured oil. In a slightly different form, this equation with some errors in the dimension is given in the paper. Analysis of experimental data on the filtration of anomalous oils allowed us to approximate the ratio τ_0/τ in the form

$$\ln\frac{\tau}{\tau_0} = \alpha\left(\frac{\text{grad}\,P}{(\text{grad}\,P)_0}\right)^{n} \tag{7.36}$$

Expression (7.36) can be considered as a new rheological equation describing the course of anomalous oils.

Obviously, the index n, depending on the temperature and properties of the formation, characterizes the steepness of the section BC. To date, many concepts and models have been put forward to describe the shear flow of disperse systems, resulting in a wide variety of rheological dependencies of effective viscosity on shear stress τ and shear rate $\dot{\gamma}$.

Using the experimental data and equations (7.35) and (7.36), we represent the filtration rate in the following form

$$V = K_2(T)\left(1 - \exp\left(-\alpha_2(T)(z/z_0)^{6}\right)\right)z \tag{7.37}$$

where $\alpha_2 = 0.1422\exp(-0.0247T)$, $K_2(T) = 1.4\times10^{-5}\exp(0.0364T)$, $z = \text{grad}\,P$, $z_0 = (\text{grad}\,P)_0$, $K_2 = k/\eta^{*}$ is the mobility of oil. Comparison of the calculated (7.37) and experimental values of the filtration rate of anomalous oils for different deposits is presented in Figure 7.12.

The values of the coefficients entering into equation (7.37) as a function of temperature are given in Table 7.2.

The dependence of the initial pressure gradient on temperature is given in the form

$$(\text{grad}\,P)_0 = 2.197\times10^{-4} + \frac{0.3275}{T} \tag{7.38}$$

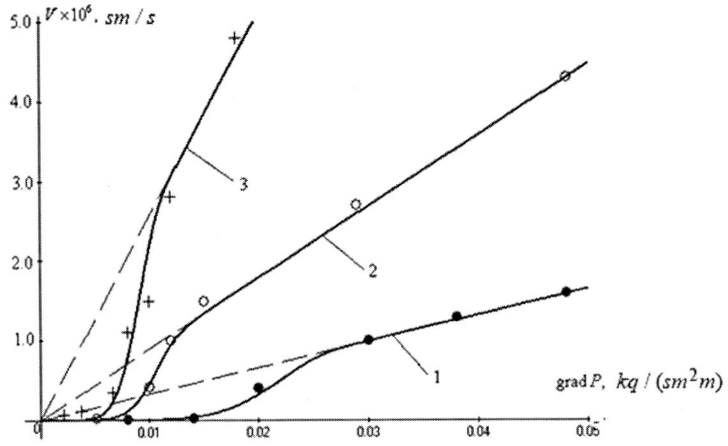

FIGURE 7.12 Change in the filtration rate of structured oils for different temperatures: (1) T = 24°C; (2) T = 50°C; and (3) T = 80°C.

TABLE 7.2
The Coefficients of Equation (7.37)

α_2	K_2	T°C	(grad P)$_0$
0.08	0.000033	24	0.0140
0.04	0.0000895	50	0.00625
0.02	0.000245	80	0.00470

TABLE 7.3
The Coefficients of Equation (7.39)

α_2	$K_1(T)$	T°C	(grad P)$_0$
0.025	0.000045	22	0.018
0.012	0.000075	50	0.008

The equation of the filtration rate is represented by the following formula

$$V = K_1(T)\left(1-\exp\left(-\alpha_1(T)\left(z/z_0\right)^6\right)\right)z, \tag{7.39}$$

where $\alpha_1 = 0.0445\exp(-0.0262T)$, $K_1 = 3.01\times10^{-5}\exp(0.01824T)$, $z = \partial P/\partial x$. Table 7.3 provides the values of the coefficients appearing in this equation.

The dependence of the gradient of the initial pressure on temperature is given in the form

$$(\text{grad}\,P)_0 = 1.4285\times10^{-4} + \frac{0.3928}{T} \tag{7.40}$$

The change in the effective viscosity of anomalous oil from the pressure gradient on the basis of experimental data is determined by the empirical formula

$$\eta^* = (\eta_0 - \eta_\infty)\exp(-30z^6) + 28\exp(-26.65z) + \eta_\infty \tag{7.41}$$

where η_0, η_∞ are the initial and final viscosity of the oil.

The solution of equation (7.41) is compared with the experimental data at temperature $T = 24°C$ in Figure 7.13.

It follows from Figure 7.13, at low-flow velocities, that the effective viscosity of the anomalous oil depends on the shear rate or on the pressure gradient, with $\tau_0 < \tau \le \tau_P$ and the effective viscosity decreasing from the maximum value η_0 to the minimum η_∞ and then stabilizing.

Figure 7.14 shows the dependence of the mobility of anomalous oil on the pressure gradient, taking into account (7.41), calculated from the expression

$$k/\eta^* = \frac{0.0310}{\eta^*} \tag{7.42}$$

As follows from Figure 7.14, the ratio k/η^* or mobility of oil at shear stress values $\tau \le \tau_0$ increases very slowly and remains practically constant $\tau_0 < \tau \le \tau_P$; when the mobility of oil intensively increases to

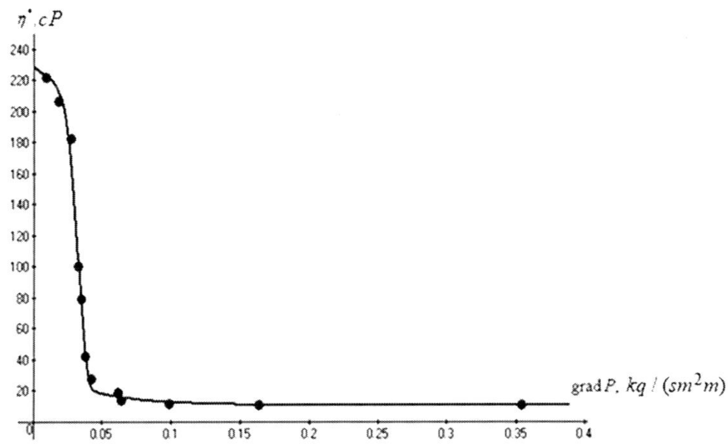

FIGURE 7.13 Dependence of viscosity on the pressure gradient.

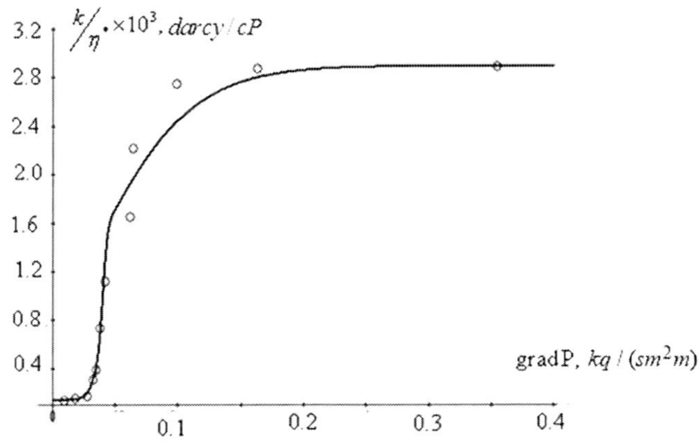

FIGURE 7.14 Dependence of the mobility of oil on the pressure gradient.

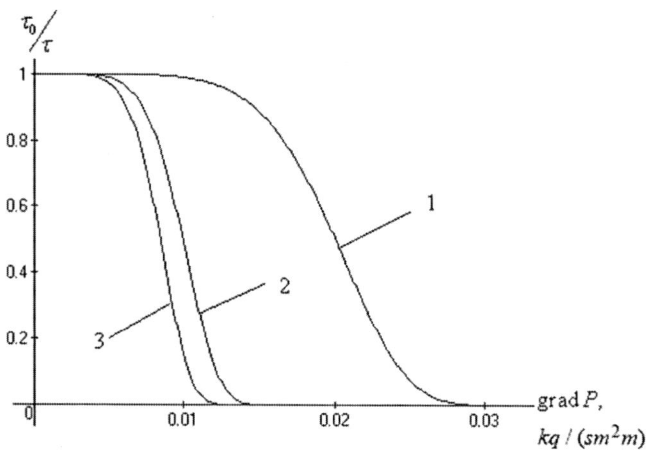

FIGURE 7.15 Dependence of the shear stress on the pressure gradient at temperatures: (1) T = 24°C; (2) T = 50°C; and (3) T = 80°C.

a maximum value, the transition from the minimum to the maximum oil mobility value occurs in a narrow range of the pressure gradient change; and the mobility of the oil stabilizes only at the values of the pressure gradient, which corresponds to the ultimate destruction of the structure $\tau > \tau_P$.

In Figure 7.15, theoretical dependences of the change in the shear stress on the pressure gradient is calculated from formula

$$\frac{\tau_0}{\tau} = \exp\left(-\alpha\left(\frac{\text{grad}\,P}{(\text{grad}\,P)_0}\right)^6\right) \tag{7.43}$$

In concentrated dispersed systems, such as anomalous oils with a content of solid particles of various types and nature (clay, sand, limestone) and asphaltene particles, the concentration of particles, their shape, and size have a significant influence on the value of the effective viscosity.

Thus, the rheological properties of anomalous oils are formed by the volume concentration of particles, the nature of the interaction forces between the particles (molecular and hydrodynamic forces), and the structure of the aggregates being formed.

7.4 THE FLOW A LAYER OF OIL PRODUCTS FROM A VERTICAL SURFACE

Film flows of rheological complex fluids are found in many branches of chemical and oil-refining technology: during transportation of petroleum products, during the emptying of reservoirs, during the collection and storage of oil sludge and oil products, etc. Therefore, it is important to determine the characteristics of such films (thickness, flow, runoff time) depending on the rheological properties of the liquid. In most cases, such liquids, which are in a viscous-flow state, have pronounced non-Newtonian properties: nonlinear viscosity, plasticity, and viscoelasticity.

Important parameters for the flow of liquids is the determination of the thickness of the layer in time, the spreading rate, and the angle of wetting of the surface, depending on the properties of the oil and the medium. The nonlinearity of rheological properties and their change in time leads to certain difficulties in solving this problem. For viscous liquids with nonlinear rheological characteristics, the possibility of retarding wetting and draining is usually associated with the release of a new phase in the volume of the liquid as a result of physical and chemical transformations (volumetric evaporation or chemical transformation) and, as a consequence, an increase in density and effective viscosity. Many complex rheology-stable (rheological characteristics of which do not depend on time) fluids under one-dimensional shear have a flow curve different from Newtonian, and such liquids belong to the class of nonlinear viscous, pseudoplastic (polymers, oil products), and dilatant (highly concentrated suspensions, emulsions).

In contrast to the considered variants of the flow of Newtonian and especially non-Newtonian fluids with a weakly pronounced nonlinearity rheological properties (in the absence of limiting shear stress), in structured disperse systems, the nonlinearity and, accordingly, the range of its variation depending on the rate of deformation or shear stress is significant, which in itself qualitatively changes the mechanism and kinetics of runoff and deceleration of this process in time. In viscous liquids with a slightly pronounced non-Newtonian character, there is no dependence of the effective viscosity on the shear stress, and for this reason the nonlinearity of the viscosity affects only the kinetics of the runoff and the possible change in the contact angle. Thus, the spreading of structured disperse systems over a solid surface is determined by their structural and rheological properties and by the laws governing their variation at the beginning and during spreading, in particular, under the influence of external influences (intrinsic weight, stresses from external forces) and fluid interaction with a solid base and processes occurring at the liquid–solid interface and on the free surface. Consequently, the flow of oil sludge and petroleum products can be attributed to the case of a spreading of a liquid with slightly pronounced non-Newtonian characteristics or a weak nonlinearity of rheological properties. In solving such problems, let us assume that the flow is steady, laminar, and non-volatile.

The flow of a viscoelastic fluid along the lateral surface of a circular cylinder is considered [97], where a numerical solution of the flow problem is obtained and the influence of the flow parameters and rheological properties of the fluid on the velocity and temperature profiles and on the thickness of the flowing film is shown.

Let us consider the steady flow of a layer of non-Newtonian power liquids from the vertical surface of cylindrical devices of large diameter (reservoirs). Since the diameter of the apparatus is greater than the thickness of the stacked layer $d_T >> \Delta$, the cylindrical surface can be considered as flat, for which the hydrodynamic equations are represented in the form

$$\frac{\sigma}{\rho_c}\frac{d^3\Delta}{dx^3} + \frac{k}{\rho_c}\frac{\partial}{\partial y}\left(\left|\frac{\partial V_x}{\partial y}\right|^{n-1}\frac{\partial V_x}{\partial y}\right) + g = 0$$

$$\frac{\partial V_x}{\partial y} + \frac{\partial V_y}{\partial x} = 0, \tag{7.44}$$

$$y = 0, \quad \frac{\partial V_x}{\partial y} = 0, \quad V_x = 0 \tag{7.45}$$

Using the boundary conditions (7.44), we obtain the solution of the first equation in the form

$$V_x = \left(\frac{\sigma}{k}\frac{d^3\Delta}{dx^3} + \frac{g\rho_c}{k}\right)^{1/n}\frac{n}{n+1}y^{n+1/n} \tag{7.46}$$

It should be noted that the capillary forces due to the curvature of the surface are small for cylindrical devices of large radius. If the dimensionless Gouger number $\text{Go} = d_T\left(\frac{\rho_c g}{8\sigma}\right)^{1/2} > 1.8$, then the capillary forces can be neglected.

Neglecting the capillary flow, that is, first term in (7.46), then we have the wave character of the film flow

$$V_x = \frac{n}{n+1}\left(\frac{g\rho_c}{k}\right)^{1/n} y^{n+1/n} \tag{7.47}$$

Equation (7.47) determines the distribution of the rate of flow along the thickness of the layer, depending on the rheological properties of the fluid and is the equation for the flow of a layer of a non-Newtonian power fluid $\tau = k|\dot{\gamma}|^{n-1}\dot{\gamma}$ and coincides with the velocity profile for Oswald–de Ville-type power liquids, where $k\left[\frac{Ns^n}{m^2}\right]$ is the liquid consistency index. It should be noted that oil sludge and oil emulsions with water and solid impurity contents show slightly pronounced non-Newtonian properties $k = 0.03 - 0.3\,{}^{Ns^n}\!/_{m^2}$, $n = 0.7 - 1.0$. Using condition

$$V_y = -\frac{\partial}{\partial x}\int_0^{\Delta} V_x dy,\quad V_y \approx \frac{d\Delta}{dt} \tag{7.48}$$

and substituting expression (7.47) in (7.48), we define the equation for the non-empty flow of a power fluid in the form

$$\begin{gathered}\frac{\partial\Delta}{\partial t} + \left(\frac{\Delta}{\Delta_0}\right)^{\frac{n+1}{n}} V_0 \frac{\partial\Delta}{\partial x} = 0 \\ \Delta(x,t)\Big|_{t=0} = \Delta_0,\quad \Delta(x,t)\Big|_{x=0} = \Delta(t)\end{gathered} \tag{7.49}$$

Here, it is assumed that the maximum velocity of runoff on the surface of the film is

$$V_0 = \frac{n}{n+1}\left(\frac{\rho_s g}{k}\right)^{1/n} \Delta_0^{\frac{n+1}{n}} \tag{7.50}$$

A particular solution of equation (7.49) can be represented in the form

$$\Delta(x,t) = \Delta_0\left(\frac{x}{V_0 t}\right)^{\frac{n}{n+1}} \tag{7.51}$$

substitution of which into equation (7.49) transforms the latter into an identity.

Analogously to (7.51), the general approximate solution, taking into account the volume of the flowing fluid layer, is represented in the form

$$\Delta(x,t) \approx \left(\frac{h-x}{V_0 t}\right)^{n/n+1} \exp\left[-\frac{n+1}{2n+1}\left(\frac{h}{V_0 t}\right)^{n/n+1}\frac{x}{h}\right] + \Delta_0(t) \tag{7.52}$$

If oil sludges and oil products exhibit viscoplastic properties, which characterize them under long-term storage, that is, properties of Bingham fluids, then the maximum rate of flow of the layer in a similar way can be determined by formula

$$V_0 = \frac{\rho_s g \Delta_0^2}{2\eta_s}\left(1 - \frac{\tau_0}{\rho_s g \Delta_0}\right)^2 \tag{7.53}$$

where τ_0 is the yield strength of the layer. Using equation (7.53), the change in the thickness of the layer is presented in a special form, as

$$\Delta = \frac{\rho_s g \Delta_0}{\rho_s g \Delta_0 - \tau_0}\left(\frac{2\eta_s}{\rho_s g}\right)^{1/2}\sqrt{\frac{x}{t}} \tag{7.54}$$

Thus, the expressions 7.52 through 7.54 allow us to determine the thickness of the flowing layer for viscous, pseudoplastic and viscoplastic liquids.

7.5 MASS AND HEAT TRANSFER IN NON-NEWTONIAN FLUIDS

The analysis and solution of problems of mass transfer and heat transfer during the flow of non-Newtonian fluids is hampered by a more complex flow velocity profile and in most cases is not amenable to an analytical solution, since, depending on the value, the equations of mass and heat transfer become nonlinear or fractional-nonlinear. The flow of non-Newtonian fluids in pipes and channels, as well as particles in non-Newtonian fluids, has been considered in many works. Since non-Newtonian fluids usually have high friction stress values, turbulent flow regimes are not characteristic for them [3,21].

Experimental studies of heat-mass transfer in the flow of various types of non-Newtonian fluids in the flow around different surfaces are given in the papers [98–104]. Both in the Newtonian fluid and in the non-Newtonian fluid, it is possible to adhere to analogies between the transfer of momentum, heat, and mass, that is, you can write

$$\begin{gathered}
\tau = \eta\left|\frac{\partial V_x}{\partial y}\right|^{n-1}\frac{\partial V_x}{\partial y} = k\frac{\partial V_x}{\partial y}, \qquad k = \eta\left|\frac{\partial V_x}{\partial y}\right|^{n-1} \\
q = a_0\left|\frac{\partial T}{\partial y}\right|^{n-1}\frac{\partial T}{\partial y} = a_T\frac{\partial T}{\partial y}, \qquad a_T = a_0\left|\frac{\partial T}{\partial y}\right|^{n-1} \\
j = D_0\left|\frac{\partial C}{\partial y}\right|^{n-1}\frac{\partial C}{\partial y} = D\frac{\partial C}{\partial y}, \qquad D = D_0\left|\frac{\partial C}{\partial y}\right|^{n-1}
\end{gathered} \tag{7.55}$$

Here, a_T is the coefficient of thermal diffusivity for a non-Newtonian fluid, the dimension $\left[\mathrm{m}^2/\mathrm{s}\right]$ and a_0 is $\left[\frac{m^{n+1}}{sK^{n-1}}\right]$, D is the diffusion coefficient for a non-Newtonian fluid, the dimension $\left[\mathrm{m}^2/\mathrm{s}\right]$ and D_0 is $\left[\frac{m^{4n-2}}{s\,kg^{n-1}}\right]$ if the concentration of the substance is measured kg/m^3. It is obvious that the theoretical justification of these relations is very difficult, although for the solution of heat transfer problems these expressions are widely used [100,105–108].

7.5.1 Mass Transfer in a Non-Newtonian Fluid in Flow Past a Plan Plate

Let us consider the mass transfer in the flow past a planar infinite plate by a non-Newtonian power-law fluid, described within the boundary layer, by the equation

$$\begin{gathered}
V_0\frac{\partial C}{\partial x} = D\frac{\partial^2 C}{\partial y^2} \\
x = 0,\ C = 0; \quad y = 0,\ C = C_s
\end{gathered} \tag{7.56}$$

The solution of equation (7.56) with given boundary conditions is represented in the form

$$C = C_s \operatorname{erfc}\left(\frac{y}{2}\right)\sqrt{\frac{V_0}{D\,x}} \tag{7.57}$$

where $\operatorname{erfc}\left(y/2\right) = 1 - \operatorname{erf}\left(y/2\right)$, $\operatorname{erf}\left(y/2\right) = \frac{2}{\sqrt{\pi}}\int_0^{y/2}\exp\left(-z^2\right)dz$ is the error integral.

Assuming that the expression for the maximum velocity for power liquids is expressed as

$$V_0 = \frac{n}{n+1}\left(\frac{\rho_c g}{k}\right)^{1/n} h^{\frac{n+1}{n}} \tag{7.58}$$

and for a Bingham fluid in the form

$$V_0 = \frac{\rho_c g h - \tau_0}{2\rho_c g \eta} \tag{7.59}$$

we define the mass flow to the surface, respectively, for the power-law fluid and the Bingham fluid, in the form

$$j = -D\frac{\partial C}{\partial y}\Big|_{y=0} = C_s\left[\frac{n}{n+1}\left(\frac{\rho_c g}{k}\right)^{1/n} h^{\frac{n}{n+1}}\frac{D}{\pi x}\right]^{1/2}$$
$$j = C_s\left[\frac{\rho_c g h - \tau_0}{2\rho_c g \eta}\frac{D}{\pi x}\right]^{1/2} \tag{7.60}$$

The Sherwood number for mass transfer in a power fluid and Bingham fluid is determined, respectively, in the form

$$\text{Sh} = \frac{\beta_L x}{D} = \left[\frac{n}{n+1}\left(\frac{\rho_c g}{k}\right)^{1/n} h^{\frac{n}{n+1}}\frac{x}{\pi D}\right]^{1/2} = \left(\frac{\text{Pe}_\text{S}}{\pi}\right)^{1/2}$$
$$\text{Sh} = \left[\frac{\rho_c g h - \tau_0}{2\rho_c g \eta}\frac{x}{\pi D}\right]^{1/2} = \left(\frac{\text{Pe}_\text{B}}{\pi}\right)^{1/2} \tag{7.61}$$

where $\text{Pe}_\text{S} = \frac{n}{n+1}\left(\frac{\rho_c g}{k}\right)^{1/n} h^{\frac{n}{n+1}}\frac{x}{D}$, $\text{Pe}_\text{B} = \frac{\rho_c g h - \tau_0}{2\rho_c g \eta}\frac{x}{D}$ – respectively, the Peclet numbers for the power-law liquid and for the Bingham fluid.

Using the integral equations of the boundary layer for the flow of non-Newtonian fluid on a flat surface, the Sherwood number can be defined as [104]

$$\text{Sh} = \frac{3}{2}\left[\frac{30(n+1)F(n)}{2n+1}\right]^{1/3}\text{Re}_x^{(n+2)/3(n+1)}\,\text{Sc}^{1/3}$$
$$\text{Re}_x = \frac{\rho V_0^{2-n}x^n}{k},\quad \text{Sc} = \frac{k}{\rho D}\left(\frac{V_0}{x}\right)^{n-1},\quad F(n) = \left[\frac{280}{39}(n+1)\left(\frac{3}{2}\right)^n\right]^{1/n+1} \tag{7.62}$$

For a Newtonian fluid $n=1$, we have $F(n)=4.64$, $\mathrm{Sh}=6.791\mathrm{Re}_x^{1/2}\mathrm{Sc}^{1/3}$, and the thickness of the dynamic boundary layer is determined in the form $\delta/x = F(n)\mathrm{Re}_x^{1/(1+n)}$.

In the flow of a power non-Newtonian fluid on a flat surface, when the velocity profile is approximated by a cubic parabola, the mass-transfer coefficient is determined as [109]

$$\beta_L = \frac{9(n+1)}{2(n+1)}\left[\frac{30F(n)(n+1)}{2n+1}\right]^{-1/3} D^{2/3}\left(\frac{k}{\rho_s}\right)^{-1/3(n+1)} V^{1/n+1} L^{\frac{n+2}{3(n+1)}} \tag{7.63}$$

For the Newtonian fluid $(n=1)$, equation (7.63) is transformed to the form

$$\beta_L = 0.664 D^{2/3}\nu^{-1/6}V^{1/2}L^{-1/2} \tag{7.64}$$

The Sherwood number for a porous layer of spherical particles is defined as

$$\mathrm{Sh} = \frac{9(n+1)}{2(n+1)}\left[\frac{30A(n+1)}{2n+1}\right]^{-1/3}\left(\frac{1+3n}{1+n}\right)^{\frac{1}{n+1}} 2^{\frac{1-n}{6(n+1)}}\varepsilon^{-\frac{1}{n+1}}\mathrm{Sc}^{1/3}\,\mathrm{Re}^{\frac{n+2}{3(n+1)}} \tag{7.65}$$

Here, $\mathrm{Sc} = \dfrac{a^{1-n}\left(k/\rho_c\right)}{DV_0^{1-n}}$ is Schmidt number, $\mathrm{Re} = \dfrac{a^n V_0^{2-n}}{k/\rho_c}$ is Reynolds number, a is particle diameter, ε is layer porosity, and V_0 fluid velocity in the layer. When a Newtonian fluid $(n=1)$ flows through a layer of particles with porosity ε, equation (7.65) is transformed to the form

$$\mathrm{Sh} = 0.938\,\varepsilon^{-1/2}\,\mathrm{Sc}^{1/3}\,\mathrm{Re}^{1/2} \tag{7.66}$$

Mass transfer during the flow of a film of a non-Newtonian liquid on a plane surface is given in the paper [110] described by the equation

$$\left(1-\xi^{\frac{n}{n+1}}\right)\frac{\partial\theta}{\partial z} = \frac{\partial^2\theta}{\partial\xi^2}$$

$$\theta = \frac{C_s - C}{C_s - C_0},\ \xi = \frac{y}{\delta},\ z = \frac{xD}{V_s\delta^2},\ V = V_s\left[1-\left(\frac{y}{\delta}\right)^{\frac{n}{n=1}}\right],\ V_s = \frac{n}{n+1}\left(\frac{\rho_c g\delta}{k}\right)^{1/n}\delta \tag{7.67}$$

$$\xi=0,\ \theta=0;\ \xi=1,\ \frac{\partial\theta}{\partial\xi}=0,\ z=0,\ \theta=1$$

The solution of equation (7.67) with given boundary conditions can be sought by the method of separation of variables and finally presented in the form

$$\theta = \sum_{m=1}^{\infty} A_m Y_m(\xi)\exp\left(-\mu_m^2 z\right) \tag{7.68}$$

where μ_m^2 the eigenvalues, and Y_m are Eigen functions, which are found from the solution of the Sturm–Louisville equation

$$\frac{d^2Y}{d\xi^2} + \mu^2\left(1-\xi^{\frac{1+n}{n}}\right)Y = 0,\quad Y(0)=0,\ Y'(1)=0 \tag{7.69}$$

Coefficients of the series are defined as

$$A_m = \frac{\int_0^1 \left(1-\xi^{\frac{n+1}{n}}\right) Y_m(\xi)\,d\xi}{\int_0^1 \left(1-\xi^{\frac{n+1}{n}}\right) Y_m^2(\xi)\,d\xi} \tag{7.70}$$

In the final analysis, the solution is represented as a series of

$$Y = \sum_{i=0}^{\infty} a_i \xi^i,\ a_1 = 1,\ a_i = \frac{\mu^2}{i(i+1)}\left(-a_{i-2} + a_{i-5}\right) \tag{7.71}$$

The Sherwood number is defined as

$$\mathrm{Sh} = \frac{\left(\frac{d\bar{\theta}}{d\xi}\right)\Big|_{\xi=0}}{\bar{\theta}} = \frac{1+n}{1+2n}\frac{\sum_{m=1}^{\infty} A_m Y'(0)\exp\left(-\mu_m^2 z\right)}{\sum_{m=1}^{\infty}\frac{A_m Y'(0)\exp\left(-\mu_m^2 z\right)}{\mu_m^2}} \tag{7.72}$$

and the average value of the number $\overline{\mathrm{Sh}}$ in the form

$$\overline{\mathrm{Sh}} = -\frac{1+n}{1+2n}\frac{1}{z}\ln\bar{\theta},\qquad \bar{\theta} = \frac{1+n}{1+2n}\sum_{m=1}^{\infty}\frac{A_m Y'(0)\exp\left(-\mu_m^2 z\right)}{\mu_m^2} \tag{7.73}$$

Numerical solutions of these equations for different values of y allow us to estimate the change $n = 2.5;\ 1.25;\ 1/2;\ 1/3;\ 1/4$ in $\mu_1^2 = 5.12;\ \mu_2^2 = 39.66;\ \mu_3^2 = 106.25;\ \mu_4^2 = 204.85$, the average value of concentration, and number $\overline{\mathrm{Sh}}$ (Figure 7.16). Obviously, this series converges quickly, and the solutions are of a very approximate nature, which makes it possible to obtain approximate values of the average Sherwood number.

The results of the analysis of the solutions presented in this section show that for the same tasks, depending on the assumptions and the solution method, we have different formulas for determining the number $\mathrm{Sh} = \mathrm{F}(\mathrm{Re}, \mathrm{Sc}, \mathrm{n})$.

7.5.2 Mass Transfer for a Flow of a Power Non-Newtonian Fluid in Pipes

The velocity of a laminar flow of a power non-Newtonian fluid in tubes of circular cross section is given by

$$V = V_m \frac{3n+1}{n+1}\left[1-\left(\frac{r}{R}\right)^{\frac{n+1}{n}}\right] \tag{7.74}$$

Here, $V_m = Q/\pi R^2$ is the average flow velocity over the section, Q is the volumetric flow rate of the liquid, and R is the radius of the tube. Consider the mass transfer in a power non-Newtonian fluid,

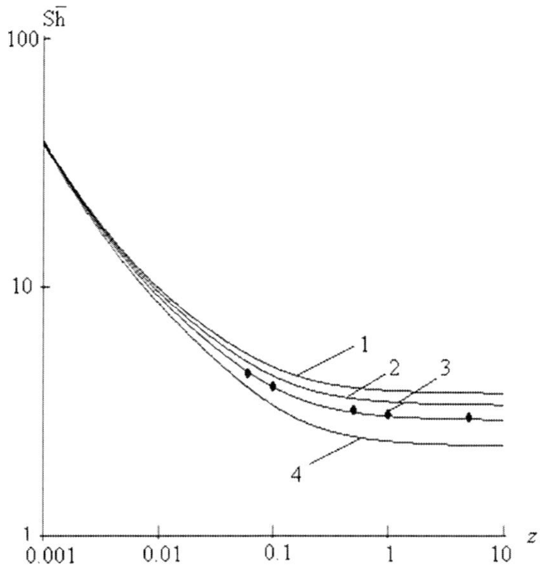

FIGURE 7.16 Dependence of the average of the numbers $\overline{\mathrm{Sh}}$ on n equal to: (1) ∞; (2) 2.5 (dilatant fluid); (3) 1 (Newtonian fluid); and (4) 1/4 (pseudoplastic fluid).

taking the analogy between the velocity and concentration profile, using which, for the diffusion coefficient, we can write

$$D = D_0 \left| \frac{\partial C}{\partial r} \right|^{n-1} \tag{7.75}$$

where D_0 us there is a certain coefficient. Then the mass flux per unit surface is defined as

$$j = -D_0 \left| \frac{\partial C}{\partial r} \right|^{n-1} \frac{\partial C}{\partial r} \Big|_{r=R} \tag{7.76}$$

If the mass flow to the surface of the pipe is constant along its entire length $\partial C/\partial x = \partial C_m/\partial x$, it can be assumed that C_m is the average concentration over the cross section of the pipe and is determined taking into account how $V_m = Q/\pi R^2 = \int_0^R V r dr / \pi R^2$

$$C_m = \frac{\int_0^R 2\pi V C r dr}{\int_0^R 2\pi V r dr} = \frac{2}{R^2 V_m} \int_0^R V C r dr \tag{7.77}$$

Then, taking into account expressions (7.75) and (7.77), the equation of mass transfer in cylindrical coordinates is written in the form

$$V_m \frac{3n+1}{n+1}\left[1-\left(\frac{r}{R}\right)^{\frac{n+1}{n}}\right]\frac{\partial C_m}{\partial x} = \frac{D_0}{r}\frac{\partial}{\partial r}\left(\left|\frac{\partial C}{\partial r}\right|^{n-1} r\frac{\partial C}{\partial r}\right)$$

$$r = 0,\quad \frac{\partial C}{\partial r} = 0;\quad r = R,\quad C = C_w \tag{7.78}$$

By twice integrating equation (7.78) for given boundary conditions, the solution can be represented in the form

$$C = C_w - \frac{1}{2^{\frac{n+1}{n}}}\frac{n}{n+1}R^{\frac{n+1}{n}}\left(\frac{V_m}{D_0}\frac{dC_m}{dx}\right)^{1/n}\left[\left(\frac{3n+1}{n+1}\right)^{\frac{n+1}{n}}\left(\left(1-\frac{2n}{3n+1}\left(\frac{r}{R}\right)^{\frac{n+1}{n}}\right)^{\frac{n+1}{n}}\right)-1\right] \tag{7.79}$$

For a Newtonian flow, from this equation we obtain

$$C = C_w - \frac{1}{2}R^2\frac{V_m}{D_0}\frac{dC_m}{dx}\left[\frac{3}{4}+\frac{1}{4}\left(\frac{r}{R}\right)^4-\left(\frac{r}{R}\right)^2\right] \tag{7.80}$$

The mass flow per unit surface is

$$j_w = -D\frac{\partial C}{\partial r}\Big|_{r=R} = D\left(\frac{R}{2}\frac{V_m}{D_0}\frac{dC_m}{dx}\right)^{1/n} \tag{7.81}$$

Substituting (7.79) into (7.77) after simple but laborious transformations, we obtain

$$C_w - C_m = -\frac{1}{2^{1/n}}\frac{n(3n+1)}{(n+1)^2}R^{\frac{n+1}{n}}\left(\frac{V_m}{D_0}\frac{dC_m}{dx}\right)^{1/n}\mathrm{I}(n),$$

$$\mathrm{I}(n) = \int_0^1\left[\left(\frac{3n+1}{n+1}\right)^{\frac{n+1}{n}}\left(1-\frac{2n}{3n+1}\eta^{\frac{n+1}{n}}\right)^{\frac{n+1}{n}}-1\right]\left(1-\eta^{\frac{n+1}{n}}\right)\eta\, d\eta,\quad \eta = r/R \tag{7.82}$$

Using (7.82), we define the generalized Sherwood number in the form

$$\mathrm{Sh} = \frac{j_w}{(C_w - C_m)}\frac{2R}{D} = \frac{2(n+1)^2}{n(3n+1)\mathrm{I}(n)} \tag{7.83}$$

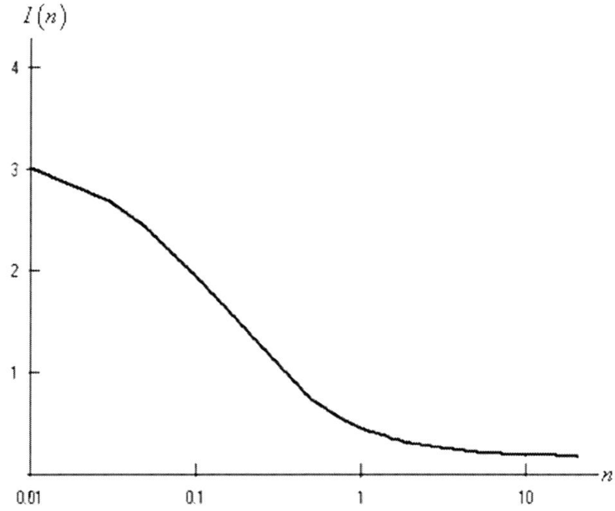

FIGURE 7.17 The value of the integral $I(n)$ at $j_w = \text{const}$.

To determine the number Sh, it is necessary to determine the numerical value of the integral $I(n)$ (Figure 7.17). The distribution of the dimensionless concentration of matter along the radius of the tube $\theta = \frac{C - C_w}{C_m - C_w}$ is defined as

$$\theta = \frac{n+1}{2(3n+1)} \frac{\left(\frac{3n+1}{n+1}\right)^{\frac{n+1}{n}} \left[1 - \frac{2n}{3n+1}\left(\frac{r}{R}\right)^{\frac{n+1}{n}}\right] - 1}{\mathrm{I}(n)} \tag{7.84}$$

Figure 7.18 shows the dependence of the generalized number on n, and with (shown in dashed lines) this value corresponds to the Newtonian fluid Sh = 4.464 for $j_w = \text{const}$.

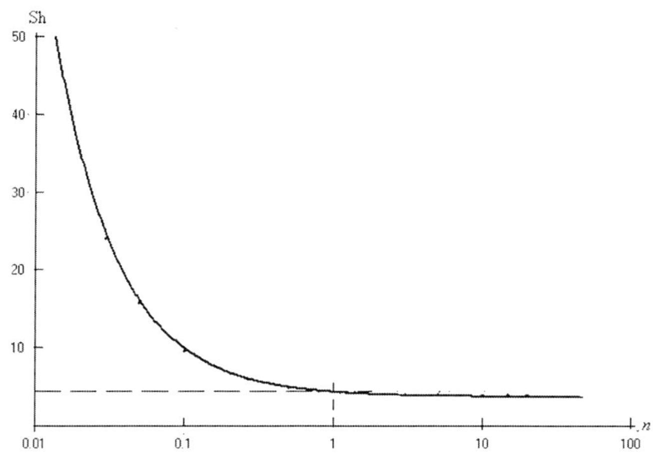

FIGURE 7.18 Dependence of a number Sh on n.

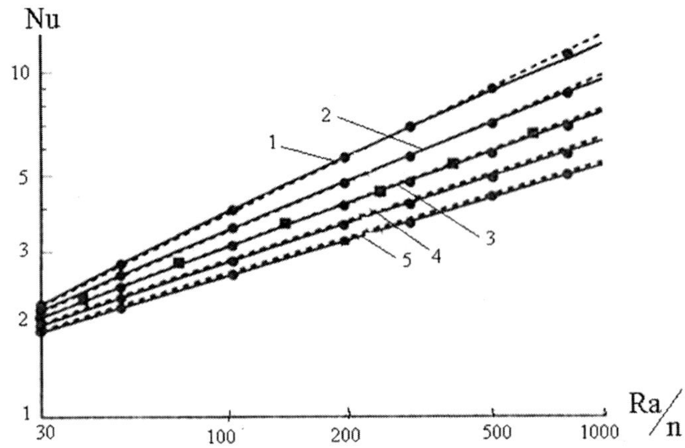

FIGURE 7.19 Dependence of the Nusselt number on the Rayleigh number equal to: (1) 0.6; (2) 0.8; (3) 1.0; (4) 1.2; and (5) 1.4 (the point is a numerical solution, the dashed line is [7.67]).

7.5.3 Heat Transfer in a Non-Newtonian Fluid

At present, there are many studies on the heat transfer in non-Newtonian fluids [111]. And most of the research is reduced to the development of empirical or semi-empirical dependencies for determining the heat transfer coefficient on the basis of experimental data. In particular, the heat transfer with free convection for a flow of a power non-Newtonian fluid through an inclined porous plate is determined by the following theoretical equation [105]

$$\mathrm{Nu} = \frac{n(11-3n)}{16}\mathrm{Ra}^{\frac{2}{3n+1}}, \qquad \mathrm{Ra}/n > 30 \tag{7.85}$$

Here, Ra = GrPr is the Rayleigh number, which characterizes free convection. A comparison of the calculated expression (7.85) with the experimental data is shown in Figure 7.19.

For $n = 1$, that is, for the Newtonian fluid, the experimental data corresponding to the values of the number are given.

$$\mathrm{Nu} = 0.5\mathrm{Ra}^{1/2} = 0.5(\mathrm{Gr\,Pr})^{1/2} \tag{7.86}$$

In this paper [109], the heat transfer during flow around a circular cylinder by a non-Newtonian fluid is described by expression

$$\mathrm{Nu} = 0.72n^{-0.4}\,\mathrm{Re}_t^{\frac{1}{n+1}}\,\mathrm{Pr}^{1/3} \tag{7.87}$$

and the mass transfer during flow around the cylinder is defined as

$$\mathrm{Sh} = 0.785\,\mathrm{Re}_t^{1/2}\,\mathrm{Sc}_t^{1/3}, \qquad 10 < \mathrm{Re}_t < 25000 \tag{7.88}$$

Here, $\mathrm{Re}_t = \frac{d_T^n \rho_c V^{2-n}}{k}$, $\mathrm{Sc} = \frac{Vd_T}{x}\mathrm{Re}_T^{-\frac{2}{n+1}}$. In work to solve a joint heat exchange and mass transfer problem with free convection of a non-Newtonian fluid in a porous medium, the following equations are given

$$\mathrm{Nu} = \frac{\alpha x}{\lambda} = \left[\mathrm{Pe}^{\frac{n+1}{2n+1}} + \mathrm{Ra}^{\frac{n}{2n+1}}\right] / \theta(\xi, 0)$$

$$\mathrm{Sh} = \frac{\beta_L x}{D} = \left[\mathrm{Pe}^{\frac{n+1}{2n+1}} + \mathrm{Ra}^{\frac{n}{2n+1}}\right] / C(\xi, 0)$$

$$\xi = \frac{V_0 x c_p \rho_c}{\lambda}\left[\mathrm{Pe}^{\frac{n+1}{2n+1}} + \mathrm{Ra}^{\frac{n}{2n+1}}\right] f(\xi, \eta) \tag{7.89}$$

$$\eta = \frac{y}{x}\left[\mathrm{Pe}^{\frac{n+1}{2n+1}} + \mathrm{Ra}^{\frac{n}{2n+1}}\right]$$

The problem is solved by a numerical method for determining the dimensionless temperature $\theta(\xi, 0)$ and the dimensionless concentration $C(\xi, 0)$ and, accordingly, the numbers Sh and Nu.

In the literature [106, 107, 112–117], it is possible to meet a number of purely private numerical solutions of various problems associated with mass exchange and heat transfer in the flow of a non-Newtonian fluid. However, numerical solutions for mass and heat transfer, with the exception of calculating temperature profiles and concentrations, do not always allow one to uniquely obtain equations and expressions for calculating mass and heat transfer coefficients or numbers Sh and Nu.

In the flow of Bingham liquids (Table 7.1), the thickness of the hydrodynamic boundary layer is estimated as [116]

$$\delta \sim \frac{\nu_c}{2\sqrt{g\beta\,\Delta TL}}\left[\mathrm{Bn} + \sqrt{\mathrm{Bn}^2 + 4\left(\mathrm{Ra}/\mathrm{Pr}\right)^{1/2}}\right] \tag{7.90}$$

Here, $\mathrm{Bn} = \frac{\tau_0}{\eta_c}\sqrt{\frac{L}{g\beta\Delta T}}$ is the Bingham number, $\mathrm{Ra} = \mathrm{GrPr} = \frac{\rho_c^2 g c_p \beta \Delta T L^2}{\eta_c \lambda}$, λ is the coefficient of thermal conductivity of the liquid layer, and ΔT is the temperature difference. Using this equation, the average Nusselt number is estimated as

$$\overline{\mathrm{Nu}} \sim \frac{2\mathrm{Ra}^{1/2}}{\mathrm{Pr}^{1/2}\left(\mathrm{Bn} + \sqrt{\mathrm{Bn}^2 + 4\left(\mathrm{Ra}/\mathrm{Pr}\right)^{1/2}}\right)} f(\mathrm{Bn}, \mathrm{Pr}) \tag{7.91}$$

where $f(\mathrm{Bn}, \mathrm{Pr}) \sim \delta/\delta_T$ is the function depending on the numbers Bn and Pr, which is equivalent to the ratio of the thickness of the dynamic and thermal δ_T boundary layers. As follows from this expression, the Nusselt number increases with decreasing Bingham number $\mathrm{Bn} \to 0$ and with increasing Rayleigh number. When this expression is simplified to the form

$$\overline{\mathrm{Nu}} \sim \left(\frac{\mathrm{Ra}}{\mathrm{Pr}}\right)^{1/4} f(\mathrm{Bn}, \mathrm{Pr}) \tag{7.92}$$

The effective viscosity of a Bingham liquid is estimated as [116]

$$\frac{\eta_{эф}}{\eta} \sim 1 + \mathrm{Bn}\left(\frac{\mathrm{Bn}}{2\mathrm{Gr}^{1/2}} + \frac{1}{2\mathrm{Gr}^{1/2}}\sqrt{\mathrm{Bn}^2 + 4\mathrm{Gr}^{1/2}}\right) \tag{7.93}$$

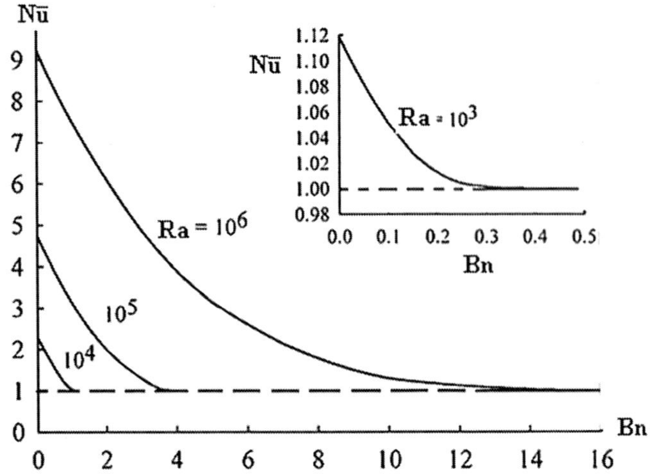

FIGURE 7.20 Dependence of the average number on the numbers Bn and $\mathrm{Ra}(\mathrm{Pr}=7)$.

In particular, for a Newtonian flow, the solution of this type of problem makes it possible to obtain empirical correlation for the average Nusselt number [116] in the form

$$\overline{\mathrm{Nu}} = 0.18\left(\frac{\mathrm{Ra\,Pr}}{0.2+\mathrm{Pr}}\right)^{0.29} \tag{7.94}$$

Figure 7.20 shows the influence of the numbers Ra and Pr on the average Nusselt number.

In the general case, in practical problems, the dependence of the average number on the numbers is determined in the form

$$\overline{\mathrm{Nu}} = \mathrm{bRa}^{m}\left(\frac{\mathrm{Pr}}{1+\mathrm{Pr}}\right)^{n} \tag{7.95}$$

where b, m, n are the coefficients determined experimentally.

7.5.4 Mass Transfer upon Evaporation from the Film Surface of Non-Newtonian Fluid

In the volume of the layer of oil sludge and petroleum products flowing from the vertical surface, evaporation of light fractions, partial oxidation of petroleum products with formation of tarry compounds, and asphaltenes can occur with a certain temperature at a certain temperature, leading to an increase in the effective viscosity and density of the liquid and, accordingly, to the deceleration of the flow rate from the surface. For large numbers $\mathrm{Pe} = {}^{V_0 h}\!/_{\mathrm{D}}$, when diffusion along the film can be neglected in the boundary-layer approximation and at a constant temperature, the distribution of the concentration inside the film with allowance for only transverse diffusion for steady-state evaporation is described by the following equation

$$V_0 \frac{\partial c}{\partial x} = D \frac{\partial^2 c}{\partial y^2} \tag{7.96}$$

$$x = 0,\ \ c = 0;\ \ y = \delta,\ \ c = c_s$$

Here, c is the concentration of vapor on steam, c_s the concentration of vapor on the surface of the layer, D is the diffusion coefficient, h is the height of the layer, and $V_0 = \frac{n}{n+1}\left(\frac{\rho_s g}{k}\right)^{1/n}\delta_0^{\frac{n+1}{n}}$ is the velocity of flow of the layer. Restricting ourselves to the principal term of the expansion in equation (7.96), the fluid velocities near a free surface V_0 can be regarded as the maximum flow velocity of the film. Introducing a dimensionless coordinate $\xi = \frac{y}{2}\sqrt{\frac{V_0}{Dx}}$, equation (7.96) becomes an ordinary differential equation of the form

$$\frac{d^2c}{d\xi^2} - 2\xi\frac{dc}{d\xi} = 0 \tag{7.97}$$

The final solution with given conditions is presented in the form

$$c(x, y) = c_s\left[1 - \text{erf}\left(\frac{\delta_0 - y}{2}\left(\frac{n\rho_s g}{(n+1)kDx}\delta_0^{\frac{n+1}{n}}\right)^{1/2}\right)\right] \tag{7.98}$$

The flux of vapor through a unit of the surface of the layer is determined as

$$j = -D\left.\frac{\partial c}{\partial y}\right|_{y=\delta} = c_s\left(\frac{V_0 D}{\pi x}\right)^{1/2} \tag{7.99}$$

Total steam flow per unit time is determined as

$$I = 2\pi(R - \delta_0)\int_0^h j\,dx = 4\pi^{1/2}(R - \delta_0)c_s(V_0 Dh)^{1/2} \tag{7.100}$$

where R, h are the radius of the reservoir and the height of the layer. Using expression (7.100), we define the Sherwood number in the form

$$\text{Sh} = \frac{\beta_L\delta_0}{D} = 4\pi^{1/2}\left(\frac{R - \delta_0}{h}\right)\text{Pe}_s^{1/2} \approx 7.08\,\text{Pe}_s^{1/2}\left(\frac{R - \delta_0}{h}\right) \tag{7.101}$$

If we assume that $R >> \delta$, then this equation is rewritten in the form

$$\text{Sh} \approx 7.08\,\text{Pe}_s^{1/2}\left(\frac{R}{h}\right) \tag{7.102}$$

where $\text{Pe}_s = \frac{n}{n+1}\left(\frac{\rho_s g}{k}\right)^{1/n}\delta_0^{\frac{n+1}{n}}\frac{h}{D}$ is the Peclet number for the pseudoplastic liquid, and h is the height of the layer. Figure 7.21 shows the nature of the change Pe in the number of n.

The thickness of the diffusion layer can be defined as $\delta_D = \frac{Dc_s}{j} = \left(\frac{\pi}{\text{Pe}_s}\frac{x}{h}\right)^{1/2}$, that is, is inversely proportional to Pe_s and increases in proportion to the height of the layer. For viscoplastic fluids, mass transfer will also be described by equation (7.96), where the number Pe_s is determined by the following equation: $\text{Pe}_s = \frac{\rho_c g\delta_0^2}{2\eta_s}\left(1 - \frac{\tau_0}{\rho_s g\delta_0}\right)^2\frac{h}{D}$. In this paper [8], the number for the thickness of the diffusion layer can be defined as $\delta_D = \frac{Dc_s}{j} = \left(\frac{\pi}{\text{Pe}_s}\frac{x}{h}\right)^{1/2}$, that is, the inversely proportional mass transfer Sh of bubbles and droplets in a pseudoplastic liquid $0.6 \le n \le 1.0$ is expressed by a semi-empirical formula of the form.

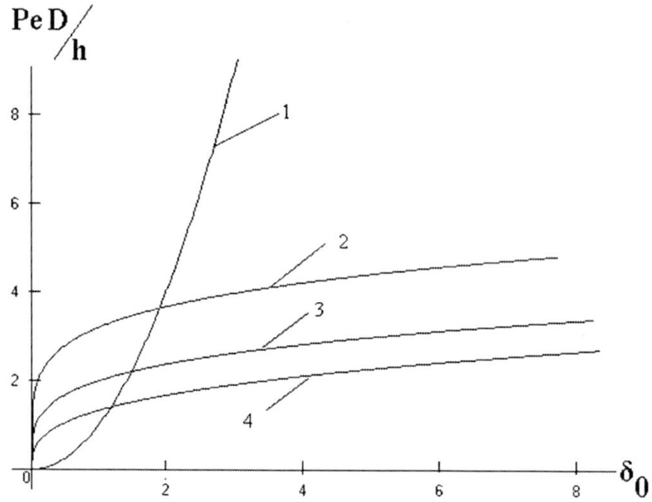

FIGURE 7.21 Dependence of the number Pe on the thickness of runoff for different $\rho_s g/k = 2.0$ values n equal to: (1) $n = 1$; (2) $n = 1/2$; (3) $n = 1/3$; and (4) $n = 1/4$.

As follows from these equations, the Sherwood number or mass-transfer coefficient at close to 1 depends on the properties of the pseudoplastic fluid and on the Peclet number in the power of ½, that is, with an increase in the flow velocity, the mass-transfer efficiency increases due to the renewal of the surface of the layer.

$$\mathrm{Sh} = \left[(0.497 - 0.284n)\mathrm{Pe}\right]^{1/2} \tag{7.103}$$

7.6 DRAG COEFFICIENT OF PARTICLE IN A NON-NEWTONIAN FLUID

Non-Newtonian fluids differ from ordinary fluids in that their viscosity changes with increasing shear rate, and the shear rate index characterizes the degree of non-Newtonian behavior of the material. When $n < 1$, a liquid exhibits pseudoplastic properties, and $n > 1$ is with dilatant properties. These conditions impose special conditions on the description of the hydrodynamics of the flow of non-Newtonian fluids and are reflected in the description of the drag coefficients of particles in similar liquids.

Theoretical and experimental studies of hydrodynamics and the drag coefficient of solid particles, droplets, and bubbles in a Newtonian fluid are given in [4], and in a power non-Newtonian fluid are given in the works [3,117–127]. Using asymptotic methods [3,127], we obtained a formula for calculating the bubble drag coefficient in a power non-Newtonian fluid $(n < 1)$ for small numbers $\mathrm{Re}_d < 1$

$$C_D = \frac{16}{2^n \mathrm{Re}_d} X(n), \qquad X(n) = 2^{n-1} 3^{\frac{n-3}{2}} \frac{13 + 4n - 8n^2}{(2n+1)(n+2)} \tag{7.104}$$

where $X(n)$ is the parameter characterizing the rheological properties of the flow and depending on the exponent n, $\mathrm{Re}_t = \frac{d_T^n \rho_c V^{2-n}}{k}$. Calculations using formula (7.104) show that for pseudoplastic liquids, the resistance coefficient is higher, and for dilatant fluids below the corresponding resistance coefficients when the flow of a Newtonian fluid flows around the bubble. In addition to expression (7.104), Chabra [117] gives experimental values and various dependences on

$$X(n) = 2^n 3^{\frac{n-3}{2}} \left[1 - 3.83(n-1)\right], \quad 0.7 \le n \le 1;$$

$$X(n) = \frac{2}{3}\left(3\gamma^2\right)^{\frac{n-3}{2}} \frac{13 + 4n - 8n^2}{(2n+1)(n+2)}, \quad \gamma > 10;$$

$$X(n) = 3^{-\frac{n+3}{2}} \left[\frac{2(2n+1)(2-n)}{n^2}\right]^2; \tag{7.105}$$

$$X(n) = 2^n 3^{\frac{n-3}{2}} \frac{1 + 7n - 5n^2}{n(n+2)}, \quad n < 1.$$

The second formula (7.105) characterizes the behavior of droplets in a non-Newtonian fluid. The preceding expressions, obtained theoretically, do not allow solving this problem for the general case.

Satisfactory dependence $X(n)$ on n for solid particles in non-Newtonian fluid for a sufficiently large range of variation $0.1 \le n \le 1.8$ using experimental data can be represented in the form

$$X(n) = \frac{7}{450}(5n+9)^2 \exp\left[-n^{\frac{3}{2}}\left(1 + \frac{n^{\frac{3}{2}}}{10}\right)\right] \tag{7.106}$$

From the formula (7.106) it follows that, for $n = 1$, the value $X(n) \approx 1$. Comparison of calculated values $X(n)$ with experimental values is given in Figure 7.22.

The paper [122] also gives various formulas characterizing the deformation of bubbles in a non-Newtonian fluid

$$\begin{aligned} Y_K &= \frac{b_0}{a_0} = 0.0628\delta^{0.46}, && 20 \le \delta \le 100, \\ Y_K &= 1.4, && \delta \le 4, \\ Y_K &= 6.17\delta^{-1.07}, && 4 \le \delta \le 20, \\ \delta &= \mathrm{Re}_t \mathrm{Mo}_t^{0.078}, && \mathrm{Mo}_n = \mathrm{We}^{n+2}\mathrm{Fr}^{2-3n}\mathrm{Re}_t^{-4}, \quad 0.64 \le n \le 0.9 \end{aligned} \tag{7.107}$$

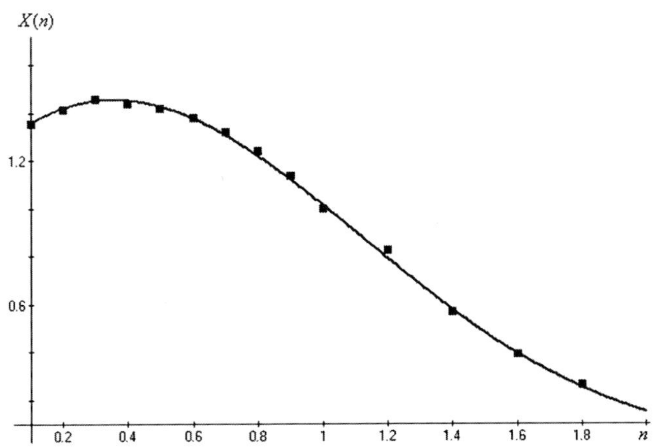

FIGURE 7.22 Dependence $X(n)$ on n.

TABLE 7.4
Drag Coefficients of Particles in a Non-Newtonian Fluid

№	Formulas		References
1	2		3
1.	$C_D=\frac{24}{\mathrm{Re}_t}\left[1+0.1118\left(k_2\,\mathrm{Re}_t\right)^{0.654}\right]+\frac{0.4305k_2}{1+\frac{3305}{k_2\,\mathrm{Re}_t}},\quad \mathrm{Re}_t=k_1\,\mathrm{Re},$ $k_1^{-1}=0.33+0.67\varphi^{-1/2},$ $\log k_2=1.815\left(-\log\varphi\right)^{0.574},\quad 10^{-3}\le \mathrm{Re}_d\le 10^5,\quad 0.1<\varphi<1$	(7.108)	[128]
2.	$C_D=\frac{24}{\mathrm{Re}_t}\left(1+0.173\,\mathrm{Re}_t^{0.657}\right)+\frac{0.413}{1+16300\,\mathrm{Re}_t^{-1.09}},\quad \mathrm{Re}_t<1600,$ $0.16\le n\le 1$	(7.109)	[129]
3.	$C_D=\frac{24}{\mathrm{Re}_t}\left(1+0.24\,\mathrm{Re}_t^{0.5}\right)+\frac{1+0.0000238\,\mathrm{Re}_t}{1+370\,\mathrm{Re}_t^{-1}},\quad \mathrm{Re}_t<2.5\times10^4$	(7.110)	[130]
4.	$C_D=\frac{24}{\mathrm{Re}_t}\left[1+8.1723\exp(-4.066\,\varphi)\,\mathrm{Re}_t^{0.0964+0.557\varphi}\right]+\frac{73.69\,\mathrm{Re}_t\exp(-5.075\,\varphi)}{\mathrm{Re}_t+5.378\exp(6.212\,\varphi)}$	(7.111)	[130]
5.	$C_D=\begin{cases}\frac{35.2\left(2^n\right)}{\mathrm{Re}_t^{1.03}}+n\left[1-\frac{20.9\left(2^n\right)}{\mathrm{Re}_t^{1.11}}\right], & 0.2\left(2^n\right)\le \mathrm{Re}_t\le 24\left(2^n\right)\quad 0.38\le n\le 1\\ \frac{37\left(2^n\right)}{\mathrm{Re}_t^{1.1}}+0.25+0.36n, & 24\left(2^n\right)\le \mathrm{Re}_t\le 100\left(2^n\right)\end{cases}$	(7.112)	[131]
6.	$C_D=\frac{32.5}{\mathrm{Re}_t}\left(1+2.5\,\mathrm{Re}_t^{0.2}\right),\quad \mathrm{Re}_t<150,\ 0.77\le n<1,\quad 0.35\le\varphi<0.7$	(7.113)	[132]
7.	$C_D=\frac{16}{\mathrm{Re}_t}\left(1+0.173\,\mathrm{Re}_t^{0.657}\right)+\frac{0.431}{1+16300\,\mathrm{Re}_t^{-1.09}},\ \mathrm{Re}_t<1000,\ 0.16\le n<1$	(7.114)	[133]
8.	$C_D=\frac{24}{\mathrm{Re}_t}X^*(n),\quad X^*(n)=X(n)=3^{2n-3}\frac{n^2-n+3}{3^{3n}},\quad \mathrm{Re}_t<10^{-5}$ $X^*(n)=X(n)+\frac{1-n^2}{3n+1}\log\left(10^3\,\mathrm{Re}_t\right),\qquad 10^{-5}\le \mathrm{Re}_t<10^{-3}$ $X^*(n)=X(n)+\frac{4n^4}{24\,\mathrm{Re}_t^{\frac{n-3}{3}}},\quad 10^{-3}\le \mathrm{Re}_t<10^3,\quad 0.56\le n\le 0.89$	(7.115)	[126]

In Table 7.4, many formulas are given for calculating the drag coefficients of solid spherical particles, droplets, and bubbles in a non-Newtonian fluid.

In this table, formulas (7.83), (7.98), and (7.101) describe the coefficient of resistance of a solid spherical particle; formulas (7.85), (7.96), and (7.100) refer to irregularly shaped particles with different shape factors; and (7.104) describes the coefficient of resistance of gas bubbles in a non-Newtonian fluid.

For small values of the number $5\le \mathrm{Re}_d\le 25$, the coefficient of bubble resistance in a non-Newtonian fluid can be determined in the form of [118,133–135]

$$C_{DG}=\frac{K(n)}{\mathrm{Re}_t}\tag{7.116}$$

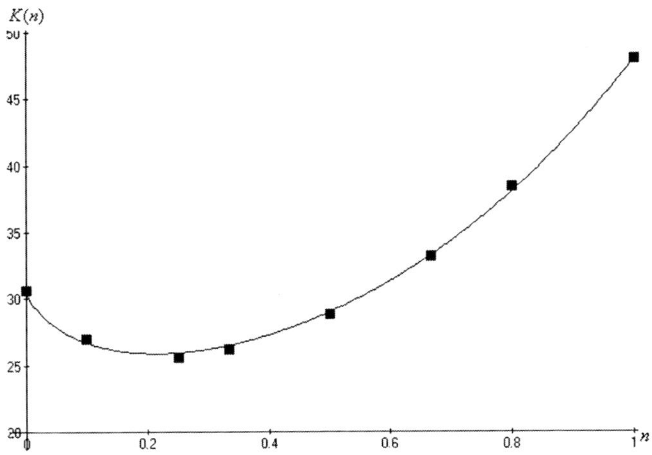

FIGURE 7.23 Dependence $K(n)$ on n.

Here, $K(n)$ is the coefficient depends on n. Using the experimental data given in the paper [118], in this paper we propose the following formula

$$K(n) = 30.6\frac{\exp(2.53n)}{\left(n^{\frac{6}{7}}+1\right)^3} \tag{7.117}$$

As follows from Figure 7.23, the coefficient $K(n)$ varies significantly with respect $0 \le n \le 1$ to u and, passing through a minimum at $n = 1$, tends to $K(n) \approx 48$, which characterizes the bubble resistance coefficient in the Newtonian fluid proposed by Levich [6] for small numbers Re_d.

The set of experimental data for the drag coefficient of particles in a non-Newtonian fluid collected from literature sources for $\mathrm{Re}_d < 1000$ is given in the work [107] and in Figure 7.24.

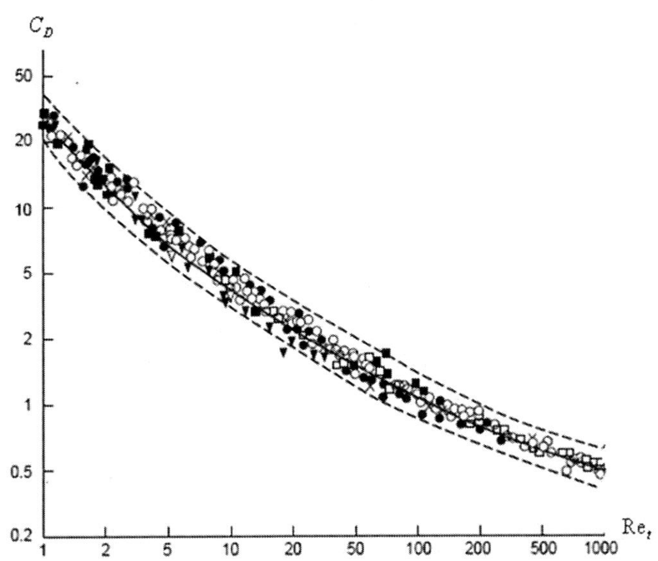

FIGURE 7.24 Drag coefficient of solid particles in a non-Newtonian liquid (solid line is the drag coefficient in Newtonian fluid; the dashed lines correspond to 30% threshold of spread of experimental data).

In this paper, we propose an equation for determining the coefficient of resistance of solid particles in a non-Newtonian fluid for a wide range of variation in the number $0.1 < \mathrm{Re}_t < 10^5$ [24].

$$C_D = \frac{24X(n,\mathrm{Re}_t)}{\mathrm{Re}_t}\left[1+\mathrm{Re}_t^{\frac{2}{3}}\xi(\mathrm{Re}_t)\right]+0.3\left(1-\exp\left(-0.2\times10^{-19}\,\mathrm{Re}_t^4\right)\right)$$

$$X(n,\mathrm{Re}_t) = \frac{5\cdot 6^{n-1}}{1+4n}\left[1+(1-n)^{\frac{1}{3}}\frac{\mathrm{Re}_t^{\frac{3}{2}}}{1+(1-n)^{\frac{1}{3}}\mathrm{Re}_t^{\frac{3}{2}}}\right],\quad 0.8 < n \le 1,\ 0.1 \le \mathrm{Re}_t \le 10^5 \tag{7.118}$$

$$\xi(\mathrm{Re}_t) = \frac{\mathrm{Re}_t^2}{5+8\mathrm{Re}_t^2}+\frac{\mathrm{Re}_t}{6.8\times10^6\,\mathrm{Re}_t^{-\frac{5}{4}}+32\mathrm{Re}_t^{\frac{2}{3}}+952.8\times10^{-13}\,\mathrm{Re}_t^{\frac{10}{3}}}$$

For values $\mathrm{Re}_t < 1000$, this equation simplifies to the form

$$C_D \approx \frac{24X(n,\mathrm{Re}_t)}{\mathrm{Re}_t}\left(1+0.125\mathrm{Re}_t^{\frac{2}{3}}+2.51\times10^{-5}\,\mathrm{Re}_t^2\right) \tag{7.119}$$

Comparison of the calculated values of the coefficient of resistance according to equation (7.118) with the experimental data given in the paper [133–138] is given in Figure 7.25.

As follows from this figure and calculations using formula (7.118), a decrease in the coefficient of resistance of solid particles in viscoplastic liquids is less than in Newtonian.

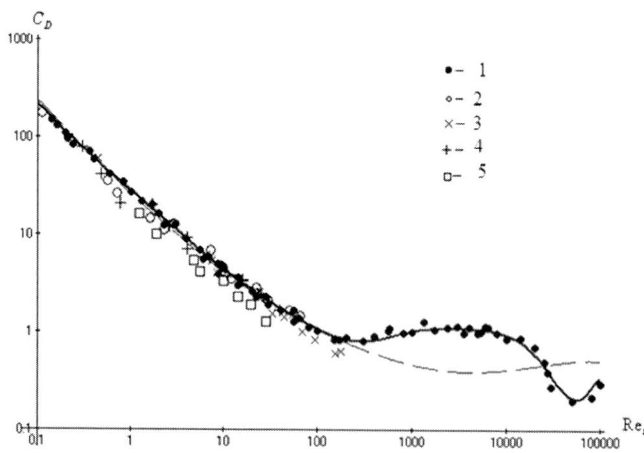

FIGURE 7.25 Dependence of the drag coefficient of a solid particle in a non-Newtonian fluid on the number according to the data of various authors (the dotted line is the coefficient of resistance of a particle in a Newtonian liquid): (1) $n = 0.84 - 0.86$ [135]; (2) $n = 0.75 - 0.90$ [133]; (3) $n = 0.75 - 0.92$ [133]; (4) $n = 0.56 - 0.75$ [70]; and (5) $n = 0.73$ [136].

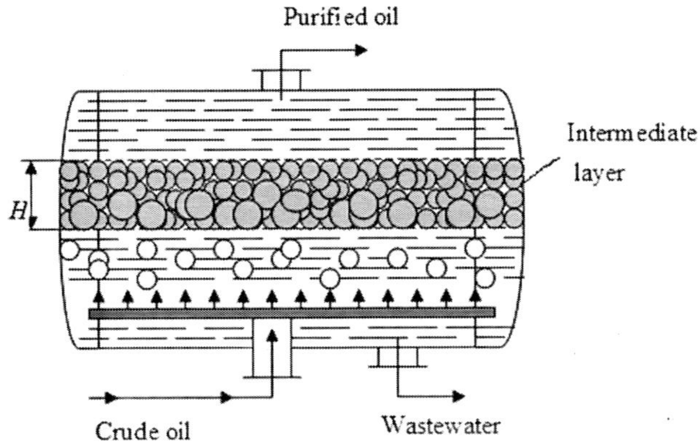

FIGURE 7.26 The structure of the oil emulsion in the apparatus.

7.7 SEPARATION OF OIL EMULSIONS AND THEIR RHEOLOGY

The processes of separation of oil emulsions are the most important stage for the preparation and cleaning of crude oil from water, mineral salts, and various associated admixtures that oil contains. The processes of oil emulsions, which are aimed at the complete reduction in the aggregative and kinetic stabilities of the emulsions, are carried out in different ways: in the gravitational (settling systems and other modifications), in the centrifugal, electrical, and magnetic fields, and also by the use of filtration through the solid and liquid layers, applying of microwave and membrane technologies The oil emulsions are in the majority of applications. Where sizes of drops and their concentrations are small, there are the electrostatic field and membrane technologies. The small size droplets ($a \leq 200$ μm) in the settlers in accordance with the terms of the dynamic equilibrium create the suspended layer of a filter bed, which mostly holds the fine-dispersed droplets and mechanical admixtures in its volume. The experimental and theoretical researches prove the very complex, in terms of dispersed composition (exfoliation) and running phenomena (coalescence, deformation, and breaking). The intermediate layer is a dynamic layer, as far as its structure (drops dispersion) and thickness are not constant. The intermediate layer occupies 10%–20% of the total volume of the apparatus (Figure 7.26).

Changing of the structure, dispersed composition, and drop geometry by time favors the changing of the porosity of the intermediate layer and the origin of the turbulent flow in the porous channels. A lot of works are dedicated to the theoretical and experimental researches of the mechanism of formation, stabilization, and destruction of oil emulsions as heterogeneous mediums, but many problems associated with phenomena, which occur on the interface of oil and water, with coalescence and breakup of water drops, exfoliation, and sedimentation. Oil emulsions are polydisperse mediums with water droplets the sizes of 1–150 μm, although they may contain some coarsely dispersed (150–1000 μm) and colloidal particles (0.001–1 μm).

Such a spread of sizes exercise a significant influence on the breakup, separation, and drop sedimentation mechanisms in the oil emulsions. The mechanism of breakup and coalescence of drops in the oil emulsions can be subdivided into the following stages:

1. Deformation and breakup of adsorption casings on the oil and water interface in the volume of stream using surface-active substances under defined temperatures (60°C–70°C) and pressures;
2. Approaching and collision of the drops of different sizes with interfacial film formation. It should be noted that the drops transport in the polydisperse medium basically defined by the hydrodynamic conditions and turbulence of flow. In the isotropic turbulence environment, the frequency of drop collision depends on specific dissipation energy of the turbulent flow and properties of the medium and dispersed phase. Interfacial film of a circular

section is formed as a result of collision and fixing of two drops with a_1 and a_2 sizes and radius of that film can be determined as [8]:

$$R_K = \left[\frac{3\pi}{4} P_m \left(k_1 + k_2\right) a_r\right]^{1/3} \tag{7.120}$$

where R_K is radius of interfacial film; P_m is maximal compression pressure; k_1, k_2 are coefficients of elasticity of two drops; $a_r = {}^{a_1 a_2}\!/\!{}_{(a_1+a_2)}$ is average size of the drops; and a_1, a_2 are the two drops sizes.

3. Thinning down and rupture of the interfacial film with the following coalescence and agglomeration of the drops. Rupture of interface film helps to junction the smaller droplets to the droplets with bigger size. It is important to note that as a result of transport of the oil emulsion in the pipes, the rate of breaking of the droplets becomes much higher than the coalescence rate, so the oil emulsion is characterized by higher spread of droplet sizes and by polydispersity of the medium; and
4. The sedimentation of the droplets and extraction of the dispersed phase as a continuous phase (exfoliation).

Another important factor affecting the effectiveness of the separation of oil emulsions are conditions of thinning and rupture of interface film and the rate of coalescence associated with breakup of adsorbed film on the surface of the drops with demulsifier participation.

Based on that, the purpose of this work is researching: (1) the structure of intermediate layer on basis of theoretical model of a liquation, (2) affection of asphaltenes on the formation of adsorptive films, (3) issues associated with thinning and breakup of the interface film, and (4) evolution of the function of the droplet distribution by sizes and by time taking into account the coalescence and the breaking of droplets.

7.7.1 Influence of Asphaltenes on the Separation of Oil Emulsions

The structurally mechanical stability of emulsion systems are associated with formation of adsorbed layers on the oil and water interface. These layers consist of asphaltenes, gums, paraffins, mineral salts, and solid particles, that is, from natural surface active substances. Initiation and formation of the adsorption layer on the water droplet surface with elastic and viscous properties lead to the stabilizing of oil emulsions. Therefore, the stability of oil emulsions is the result of a physical barrier, which hampers the breaking of the film when the collision energy between droplets is not enough for the destruction of the adsorption layer. The mechanism of formation of the adsorptive films on the surface is determined by the following stages:

1. The diffusion transfer of the mass of matter (asphaltenes) from the oil volume to the surface of water droplets in the boundary layer is described by an equation of the form

$$V_r \frac{\partial C}{\partial r} + \frac{V_\theta}{r} \frac{\partial C}{\partial \theta} = D \frac{\partial^2 C}{\partial r^2} \tag{7.121}$$

where C is the concentration of asphalt–tarry substances, which diffuse to the surface of a drop of water. The solution of this equation when introducing complex transformations and assumptions with boundary conditions

$$r \to \infty,\ C = C_0,\quad r = a,\ C = C_S \tag{7.122}$$

is represented as [6]

$$C = \frac{2(C_0 - C_S)}{\sqrt{\pi}} \int_0^N \exp(-z^2)dz + C_S \tag{7.123}$$

Here, $N = \frac{aV_0 \sin^2\theta\, y}{2\sqrt{DV_0 a^3 \left(2/3 - \cos\theta + \cos^3\theta/3\right)}}$, y is the distance to the surface of the drop of water, and V_0 is the value of the velocity at the surface of the drop. The diffusion flux density on the surface of a drop is determined from the following expression

$$J = D\left(\frac{\partial C}{\partial y}\right)\bigg|_{y \to 0} = \left(\frac{DV_0}{a}\right)^{1/2} \sqrt{\frac{3}{\pi} \frac{(1+\cos\theta)^2}{2+\cos\theta}} (C_0 - C_S) \tag{7.124}$$

It follows from this expression that the density of the diffusion flux is proportional to the square root of the velocity of the liquid (oil) flow on the surface of the drop and is inversely proportional to the root of the droplet size. Thus, as the water droplet size grows, the diffusion flux on its surface decreases, and the distribution of the concentration of asphalt–tar content on the surface depends on the angle, that is, is uneven. From this expression, we obtain the maximum value of the density of the diffusion flux on $\theta = 0$ the surface of the drop in the form

$$J = \left(\frac{2DV_0}{\pi a}\right)^{1/2} \tag{7.125}$$

The mass flux to the surface of a moving drop per unit time for small numbers $\mathrm{Re} = \frac{Ua_r}{\nu_c} << 1$ is defined as

$$I = 2\pi a^2 \int_0^{\circ} J \sin\theta d\theta \tag{7.126}$$

Having expressed the velocity on the surface of the drop in the form $V_0 = \frac{\eta_c}{2} \frac{U}{\eta_c + \eta_d}$, we finally obtain

$$I = \sqrt{\frac{4\pi}{3}} \left[\frac{D}{a_r} \frac{\eta_C}{\eta_C + \eta_d}\right]^{1/2} a_r^2 \Delta C \sqrt{U} \tag{7.127}$$

where η_c, η_d are the viscosity of the medium and droplets, D is the coefficient of molecular diffusion of particles, $\Delta C = C_0 - C_S$, C_0, C_S are the content of asphaltenes and resins in the volume and on the surface, and U is the velocity of the droplet.

The mass-transfer coefficient is determined using the equation

$$\mathrm{Sh} = \frac{Ia}{4\pi a^2 D(C_0 - C_S)} = \frac{2}{\sqrt{6\pi}} \mathrm{Sc}^{1/2} \mathrm{Re}^{1/2} \left(\frac{\eta_c}{\eta_c + \eta_d}\right)^{1/2} \tag{7.128}$$

The expression for estimating the mass-transfer coefficient is valid for the case, for large values of the number, this expression can give large errors, since the drop deforms and loses its sphericity. Assuming that the change in the mass of the spherical droplet as a result of the formation of the adsorption layer is determined, as, ${}^{dm}/_{dt} = I$, $m = \frac{4\pi}{3}\pi\rho_a\left[(R+\Delta)^3 - R^3\right] \approx 4\pi R^2 \Delta \rho_a$, $\Delta \leq R$, $\Delta \leq R$, then the thickness of the layer is determined in the form

$$\frac{\Delta}{a_r} = 0.65\left[\frac{1}{Pe}\frac{1}{1+\gamma}\right]^{1/2}\frac{\Delta C}{\rho_a}St \tag{7.129}$$

where Δ is the thickness of the adsorption layer on the surface of a water droplet, $\gamma \approx 0.04$, $\mathrm{Pe} = Ua_r/D$ for the oil emulsion this value is the very small Peclet number, $\mathrm{St} = Ut/a_r$ is the Strouhal number, ρ_a is the density of the adsorbed layer, and a_r is the average size of the water droplets. From equation (7.129) it follows that the thickness of the adsorbed layer depends on the diffusion coefficient of the asphaltene particles to the drop surface, the size and mobility of the surface of the drops, and on the concentration of asphaltenes in the flow volume. For the values of, $\mathrm{Pe} = 10^4 - 10^5$, $\gamma = 0.8$, $D \approx 10^{-10} - 10^{-9}\,{}^{\mathrm{m}^2}/_{\mathrm{s}}$, ${}^{\Delta C}/_{\rho_a} \approx 10^{-5}$, $St = 10^4 - 10^5$ from the equation, we estimate ${}^{\Delta}/_{a_r} \approx 0.1 - 0.15$. The large values of the number Pe, which are a consequence of the small values of the diffusion coefficient of particles in the liquid, in some cases determine the predominance of convective transport of matter over the diffusion one. Further compaction of the adsorption layer under the influence of external perturbations and chemical transformations promotes an increase in the density of the layer and the "aging" of the emulsions. Despite the insignificant thickness of the adsorption layer in comparison with the droplet size, their strength on the surface of droplets for various oils varies between 0.5 and 1.1 N/m^2.

2. Substance adsorption on the surface of the drops; and
3. Desorption and disruption of the adsorptive layer with surface-active substances participation. If the adsorption and desorption rate is small in comparison with feed rate of the substance to the surface of the droplet, then the adsorptive layer formation process is limited by the adsorption and desorption processes. Assuming that concentration of the adsorbed substance in volume C_0 and on the surface Γ. By analogy with a derivation of a Langmuir equation, assuming that adsorption rate of the substance on the surface of the droplet is equal to $W_A = \beta C_0\left(1 - {}^{\Gamma}/_{\Gamma_\infty}\right)$ and desorption rate is equal to $W_D = \alpha\Gamma$, then in the equilibrium condition $(W_A = W_D)$ we have

$$\Gamma = \frac{KC_0}{1 + K_0 C_0} \tag{7.130}$$

where α, β are the some constants that depend on temperature, and $K = {}^{\beta}/_{\alpha}$, $K_0 = {}^{\beta}/_{\alpha\Gamma_\infty}\, a$, Γ_∞ is the maximum saturation of the droplet surface. Equation (7.130) matches very well with many experimental data for the oils from the different oilfields. Figure 7.27 shows the adsorption isotherms of asphaltenes (T = 40°C) on the surface of the water drops for North-Caucasian oils and data calculated by equation (7.130), where $K = 55$, $K_0 = 0.5$.

For the adsorbed films breakup in the flow volume, the different demulsifies (surface-active substances), which are characterized by high surface activity during adsorption, are used. Adsorbed films breakup mechanism consists of diffusion transfer of demulsified to the film surface with further adsorption and penetration to the film volume, defects and cracks formation in its structure, change in surface tension, and reduction of the strength properties, which qualitatively changes the rheological properties of the films on the oil and water interface. Further oil emulsion separation is determined by the collision frequency of the drops, their fixing on the surface, thinning, and breakup of the interfacial film.

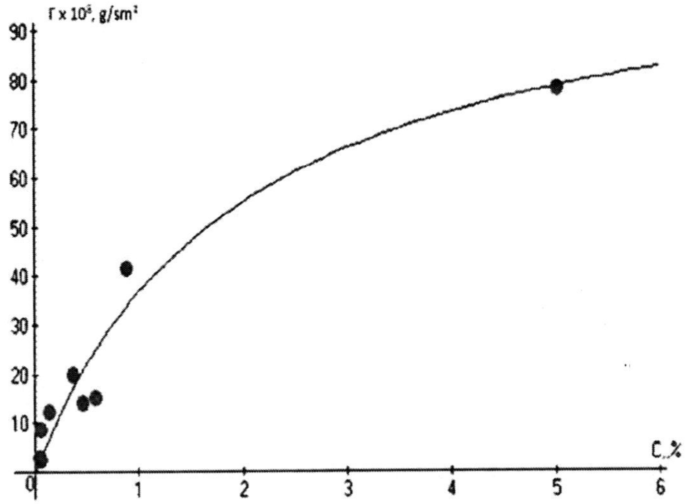

FIGURE 7.27 Dependence of the concentration of adsorbed substance on the surface on the concentration of asphaltenes.

7.7.2 Viscosity of Oil Emulsions

Section 1.6 lists many formulas for calculating the effective viscosity of oil-dispersed systems. One of the important rheological parameters of emulsions is their dynamic viscosity, depending on the volume fraction, size, and shape of the droplets, on the ratio of the droplet viscosity to the viscosity of the medium $\lambda = \eta_d/\eta_c$ (droplet surface mobility), on the shear stress in concentrated emulsions (Figure 7.28), etc.

These curves are described by equations:

$$\begin{aligned} &\eta(\tau) = \eta_\infty + (\eta_0 - \eta_\infty)\exp(-k\tau) \\ &\eta_0(T) = 15.6 - 0.96T + 0.0134T^2, \\ &\eta_\infty(T) = 3.62 - 0.22T + 0.003T^2 \\ &k = 0.015 \end{aligned} \tag{7.131}$$

In addition to these parameters, the viscosity of oil emulsions is significantly influenced by the physical phenomena of coalescence and crushing of droplets that change the structure of the dispersed medium. The possibility of taking into account various physical phenomena in the rheology of oil emulsions is a complex task, in connection with which the viscosity of such systems is expressed by empirical relationships.

The paper [136] provides a formula for calculating the viscosity of an oil emulsion for various types of oil

$$\frac{\eta_э}{\eta_н} = \exp(5\varphi)\left(1 - 3\varphi + b\varphi^2\right) \tag{7.132}$$

Here, $\eta_э$, $\eta_н$ are the dynamic viscosities of the emulsion and oil, φ is the volume fraction of the drops, b is the emulsion type factor, with $b = 7.3$ for highly concentrated emulsions; $b = 5.5$ for concentrated emulsions; $b = 4.5$ for medium concentration emulsions; $b = 3.8$ for dilute emulsions; and

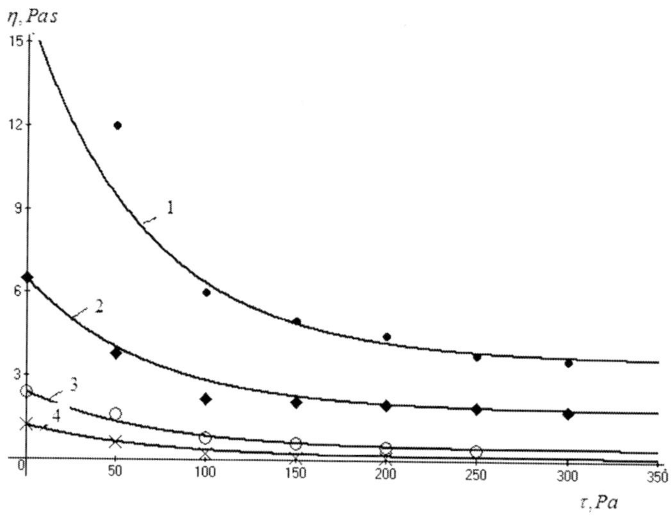

FIGURE 7.28 Dependence of the effective viscosity of oil emulsions on the shear stress at temperatures: (1) 0°C; (2) 10°C; (3) 20°C; and (4) 50°C.

$b = 3.0$ for very dilute emulsions. The calculated curves for the viscosity of the oil emulsion versus the volume content of water are given in Figure 7.29.

It is important to note that, in addition to the above factors, the viscosity of emulsions is associated with the presence of deformable droplets and bubbles in them, and at large droplet concentrations, with the formation of coagulation structures leading to rheological properties. The paper [137]

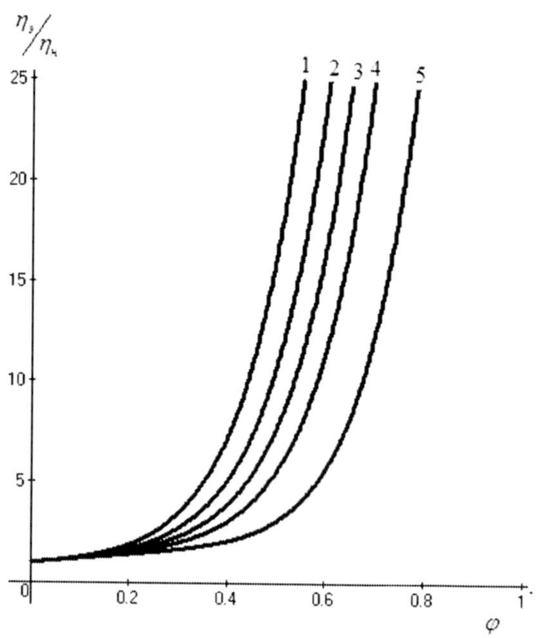

FIGURE 7.29 Dependence of viscosity of oil emulsion on water content: (1) highly concentrated emulsions, $b = 7.3$; (2) concentrated emulsions, $b = 5.5$; (3) for medium emulsion concentrations, $b = 4.5$; (4) for diluted emulsions, $b = 3.8$; and (5) for highly diluted emulsions, $b = 3.0$.

considers possible options for calculating the viscosity of emulsions, taking into account the structural change. If we introduce the relaxation time of the stress in the form

$$\tau_p = \frac{\eta_c R}{\sigma} \frac{(2\lambda+3)(19\lambda+16)}{40(\lambda+1)} \tag{7.133}$$

then the viscosity of the emulsions can be calculated from formula

$$\eta_э = \frac{\eta_c}{1+(\tau_p\dot{\gamma})^2}\left[1+\frac{1+2.5\lambda}{1+\lambda}\varphi+\left(1+\varphi\frac{5(\lambda-1)}{2\lambda+3}\right)\left((\tau_p\dot{\gamma})^2\right)\right] \tag{7.134}$$

For small values $\tau_p\dot{\gamma} << 1$ of the quantity, this equation takes the following form

$$\frac{\eta_э}{\eta_c} = 1+\frac{1+2.5\lambda}{1+\lambda}\varphi \tag{7.135}$$

and, for $\lambda \to \infty$, that is, for solid particles is $\eta_э/\eta_A = 1+2{,}5\varphi$. The paper gives the following formula for calculating the viscosity of mono-disperse emulsions depending on the droplet size and their volume fraction

$$\frac{\eta_э}{\eta_c} = 1+\frac{\eta_c+2.5\eta_d+(\eta_{ds}+\eta_{di})/a}{\eta_c+\eta_d+0.4(\eta_{ds}+\eta_{di})/a}\varphi \tag{7.136}$$

Here, η_{ds} is interphase shear viscosity, η_{di} is dilatant viscosity, and a is droplet size.

The rheology of oil emulsions is determined by their structure; since the emulsions are thermodynamically unstable disperse media formed by two mutually insoluble liquids in one another, where one phase is suspended in the form of fine droplets. The droplets of the dispersed phase of small dimensions have a spherical shape corresponding to the minimal surface and the smallest energy for a given volume. The shape of the large droplets differs from the spherical, which affects the effective viscosity. The presence in the oil emulsion of various particles of the solid phase, asphaltenes, and resinous substances capable of creating coagulation structures and aggregates can have a significant effect on the rheological models. A large spread of water droplet sizes and their time variations can be taken into account using the evolution of the droplet distribution function in size and time, as will be discussed in the next chapter.

8 Aggregate and Sedimentation Unstable Dispersion Systems

Aggregative unstable disperse systems are characterized by the inconsistency of the state of the medium, caused by a continuous change in the physical properties of the particles, that is, change in the volume and size of particles as a result of their interaction, collision, coalescence, and breaking at a certain concentration of particles in a closed volume. The problems of coalescence, crushing, agglomeration, and precipitation are special types of mass-transfer problems associated with the diffusion transfer of particles to the phase interface. As a result of these physical phenomena, new particles and aggregates are formed, which changes the number of particles per unit volume, their distribution by size, and the entire spectrum of dimensions. The change in the number and size of particles leads to a change in the interphase surface, which significantly affects mass transfer and heat exchange in the medium. The change in the particle size also affects the sedimentation instability of the dispersed medium due to the precipitation of large particles from the volume of the dispersed system. The sedimentary instability of the system is the result of phase separation and separation as a result of precipitation or emergence of the dispersed phase.

8.1 FUNDAMENTALS OF THE THEORY OF COALESCENCE AND BREAKING OF DROPLETS AND BUBBLES IN ISOTROPIC TURBULENT FLOW

The physical phenomena of coalescence and breaking of droplets and bubbles in a turbulent flow form the basis of many processes of chemical, oil refining, food, and pharmaceutical technology and are associated mainly with changes in the spectrum and number of particles per unit volume that determine the magnitude of the interphase surface of mass and heat transfer. The coalescence of droplets and bubbles associated with their collision and subsequent enlargement is widely used in the processes of separation and phase separation (deposition, ascent) in emulsions, suspensions, and other multiphase media. The breaking of droplets and bubbles, necessary to increase the interfacial surface, is widely used in mass-transfer processes of liquid extraction, absorption, gas–liquid reactors, spraying and combustion processes, etc. The essence of the processes of coalescence and breaking of droplets and bubbles consists in the loss of aggregative and, in some cases, sedimentation stability of the dispersed system as a whole, under the influence of external forces or spontaneously because of the desire to reduce excess surface energy. Theoretical and experimental studies of the processes of coalescence, breaking, and deformation of droplets and bubbles are given in many works [4,6,8,130–150]. The first theoretical studies of coalescence phenomena of two-particle collisions of drops of very small sizes in the Brownian particle approximation in a homogeneous unbounded system on the simplest models were carried out by Smoluchowski [6,8], as a result of which the coalescence frequency $\omega = 4\pi\left(D_1 + D_2\right)\left[R_1 + R_2\right]N_0$ and time inversely proportional to the molecular diffusion coefficient of Brownian particles were determined $\tau = \left(8\pi DRN_0\right)^{-1}$. Subsequent studies of the phenomena of coalescence and breaking of droplets and bubbles flowing in pipes, in column apparatus and in mixing devices, were devoted to the experimental [139–143,146–148] and theoretical research and dynamics of these phenomena in a turbulent flow [142–153]. Various works [142–153] present the conditions, mechanisms, models, frequencies, and time of breaking and coalescence of droplets and bubbles; semi-empirical formulas for calculating the minimum and maximum particle sizes; and frequency and time of coalescence and breaking depending on the coefficient of turbulent diffusion. It is noted that in the working volume of the apparatus along with breaking there is a coalescence of bubbles, as a result of which a certain equilibrium bubble size is established, which is primarily due to the properties of the liquid and gas phases and mixing conditions. The necessary intensity of mixing is required only for the formation of a gas–liquid system and the achievement of the required gas content in the volume. Some studies have focused on specific cases of coalescence, in particular, heterogeneous hydrodynamic coagulation in agitators, that is, adsorption coagulation [154], coalescence of a low-boiling disperse phase in a turbulent flow, and coagulation of particles in an inhomogeneous field [156]. The main characteristics of these processes are the frequency of coalescence (or collision) and breaking determined in the isotropic turbulent flow by energy dissipation and the scale of turbulence, as well as the properties and dimensions of the particles themselves and the medium (density, viscosity, surface tension), and in most cases the state of the dispersed system is determined by minimum and maximum sizes of droplets and blisters. The literature [157] analysis of the processes of breaking droplets and bubbles in a turbulent flow is given, in which the main attention is paid to the analysis of various mechanisms of breaking and to the problems of determining the frequency, its dependence on the specific dissipation of energy, the properties of the medium, and the particle sizes. The minimum particle size characterizes such a state of the disperse system, which is more prone to coalescence, and the maximum dimensions characterize the state of the system more prone to deformation and breaking of droplets and bubbles. Many works are devoted to the processes of coalescence and fragmentation in turbulent flow [155–157], where the following are studied: the influence of isotropic turbulence parameters on the processes of coalescence and fragmentation of droplets and bubbles in gas–liquid and liquid–liquid systems; the influence of the properties of the medium and drops on the course of processes and on the distribution of drops [161]; the effect of particle concentration on coalescence and crushing in multiphase media [162]; Others are problems of calculating these processes [165–167]. Another characteristic of the dispersed system in which coalescence and crushing processes occur is the rate of change in the size and number of particles, in other words, the evolution of the distribution function in size and time, theoretical and experimental studies of which are given in the works [40,167–170]. Important characteristics of an isotropic turbulent flow are the Kolmogorov scale of turbulence λ_0 and energy dissipation on the surface of a drop E_R, defined in the form [139–143,146–148]

$$\lambda_0 = \left(\frac{\nu_C^3}{\varepsilon_R}\right)^{1/4}, \quad E_R \sim \frac{\pi a^3}{6}\tau_P \sim \frac{\pi a^2 \eta_c}{6}\sqrt{\frac{\rho_c\left(\varepsilon_R a\right)^{2/3}}{\rho_d}} \tag{8.1}$$

where $\tau_P = \frac{\eta_d \bar{U}}{a} \sim \frac{\eta_d\left(\varepsilon_R a\right)^{1/3}}{a}$ is the viscous stress on the surface of the drop, and $\bar{U}^2(r) = C_0\left(\varepsilon_R r\right)^{2/3}$ is the root-mean-square (RMS) velocity of the turbulent flow. Thus, the kinetics of breaking and coalescence of droplets in a turbulent flow is determined by the statistics of pressure pulsations in the flow zones with the maximum energy dissipation rate and can be described starting from the "two-thirds" law and the normal distribution law of turbulent pulsations.

In a turbulent flow, a turbulent number $\mathrm{We} = \rho_c \left\langle \bar{U}^2(a_e) \right\rangle a_e / \sigma$ [141] is introduced, where $\left\langle \bar{U}^2(a_e) \right\rangle$ is the RMS velocity of the turbulent flow and a_e is the equivalent particle size. Another form of expression for the Weber number in an isotropic turbulent flow is the formula: $\mathrm{We}_T = 2\rho_c \varepsilon_R^{2/3} a_e^{5/3} / \sigma$. The ratio of the surface energy of the droplet $\left(E_\sigma \sim \pi a^2 \sigma / a \sim \pi a \sigma\right)$ to the energy of the turbulent flow $\left(\bar{E}_T \sim \pi a^2 \left(\Delta P_T\right), \Delta P_T = C_1 \rho_c \left(\varepsilon_R a\right)^{2/3}\right)$ characterizes its deformation and the efficiency of the crushing process

$$\frac{E_\sigma}{\bar{E}_T} \sim \frac{\sigma}{\rho_c \varepsilon_R^{2/3} a^{5/3}} = \mathrm{We}_T^{-1} \tag{8.2}$$

It should be noted that in the processes of coalescence and crushing of droplets and bubbles in an isotropic turbulent flow, the turbulent diffusion coefficient plays an important role, which is determined for a different range of values of the scale of turbulent pulsations λ in the form [6]

$$\lambda > \lambda_0, \quad D_T = \alpha_0 \left(\varepsilon_R \lambda\right)^{1/3} \lambda; \quad \lambda < \lambda_0, \quad D_T = \alpha_0 \left(\varepsilon_R / \nu_C\right)^{1/2} \lambda^2 \tag{8.2a}$$

At the same time, taking into account the degree of entrainment of particles by the pulsating medium for the coefficient of turbulent diffusion of particles, we can write [12]: $D_{TP} \approx \mu_p^2 D_T$. where μ_p^2 is the degree of entrainment of particles by a turbulent flow, depending on the particle size, and with their growth it can be assumed that $\mu_p^2 \to 0$. In a broader sense, in the work [48], on the basis of available experimental studies, empirical formulas are proposed for determining the diffusion coefficients of particles as a function of the dynamic flow velocity and deposition rate, and so on.

8.2 THE MAXIMUM AND MINIMUM SIZES OF DROPLETS AND BUBBLES

Dispersed systems (emulsions, suspensions) are more characterized by the polydispersity of the particle sizes, ranging from 1 µm to 200 µm, although colloidal particles and larger particles can be found in the flow. However, the state of the dispersed flow, its aggregate resilience to sedimentation and sedimentation resistance to sedimentation, as a whole, which determine the structure of the dispersion spectrum in the volume, is characterized by the minimum and maximum particle sizes. It should be noted that the processes occurring in dispersed systems are accompanied not only by collision and enlargement of colliding droplets, but also by an inverse phenomenon—a breaking caused by the fact that strongly interacting particles scatter into fragments, or cannot maintain a stable state and decay spontaneously or under the action of any perturbations on their outer surface. Thus, in dispersed systems, there is a certain size above which the droplets $a_{\max}$ are unstable, deformed, and instantly destroyed and the minimum size $a_{\min}$ determining the lower threshold of droplet stability, that is, under certain flow conditions, droplets that have reached these sizes cannot be further broken. The maximum particle size characterizes the unstable state of droplets and bubbles that depend on the hydrodynamic flow conditions of the disperse medium and, under certain conditions of turbulent flow, exhibits a propensity to decay and fragmentation of a single drop. In the literature, there are many formulas for estimating the maximum values of droplets in a turbulent flow, among which it is important to note the following:

$$a_{\max} = \frac{\sigma^{0.6}}{\rho_C^{0.4} \rho_d^{0.2}} \varepsilon_R^{-0.4}, \ [141] \ ; \ a_{\max} = 1.12 \left(\frac{\sigma}{\rho_C}\right)^{0.6} \left(\frac{\eta_C}{\eta_d}\right)^{0.1} \varepsilon_R^{-0.4}, \ [158];$$

$$a_{\max} = 0.725 \left(\sigma / \rho_C\right)^{0.6} \varepsilon_R^{-0.4}, \ [173] \ ; \ a_{\max} = \frac{\sigma \nu_c^{1/2}}{\eta_c \varepsilon_R^{1/2}} \frac{16\left(\eta_d / \eta_c\right) + 16}{19\left(\eta_d / \eta_c\right) + 16}, \ [172] \tag{8.3}$$

$$a_{\max} = \lambda_f \left(\frac{\sigma}{\varepsilon_R^{2/3} \rho_c \lambda_f^{5/3}}\right)^{0.926}, \ [173] \ ; \ a_{\max} = C^{\frac{5}{3+2\alpha}} L \left(\frac{\sigma}{\rho_c \varepsilon_R^{2/3} L^{5/3}}\right)^{\frac{3}{3+2\alpha}}$$

where λ_f is the initial length of the scale of turbulence, L is the integral scale of turbulence, α is the coefficient, and when, $\alpha = 1$ the last expression coincides with the formula proposed in the paper [173]. Calculation of these formulas under identical flow conditions gives a significant spread of the numerical values of the maximum droplet size. In order to estimate the minimum and maximum droplet sizes, we use the condition for the equality of the dynamic force acting on the drop surface $F_D = C_D(\text{Re}_d)\rho_C U^2/2$. Where $C_D(\text{Re}_d)$ is the drop and bubble drag coefficient [171–174], and $\text{Re}_d = Ua/\nu_C$ is the Reynolds number for the particle, determined by the average velocity of the turbulent flow (U) with the surface tension force $F_\sigma = 4\sigma/a_S$ [6]. It is assumed that the dynamic head acts on a certain part of the droplet surface and can therefore be related to the drag coefficient. From the condition of equality of these forces

$$C_D(\text{Re}_d)\frac{\rho_C U^2}{2} = \frac{4\sigma}{a_S}, \tag{8.4}$$

taking into account the expression for the pulsation velocity $V' = \gamma\left(\frac{\rho_d}{\rho_c}\varepsilon_R\lambda\right)^{1/3}$, we determine the expression for the maximum size of droplets or bubbles in the form [4,85–87]

$$a_{\max} = \gamma^{-6/5}\left(\frac{8}{C_D}\right)^{0.6}\left(\frac{\sigma}{\rho_C^{1/3}\rho_d^{2/3}}\right)^{0.6}\varepsilon_R^{-0.4} \tag{8.5}$$

Here the coefficient is determined by the experimental method. Starting from equation (8.5), for various regions of the number change, one can obtain a number of formulas for estimating the maximum droplet size $\left(C_D \approx 8/3\right)$

$$a_{\max} \approx 1.93\gamma^{-1.2}\left(\frac{\sigma}{\rho_C^{1/3}\rho_d^{2/3}}\right)^{0.6}\varepsilon_R^{-0.4}, \quad 2\times10^3 \le \text{Re}_d \le 5\times10^5 \tag{8.6}$$

For air bubbles in water in the region of their deformation $500 < \text{Re}_d \le 2000$, where the resistance coefficient according to the experimental data is equal $C_D \approx 14.5\text{Mo}^{1/2}\,\text{Re}_d^{4/3}$, using expression (8.5) for maximum dimensions, we obtain [4,85]

$$a_{\max} = 0.82\gamma^{-2/3}\text{Mo}^{-1/6}\left[\left(\frac{\nu_C}{U}\right)^{-4/9}\left(\frac{\sigma}{\rho_C^{1/3}\rho_d^{2/3}}\right)\right]^{1/3}\varepsilon_R^{-2/9} \tag{8.7}$$

where $\text{Mo} = \frac{g\eta_C^4}{\rho_C a^3}\frac{\Delta\rho}{\rho_C}$ is the Morton number. If the Morton number $\text{Mo} > 10^{-7}$ and $0.1 < \text{Re}_d < 100$ then, according [6] and to the experimental data [149], we obtain and the maximum bubble size is determined in the form

$$a_{\max} \approx 0.354\gamma^{-3}\left(\frac{U}{\nu_c}\right)^{3/2}\left(\frac{\sigma}{\rho_C^{1/3}\rho_d^{2/3}}\right)^{3/2}\varepsilon_R^{-1} \tag{8.8}$$

If Morton's number is $\text{Mo} < 10^{-7}$ and $0.1 \le \text{Re}_d < 100$, then we can use equation (8.8) with a coefficient of 0.192 and if $\text{Mo} < 10^{-7}$ and $100 \le \text{Re}_d < 400$ is with a coefficient equal to 0.068. In particular, if $2 < \text{Re}_d < 10^3$, then for bubbles it can be assumed that $C_D \approx 14/\text{Re}_d^{0.5}$, as a result of which expression (8.4) takes the form

$$a_{\max} \approx 0.619\gamma^{-12/7}\left[\left(\frac{U}{\nu_C}\right)^{1/2}\left(\frac{\sigma}{\rho_C^{1/3}\rho_d^{2/3}}\right)\right]^{6/7}\varepsilon_R^{-4/7}, \tag{8.9}$$

Expression (8.9) is in satisfactory agreement with the experimental data [5,6], which is shown in the work [141,142] (Figure 8.1).

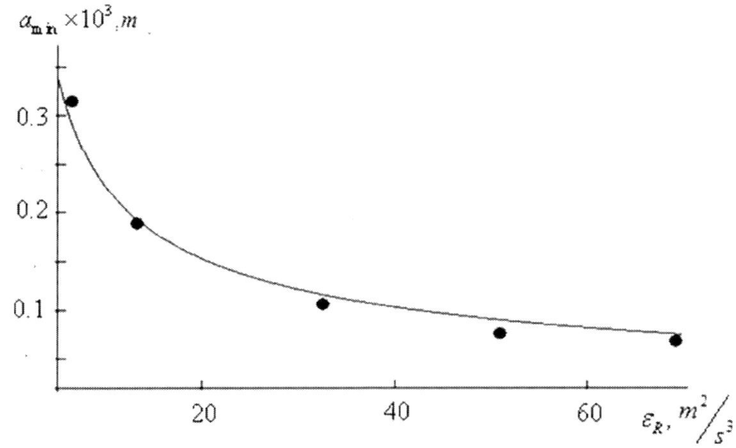

FIGURE 8.1 Change in the minimum size of bubbles as a function of the specific dissipation energy (Points are experimental data [141,142]): $\left(\mathrm{Mo} = 3\times 10^{-11},\ \rho_d = 1.29\,\mathrm{kg/m^3},\ \rho_c = 1000\,\mathrm{kg/m^3},\ \nu_c = 10^{-6}\,\mathrm{m^2/s},\ \gamma = 28,2\right)$.

In the literature, there are also many formulas using the critical Weber number $\mathrm{We}_{cr} = \frac{\rho_c \bar{U} a_{max}}{\sigma}$ to determine the maximum size of droplets and bubbles in a turbulent flow, among which

$$a_{max} = \left(\frac{\mathrm{We}_{cr}}{2}\right)^{0.6}\left[\frac{\sigma^{0.6}}{\left(\rho_C^2 \rho_d\right)^{0.2}}\right]\varepsilon_R^{-0.4},\ [141];$$

$$a_{max} = \left(\mathrm{We}_{cr}\right)^{0.6}\left[\frac{\sigma^{0.6}}{\left(\rho_C^{0.3}\rho_d^{0.3}\eta_C^{0.6}\right)}\right]\left(\frac{d_k^{1.3}}{U^{1.1}}\right),\ [142];$$

$$\left(\frac{a_{max}\rho_C U^2}{\sigma}\right)\left(\frac{\eta_C U}{\sigma}\right)^{1/2} = 38\left[1+0.7\left(\frac{\eta_C U}{\sigma}\right)^{0.7}\right],\ [167]; \tag{8.10}$$

$$a_{max} = \left(\frac{\mathrm{We}_{cr}\sigma}{2}\right)^{3/5}\rho_c^{1/5}\varepsilon_R^{-2/5},\ a_{max} = \left(\frac{\mathrm{We}_{cr}\sigma}{2}\right)^{3/5}\rho_c^{1/5}\varepsilon_R^{-2/5},\ [174,175]$$

Here We_{cr} is the critical Weber number, and d_k is the diameter of the channel. In the literature, you can find other empirical formulas for determining the maximum size of droplets

$$a_{max} = A_1\left(\sigma + A_2\eta_c\varepsilon_R^{1/3}a^{1/3}\right)^{3/5}\rho_c^{-3/5}\varepsilon_R^{-2/5},\quad A_1 \approx 1.9,\quad A_2 \approx 0.35 \tag{8.11}$$

$$\frac{\rho_c\varepsilon^{2/5}a^{5/3}}{\sigma} = A_1^{5/3}\left(1 + A_2\left(\frac{\rho_c}{\rho_d}\right)^{1/2}\frac{\eta_c\varepsilon_R^{1/3}a^{1/3}}{\sigma}\right) \tag{8.12}$$

where A_1, A_2 are the coefficients that have different values depending on the viscosity of the medium.

The minimum particle size in an isotropic turbulent flow characterizes the state of a droplet or bubble that is hydrodynamically stable to breaking under certain flow conditions and at a high concentration of particles, a tendency to intense collision and coalescence, and is also determined by condition (8.3). Taking that into account for the isotropic turbulent flow $V_\lambda = \gamma_0 \left(\frac{\rho_d}{\rho_C} \varepsilon_R \lambda \right)^{1/3}$ and $\varepsilon_R = \left({}^{\nu_c^3}\!/\!_{\lambda_0^4} \right)$, as well as the expression (8.3) for the minimal sizes of droplets and bubbles, we obtain [25]

$$a_{\min} = \frac{C_D}{8} \gamma^2 \left(\rho_c \rho_d^2 \right)^{1/3} \left(\frac{\nu_c^2}{\sigma} \right) \tag{8.13}$$

It should be noted that a simpler formula has been obtained for the minimum droplet size $a_{\min} \approx {}^{\rho_c \nu_c^2}\!/\!_{\sigma}$ [141]. Since the coefficient of resistance depends on the number, we consider the various variants of formula (8.13) [25]:

$$a_{\min} \approx \sqrt{3}\gamma \left(\frac{\nu_c^3}{U\sigma} \right)^{1/2} \left(\rho_c \rho_d^2 \right)^{1/6}, \quad \mathrm{Re}_d < 0.1, \tag{8.14}$$

$$a_{\min} \approx \sqrt{2}\gamma \left(\frac{\nu_c^3}{U\sigma} \right)^{1/2} \left(\rho_c \rho_d^2 \right)^{1/6}, \quad 0.1 \le \mathrm{Re}_d \le 100 \tag{8.15}$$

For the region $400 \le \mathrm{Re}_d \le 2\times10^3$, according to experimental data [91] for a drop of liquid in air, the following correlation can be obtained $C_D = 14.5\mathrm{Mo}^{-3/2}\,\mathrm{Re}_d^{4/3}$, taking into account which one can write

$$a_{\min} = 0.168\gamma^{-2/3}\mathrm{Mo}^{-3/2} \left(\frac{\sigma^3}{U^4 \rho_c \rho_d^2 \nu_c^2} \right) \tag{8.16}$$

For air bubbles in water, for $\mathrm{Re}_d > 2\times10^3$, the value of the coefficient of resistance is set at a level equal to $C_D = {}^8\!/\!_3$ and

$$a_{\min} = \frac{1}{3} \gamma^2 \left(\frac{\nu_c^2}{\sigma} \right) \left(\rho_c \rho_d^2 \right)^{1/3} \tag{8.17}$$

In agitating devices, the minimum size of the droplets is determined as [175]

$$a_{\min} = 0.5 \left(n d_T \right)^{-1.75} \left(\frac{\sigma^3 \nu_C}{\rho_C \rho_d^2} \right)^{1/4} \tag{8.18}$$

where n is the angular frequency of the stirrer, and d_T is the diameter of the stirrer. As follows from this equation, the minimum droplet size is determined by the stirrer characteristics: the angular velocity of rotation and the size of the agitator, as well as the properties of the medium and the droplet. Table 8.1 compares the experimental data [175,176] with the calculated values of the minimum oil droplets in water by formula (8.18).

The above correlations for calculating the minimum and maximum sizes of deformable particles (droplets, bubbles) can be used to solve problems of coalescence, breaking, and separation of disperse systems. Theoretical and experimental studies of the state of deformable particles in a turbulent flow [177–179] made it possible to propose various formulas for estimating the maximum and minimum sizes of dispersed inclusions, and, as studies have shown, these dimensions depend primarily on the specific dissipation of energy, on the properties of the medium and particles, on

TABLE 8.1
Comparison of the Experimental [176] and Calculated Values of the Minimum Sizes of Oil Droplets in Water: $\sigma = 72.2\times 10^{-3}$ N/m, $\rho_C = 1000$ kg/m³, $\rho_d = 850$ kg/m³, $\nu_C = 10^{-6}$ m²/s, $d_T = 0.027$ m

$\mathrm{Re}_d = \frac{n d_T a_m}{\nu_C}$	$n, \sec^{-1}$	$U = n d_T$ m/s	$a_{min} \times 10^6$ m experience	$a_{min} \times 10^6$ m (8.18)
22.50	16.67	0.450	53.0	54.2
19.44	20.0	0.540	39.0	39.4
16.87	25.0	0.675	26.0	26.6
14.58	30.0	0.810	19.0	19.4
13.23	35.0	0.945	15.0	14.7
11.25	41.67	1.125	10.5	10.9

Source: Pilch, M. and Erdman, C.A. *Int. J. Multiph. Flow*, 13, 741–752, 1987.

the resistance coefficient particles, and in some cases also from the critical value of the Weber number We_{cr} [91,42]. The use of the coefficient of resistance in the equation of equilibrium (8.4) allows us to expand the range of adequate description of the maximum and minimum particle sizes. Comparing the above formulas with various experimental measurements, it can be noted that the maximum droplet size is proportional to the specific dissipation of energy $\varepsilon_R^{-0.4}$ only for the region of developed turbulence at $\mathrm{Re}_d \geq 2\times 10^3$. In other cases, the degree of specific dissipation of energy can vary significantly.

In this paper [181], using experimental data [8,91], an expression is proposed $\mathrm{Re}_d\,\mathrm{Mo}^{1/6} < 7$ in which bubbles do not undergo significant deformation, where the relationship between numbers Mo, We, and Re_d is expressed as $\mathrm{Mo} = \frac{4}{3} C_D \mathrm{We}^3 \mathrm{Re}_d^{-4}$.

8.3 DEFORMATION OF DROPS AND BUBBLES

The deformation of drops and bubbles, first of all, is characterized by a violation of the balance of external and surface forces acting on a particle in a turbulent flow. In the simplest case, with insignificant gravitational forces and drag forces, such forces are forces of hydrodynamic head and surface tension. The pressure forces are proportional to the high-velocity head, $F_D \sim \frac{\rho_c U}{2}$ and the surface tension force is the capillary pressure $F_\sigma = \frac{2\sigma}{a_e}$. If $\mathrm{We} \sim \frac{F_D}{F_\sigma} << 1$, then at small numbers $\mathrm{Re}_d << 1$, the droplets and bubbles have a strictly spherical shape. Under the condition $F_D \geq F_\sigma$ or $\mathrm{We} \geq 1$, $\mathrm{Re}_d > 1$ either the surface of the drop loses its stability and it deforms by taking the shape of an oblate ellipsoid of revolution at the beginning, and with a further increase in the number and assumes various configurations up to a stretched filament that cannot be theoretically investigated and described (Figure 8.2). It should be noted that in dispersed systems there is some maximum size a_{max} above which the drops are unstable, deformed, and instantly destroyed, and the minimum

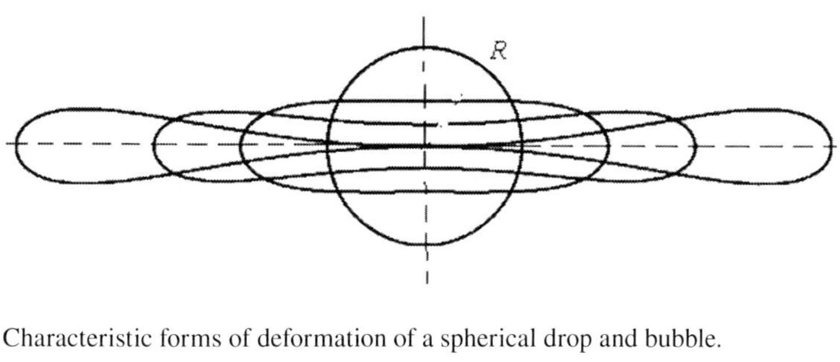

FIGURE 8.2 Characteristic forms of deformation of a spherical drop and bubble.

size $a_{\min}$ determining the lower threshold of droplet stability, that is, under certain flow conditions, droplets that have reached these sizes cannot be further broken up. The maximum particle size characterizes the unstable state of droplets and bubbles that depend on the hydrodynamic flow conditions of the disperse medium and, under certain conditions of turbulent flow, exhibits a tendency to deform the shape and fragmentation of a single drop.

The volume deformation of droplets and bubbles is based on a three-dimensional model and reduces to a change in the shape of a spherical particle to an ellipsoidal one. And the drop undergoes simultaneous stretching and compression with a constant volume. In the literature there is a large number of empirical formulas describing the deformation of drops and bubbles. Thus, the paper [141] gives the following formula $\chi = \sqrt{1 - 9/16\,\mathrm{We}}$, in [149] is $\chi = 1 - \mathrm{We}/9$, in [182] is $\chi = 1 + 0.09/\mathrm{We}^{0.95}$, in [183] is $\chi = 1 - 0.11\mathrm{We}^{0.82}\psi(\rho_c, \gamma, \mathrm{We})$ and etc., which describe the droplet behavior at a value $\mathrm{We} \le 1.01 - 1.13$. In comparison with multidimensional deformation, the volume deformation is the simplest case with the preservation of a certain symmetry of the shape (Figure 8.2).

It is important to note that for any deformation of the droplet shape, the surface area of the particle increases with the volume of liquid in the drop constant, which is an important factor for increasing the interfacial surface.

In the papers [6,141], the fluctuation frequency of oscillations of the drop surface using the Rayleigh equation as a result of the effect of a turbulent pulsation of a certain frequency on the surface of droplets and bubbles on the shape change is defined as

$$\omega(k) = \left[\left(\frac{2\sigma}{\pi^2 \rho_c a^3}\right)\left(\frac{(k+1)(k+2)k(k-1)}{(k+1)\rho_d/\rho_c + 1}\right)\right]^{1/2} \tag{8.19}$$

where k is the wave number. From $k = 2$ in this formula, it is possible to obtain formulas for determining the frequency corresponding to the fragmentation of bubbles ($\rho_d << \rho_c$) $\omega(a) = \frac{2\sqrt{6}}{\pi}\left(\frac{\sigma}{\rho_c a^3}\right)^{1/2}$ and drops $(\rho_d >> \rho_c)$ $\omega(a) = \frac{4}{\pi}$ As a result, for small deformations, the shape of the droplet is determined by the superposition of linear harmonics

$$r(t,\theta) = R\left[1 + \sum_k A_k \cos(\omega_k t) P_k(\cos\theta)\right] \tag{8.20}$$

where $P_k(\cos\theta)$ is the Legendre functions, A_k is the coefficients of the series, defined as $A_k = A_{k0}\exp(-\beta_k t)$, β_k is the attenuation coefficient, defined in the form

$$\beta_k = \frac{(k+1)(k-1)(2k+1)\eta_d + k(k+2)(2k+2)\eta_c}{\left[\rho_d(k+1) + \rho_c k\right]R^2} \tag{8.21}$$

In this paper [184], the expression (8.20) for small numbers Re_d is represented in the form

$$r(t,\theta) = R\left[1 + \xi(\cos\theta)\right] \tag{8.22}$$

where $\xi(\cos\theta) = \lambda_m \mathrm{Ac}\,\mathrm{Re}_d^2\, P_2(\cos\theta) - \frac{3}{70}\lambda_m \frac{11+10\gamma}{1+\gamma}\mathrm{Ac}^2\,\mathrm{Re}_d^3\, P_3(\cos\theta) + \ldots$. R is the radius of a spherical particle. Here $\lambda_m = \varphi(\gamma)$ and in particular for gas bubbles in the water environment $\lambda_m = 1/4$, for water droplets in the air $\lambda_m \to 5/48$, etc. Given that $\mathrm{We} = \mathrm{Ac}\,\mathrm{Re}_d^2$, we get

$$\xi(\cos\theta) = \lambda_m \mathrm{We}\, P_2(\cos\theta) - \frac{3}{70}\lambda_m \frac{11+10\gamma}{1+\gamma}\mathrm{We}^2\,\mathrm{Re}_d^{-1}\, P_3(\cos\theta) + \ldots. \tag{8.23}$$

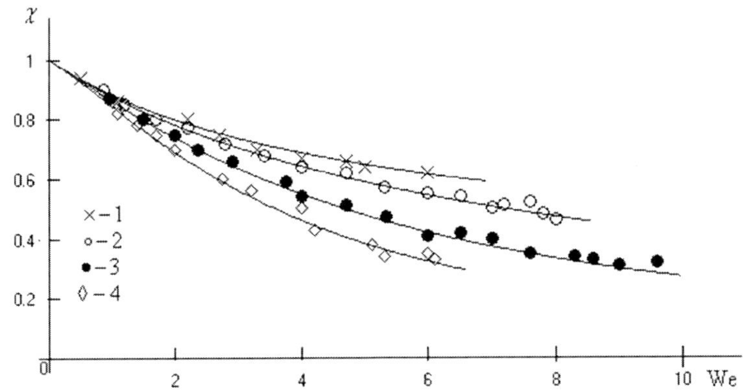

FIGURE 8.3 Dependence of the degree of deformation on the number for different numbers Mo, equal to: (1) 7; (2) 1.4; (3) 0.023; and (4) 0.0001.

As a parameter characterizing the deformation of droplets and bubbles, we consider the ratio of the semi-major axis of the ellipsoid to the large one, that is, $\chi = a/b$. Proceeding that for $\theta = 90°$, $P_2(\cos 90°) = -0.5$, $P_3(\cos 90°) = 0$, from equations (8.22) and (8.23) we can write $a = R(1+\alpha_0 \mathrm{We})$ and $\theta = 0°$, $P_2(\cos 0) = 1.0$, $P_3(\cos 0) = -1$ define $b = R(1+\beta_0 \mathrm{We} + \beta_1 \mathrm{We}^2)$, where $\alpha_0 = 0.5\lambda_m$, $\beta_0 = \lambda_m$, $\beta_1 = \frac{3}{70}\lambda_m \frac{11+10\gamma}{1+\gamma}\mathrm{Re}_d^{-1}$.

Then $\chi = a/b$, assuming that we finally obtain the expression for the dependence of the degree of deformation on the number in the form

$$\chi = \frac{1+\alpha_0 \mathrm{We}}{1+\beta_0 \mathrm{We} + \beta_1 \mathrm{We}^2} \tag{8.24}$$

This expression is typical for describing small deformations of drops. As a result of the use of experimental studies on the deformation of the shape of bubbles in a liquid medium [149] with different numbers and to extend the range of application of equation (8.6), the following expression is proposed

$$\chi = \frac{1+0.06\mathrm{We}}{1+0.2\mathrm{We} + \beta_1 \mathrm{We}^2} \tag{8.25}$$

where $\beta_1 = 0.005(2 - \ln \mathrm{Mo})$ with a correlation coefficient equal to $r^2 = 0.986$. In Figure 8.3, comparison of calculated values according to equation (8.25) with experimental data is given [149].

It follows from Figure 8.3 that expression (8.25) describes quite satisfactorily the deformation of drops and bubbles for the region of variation of $\mathrm{We} < 10$ and $10^{-4} \le \mathrm{Mo} \le 7$.

Based on the analysis of the experimental data [149], and the elements of the theory of elasticity, it is established that the relative volume deformation of droplets and bubbles is proportional to the volume of the drop, which can be written in the following form

$$\frac{\Delta\chi}{\chi - \chi_s} = -K_s \upsilon \, \Delta a \tag{8.26}$$

where K_s is the coefficient of volumetric elasticity, the steady-state value of the droplet deformation parameter up to the ellipsoidal shape. Passing to the limit, we have

$$\lim_{\Delta a \to 0} \frac{\Delta\chi}{\Delta a} = \frac{d\chi}{da} = -K_s \upsilon (\chi - \chi_s) \tag{8.27}$$

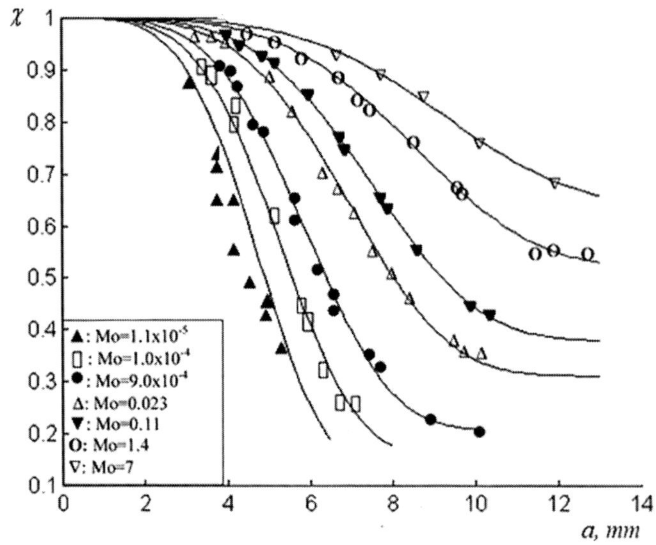

FIGURE 8.4 Comparison of experimental and calculated values of the degree deformation of drops and bubbles at different numbers Mo.

If the volume of the drop is constant, this equation is written in the form

$$\frac{d\chi}{da} = -\frac{\pi a^3}{6} K_s (\chi - \chi_s), \qquad \chi(a)|_{a\to 0} = 1 \tag{8.28}$$

The solution of equation (8.28) with the initial conditions is represented in the form

$$\chi = \chi_s + (1 - \chi_s)\exp(-\beta_s a^4) \tag{8.29}$$

where $\beta_s = \pi/24K_s$ is the coefficient determined on the basis of the experimental data.

Using experimental studies [149], the following approximations are obtained for the coefficients β_s and χ_s:

$$\beta_s = 1.475 \times 10^{-4} \mathrm{Mo}^{1/3}, \quad \chi_s = 0.5\mathrm{Mo}^{1/8} \tag{8.29a}$$

Figure 8.4 shows the dependence on the droplet size for different numbers Mo.

As follows from Figure 8.4, equation (8.29) describes quite satisfactorily the deformation of bubbles in a large range of the number Mo change, due to the introduction of empirical dependences (8.29a). Figure 8.5 shows the deformation surface of droplets as a function of size and number Mo.

8.4 FRAGMENTATION OF DROPLETS AND BUBBLES IN AN ISOTROPIC TURBULENT FLOW

The fragmentation of droplets and bubbles in an isotropic turbulent flow is an important factor for increasing the interfacial surface and the rate of heat and mass transfer in disperse systems. The mechanism of fragmentation of deformable particles is determined by many factors, among which it is important to note the following: (1) the effect of turbulent pulsations of a certain frequency on the surface of droplets and bubbles on the change in shape. Figure 8.5 shows the scheme of fragmentation of a single drop in the primary and secondary stages; (2) the boundary instability on the surface of the drop, determined by the turbulence of the boundary layer or the general

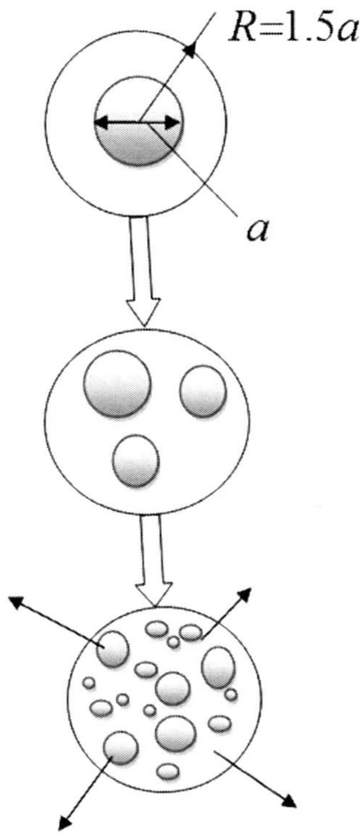

FIGURE 8.5 Schematic representation mechanism of fragmentation of single drops in the imaginary sphere.

instability as a result of reaching the size of the drop of the maximum value $a \geq a_{max}$; (3) the action of the external environment, in which the fragmentation of the droplets is defined as the equilibrium between external forces from the continuous phase (dynamic head) and the forces of surface stress, which resist the destruction of the drop. It should be noted that this condition can also characterize the deformation of the shape of droplets and bubbles; and (4) as a result of mutual elastic collision of drops with intensive mixing of the system. It is important to note that not every collision of droplets and bubbles leads to their merging and coalescence, and in an elastic collision, the drop can decay into fragments, thereby changing the spectrum of the size distribution, although there are no papers indicating the number of particles formed as a result of such decay.

A general overview of the fragmentation of droplets and bubbles is given in the paper [157], where problems related to the fragmentation frequency and the character of the particle size distribution function are considered, although the analysis of the maximum and minimum dimensions and the characteristic features of the effect of secondary crushing processes on the change in the multimodal droplet distribution function are not considered. Despite the many mechanisms for the fragmentation of droplets and bubbles, an important parameter characterizing this process is the frequency of fragmentation in a turbulent flow, the definition of which is devoted to many works [10,16,139,150,162,168,175,181]. In this paper [139], based on the analysis of the surface energy and the kinetic energy of the turbulent flow for the frequency of fragmentation of drops, the following expression is proposed

$$\omega(a) = C_1 a^{-2/3} \varepsilon_R^{1/3} \exp\left(-\frac{C_2 \sigma}{\rho_c \varepsilon_R^{2/3} a^{5/3}}\right) \tag{8.30}$$

The fragmentation of droplets in a turbulent flow in the absence of external mechanical energy is carried out for droplets whose dimensions are $a>a_{max}$. For turbulence conditions $\lambda>\lambda_0$, the equation for the change in the number of particles, taking into account their turbulent diffusion, is represented in the form

$$\frac{\partial N}{\partial t}=\frac{1}{r^2}\frac{\partial}{\partial r}\left(r^2 D_{TP}\frac{\partial N}{\partial r}\right) \tag{8.31}$$

$$t=0,\ r>R,\ N=N_0;\ t>0,\ r=R,\ N=0;\ t>0,\ r\to\infty,\ N=N_0$$

We give expressions for the turbulent diffusion of particles [6,145]

$$\lambda>\lambda_0,\qquad D_T=\alpha_0\mu_p^2\left(\varepsilon_R\lambda\right)^{1/3}\lambda$$
$$\lambda<\lambda_0,\qquad D_T=\alpha_0\mu_p^2\left(\varepsilon_R/\nu_C\right)^{1/2}\lambda^2 \tag{8.32}$$

where N is the concentration of particles per unit volume, λ is the scale of turbulent pulsations, and λ_0 the Kolmogorov scale of turbulence. Then equation (8.31) can be rewritten as

$$\frac{\partial N}{\partial t}=\frac{\alpha}{r^2}\frac{\partial}{\partial r}\left(\mu_R^2\varepsilon_R^{1/3}r^{10/3}\frac{\partial N}{\partial r}\right) \tag{8.33}$$

solution, which will be presented in the form

$$N(r,t)=\sum_{n=1}^{\infty}A_n J_2\left[\mu_n\left(r/R\right)^{1/3}\right]\exp(-\mu_n^2 t) \tag{8.34}$$

where $A_n=\frac{2}{R^2}\frac{\int_0^R N_0 J_2\left[\mu_n\left(r/R\right)^{1/3}\right]rdr}{J_1^2(\mu_0)}$, μ_n are the eigenvalues determined from the solution $J_2(\mu_n)=0$ of both $\mu_n=q_n\left(\varepsilon_R^{1/3}\mu_R^2\alpha/3R^{2/3}\right)$, $J_1(\mu_n)$, $J_2(\mu_n)$, are the Bessel functions of the first and second kind, and q_n is the coefficients characterizing the series (8.34). Since the series converges rapidly, we restrict the first term to the frequency of the fragmentation in the form [4,85–87]

$$\omega=4\pi R^2 D_{TP}\frac{\partial N}{\partial r}\Big|_{r=R}=C_1\varphi_0\left(\frac{\varepsilon_R}{a^2}\right)^{1/3}\exp\left[-C_2\left(\frac{\varepsilon_R}{a^2}\right)^{1/3}t\right] \tag{8.35}$$

where φ_0 is the volume fraction of particles at the initial instant of time. Time of fragmentation of drops can be taken in the form, although in works it is defined as $t\sim\sigma/(\rho_c\varepsilon_R a)$, and in work [6] is $t\sim\rho_c a\nu_c/\sigma$. Then, taking into account this equation and the reasoning given, the frequency of fragmentation is defined in the form

$$\omega(a)\approx C_1\varphi_0\left(\frac{\varepsilon_R}{a^2}\right)^{1/3}\exp\left(-C_2\frac{\sigma}{\rho_c\varepsilon_R^{2/3}a^{5/3}}\right)=C_1\varphi_0\left(\frac{\varepsilon_R}{a^2}\right)^{1/3}\exp\left(-C_2\mathrm{We}_T^{-1}\right) \tag{8.36}$$

For the case $\lambda<\lambda_0$, that is, for viscous flow using the second equation (8.31), the crushing frequency can be determined by the following expression

$$\omega(a)=C_{01}N_0a^3\left(\frac{\varepsilon_R}{\nu_c}\right)^{1/2}\exp\left[-C_{02}\frac{\sigma}{\left(\nu_c\varepsilon_R\right)^{1/2}a\rho_c}\right] \tag{8.37}$$

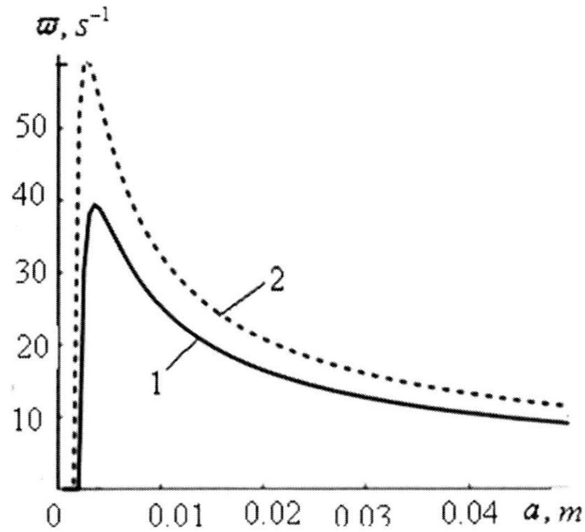

FIGURE 8.6 Dependence of the frequency of crushing of bubbles on their sizes for different energy dissipation values equal to: (1) $\varepsilon_R = 5.0\,\mathrm{m}^2 s^{-3}$ and (2) $\varepsilon_R = 10.0\,\mathrm{m}^2 s^{-3}$.

As follows from this equation, the frequency of fragmentation of droplets and bubbles in a viscous region or in a liquid medium is inversely proportional to the viscosity of the medium $\sim \nu_c^{-1/2}$, that is, with increasing viscosity, the crushing frequency decreases.

In the case of a dispersed system with a volume fraction of particles, the frequency of crushing is determined in the form

$$\omega(a) = C_{10}\frac{\varepsilon_R^{1/3}}{a^{2/3}(1+\varphi)}\exp\left(-C_{11}\frac{(1+\varphi)^2\sigma}{\rho_d a^{5/3}\varepsilon_R^{2/3}}\right) = C_{10}\frac{\varepsilon_R^{1/3}}{a^{2/3}(1+\varphi)}\exp\left(-C_{11}\frac{(1+\varphi)^2}{\mathrm{We}_T}\right) \tag{8.38}$$

Below are characteristic curves for the frequency of fragmentation of droplets depending on the equivalent size and specific dissipation of energy.

It follows from Figure 8.6 that with increasing specific dissipation of energy or intensity of turbulence, the frequency of fragmentation increases. The above studies on the deformation and fragmentation of droplets and bubbles are important for calculating the change in the number of particles per unit volume and interfacial surface in order to improve mass and heat transfer in liquid–liquid and gas–liquid systems. Moreover, in practical problems in polydisperse systems this problem should be related to the evolution of the droplet distribution function in size and time, which gives a broader interpretation of this problem. As follows from (8.36), the frequency of fragmentation of drops as $t \to \infty$ tends to zero; $\omega_D \to 0$, that is, the droplet size in the stream tends to minimum. However, many experimental data on fragmentation of droplets in a turbulent flow show that when the droplet size is set at a minimum size, which corresponds to the aggregative stability of the system. Taking this condition into account, the change in the average mass of the drop can be represented as:

$$\frac{dm}{dt} = -\omega_D m \tag{8.39}$$

where m is the reduced average mass of drops with respect to the minimum size. Setting for a spherical drop $m = \frac{\pi}{6}\rho_d(a - a_{\min})^3$, for an average droplet size, we obtain:

$$\frac{da}{dt} = -q_R(a - a_{\min}), \qquad t = 0,\ a = a_0 \tag{8.40}$$

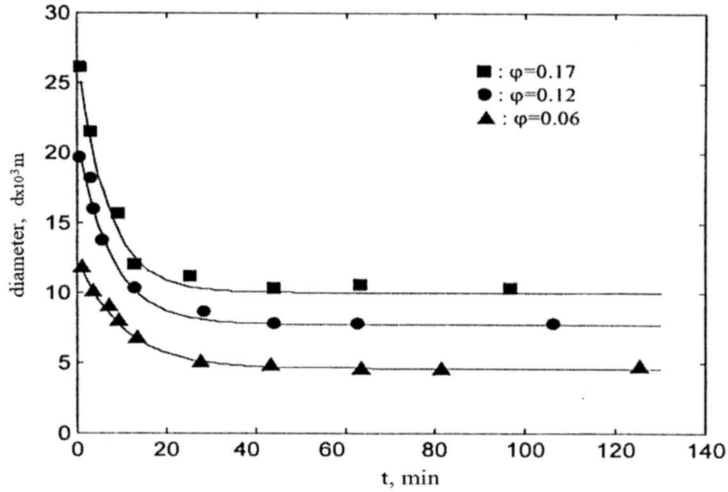

FIGURE 8.7 The change in the average droplet size in time, depending on the volume fraction.

where $q_R = {}^{\omega D}\!/_{3}$. Experimental studies of crushing oil droplets in an aqueous medium using mixing devices, taking into account $\varepsilon_R \sim \omega^3 D_M^2$, $K_R \sim 1/t$, yielded the following approximate expressions: $a_{\min} = 1.695.10^4\left(\omega D_M\right)^{-1.75}$, $a_0 = 1.75.10^4\left(\omega D_M\right)^{-1.5}$, $q_R = 0.0125\left({}^{\omega}\!/_{t}\right)^{1/2}$. Figure 8.7 compares the experimental data on the change in the average size of oil droplets in water with the calculated values (8.40), depending on their volume fraction and time t.

As follows from expression (8.36), the frequency of fragmentation depends on the volume fraction of the droplets in the flow, and using the experimental data in expression (8.40) for a_0, $a_{\min}$, and q_R, the following approximations can be obtained

$$a_0 = 104\varphi^{3/4}, \quad a_{\min} = 38\varphi^{3/4}, \quad q_R = 0.305\varphi^{2/5} \tag{8.41}$$

where φ is the volume fraction of droplets in the stream.

Figure 8.7 compares the experimental values of the average droplet size with respect to time and calculated from the solution (8.40) $a = a_{\min} + \left(a_0 - a_{\min}\right)\exp\left(-q_R t\right)$, for different values of $\varphi = 0.06$; 0.12; 0.17. As follows from Figure 8.7, with an increase in the volume fraction of droplets, their minimum size is set at a higher level. It should be noted that equation (8.40) describes quite satisfactorily the experimental data on the fragmentation of droplets in a turbulent gas flow [144] and in the case of crushing in mixing devices [10,11,150,175]. In the literature there are many empirical and semi-empirical formulas for determining the frequency of fragmentation

$$\begin{aligned}
&\omega\left(a\right) = C_3 a^{-2/3}\varepsilon_R^{1/3}\left(\frac{2}{\sqrt{\pi}}\right)\Gamma\left(\frac{3}{2}\frac{C_4\sigma}{\rho_c\varepsilon_R^{2/3}a^{5/3}}\right), \quad [157,187-190]\\
&\omega\left(a\right) = K_0\frac{\sqrt{8.2\varepsilon_R^{2/3}a^{2/3} - {}^{\sigma}\!/_{\rho_c a}}}{a}, \quad [168]\\
&\omega\left(a\right) = C_5\varepsilon_R^{1/3}\,erfc\left(\sqrt{C_6\frac{\sigma}{\rho_c\varepsilon_R^{2/3}a^{5/3}} + C_7\frac{\eta_d}{\sqrt{\rho_c\rho_d}\,\varepsilon_R^{1/3}a^{4/3}}}\right), \quad [157,191-193]\\
&\omega\left(a\right) = \frac{a^{5/3}\varepsilon_R^{19/15}\rho_c^{7/5}}{\sigma^{7/5}}\exp\left(-\frac{\sqrt{2}\sigma^{9/5}}{a^3\rho_c^{9/5}\varepsilon_R^{6/5}}\right), \quad [26,66];
\end{aligned} \tag{8.42}$$

The last equation determines the frequency of fragmentation of the droplets in the mixing devices and depends on the mixing parameters. For multiphase systems with a volume fraction of drops, the frequency of their fragmentation can be determined in the form [139,168]

$$\omega(a) = C_{10} \frac{\varepsilon_R^{1/3}}{a^{2/3}(1+\varphi)} \exp\left(-C_{11} \frac{(1+\varphi)^2 \sigma}{\rho_d a^{5/3} \varepsilon_R^{2/3}} \right) \tag{8.43}$$

The rate of fragmentation of drops in an isotropic turbulent flow is characterized by a rate constant, defined in the form [179]

$$\begin{aligned} \mathrm{Re}_d < 1, \qquad & k_R = A_0 \frac{\varepsilon_R^{1/3}}{a^{2/3}} \exp\left(-\frac{A_1 \sigma}{\rho_c \varepsilon_R^{2/3} a^{5/3}} \right), \\ \mathrm{Re}_d > 1, \qquad & k_R = A_0 \frac{\rho_c a^{2/3} \varepsilon_R^{1/3}}{\eta_c} \exp\left(-\frac{A_1 \sigma}{\rho_c \varepsilon_R^{2/3} a^{5/3}} \right) \end{aligned} \tag{8.44}$$

In principle, the expression given in parenthesis characterizes the ratio of surface energy $\left(E_\sigma \sim \pi a^2 \sigma / a \sim \pi a \sigma\right)$ to the energy of the turbulent flow $\left(\bar{E}_T \sim \pi a^2 (\Delta P_T),\ \Delta P_T = C_1 \rho_c (\varepsilon_R a)^{2/3}\right)$ and characterizes the efficiency of the crushing process

$$\frac{E_\sigma}{\bar{E}_T} \sim \frac{\sigma}{\rho_c \varepsilon_R^{2/3} a^{5/3}} \tag{8.45}$$

Analyzing equations (8.42 through 8.44), it can be noted that the frequency of breaking in an isotropic turbulent flow for a region $\lambda > \lambda_0$ is determined mainly by the turbulence parameters (specific energy dissipation, turbulent pulsation scale), the density of the medium, surface tension, and for viscous flow $\lambda < \lambda_0$—additionally and viscosity of the medium. It is important to note that the crushing of droplets and bubbles in an isotropic turbulent flow is preceded by a deformation of their shape, and for large numbers and sufficiently small numbers can take forms that cannot be described. The equilibrium condition between the surface forces and the external forces of the turbulent flow can also characterize the initial conditions for deformation of the particles. Experimental studies of the deformation of bubbles for different numbers We, Mo, and Re_d are given in the works [149,173], and theoretical studies are given in the works [21,24,91,180].

8.5 COALESCENCE OF DROPLETS AND BUBBLES IN THE FLOW

The coalescence of droplets and bubbles plays an important role in the course of various technological processes of chemical technology, and primarily in the reduction of the interfacial surface, in the separation and separation of particles of different sizes, accompanied by their deposition or ascent. The mechanism of the coalescence of droplets and bubbles is determined by the following stages: (1) mutual collision of particles with a certain frequency in a turbulent flow; (2) the formation of an interphase film between two drops and its thinning; (3) rupture of interfacial film and drainage of liquid from one drop to another, fusion and formation of a new drop. Mutual collisions of particles in the volume of the flow occur for various reasons: (a) due to convective Brownian diffusion of the finely dispersed component of the particles to the surface of a larger particle, characteristic mainly for laminar flow at low Reynolds numbers; (b) due to turbulent flow and turbulent diffusion; (c) due to the presence of additional external fields (gravitational, electric, electromagnetic, etc.). If the Kolmogorov scale of turbulence λ_0 is less or comparable to the droplet size in the region of viscous flow, then the process is accompanied by a turbulent wander, similar to Brownian, resulting in the appearance of turbulent diffusion. However, turbulent diffusion can be characteristic of large particle sizes at distances large, owing to the high intensity of turbulent pulsations and the inhomogeneity of the hydrodynamic field; (d) due to the effect of meshing as a result of convective small particles in the vicinity of the incident large particle. As a result of the deposition or emergence of large particles, due to the formation of a hydrodynamic trace, the capture of small particles by large particles increases substantially, which leads to gravitational coalescence if they fall along lines close to the center line. For the coalescence of droplets, an important role is played by the capture coefficient, which determines the deviation of the actual capture cross section from the geometric [40]

$$\vartheta = \frac{I}{\pi \left(L+R\right)^2 N_0 V_\infty} \tag{8.46}$$

where ϑ is the capture coefficient, the mass flux to the surface of the extracted particle, L is the characteristic scale of the distance, and V_∞ is the velocity of the medium unperturbed by the sphere of flow. The connection between the Sherwood number and the capture coefficient for convective diffusion has the form

$$\text{Sh} = \frac{1}{2}\text{Pe}\,\vartheta\left(1+\frac{R}{L}\right)^2 \tag{8.47}$$

The capture ratio is defined as [40]

$$\vartheta \approx \frac{4}{\text{Pe}}\left(1-\frac{N_L}{N_0}\right)b_*\left(1+b_0\text{Pe}\right),\ \text{Pe}<<1;\ \vartheta \approx \left(1-\frac{N_L}{N_0}\right)\text{Pe}^{-2/3}\left(1+0.738\text{Pe}^{-1/3}\right),\ \text{Pe}>>1 \tag{8.48}$$

Here N_L is the concentration of particles on the surface of a sphere of radius $r = L$, b_i, is the coefficients; (e) due to the inhomogeneity of the temperature and pressure fields, which contribute to the appearance of forces proportional to the temperature and pressure gradients and acting in the direction of decreasing these parameters. For the finely dispersed component of the dispersed flow, as a result of these forces, their migration is due to thermal diffusion and barodiffusion, which also contributes to their collision and coalescence; (f) in addition to these phenomena, physical phenomena (evaporation, condensation of droplets), accompanied by the appearance of a hydrodynamic repulsive force (the Fassi effect), evaporating droplets due to evaporation (Stefan flow), or the condensation growth of a droplet are caused by the appearance of a force acting in the opposite direction. Despite the various mechanisms, the coalescence of droplets and bubbles in the turbulent flow forms the basis for calculating the phenomena occurring in dispersed gas–liquid systems (columns, gas–liquid reactors) and liquid–liquid (mixing devices) in chemical technology processes [10,11,140,143,145,151,152,159,162,165,194], since these phenomena determine the magnitude interphase exchange surface.

The coalescence of droplets and bubbles is characterized by the following stages: (1) the approach and collision of droplets of different sizes in a turbulent flow with the formation of an interfacial film between them. It should be noted that the transfer of droplets in a poly disperse medium is determined mainly by the hydrodynamic conditions and the intensity of turbulence in the flow. Under conditions of isotropic turbulence, the collision frequency of droplets depends on the specific energy dissipation of the turbulent flow and on the properties of the medium and the dispersed phase [10,139,165,186,194]. As a result of the collision and fixation of two drops with dimensions a_1 and a_2 between them, an interfacial film of circular cross section is formed (Figure 8.8), whose radius can be determined in the form

$$R_K = \left[\frac{3\pi}{4} P_m\left(k_1+k_2\right) a_r\right]^{1/3} \tag{8.49}$$

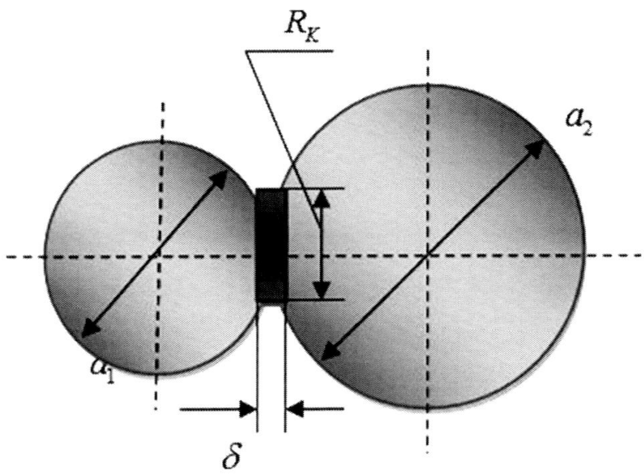

FIGURE 8.8 Collision of two drops with the formation of interfacial film.

where P_m is the maximum compressive pressure, k_1, k_2 are the elasticity coefficients of two drops, $a_r = {a_1 a_2}/{(a_1+a_2)}$ is the average droplet size, and a_1, a_2 are the diameters of two drops. In this paper [139], the expressions for the hydrodynamic compression pressure in a turbulent flow are defined $P_m = \pi\, a_r^2 \rho_C \bar{U}^2$, as for isotropic turbulence is $P_m \sim \rho_c \varepsilon_R^{2/3} \frac{(a_1 a_2)^2}{(a_1+a_2)^{4/3}}$.

(2) Thinning and rupture of the interfacial film under the action of external forces and deforming (wedging) stress with subsequent drainage of liquid from one drop to another, coalescence, and enlargement of droplets. The rupture of the interfacial film promotes the coalescence of smaller droplets into larger ones, which contributes to a reduction in the total number of droplets in the volume and to a violation of sedimentation stability as a result of the dropping of large particles from the general spectrum of sizes and the nature of their distribution.

When two drops are fixed, as a result of their collision, the interfacial film formed under the action of various kinds of forces (gravitational, dynamic, capillary, molecular, etc.) becomes thinner to a certain critical thickness and breaks with the further merging (Figure 8.8).

In principle, the interfacial surface formed between the two drops has a finite curvature, depending on the radii of the droplets and the alternating surface tension. Assuming that the plane flow of an insignificant curvature of the circular cross section is laminar, the equation of momentum transfer in cylindrical coordinates is written in the form [194]

$$-\frac{\partial P}{\partial r} + \frac{\eta}{gr^2}\frac{\partial^2 V_r}{\partial \theta^2} + \frac{\eta}{g}\frac{\partial^2 V_r}{\partial x^2} = 0$$

$$-\frac{1}{r}\frac{\partial P}{\partial \theta} + \frac{2\eta}{gr^2}\frac{\partial V_r}{\partial \theta} = 0 \qquad (8.50)$$

$$\frac{\partial V_x}{\partial x} + \frac{1}{r}\frac{\partial (rV_r)}{\partial r} = 0$$

where P is the pressure in the film, the components of the flow velocity in the film, and θ polar angle. One of the important boundary conditions for the solution of these equations is

$$x = \delta, \quad -\eta\frac{\partial V_r}{\partial r} = \frac{d\sigma}{dr} + \frac{1}{R_K \sin\theta}\frac{\partial \sigma(\cos\theta)}{\partial \theta} \qquad (8.51)$$

Determining the presence of convective flow in a liquid film according to the Marangoni effect [195–198], we note that this problem, taking into account the presence of surfactants, phase inversion and the Marangoni effect for the one-dimensional pressure distribution in the film has been solved numerically [196]. The Marangoni effect can be considered as a thermocapillary flow due to temperature changes in the film and convective flow due to a change in concentration or surface tension. As a result of simple, time-consuming transformations of the system of equations (8.50), we obtain an equation for changing the thickness of the interphase film in time in the form [194,198]

$$\frac{d\delta}{dt}=\frac{2g\Delta P}{3\pi\eta R_K^4}\delta^3+\frac{2\delta^2}{3\eta R_K^2\sin\theta}\frac{\partial\sigma}{\partial\theta} \tag{8.52}$$

To ΔP determine the following expression $\Delta P=\left(P_D+P_K\right)\pi R_K^2+\Pi$, P_D, P_K, is where the dynamic and capillary pressure $\left(P_K={}^{2\sigma}\!/_{g\delta}\right)$ acting in the film, Π is the wedging pressure, determined for both the spherical droplet $\Pi=-\frac{AR_K^2}{6\delta^2}$ and the deformable droplets $\Pi=-\frac{AR_K^2}{6\delta^3}$, and A is the Van der Waals-Gammaker constant ($A\sim10^{-21}J$) [197,199]. Taking into account the foregoing, by transforming (8.51), the equation of thinning of the interphase film can be represented in the form

$$\frac{d\delta}{dt}=b_1\delta^3+b_2\delta^2-b_3\delta\,,\; t=0,\;\delta=\delta_0 \tag{8.53}$$

where $b_1=\frac{2gP_D}{3\eta R_K^4}$, $b_2=\frac{2}{3\eta R_K^2}\left(2\sigma+\frac{1}{\sin\theta}\frac{\partial\sigma}{\partial\theta}\right)$, and $b_3=\frac{1}{9}\frac{Aa_rg}{\pi\eta R_K^4}$, Various analytical solutions of the given equation are proposed in the works:

1. For thin films one can assume that $P_D << P_K-\Pi/\pi R_K^2$. In this case, the solution of equation (8.53) is represented as

$$\delta(t)=\frac{\delta_0\exp\left(b_3t\right)}{1+\beta_1\delta_0\left(\exp\left(b_3t\right)-1\right)} \tag{8.54}$$

where $\beta_1=\frac{b_2}{b_3}=\frac{6\pi R_K^2}{Aa_rg}\left(2\sigma+\frac{1}{\sin\theta}\frac{\partial\sigma}{\partial\theta}\right)$. For very thin films, solution (8.53) is represented as

$$\delta(t)\approx\delta_0\exp\left(-\beta_3t\right) \tag{8.55}$$

where $\beta_3=\frac{2}{3}\frac{P_Dg\delta_0}{\eta_cR_K^2}$. In the case of deformable drops, we have

$$\delta(t)\approx\delta_0-\beta_2t,\;\beta_2=\frac{Ag}{9\pi\eta\delta_0R_K^2} \tag{8.56}$$

2. For thick films we can assume that $(P_D+P_K)>>{}^{\Pi}\!/_{\pi R_K^2}$. It is important to note that the main forces that determine the rupture of interfacial film at large thicknesses are forces due to velocity pulsations, that is, hydrodynamic forces. The solution (8.53) is represented in the form

$$\left(P_D+P_K\right)>>{}^{\Pi}\!\Big/_{\pi R_K^2},\; P_D>>P_K,\;\delta(t)=\frac{\delta_0}{1+b_2\delta_0t} \tag{8.57}$$

The preceding particular solutions (8.54 through 8.57) can be used in practical calculations of the thickness of the interfacial film for particular cases. As follows from equation (8.54), the Marangoni effect is a partial correction to the coefficient of surface tension in the coefficient b_2, although it can have a significant effect on the nature of the flow and on the velocity distribution in the interfacial film. The presence of two-dimensional pressures and the complexity of its distribution on the

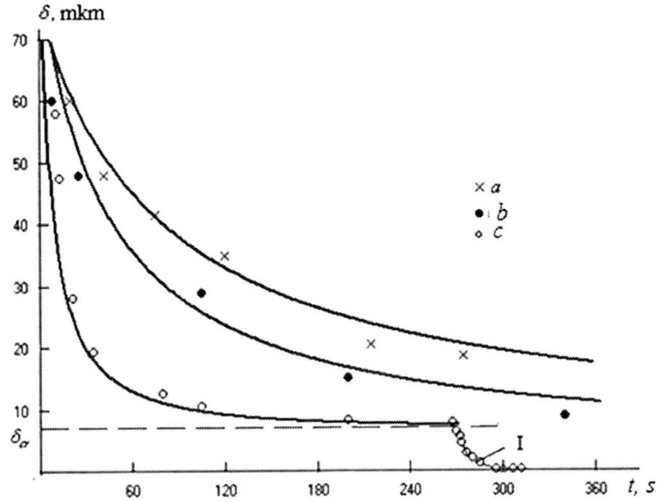

FIGURE 8.9 Comparison of the calculated (8.54) and experimental values of the thickness of the interfacial film versus time at different concentrations (g/L) of the demulsified: (a) = 0.2; (b) = 0.5; and (c) = 1.0 (I—calculation by the equation [8.55]).

surface of a liquid film, taking into account equations (8.50), show that when it is thinned, the presence of the Marangoni effect to some extent helps stabilize the coalescence processes in the liquid–liquid system, that is, has a retarding effect of tearing the film. As follows from equation (8.54) and the formula for determining the coefficient, the Marangoni effect contributes to the temporary stabilization of the interfacial film, since, at any point where the film becomes thinner due to the influence of external forces, a local increase in surface tension appears, counteracting thinning. The process of thinning and rupturing the film has a random (spontaneous) character and, as noted in the paper [200], the probability of its rupture is inversely proportional to its thickness. In the destruction of the interfacial film, an important role is played by the hydrodynamic forces that generate turbulence and, above all, high-frequency turbulent pulsations (P_D) that promote mechanical weakening of the interphase film and intermolecular bonds between its components, a decrease in the strength and destruction of the film as a result of their deformation (local stretching and compression) conditions of mutual effective collision due to an increase in the collision frequency. In the case of a one-dimensional pressure distribution in a film with the participation of surfactants, this problem is also solved in the work [201,202]. Experimental data on thinning of the interfacial film are given in the works [201,203]. Figure 8.9 compares the experimental and calculated values by equation (8.54) of thinning the film thickness, and after reaching the thickness of the critical value, the calculation by formula (8.55) is the most acceptable.

After the rupture of the interfacial film, the liquid drains from one drop to the other, the velocity of which is determined in the works [194,204].

8.5.1 Coalescence of Drops in an Isotropic Turbulent Flow

The coalescence of droplets and bubbles is described by the diffusion mass transfer equations in an isotropic turbulent flow. Turbulent diffusion of particles characterizing their pulsating motion is determined by the following expression [12]:

$$D_{TP} = \mu_R^2 D_T \tag{8.58}$$

where μ_R^2 is the degree of entrainment of particles by turbulent pulsations, which depends on the droplet size and the frequency of turbulent pulsations. If we select a drop and draw an imaginary sphere around it with a radius $R = a$, then we can assume that any drop passing through this sphere necessarily encounters a central drop. In this case, the frequency of collisions is analogous to the particle flux per unit time and can be set equal to the specific flux of particles flowing through this surface:

$$w = \pm\pi\, a^2 D_{TP} \left.\frac{\partial N}{\partial r}\right|_{r=R} \tag{8.59}$$

The minus sign corresponds to crushing, and the plus sign indicates the coalescence of the drops and their transfer through the surface. The coalescence process can be considered as a mass-exchange process, and therefore, the change in the number of drops for $w = \pm\pi\, a^2 D_{TP} \left.\frac{\partial N}{\partial r}\right|_{r=R}$, taking into account the expressions for the diffusion coefficients Pe << 1, can be written as:

$$\frac{\partial N}{\partial t} = \frac{\alpha}{r^2}\frac{\partial}{\partial r}\left(\mu_R^2 \varepsilon_R^{1/3} r^{10/3} \frac{\partial N}{\partial r}\right) \tag{8.60}$$

With boundary conditions

$$t = 0,\ r > R,\ N = N_0,\ t > 0,\ r = R,\ N = 0,\ r \to \infty,\ N = N_0 \tag{8.61}$$

The solution (8.60) with the boundary conditions (8.61) is represented as [9,11]

$$N(r,t) = \sum_{n=1}^{\infty} A_1 J_2\left[\mu_n \left(r/R\right)^{1/3}\right] \exp\left(-\mu_n^2 t\right) \tag{8.62}$$

where $A_n = \frac{2}{R^2}\frac{\int_0^R N_0 J_2\left[\mu_n^2 (r/R)^{1/3}\right] r\,dr}{J_1^2(M_n)}$, μ_n are the eigenvalues determined from the solution of the equation $J_2(q_n) = 0$, as $\mu_n = q_n(\varepsilon_R^{1/3}\mu_R^2\alpha/3R^{2/3})$, $J_1(\mu_n)$, $J_2(\mu_n)$, are Bessel functions of the first and second order. Since the series (8.62) converges rapidly, it is sufficient to use the first term and determine the frequency of coalescence in the form:

$$\omega_k = -D_{TP}\left.\frac{\partial N}{\partial r}\right|_{r=R} \approx C_1\left(\frac{\varepsilon_R}{a^2}\right)^{1/3} \exp\left(-t/T_1\right) \tag{8.63}$$

where $C_1 = 16\mu_R^2\alpha\phi_0$, $C_2 = 8.5\mu_R^2\alpha$, are the coefficients, $\varphi_0 = N_0\frac{\pi a^3}{6}$ is the initial volume fraction of drops in the stream, N_0 is the initial number of drops per unit volume, and $T_1 = C_2^{-1}(a^2/\varepsilon_R)^{1/3}$ is the total coalescence time of the drops.

The frequency of coalescence of droplets at very small values of time $t/T_1 < 1$ and at high coalescence rates assumes a steady-state value equal to $\omega_k \sim (\varepsilon_R/a^2)^{1/3}$.

For the case $\lambda < \lambda_0$, using the expression (8.61) in equation (8.60), we obtain the following solution

$$N(r,t) = N_0\left(1 - \left(R/r\right)^3\right)\exp\left(-\alpha\left(\varepsilon_R/\nu_C\right)^{1/2} t\right) \tag{8.64}$$

and the collision frequency in the form

$$\omega_k = 72\varphi_0\alpha\mu_R^2\left(\varepsilon_R/\nu_C\right)^{1/2}\cdot\exp\left(-\alpha\left(\varepsilon_R/\nu_C\right)^{1/2}t\right) \tag{8.65}$$

It follows from equation (8.65) that for $\lambda<\lambda^0$ the viscosity of the medium exerts a significant influence on the coalescence frequency, and the higher the viscosity, the lower the frequency, $\varepsilon_R \sim (\varepsilon_R/\nu_c)^{1/2}$. It should be noted that the particle diffusion coefficient is significantly influenced by their size and flow direction. Thus, for example, the time of their relaxation and the rate of precipitation affect the diffusion coefficient of the particles, as shown in [4]. Work [156] considers the coagulation of particles in an inhomogeneous field, although the solutions given above correspond to a homogeneous symmetric distribution of the droplet concentration.

The theory and description of the physical phenomena of coalescence and crushing of droplets and bubbles in an isotropic turbulent flow is a deterministic-stochastic problem, including the solution of problems of deterministic nature (hydrodynamic problems in a turbulent flow) and stochastic nature, associated mainly with random abrupt changes in particle sizes and their distribution in size and time. At the same time, the problems of coalescence and fragmentation can be considered from the standpoint of the theory of "birth and death" of particles [205] and the theory of catastrophes [206], since the basis of these processes is the formation of new particles associated with the disappearance or "death" of old particles, the emergence or "birth" of new particles, and are characterized by the loss of aggregative and sedimentation stability of the dispersed medium. As follows from formulas 8.60 through 8.63, the coefficient of surface tension and energy dissipation and the physical properties of the medium and particles are important parameters that provide the aggregative stability of a liquid–liquid or liquid–gas dispersion medium to crushing, deformation, and coalescence. In an isotropic turbulence flow, the ratio is σ/ε_R, since these parameters are the main ones in estimating the maximum (8.6 through 8.8) and minimum (8.14 through 8.20) droplet sizes, in calculating the frequencies of crushing and coalescence. In some studies, the value of the maximum size of droplets and bubbles is associated with the critical Weber number as follows: $a_{\max} \sim \mathrm{We}_{\mathrm{cr}}^{3/5}$, where $\mathrm{We}_{\mathrm{cr}} = 1.18$ [171], $\mathrm{We}_{\mathrm{cr}} = 2.6$ [206], $\mathrm{We}_{\mathrm{cr}} = 2.7-7.8$ [207]. Although in most cases the formulas given in the literature for estimating the minimum and maximum sizes of droplets and bubbles are formal in nature, since their comparison with the experimental values gives a big error, however, for a certain selection of unknown coefficients (for example, the parameters in equations (8.6 through 8.8) and (8.14 through 8.20)), it is possible to achieve satisfactory convergence in these formulas for the calculated and experimental values.

The imbalance between the phenomena of coalescence and breaking, associated with changes in the regime and external parameters, contributes to a shift in the process in one direction or another. However, for sufficiently large values of the angular velocity of mixing, the process shifts both toward breaking and coalescence, which is associated with an increase in the specific dissipation of energy $\varepsilon_R \sim n^3$ and the frequency of fragmentation $\omega(a) \sim n^{1/2}$ [11,175]. An increase in the rate of fragmentation of droplets contributes to an increase in the number of drops per unit volume and, correspondingly, to an increase in the probability of collision and enlargement of the drops. This fact is clearly observed in the processes of liquid extraction, separation, and separation of emulsions [208]. In practice (in particular, in the oil–water system), the phenomena of coalescence and fragmentation of drops can be complicated by the presence of various films on the surface of droplets that inhibit the process [202]. The problems of coalescence and fragmentation of drops, characterized by a complex random spasmodic behavior, conceal many complexities and subtleties. A deeper analysis of these phenomena on the basis of mathematical regularities of transport phenomena allows us to standard way calculate such systems in a certain approximation, as continuous ones, with an infinitesimal jump.

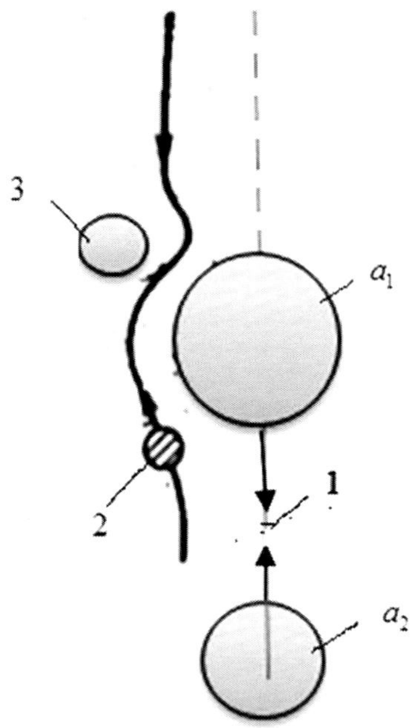

FIGURE 8.10 Scheme of gravitational coagulation: (1) inertial deposition of particles; (2) capture of small particles large; and (3) attraction to the trail of a large particle.

8.5.2 Gravitational Coagulation of Particles

Gravitational coagulation is carried out as a result of inertial phenomena, sedimentation by capture and pulling up of small particles by large ones. The inertial forces act in the direct collision of relatively large (more than 0.1 μm) polydisperse particles. Most often, fine particles are precipitated by gripping and pulling while moving along curved trajectories. A special case of coagulation in dispersed media is gravitational coagulation, which occurs as a result of the difference in the rates of deposition of particles in a polydisperse system. As a result of gravitational coagulation, larger particles are formed. Gravitational coagulation plays an important role in the process of separation of suspensions and emulsions by sedimentation, as a result of which solid aggregates are formed, and the rate of precipitation of such aggregates increases with the growth of their sizes. Intensive turbulence, at which the rate of dissipation of turbulent energy becomes equal to a few tenths, can ensure the collision efficiency of particles of the same order as gravitational coagulation. For coarse-dispersed emulsions and suspensions, the diffusion process proves to be very slow and cannot by itself ensure the convergence of particles. The latter mechanism also plays an important role in the enlargement of gas bubbles rising in the liquid column. To assess the influence of various factors that determine the rate of gravitational coagulation, let us consider the motion of one isolated particle relative to the total flow carrying other (smaller) particles moving along with the flow (Figure 8.10).

The size of a small particle is assumed to be so large as to neglect Brownian motion. For arbitrary values of the number in the absence of hydrodynamic interaction between particles, the equation for the motion of a single particle in the gravitational field is represented as

$$\frac{dV_1}{dt}=\gamma\frac{dV_1}{dt}-\beta\left(1+\zeta\left(\mathrm{Re}_d\right)\right)\left(V_1-U\right)+g,\ \ \gamma=\frac{3}{2}\frac{\rho_c}{\rho_d},\ \ \beta=\frac{1}{\tau_{p1}},\ \tau_{p1}=\frac{1}{18}\frac{\Delta\rho a_1^2}{\eta_c} \tag{8.66}$$

The function $\zeta(\mathrm{Re}_d)$ characterizes the degree of deviation of the motion of particles from the Stokes regime.

At moderate values of the inertia index, the first term on the right-hand side is relatively small and can be neglected, as a result of which the given equation for the first and second particles (Figure 8.10) for Stokes particles $\mathrm{Re}_d < 1$ transforms to the form

$$\frac{dV_1}{dt} = -\beta_1(V_1 - U) + g,\ t = 0,\ V_1 = V_{10}$$
$$\frac{dV_2}{dt} = -\beta_2(V_2 - U) + g,\ t = 0,\ V_2 = V_{20} \tag{8.67}$$

The solution of these equations is represented in the form

$$V_1 = (U + g\tau_{p1})\left(1 - \exp\left(-{t}/{\tau_{p1}}\right)\right) + V_{10}\exp\left(-{t}/{\tau_{p1}}\right)$$
$$V_2 = (U + g\tau_{p2})\left(1 - \exp\left(-{t}/{\tau_{p2}}\right)\right) + V_{20}\exp\left(-{t}/{\tau_{p2}}\right) \tag{8.68}$$

These solutions refer to the case of capture by a large particle of a small particle along the path of their succession, located on one axis. The exponential term representing the "acceleration" process of a particle tends to zero at $t >> \tau_p$, which leads to stationary solutions $V_1 = V_{10},\ V_2 = V_{20}$. The ratio of the relaxation times of a part for equal values of the density and viscosity of the medium is determined as

$$\frac{\tau_{p1}}{\tau_{p2}} = \frac{a_1^2}{a_2^2} \tag{8.69}$$

The capture coefficient is chosen, starting from the condition in the cylinder with the axis located at the center of the large drop and the radius equal to R_z, and the capture of the particle takes place. Then there is a value $R_z = R_{z0}$ at which a small particle with a large particle will bypass it, if $R_z < R_{z0}$. Having determined the value R_{z0}, the coefficient of capture of large particles by small particles (Figure 8.10) can be expressed by the formula

$$\varepsilon_z = \frac{R_{z0}^2}{R_1^2} \tag{8.70}$$

The paper presents a lot of experimental data, borrowed from the literature, devoted to the dependence of the capture coefficient on the Stokes number. Such curves, depending on the number, are shown in Figure 8.11.

The experimental data are well described by the following relationship

$$\varepsilon_z = A(\mathrm{Re}_d)\mathrm{erf}\left(B(\mathrm{Re}_d\, k_0)\mathrm{St}\right) \tag{8.71}$$

where $A(\mathrm{Re}_d) = 0.65 + 0.004\,\mathrm{Re}_d$, $\mathrm{St} = \frac{1}{18}\frac{\rho_d a_1^2}{\eta_c a_2^2}\upsilon(a_1)$, $erf(x) = 1 - \exp(-x - \frac{\sqrt{\pi}}{2}x^2)$, is the probability integral. For $\mathrm{Re}_d \to \infty$ and large numbers St, the capture coefficient is $\varepsilon_z \to 1$. The change in the number of particles in the capture of small particles by precipitating coarse particles is expressed by the equation

$$\frac{dN}{dt} = -K_g N N_k \tag{8.72}$$

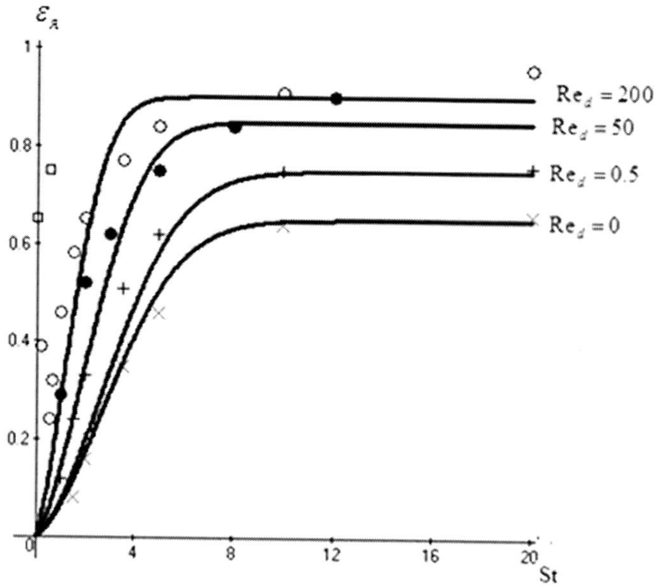

FIGURE 8.11 Dependence of the capture coefficient of particles on the Stokes number.

Here, N is the number of small particles per unit volume, N_k is the number of large particles per unit volume, and K_g is the coefficient of gravitational coagulation of particles, equal to the volume from which a large particle extracts small particles per unit time. If the number of large particles is constant, the solution of this equation will be written as

$$N = N_0 \exp\left(-\int K_g N_k dt\right) \tag{8.73}$$

After the collision and gravitational coagulation in the case of the merging of these droplets, a new drop is formed, the size being equal to $a_s = (a_1^3 + a_2^3)^{1/3}$. The sedimentation rate of a drop in size a_s can be determined from the solution of equation

$$\frac{dV_s}{dt} = -\beta_s\left(1 + \zeta\left(\mathrm{Re}_d\right)\right)\left(V_s - U\right) + g, \qquad t = 0,\ V_s = V_{s0} \tag{8.74}$$

where $\beta_s = \tau_p^{-1}$. When large and small drops merge, sufficiently large droplets can be obtained, the rate of subsidence of which differs from the Stokes mode in view of the nonlinear dependence $\zeta\left(\mathrm{Re}_d\right)$ on the number Re_d. If you enter a number $\mathrm{Re}_d = \frac{a_s|V_s - U|}{\nu_c}$, then the equation can be written in the form

$$\frac{d\,\mathrm{Re}_d}{dt} = -\beta_s\left(1 + \zeta\left(\mathrm{Re}_d\right)\right)\mathrm{Re}_d + \frac{g a_s}{\nu_s} \tag{8.75}$$

It should be noted that gravity coalescence is characteristic for the processes of separation of water droplets from oil in oil emulsion, and coalescence after the impact of droplets occurs according to the above mechanism, that is, thinning of the interphase film, its rupture, and the merging of droplets.

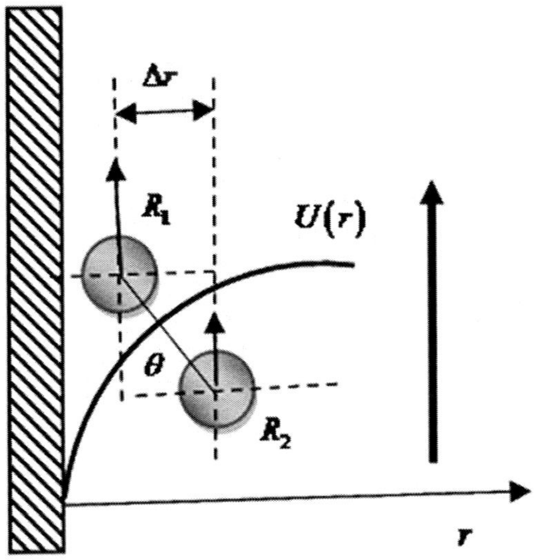

FIGURE 8.12 Scheme of gradient coagulation.

8.5.3 Gradient Coagulation of Particles

Gradient coagulation of particles is characteristic for laminar flow or for the boundary layer of a turbulent flow. Gradient coagulation is a consequence of the flow of particles in a gradient flow field, where the presence of a velocity gradient can lead to collisions between two particles of size R_1 and R_2 moving with different velocities at a distance from each other (Figure 8.12).

Consider the following assumptions:

- The particle sizes are such that the effect of diffusion can be neglected;
- There is no hydrodynamic interaction of the particles, leading to the curvature of their trajectories;
- There is no influence of external fields of non-hydrodynamic nature on particle motion;
- Magnus lift forces do not act on drops in the gradient field; and
- The particle sizes are such that they follow laminar lines currents of fluid.

Consider two particles in a gradient laminar flow and moving parallel to the wall at a velocity $V(r)$ at a distance Δr from each other. Assuming that $V(r) \approx \frac{\partial V}{\partial r}\Delta r$, we define the conditions for the collision of these particles in the form ($R_k = R_1 + R_2$ is the coagulation radius)

$$\Delta r \le \left(R_1 + R_2\right)\sin\theta \tag{8.76}$$

The number of collisions per unit time is

$$I = N_0 \int V(r) 2\left(R_1 + R_2\right)\cos\theta d\left(\Delta r\right) = \frac{4}{3} N_0 \frac{\partial V}{\partial r}\left(R_1 + R_2\right)^3 \tag{8.77}$$

Flow velocity distribution for laminar flow

$$V(r) = V_{\max}\left(1 - \frac{r^2}{R_0^2}\right) \tag{8.78}$$

where V_{max} is the maximum value of the flow velocity at the center of the pipe, R_0 is the radius of the pipe. Then, assuming that,

$$\left.\partial V \big/ \partial r\right|_{r = R_1 + R_2} = -2V_{max} \frac{R_1 + R_2}{R_0^2} \tag{8.79}$$

the number of collisions per unit time will be determined as

$$I = -\frac{8}{3} N_0 V_{max} \frac{(R_1 + R_2)^4}{R_0^2} \tag{8.80}$$

For particles of the same size $R_1 = R_2 = R$, we have

$$I = -\frac{128}{3} N_0 V_{max} \frac{R^4}{R_0^2} \tag{8.81}$$

In these expressions, the minus sign indicates an increase in the flow velocity as it approaches the center of the tube, and therefore, in calculating the number of collisions, an absolute value should be taken. Thus, the number of collisions increases with increasing particle size and flow velocity. Comparing the number of collisions with Brownian coagulation, we have

$$\beta = \frac{I_{grad}}{I_{Br}} = \frac{32}{3\pi} \frac{V_{max}}{DR_0^2} R_k^3 = \frac{16}{3} \mathrm{Pe} \left(\frac{R_k}{R_0} \right)^3 \tag{8.82}$$

Here, $\mathrm{Pe} = 2R_0 V_{max} / D$ is the Peclet number. Thus, if the number $\mathrm{Pe} \gg 1$, then gradient coagulation prevails over Brownian particle coagulation.

The solution of the problem of gradient coagulation of particles in a turbulent flow presents some difficulties, since curvatures of particle trajectories are possible in the volume of a turbulent boundary layer. This is primarily due to the complex structure of the turbulent boundary layer, consisting of: (1) from the viscous sublayer region $y_+ \le 5$; (2) from the transition layer, $5 \ge y_+ \le 10$; and (3) from the region of developed turbulence $y_+ > 10$.

Analogously to the foregoing, let us consider the gradient coagulation of particles in the viscous sublayer of a turbulent boundary layer,

$$U = \frac{U_*^2}{\nu_c} y, \quad \Delta y \le (R_1 + R_2) \sin\theta, \quad \frac{\partial U}{\partial y} = \frac{U_*^2}{\nu_c} \tag{8.83}$$

Then the total flux of particles is defined as

$$I = \frac{4}{3} N_0 \frac{U_*^2}{\nu_c} (R_1 + R_2)^3 \tag{8.84}$$

For particles of the same size $R_1 = R_2 = R$, this expression is converted to the form

$$I = \frac{32}{3} N_0 \frac{U_*^2}{\nu_c} R^3 \tag{8.85}$$

As follows from these formulas, in the viscous sublayer with increasing viscosity of the liquid, the collision frequency decreases. Obviously, these expressions can be valid provided that $R_1 + R_2 < y_+$.

8.6 DRAG COEFFICIENTS OF SOLID PARTICLES, DROPLETS, AND BUBBLES IN THE FLOW

One of the important parameters that determine the migration of particles in the flow is the drag force arising as a result of friction between the surface of the particle and the flow around it. The value of the drag coefficient for spherical particles in the classical sense is determined in the form

$$C_D = \frac{F_{TP}}{\left(\rho_c U_\infty^2 / 2\right)\left(\pi a^2 / 4\right)} \quad (8.86)$$

where C_D is the drag coefficient of particle, F_{TP} is friction force, and U_∞ is flow velocity far from the particle. The frictional force F_{TP} depends on the configuration of the streamlined surface, the nature of the hydrodynamic flow, the particle size, and the properties of the medium, and also the nature of the mass exchange processes occurring on the particle surface (evaporation condensation, crystallization, etc.). Numerous experimental studies to determine the coefficient of resistance led to the so-called standard Rayleigh curve for a single solid sphere moving at a constant velocity in a stationary isothermal fluid. For bodies of complex configurations, the theoretical definition of the resistance coefficient presents great difficulties, and in most cases its evaluation is carried out on the basis of experimental studies and empirical correlations. This primarily applies to the determination of the drag coefficients of deformable particles (drops, bubbles) for large values of the number, with the corresponding flow patterns (separation of the boundary layer from the surface, hydrodynamic trail, resistance crisis, etc.) and for single particles characterized by nonsphericity of the surface. Particle resistance coefficients depend on the nature of their hydrodynamic flow around depending on the number and their configuration. Figure 8.13 shows the characteristic resistance curves for solid particles of different shapes.

In chemical technology, there are many processes involving or with the formation of dispersed particles, where the particle sizes in the flow vary from zero to some limiting value (crystallization, condensation, swelling, granulation), and vice versa, where the particles can completely disappear as a result of physical and chemical transformations (dissolution, sublimation, evaporation).

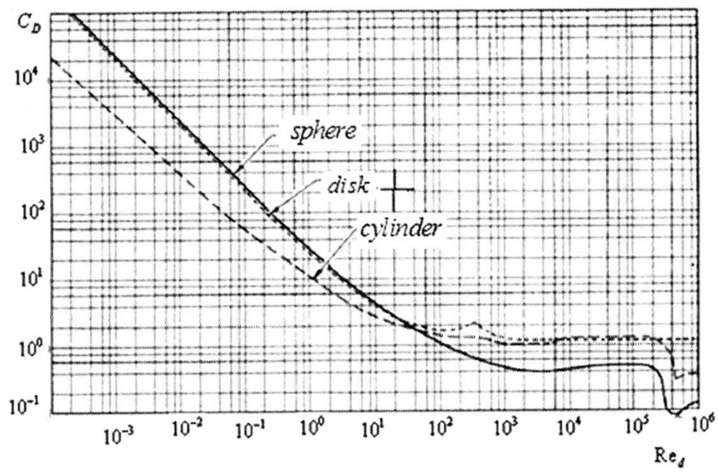

FIGURE 8.13 Characteristic curves of the change in the resistance coefficients spherical, cylindrical particles and a disk.

TABLE 8.2
Numbers for Similarity and the Relationship between Them

	Number	Expression	Relationship
1	Archimed	$\mathrm{Ar} = \frac{ga^3}{\nu_c^2}\frac{\Delta\rho}{\rho_c}$	$\mathrm{Mo} = \frac{g\eta_c^4}{\rho_c\sigma^3}\frac{\Delta\rho}{\rho_c}$ $\mathrm{We} = \frac{\rho_c U^2 a}{\sigma}$
2	Morton	$\mathrm{Bo} = \frac{ga^2}{\sigma}\Delta\rho$	$\mathrm{Ac} = \frac{\rho_c \nu_c^2}{\sigma a}$
3	Weber	$\mathrm{Fr} = \frac{U^2}{ag}$	$\mathrm{Ar} = \mathrm{BoAc}^{-1}$ $\mathrm{Ar} = \mathrm{MoAc}^{-3}$
4	Bond	$\mathrm{Mo} = \mathrm{We}^3\mathrm{Re}_d^{-4}\mathrm{Fr}^{-2} = \mathrm{BoWe}^2\mathrm{Re}_d^{-4}$	$\mathrm{We} = \mathrm{Re}_d^{4/3}\mathrm{Mo}^{1/3}\mathrm{Fr}^{-1/3} = \mathrm{AcRe}_d^2$
5	Acrivos	$\mathrm{We} = \mathrm{Re}_d^2(\mathrm{MoBo})^{-1/2}$	$\mathrm{Bo} = \mathrm{ArAc} = \mathrm{WeFr}^{-1}$
6	Froud	$\mathrm{Ac} = \mathrm{WeRe}_d^{-2}$	$\mathrm{Fr} = \mathrm{WeBo}^{-1}\frac{\Delta\rho}{\rho_c}$

Along with these there are mass-exchange processes of agglomeration, coalescence and crushing, and non-stationary processes of growth or reduction of particles where their dimensions vary significantly over time in a wide range. With an increase in the droplet size, the nature of its flow significantly differs from the nature of the flow around the solid particle, since along with the physical phenomena already discussed, other effects of the pulsation of the shape and surface of the drop begin to appear, due to the mobility of the interface and the uneven distribution of pressure on it, the internal circulation of the liquid, forms, crushing, etc. In this connection, numerous expressions are given in the literature for calculating the drag coefficient of solid particles, drops and bubbles, both for small and for a large enough area of the number change [6,8,10,24–26,83,91,209–216]. For gas bubbles and liquid droplets, the determination of the coefficient of resistance for large numbers is complicated by the deformation of the shape, which depends on the values of the numbers Weber (We), Morton (Mo), Acrivos (Ac), and Bond (Bo). Table 8.2 lists various expressions for these numbers and the relationship between them. The medium's resistance to the motion of a liquid droplet in it in the general case can be represented by the relation

$$F_{SK} = \frac{1}{3}\frac{2+3\gamma}{1+\gamma}F_S\left[1+\zeta\left(\gamma,\mathrm{Ac},\mathrm{Re}_d\right)\right] \tag{8.87}$$

The change in the character of the dependence $C_D(\mathrm{Re}_d)$ over a wide range of the value of the number Re_d is due to a continuous change in the character of the hydrodynamic flow past the particle.

If we assume in the isothermal flow that the coefficient of resistance depends on the velocity, the viscosity of the flow, and the size and shape of the particles, then by introducing some equivalent size, on the basis of the dimensionality and similarity method, one can write $C_D = AU^n a^m \nu_c^k$. Taking into account the dimension of these parameters, we can determine: $n=-k$, $m=-k$. Then, in view of the arbitrariness of the coefficient A, the expression for the drag coefficient is represented as

$$C_D = A\mathrm{Re}_d^{-k} = \frac{A}{\mathrm{Re}_d}f\left(\mathrm{Re}_d\right) \tag{8.88}$$

As shown by theoretical calculations and experimental studies, for particles of spherical shape (solid particle, droplet, bubble) at small numbers (Stokes mode) $\mathrm{Re}_d \ll 1$, the drag coefficient obeys equation

$$C_D = \frac{24}{\mathrm{Re}_d} \tag{8.88a}$$

For several large values of the number Re_d, equation (8.88) can be written in the form

$$C_D = \frac{24}{\mathrm{Re}_d}\left(1+\zeta\left(\mathrm{Re}_d\right)\right) \tag{8.89}$$

where $\zeta\left(\mathrm{Re}_d\right)$ is a function of the number Re_d, obtained in most cases by empirical means on the basis of experimental studies. For drops and bubbles for large values of the number, equation (8.89) can be represented as $C_D = \frac{A}{\mathrm{Re}_d} f\left(\mathrm{Re}_d, \mathrm{Mo}, \mathrm{We}, \mathrm{Ac}\right)$.

8.6.1 Drag Coefficient of a Solid Spherical Particle

Theoretical and experimental studies of the drag coefficients for a spherical solid particle were carried out by many researchers [6,8,21,24,25] whose complete enumeration would constitute a whole bibliography. It should be noted that the Stokes solution (8.88a) is not an exact solution of the flow problem with inertial terms. The calculations carried out showed that at large distances, compared with the particle sizes, the unaccounted terms give an important correction. Ossen [8,201] proposed a method of linearization the Navier–Stokes equation, based on replacing the convective term $(U\nabla)U$ with $(U_\infty\nabla)U_\infty$, where U_∞ is the velocity of the unperturbed flow at infinity, which leads to the following result ($\mathrm{Re}_d = \frac{RU}{\nu_c}$ where R is the radius of the particle)

$$\zeta\left(\mathrm{Re}_d\right) = \frac{3}{16}\mathrm{Re}_d + \ldots \tag{8.90}$$

Using the by linearized Navier–Stokes equation, Goldstein [83] proposed the following series for a solid sphere

$$\zeta\left(\mathrm{Re}_d\right) \approx \frac{3}{16}\mathrm{Re}_d - \frac{19}{1280}\mathrm{Re}_d^2 + \frac{71}{20480}\mathrm{Re}_d^3 - \ldots \tag{8.91}$$

A study conducted by Proudman and Pearson [8,209] led to the following result

$$\zeta\left(\mathrm{Re}_d\right) = \frac{3}{16}\mathrm{Re}_d + \frac{9}{160}\mathrm{Re}_d^2 \ln \mathrm{Re}_d + \ldots \tag{8.92}$$

The expansion (8.90) from the quantitative point of view is unsuccessful, since the appearance of a term with a logarithm Re_d makes the results dependent on the method for determining the Reynolds number Re_d. The three-term expansion, independent of the method of determining the number, was obtained by Chester [210] in the form

$$\begin{gathered}\zeta\left(\mathrm{Re}_d\right) = \frac{3}{16}\mathrm{Re}_d + \frac{9}{160}\mathrm{Re}_d^2\left(\ln \mathrm{Re}_d + A\right) + O\left(\mathrm{Re}_d^2 \ln \mathrm{Re}_d\right) \\ A = \gamma_1 + \frac{5}{3}\ln 2 - \frac{323}{360} \approx 0.835\end{gathered} \tag{8.93}$$

where γ_1 is Euler's constant is equal to 0.577. Thus, the solutions of the hydrodynamic problem of flow past a spherical particle show that the Stokes law for the resistance force is only the first term in the series expansion in powers of the number Re_d.

The preceding theoretical approximate solutions are effective for very small numbers $\mathrm{Re}_d << 1$, that is, for laminar flow. For large values of the number Re_d, the theoretical solutions of the hydrodynamic problem are complicated: (1) the appearance of convective terms in the Navier–Stokes equation; (2) various hydrodynamic phenomena (separation of the boundary layer, the appearance of turbulence, the formation of a hydrodynamic track, etc.). In connection with this, for large numbers Re_d in the literature, for various regions of the Reynolds number change Re_d, there are many formulas consistent with the experimental Rayleigh curve. The appearance of the hydrodynamic trace begins at sufficiently low values Re_d. As the flow increases Re_d, the trail loses stability with respect to infinitesimal perturbations. In the interval $0.1 \le \mathrm{Re}_d \le 10$ inertial forces begin to distort symmetrical flow, the boundary layer begins to break away from the sphere in the aft region, and when $\mathrm{Re}_d \approx 20$, a ball is formed a stable vortex ring and a turbulent trace appears. When $\mathrm{Re}_d \approx 60$ the flow in the wake becomes unsteady, and with a further increase in the number $\mathrm{Re}_d \approx 500$, the flow in the wake becomes irregular, turbulent, and from the aft region with a certain frequency, increasing with, the vortex rings come off and leave in the form of a vortex path downstream. When $\mathrm{Re}_d \approx 3.10^5$, there is a crisis of resistance, and as a result, the drag coefficient drops sharply to $C_D \approx 0.1$. The typical patterns of flow around a solid spherical particle as a function of number Re_d are given below. It is important to note that, although the drag coefficient curve is differentiable and continuous, the derivatives do not represent a monotonically varying function for large regions of number variation. In particular, for values of the number Re_d corresponding to the resistance crisis, the derivatives suffer a discontinuity of the first kind: $\lim\limits_{+\mathrm{Re}_d \to 3\times10^5} C_D \approx 0.51$ and $\lim\limits_{-\mathrm{Re}_d \to 3\times10^5} C_D \approx 0.1$, although in calculating the drag coefficient the data corresponding to this region are to some extent smoothed out. At very small values of the number Re_d, the analysis of the data of numerical and experimental measurements leads to the following conclusion: $\zeta(\mathrm{Re}_d)$ at least it is a monotonically increasing function and satisfies the conditions

$$\frac{d\zeta(\mathrm{Re}_d)}{d\,\mathrm{Re}_d} \le \frac{3}{16}, \quad \zeta(\mathrm{Re}_d) \sim \mathrm{Re}_d^{\frac{2}{3}} \tag{8.94}$$

which also follows from equation (8.91). Levich's work [8], the region of flow separation from the surface of spherical droplets and bubbles, is estimated in the form

$$\theta \approx \frac{1}{\sqrt{\mathrm{Re}_d}} \tag{8.95}$$

The paper [83] gives the following empirical correlation of the hydrodynamic trace for a solid spherical particle

$$\theta \approx 180 - 42.5\left[\ln\left(\frac{\mathrm{Re}_d}{20}\right)\right]^{0.483} \tag{8.96}$$

Using the experimental data [209,211], in this paper we obtained for the region $\theta \le 180°$ a simpler formula with the correlation coefficient $r^2 = 0.995$ (Table 8.3)

$$\theta = \frac{276}{\mathrm{Re}_d^{1/7}} \tag{8.97}$$

It follows from (8.95) and (8.97) that the angle of detachment of the hydrodynamic flow from the surface of the drop and the bubble is much smaller than that of the solid particle. Due to the mobility of the surface of the droplet and the bubble, the region of flow separation from the surface occupies a very narrow region, where turbulization of the flow occurs at large numbers.

TABLE 8.3
The Nature of the Flow Around a Solid Spherical Particle

$0 < Re_d < 5$	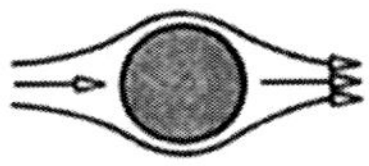	Laminar flow past a particle
$5 < Re_d < 40$	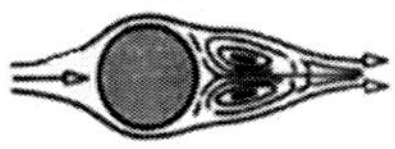	A pair of symmetric vortices from the opposite sides of the particle
$40 < Re_d < 150$		Laminar boundary layer becomes a family of periodic vortices with a certain frequency
$150 < Re_d < 3\times10^5$	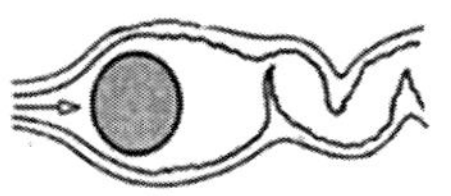	Laminar boundary layer up to the point of separation forms turbulent vortex paths
$3\times10^5 < Re_d < 3.5\times10^6$	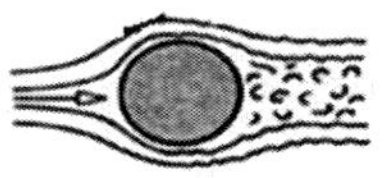	Laminar boundary layer becomes turbulent
$Re_d > 3.5\times10^6$	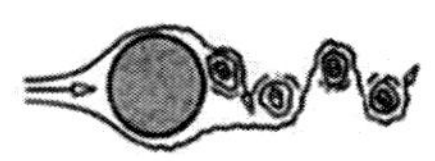	Turbulent vortex paths stabilize and narrow

Consequently, unlike a solid particle for a drop and a bubble, the point of flow separation from the surface is so low that the region of the turbulent wake covers an insignificant part of its surface. Thus, for a solid particle, the total resistance force can be considered as the resistance force acting on the surface up to the point of flow separation and having a viscous character and the drag force acting in the separation region. As noted in the paper [8], the drag force has a dissipative character, expressed in terms of the dissipation energy, and its value for small numbers corresponds to the Stokes force. Such a picture of the flow around a solid spherical body imposes significant limitations on the description of the Rayleigh curve for a large range of the number change. An analysis of the existing formulas for calculating the drag coefficient of solid spherical particles showed that for the region $\mathrm{Re}_d = \frac{Ua}{\nu_c} \le 500$ most of the equations can be reduced to the form:

a.
$$C_D = \frac{24}{\mathrm{Re}_d}\left(1 + A\mathrm{Re}_d^n\right) \tag{8.98}$$

where $A = 0.17$, $n = 2/3$ [10,12]; $A = 0.189$, $n = 0.632$ [212]; A = 0.15, $n = 0.687$ [40,91,213]; $A = 0.1538$, $n = 2/3$ [25]; $A = 0.1935$, $n = 0.6305$, $20 \le \mathrm{Re}_d \le 260$[217]; $A = 3/16$, $n = 2/3$ [10,83].

b.
$$C_D = \frac{24}{\mathrm{Re}_d}\left(1 + A\mathrm{Re}_d^m\right)^n \tag{8.99}$$

where $A = 0.065$, $m = 2/3$, $n = 3/2$, $0.5 \le \mathrm{Re}_d \le 2\times10^3$ [3]; $A = 0.27$, $m = 1$, $n = 0.43$ [3,218].

c.
$$C_D = \frac{24}{\mathrm{Re}_d} + \zeta\left(\mathrm{Re}_d\right) \tag{8.100}$$

TABLE 8.4
Determination of the Drag Coefficient of Particle for Different Areas of the Number Change Re_d

Re_d	A	n	Re_d	A	n
<0,1	24	1.0	10–100	16.8	0.60
0.1–1.0	26.9	0.95	100–800	6.1	0.38
1.0–10	26.5	0.80	800–1000	5.8	0.37

where $\zeta(\mathrm{Re}_d) = 4\mathrm{Re}_d^{-\frac{1}{3}}$, $1 \le \mathrm{Re}_d \le 100$ [21,218]; $\zeta(\mathrm{Re}_d) = 3.6\mathrm{Re}_d^{-0.313}$, $5 \le \mathrm{Re}_d \le 200$; $\zeta(\mathrm{Re}_d) = 6.48\mathrm{Re}_d^{-0.573} + 0.36$, $20 \le \mathrm{Re}_d \le 500$ [21,12]; $\zeta(\mathrm{Re}_d) = 0.03276\mathrm{Re}_d^{-0.18-0.05\log\mathrm{Re}_d}$, [21,12]; $0.01'' \mathrm{Re}_d'' 20$ [217]; $\zeta(\mathrm{Re}_d) = \frac{6}{1+\sqrt{\mathrm{Re}_d}} + 0.4$, $1 < \mathrm{Re}_d < 10^5$ [83].

d.
$$C_D = \frac{A}{\mathrm{Re}_d^n} \tag{8.101}$$

where the coefficients A and n take the following values given in Table 8.4 [12]. You can add this table to: $\mathrm{Re}_d = 10^4 - 7\times10^4$, $A = 0.16$, $n = -0.102$. A number of such approximations for the drag coefficient in the form (8.101) are given in the papers [8,10,26,40,91,214,222–231].

For large regions of variation, the numbers of formulas for calculating the drag coefficient of a spherical particle become more complex and cumbersome (Table 8.5).

The formulas for calculating the drag coefficient are suitable for relatively large values of the number Re_d and are characterized by a relative error of less than 5%. It should be noted

TABLE 8.5
Drag Coefficients of Solid Spherical Particles for a Large Range of Number Change Re_d

No.	Equations	No.	References
1.	$C_D = \frac{24}{\mathrm{Re}_d}\left[1 + 18.5\mathrm{Re}_d^{3.6} + \left(\frac{\mathrm{Re}_d}{2}\right)^{11}\right]^{1/30} + \frac{4}{9}\frac{\mathrm{Re}_d^{4/5}}{330 + \mathrm{Re}_d^{4/5}}$, $0.1 \le \mathrm{Re}_d \le 5\times10^4$	(8.102)	[25]
2.	$C_D = \left[\frac{1}{(\varphi_1+\varphi_2)^{-1} + \varphi_3^{-1}} + \varphi_4\right]^{1/10}$, $\varphi_1 = \left(\frac{24}{\mathrm{Re}_d}\right)^{10} + \left(\frac{21}{\mathrm{Re}_d^{0.67}}\right)^{10} + \left(\frac{4}{\mathrm{Re}_d^{0.33}}\right)^{10} + 0.4^{10}$, $\varphi_2 = \frac{1}{\left[\left(0.148\mathrm{Re}_d^{0.11}\right)^{-10} + 0.5^{-10}\right]}$, $\varphi_3 = \left(\frac{1.57\times10^8}{\mathrm{Re}_d^{1.625}}\right)^{10}$, $\varphi_4 = \frac{1}{\left[\left(\times10^{-17}\mathrm{Re}_d^{2.63}\right)^{-10} + 0.2^{-10}\right]}$, $\mathrm{Re}_d < 10^6$	(8.103)	[215]

(Continued)

TABLE 8.5 (*Continued*)
Drag Coefficients of Solid Spherical Particles for a Large Range of Number Change Re_d

No.	Equations	No.	References
3.	$C_D = \frac{24}{\mathrm{Re}_d}\left(1+0.15\,\mathrm{Re}_d^{0.681}\right)+\frac{0.407}{1+8710\,\mathrm{Re}_d^{-1}}, \quad \mathrm{Re}_d < 2\times10^5$	(8.104)	[219]
4.	$C_D = \frac{24}{\mathrm{Re}_d}\left(1+0.15\,\mathrm{Re}_d^{0.687}\right)+\frac{0.42}{1+42300\,\mathrm{Re}_d^{-1.16}}, \quad \mathrm{Re}_d < 2\times10^5$	(8.105)	[83,91]
5.	$C_D = 0.84\left[\alpha_1+\alpha_2\right]^{1.43}, \quad \alpha_1 = \frac{33.78}{\left(1+4.5d_f^{0.36}\right)^{0.7}\mathrm{Re}_d^{0.36}},$ $\alpha_2 = \left(\frac{\mathrm{Re}_d}{\mathrm{Re}_d+700+1000d_f}\right)^{0.28}\times\frac{1}{(d_f^4+20d_f^{20})^{0.175}},$ $d_f = \frac{d_{fh}}{\sqrt{d_{fi}d_{fm}}}$	(8.106)	[216]
6.	$C_D = \frac{24}{\mathrm{Re}_d}10^{m(\mathrm{Re}_d)}, \quad m(\mathrm{Re}_d) = 0.261\,\mathrm{Re}_d^{0.369}-0.105\,\mathrm{Re}_d^{0.431}-\frac{0.124}{1+\log^2\mathrm{Re}_d},$ $\mathrm{Re}_d < 3\times10^5$	(8.107)	[217]
7.	$C_D = \frac{24}{\mathrm{Re}_d}\left(1+0.173\,\mathrm{Re}_d^{0.657}\right)+\frac{0.413}{1+16300\,\mathrm{Re}_d^{-1.09}}, \quad \mathrm{Re}_d < 2\times10^5$	(8.108)	[218]
8.	$C_D = \frac{0.2841}{\mathrm{Re}_d^2}\left(1+\frac{9.04}{\sqrt{\mathrm{Re}_d}}\right)^2\times\lambda(\mathrm{Re}_d), \quad \lambda(\mathrm{Re}_d) = 0.96208\,\mathrm{Re}_d^2+$ $+2.73646\times10^{-5}\,\mathrm{Re}_d^3-3.93861\times10^{-10}\,\mathrm{Re}_d^4+2.47686\times10^{-15}\,\mathrm{Re}_d^5-$ $-7.15934\times10^{-21}\,\mathrm{Re}_d^6+7.43723\times10^{-27}\,\mathrm{Re}_d^7, \quad \mathrm{Re}_d < 3\times10^6$	(8.109)	[222]
9.	$C_D = \frac{24}{\mathrm{Re}_d}\left(1+0.27\,\mathrm{Re}_d\right)^{0.43}+0.47\left[1-\exp\left(-0.04\,\mathrm{Re}_d^{0.38}\right)\right], \quad \mathrm{Re}_d < 2\times10^3$	(8.110)	[223]
10.	$C_D = \begin{cases} \frac{24}{\mathrm{Re}_d}, & \mathrm{Re}_d < 0.01 \\ \left(\frac{24}{\mathrm{Re}_d}\right)\left(1+0.1315\,\mathrm{Re}_d^{0.82-0.05\log\mathrm{Re}_d}\right), & 0.01<\mathrm{Re}_d\le 20 \\ \left(\frac{24}{\mathrm{Re}_d}\right)\left(1+0.1935\,\mathrm{Re}_d^{0.6305}\right), & 20<\mathrm{Re}_d\le 260 \\ 10^{1.6425-1.1242\log\mathrm{Re}_d+0.1558\log^2\mathrm{Re}_d}, & 260<\mathrm{Re}_d\le 1500 \\ 10^{-2.4571+2.5558\log\mathrm{Re}_d-0.9295\log^2\mathrm{Re}_d+0.1049\log^3\mathrm{Re}_d}, & 1500<\mathrm{Re}_d\le 1.2\times10^4 \\ 10^{-1.9181+0.637\log\mathrm{Re}_d-0.0636\log^2\mathrm{Re}_d}, & 1.2\times10^4<\mathrm{Re}_d\le 4.4\times10^4 \\ 10^{-4.339+1.5809\log\mathrm{Re}_d-0.1546\log^2\mathrm{Re}_d}, & 4.4\times10^4<\mathrm{Re}_d\le 3.38\times10^5 \\ 29.78-5.3\log\mathrm{Re}_d, & 3.38\times10^5<\mathrm{Re}_d\le 4\times10^5 \\ 0.19\log\mathrm{Re}-0.49, & 4\times10^5<\mathrm{Re}_d\le 10^6 \\ 0.19-\frac{8\times10^4}{\mathrm{Re}_d}, & \mathrm{Re}_d>10^6 \end{cases}$	(8.111)	[91]

(*Continued*)

TABLE 8.5 (*Continued*)
Drag Coefficients of Solid Spherical Particles for a Large Range of Number Change Re_d

No.	Equations	No.	References
11.	$C_D = \frac{24}{\mathrm{Re}_d}(1+0.1806\,\mathrm{Re}_d^{0.6459}) + \frac{0.4251}{1+\frac{6880.95}{\mathrm{Re}_d}}, \quad \mathrm{Re}_d < 2\times10^4$	(8.112)	[224]
12.	$C_D = \frac{24}{\mathrm{Re}_d}\left[\frac{1+0.545\,\mathrm{Re}_d + 0.1\,\mathrm{Re}_d^{1/2}(1-0.03\,\mathrm{Re}_d)}{1+a\,\mathrm{Re}_d^b}\right],$ $a = 0.09+0.077\exp(-0.4\,\mathrm{Re}_d),\ b = 0.4+0.77\exp(-0.04\,\mathrm{Re}_d),$ $\mathrm{Re}_d < 3\times10^5$	(8.113)	[225]
13.	$C_D = \frac{12}{\mathrm{Re}_d}\left(1+0.241\,\mathrm{Re}_d^{0.687}\right) + 0.42\left(1+1.902\times10^4\,\mathrm{Re}_d^{-1.18}\right)^{-1},$ $\mathrm{Re}_d < 1.5\times10^5$	(8.114)	[10]
14.	$C_D = \begin{cases} 28.12-5.3\log\mathrm{Re}_d, & 1.7\times10^5 \le \mathrm{Re}_d \le 2\times10^5, \\ 0.1\log\mathrm{Re}_d - 0.46, & 2\times10^5 < \mathrm{Re}_d \le 5\times10^5, \\ 0.19-4\times10^4\,\mathrm{Re}_d^{-1}, & \mathrm{Re}_d > 5\times10^5 \end{cases}$	(8.115)	[10]
15.	$C_D = \frac{24}{\mathrm{Re}_d}\left(A + B\,\mathrm{Re}_d^m + C\,\mathrm{Re}_d^n\right),$ $A=1,\ B=\frac{3}{16},\ m=1,\ C=0,\quad \mathrm{Re}_d<1,$ $A=1,\ B=0.1935,\ m=0.6305,\ C=0,\quad 1<\mathrm{Re}_d<285,$ $A=1,\ B=0.015,\ m=1,\ C=0.2283,\ n=0.424,\quad 285<\mathrm{Re}_d<2000,$ $A=0,\ B=\frac{0.44}{24},\ m=1,\ C=0,\quad 2000<\mathrm{Re}_d<3.5\times10^5$	(8.116)	[220]

that equation (8.102) is obtained by the method of tangents [25], and from this equation there follow special cases for local regions of number Re_d variation that coincide with the partial equations

$$C_D = \frac{24}{\mathrm{Re}_d},\ 0.1 \le \mathrm{Re}_d \le 0.3,\ C_D = \frac{26.5}{\mathrm{Re}_d^{0.88}},\ 0.3 \le \mathrm{Re}_d \le 5,$$
$$C_D = \frac{18.6}{\mathrm{Re}_d^{0.64}},\ 5 \le \mathrm{Re}_d \le 100,\quad 100 < \mathrm{Re}_d \le 2.10^4 \tag{8.117}$$

It should be noted that in order to obtain a more precise equation (8.102), it is necessary to use a larger number of tangents corresponding to individual piecewise approximations.

The equations (8.104), (8.105), (8.108), and (8.115) assume the fulfillment of Newton's law within the limits $10^4 \le \mathrm{Re}_d \le 10^5$, that is, the proportionality of the frictional force to the square of the velocity ($C_D \approx 0.44$), which gives a significant deviation from the true experimental curve in this region. As follows from Table 8.5, to approximate the resistance curve of a spherical particle for a large region of the Reynolds number variation, it is possible to use different series or a sum of functions as the only way to solve this problem. In most cases, assuming the coefficient of resistance in the form of equation (8.3), the problem of describing the experimental data on the resistance curve of spherical particles can be reduced to a successful choice of function $\zeta(\mathrm{Re}_d)$. However, it should be remembered that the effectiveness of the developed formulas is determined by the smallest number of terms and determined coefficients.

Satisfactory compliance with the standard resistance curve gives the author's equation [26], using a series of types

$$C_D = \sum_{n=1}^{\infty} A_n \mathrm{Re}_d^{m(n)} \exp\left(-b_n \mathrm{Re}_d^{k(n)}\right) \tag{8.118}$$

in area $0{,}1 \le \mathrm{Re}_d \le 10^6$. Here A, b, m, and k are the coefficients of the series, depending on the number of the term in the series. In addition to the formulas given in this table, various correlations of the drag coefficient of spherical solid particles for different regions of the number change can be found in the works [83,91,220,221,226–228]. In this paper, we propose another equation for calculating the coefficient of resistance of a solid spherical particle for a wide range of the number $0.01 \le \mathrm{Re}_d \le 10^6$ in the form

$$C_D = \frac{24}{\mathrm{Re}_d}\left[1 + \mathrm{Re}_d^{\frac{2}{3}}\zeta(\mathrm{Re}_d)\right] + 0.22\left(1 - \exp(-1.2\times10^{-24}\mathrm{Re}_d^4\right)$$

$$\zeta(\mathrm{Re}_d) = \frac{\mathrm{Re}_d^2}{5 + 6.5\mathrm{Re}_d^2} + \frac{\mathrm{Re}_d}{3\times10^4 + 36\mathrm{Re}_d^{\frac{2}{3}} + 2.8\times10^{-13}\mathrm{Re}_d^{\frac{10}{3}}} \tag{8.119}$$

For the region $0.01 \le \mathrm{Re}_d < 800$, expression (8.119) simplifies to the form

$$C_D \approx \frac{24}{\mathrm{Re}_d}\left(1 + \frac{\mathrm{Re}_d^{\frac{8}{3}}}{5 + 6.5\mathrm{Re}_d^2}\right) \tag{8.120}$$

Figure 8.14 shows the Rayleigh curve and calculations using equation (8.119), which gives a satisfactory description within an acceptable accuracy of 8% and with a correlation coefficient equal to $r^2 \approx 0.9431$. The approximation error is achieved due to inaccurate data in the area of the resistance crisis.

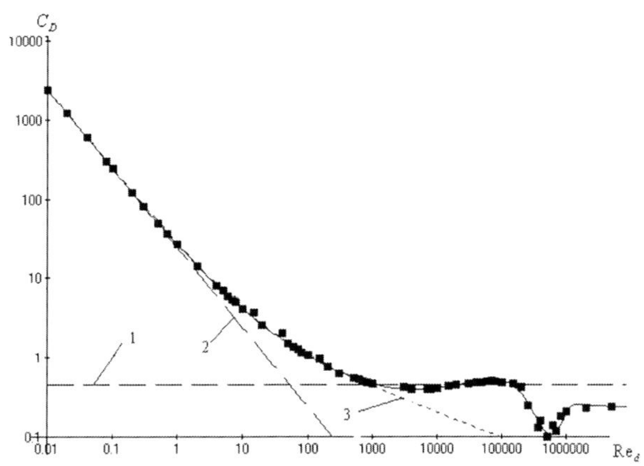

FIGURE 8.14 Drag coefficient for a solid spherical particle: (1) $C_D \approx 0.44$; (2) $C_D = \frac{24}{\mathrm{Re}_d}$; and (3) is the Schiller–Neumann equation $\left(C_D = \frac{24}{\mathrm{Re}_d}(1 + 0.15\mathrm{Re}_d^{0.687})\right)$.

It should be noted that at high flow velocities of particles, when a resistance crisis occurs, the compressibility of the fluid, accounted for by the Mach number $M = \frac{U}{c}$ (c is the velocity of sound in a given medium) [83], $M << 1$ is noticeable, and the liquid can be considered incompressible. Taking into account the fluid compressibility, the literature suggests an equation for determining the coefficient of resistance of a sphere in the form

$$C_D = \frac{24}{\mathrm{Re}_d}\left[\frac{\left(1+0.15\mathrm{Re}_d^{0.687}\right)\left(1+\exp\left[-\frac{0.427}{\mathrm{M}^{4.63}}-\frac{3}{\mathrm{Re}_d^{0.88}}\right]\right)}{1+\frac{\mathrm{M}}{\mathrm{Re}_d}\left[3.82+1.28\exp\left(-\frac{1.25\mathrm{Re}_d}{\mathrm{M}}\right)\right]}\right], \quad \mathrm{Re}_d < 3\times10^5 \qquad (8.121)$$

As follows from this equation for, it coincides with the Schiller–Neumann expression. Experimental studies show that as the number M increases, the critical value of the number Re_d at which turbulence of the boundary layer occurs increases.

8.6.2 Drag Coefficient for Particles of Irregular Shape

The spherical shape of solid particles is absolute and ideal, although in practice there is a deviation from a strict spherical shape (oval, ellipse, disk, cylinder) or irregularly shaped particles. Particles of irregular shape are characterized by a sphericity factor representing the ratio of the surface of a sphere s having the same volume as a particle of an irregular shape to the actual surface s_d of the particle, $\varphi = \frac{s}{s_d}$, where $\varphi = 0.806$ for cubic particles, $\varphi = 0.69$ for cylindrical particles, and $\varphi = 0.32$ for a disk. Numerous expressions and experimental studies for calculating the coefficient of resistance for particles of irregular shape are given in [24,32–37] (Table 8.6). The reduced formula (8.119) is suitable for calculating the coefficient of resistance of a circular disk.

The coefficient of resistance of a cylindrical body with $5 < \frac{L}{d} < 50$ and $\mathrm{Re}_d < 1$ is defined as [83]

$$C_D = \frac{4}{\mathrm{Re}_d\left[\ln\left(\frac{L}{d}\right)-0.1197\right]} \qquad (8.127)$$

It should be noted that the coefficient of resistance for a transversely streamlined cylindrical body can be determined in the form using the experimental data [226] ($\mathrm{Re}_d = \frac{UD}{\nu_c}$, D, is cylinder diameter)

$$C_D = \frac{10}{\mathrm{Re}_d^{0.778}}\left(1+\frac{\mathrm{Re}_d^{1.875}+0.368\times10^{-3}\,\mathrm{Re}_d^{2.55}}{60+6.8\mathrm{Re}_d^{1.15}+0.4\times10^{-14}\,\mathrm{Re}^{3.95}}\right)+0.36\left(1-\exp(-1.2\times10^{-24}\,\mathrm{Re}^4)\right) \qquad (8.128)$$

Figure 8.15 shows a comparison of the calculated (8.128) and experimental data of the coefficient of resistance of a cylindrical particle for $0.1 \le \mathrm{Re}_d \le 10^6$.

From equation (8.128) we have particular solutions

$$C_D \approx \frac{10}{\mathrm{Re}_d^{0.778}}, \quad \mathrm{Re}_d \le 1\ [28],$$

$$C_D \approx \frac{10}{\mathrm{Re}_d^{0.778}}\left(1+\frac{\mathrm{Re}_d^{1.875}}{60+6.8\mathrm{Re}_d^{1.156}}\right), \quad \mathrm{Re}_d \le 6.10^3, \qquad (8.129)$$

$$C_D \approx \frac{10}{\mathrm{Re}_d^{0.778}}\left(1+0.1076\mathrm{Re}_d^{0.778}\right), \quad 0.1 \le \mathrm{Re}_d \le 6.10^3$$

TABLE 8.6
Drag Coefficients for Particles of Irregular Shape

No.	Equations	No.	References
1.	$C_D = \frac{30}{\text{Re}} + \frac{67.289}{\exp(5.03\varphi)}, \quad 0.2 \le \varphi \le 1.0, \quad 0.1 \le \text{Re} \le 20$	(8.122)	[17]
2.	$C_D = \frac{24}{\text{Re}}\left(1 + b_1 \text{Re}^{b_2}\right) + \frac{b_3 \text{Re}}{b_4 + \text{Re}}, \; b_1 = \exp\left(2.3288 - 6.4581\varphi + 2.488\varphi^2\right),$	(8.123)	[224,229]
	$b_2 = 0.0964 + 0.5565\varphi,$		
	$b_3 = \exp(4.905 - 13.8944\varphi + +18.4222\varphi^2 - 10.2599\varphi^3),$		
	$b_4 = \exp\left(1.4681 + 12.258\varphi - 20.7322\varphi^2 + 15.8855\varphi^3\right).$		
3.	$C_D = \frac{24}{\text{Re}}\left[1 + 8.171\exp(-4.0655\varphi)\right]\text{Re}^{0.0964+0.5565\varphi} + \frac{73.69\,\text{Re}\exp(-5.748\varphi)}{\text{Re} + 5.378\exp(6.2122\varphi)},$	(8.124)	[224]
	$0.7 \le \varphi < 1, \quad 1 < \text{Re} \le 20$		
4.	$C_D = \frac{24}{K_1 K_2 \text{Re}}\left[1 + 0.118\left(K_1 K_2 \text{Re}\right)^{0.6567}\right] + \frac{0.4305}{1 + \frac{3305}{K_1 K_2 \text{Re}}}, \quad K_1 = \frac{1}{3} + \frac{2}{3\sqrt{\varphi}},$	(8.125)	[226,230,231]
	$K_2 = 10^{1.8148(-\log\varphi)^{-0.5743}}$		
5.	$C_D = \frac{64}{\pi \text{Re}}\left(1 + \frac{\text{Re}}{2\pi}\right), \; \text{Re} \le 0.01, \; Re = \frac{Ud_e}{\nu_c},$	(8.126)	[83,91]
	$C_D = \frac{64}{\pi \text{Re}}\left[1 + \frac{\text{Re}}{2\pi} + \frac{2\text{Re}^2}{5\pi^2}\ln\left(\frac{\text{Re}}{2}\right)\right],$		
	$\text{Re} < 1$		
	$C_D = \frac{64}{\pi \text{Re}}\left(1 + 10^m\right), \; m = -0.883 + 0.906\log\text{Re} - 0.025\log^2\text{Re},$		
	$0.01 < \text{Re} < 1.5, \quad d_e = \left(6d^2H\right)^{1/3},$		
	$C_D = \frac{64}{\pi \text{Re}}\left(1 + 0.138\text{Re}^{0.792}\right), \quad 1.5 \le \text{Re} < 133,$		
	H – disk thickness, d – disk diameter		

In an approximate version for a cylindrical body, we can write

$$C_D \approx 1.0,\; 6\cdot 10^2 \le \text{Re}_d \le 6\cdot 10^3,\; C_D \approx 1.1 - 1.4,\; 8\cdot 10^3 < \text{Re}_d < 2\cdot 10^5 \tag{8.130}$$

It follows from Figure 8.15 that the resistance crisis associated with displacement zone of separation of the boundary layer to the rear of the cylinder is due to turbulence occurs at $\text{Re}_d \ge 3\times 10^5$.

8.6.3 Drag Coefficient of Droplets and Bubbles

The motion of a liquid drop or bubble differs from the motion of a solid spherical particle of the same volume and the same mass. The fluid flowing around the drop triggers, due to friction, the circulation of the internal fluid from which the drop consists, this effect being dependent on a dimensionless

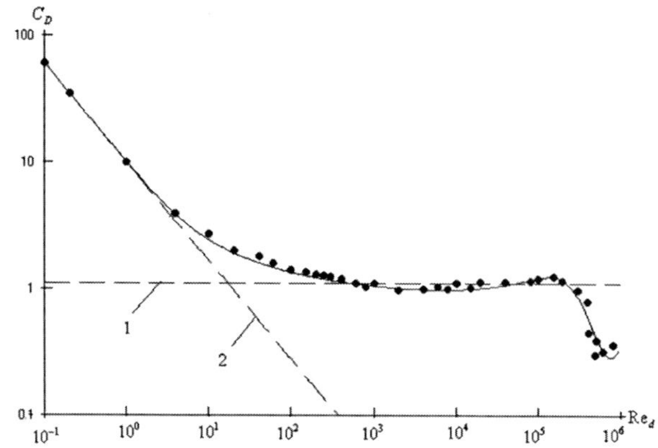

FIGURE 8.15 Drag coefficient transversely streamlined cylinder: (1) $C_D \approx 1.1$ and (2) $C_D = 10\,\text{Re}^{-0.78}$.

parameter $\gamma = \eta_d/\eta_c$ representing the ratio of the dynamic viscosity of the internal fluid to the dynamic viscosity of the external fluid and expressing the degree of surface mobility. In addition, when a droplet flows around the surface of a droplet, external forces act on the surface of the drop due to uneven pressure and tend to disrupt the spherical shape of the droplet. These forces are counteracted by the forces of surface tension, which tend to preserve the spherical shape. The presence on the surface of droplets and bubbles of adsorbed substances can have a significant effect on the forces acting on the surface, on the coefficient of resistance and on the rate of their deposition (rising).

Taking into account these peculiarities of the flow of liquid drops [232], Taylor and Acrivos, using the method of asymptotic splicing to determine the droplet drag coefficient, carried out more and theoretical analysis, as a result of which, for very small values of the number, the following formula

$$C_{DK} = \frac{8}{\text{Re}_d}\left[\frac{2+3\gamma}{1+\gamma} + \frac{\text{Re}_d}{16}\frac{2+3\gamma}{1+\gamma} + \frac{1}{40}\left(\frac{2+3\gamma}{1+\gamma}\right)^2\left(\frac{\text{Re}_d}{2}\right)^2 \ln\frac{\text{Re}_d}{2}\right] \quad (8.131)$$

Papers [210,213] modified this formula for relatively large numbers $\text{Re}_d < 6$. Large numbers Re_d in connection with the appearance of convective terms, and numerical methods for solving the Navier–Stokes equation are proposed in the work [234,235]. Theoretical and experimental studies to determine the drag coefficient, deformation, and rate of deposition (or ascent) of droplets and bubbles for moderate and large numbers Re_d, including numerical solutions, are given in the works [236–242]. In Table 8.7, some formulas are given for calculating the drop and bubble resistance coefficients. Unfortunately, most of the given bubble formulas are suitable for small and moderate numbers $\text{Re}_d < 50$ and $\text{We} < 1$, except for equations (8.134 through 8.136).

The author's equation (8.135) for determining the coefficient of resistance of gas bubbles was obtained by the method of tangents using the experimental data of Raymond and Rozant [149]. Equation (8.135) satisfies the conditions, $\text{Re}_d \le 100$, $\text{Mo} \le 7$, and $\text{Re}_d \le 1.385$ simplifies to the form (8.136), which is independent of the number. As shown by numerous experimental data on the behavior of droplets and bubbles, their deformation depends on the numbers, Re_d, Mo, and We. In particular, the number of Morton depends on the properties of the medium and the droplet and for different liquids at the normal temperature are equal to: $\text{Mo} = 1.45\times10^{-2}$ for mineral oil, $\text{Mo} = 1.75\times10^{-4}$ is water + 42% glycerin, water +13% ethyl alcohol, $\text{Mo} = 0.89\times10^{-10}$ methyl

TABLE 8.7
Drag Coefficients of Drops and Bubbles

No.	Equations	No.	References
1.	$C_{DK}=\frac{24}{\mathrm{Re}_d}\left(\frac{\frac{2}{3}+\gamma}{1+\gamma}\right),\quad \mathrm{Re}_d\ll 1$	(8.132)	[8]
2.	$C_{DK}=\begin{cases}\frac{1.68}{\mathrm{Re}_d^{1/4}}, & 0.1\le \mathrm{Re}_d\le 0.5\\ \frac{14.9}{\mathrm{Re}_d^{0.78}}, & 1\le \mathrm{Re}_d\le 10\end{cases}$	(8.133)	[91]
3.	$C_{DG}=\frac{48}{\mathrm{Re}_d}\left(\frac{1+12\mathrm{Mo}^{1/3}}{1+36\mathrm{Mo}^{1/3}}\right)+0.9\frac{\mathrm{Bo}^{3/2}}{1.4\left(1+30\mathrm{Mo}^{1/3}\right)+\mathrm{Bo}^{3/2}}$	(8.134)	[236]
4.	$C_{DG}=\begin{cases}\frac{16\zeta_0(\mathrm{Re}_d)}{\mathrm{Re}_d}\left[1+\left(\frac{\mathrm{Re}_d}{1.385}\right)^{12}\right]^{1/55}+\frac{8}{3}\frac{\mathrm{Re}_d^{4/3}\,\mathrm{Mo}^{1/3}}{24\left(1+\mathrm{Mo}^{1/3}\right)^{-1/3}+\mathrm{Re}_d^{4/3}\,\mathrm{Mo}^{1/3}},\\ \mathrm{Re}_d<100\\ \frac{16\zeta_0(\mathrm{Re}_d)}{\mathrm{Re}_d}\left(1+\left(\frac{\mathrm{Re}_d}{1.385}\right)^{12}\right)^{1/55}, \quad \mathrm{Re}_d<10,\\ \zeta_0(\mathrm{Re}_d)=1+\frac{1}{1-0.5\left(1+250\,\mathrm{Re}_d^{5}\right)^{-2}}\end{cases}$	(8.135, 8.136)	[25]
5.	$C_{DK}=\frac{1}{1+\gamma}\left[\gamma\left(\frac{24}{\mathrm{Re}_d}+\frac{4}{\mathrm{Re}_d^{1/3}}\right)+\frac{14.9}{\mathrm{Re}_d^{0.78}}\right],\quad 0.1\le \mathrm{Re}_d<10$	(8.137)	[237]
6.	$C_{DK}=\frac{48}{\mathrm{Re}_d}G(\chi)\left[1+\frac{H(\chi)}{\mathrm{Re}_d^{1/2}}\right],\quad \mathrm{Re}_d<20$	(8.138)	[238]
7.	$C_{DK}=\frac{8}{\mathrm{Re}_d}\frac{3\gamma+2}{1+\gamma}\left[1+0.05\frac{3\lambda+2}{1+\lambda}\mathrm{Re}_d\right]-0.01\frac{3\gamma+2}{1+\gamma}\mathrm{Re}_d\ln\mathrm{Re}_d,$	(8.139)	[118]
8.	$C_{DK}=1.87\left(783\gamma^2+2142\gamma+1080\right)\mathrm{Re}_d^{-0.74}\left[\left(60+20\gamma\right)\left(4+3\gamma\right)\right]^{-1},\quad 2<\mathrm{Re}_d<50$	(8.140)	[10]
9.	$C_{DK}=\left\{\left[\gamma\left(\frac{24}{\mathrm{Re}_d}+\frac{4}{\mathrm{Re}_d^{1/3}}\right)+\frac{14.9}{\mathrm{Re}_d^{0.78}}\right]+40\left(\frac{3\gamma+2}{\mathrm{Re}_d}\right)+15\gamma+10\right\}\times\left[\left(1+\gamma\right)\left(5+\mathrm{Re}_d^{2}\right)\right]^{-1},\mathrm{Re}_d\le 400$	(8.141)	[118]
10.	$C_{DK}=\begin{cases}\frac{16}{\mathrm{Re}_d}, & \mathrm{Re}_d<1.5\\ \frac{14.9}{\mathrm{Re}_d^{0.78}}, & 1.5<\mathrm{Re}_d\le 80\\ \frac{49.9}{\mathrm{Re}_d}\left(1-\frac{2.21}{\mathrm{Re}^{1/2}}\right)+1.17\times10^{-8}\,\mathrm{Re}_d^{2.615}, & 80<\mathrm{Re}_d\le 1530\\ 2.61, & \mathrm{Re}_d>1530\end{cases}$	(8.142)	[241]
11.	$C_{DK}=\frac{26.5}{\mathrm{Re}_d^{0.78}}\left[\frac{\left(1.3+\gamma\right)^2-0.5}{\left(1.3+\gamma\right)\left(2+\gamma\right)}\right],\quad \mathrm{Re}_d\le 5,\quad \gamma\le 1.4$	(8.143)	[242]
12.	$C_{DK}=\frac{16}{\mathrm{Re}_d}\left\{1+\left[\frac{8}{\mathrm{Re}_d}+\frac{1}{2}\left(1+3.31\,\mathrm{Re}_d^{-0.5}\right)\right]^{-1}\right\},\quad \mathrm{Re}_d<50$	(8.144)	[128]

alcohol, and $Mo = 3.1\times10^{-11}$ for distilled water. The second term in equation (8.135) characterizes the dependence of the bubble drag coefficient on its deformation as a function of the values of the numbers Re_d and Mo. As follows from equation (8.135), for small values of the number $Re_d < 0.4$, spherical droplets and bubbles behave like solid particles, and the drag coefficient is determined by formula (8.2a), although for $0.4 \le Re_d \le 1.385$ is $C_D = \frac{16}{Re_d}$. Thus, for the region $0.01 \le Re_d < 0.4$, we can propose the following formula

$$C_{DK} = \frac{8}{Re_d}\zeta\left(Re_d\right) = \frac{8}{Re_d}\left[1 + \frac{1}{1-0.5\left(1+250\,Re_d^5\right)^{-2}}\right] \tag{8.145}$$

where $\zeta\left(Re_d\right) \to 2$ if $Re_d > 0.4$ and $\zeta\left(Re_d\right) \to 3$, if $Re_d < 0.4$. Figure 8.16 shows a comparison of the calculated values drag coefficient of the bubble for $Re_d < 100$

$$C_{DG} = \begin{cases} \dfrac{16\zeta_0\left(Re_d\right)}{Re_d}\left[1+\left(\dfrac{Re_d}{1.385}\right)^{12}\right]^{1/55} + \dfrac{8}{3}\dfrac{Re_d^{4/3}\,Mo^{1/3}}{24\left(1+Mo^{1/3}\right)^{-1/3} + Re_d^{4/3}\,Mo^{1/3}}, \\ Re_d < 100 \\ \dfrac{16\zeta_0\left(Re_d\right)}{Re_d}\left(1+\left(\dfrac{Re_d}{1.385}\right)^{12}\right)^{1/55}, \quad Re_d < 10, \\ \zeta_0\left(Re_d\right) = 1 + \dfrac{1}{1-0.5\left(1+250\,Re_d^5\right)^{-2}} \end{cases} \tag{8.146}$$

with experimental data [149].

Using the experimental values for the minimum of the resistance curve, we can write the following expression $Re_d\,Mo^{\frac{1}{6}} \approx 7$ [175] and if $Re_d\,Mo^{1/6} > 7$, then the drop undergoes deformation. In this paper [241], this condition is expressed by the formula $We\,Re_d^{0.85} > 165$. It is noted in the

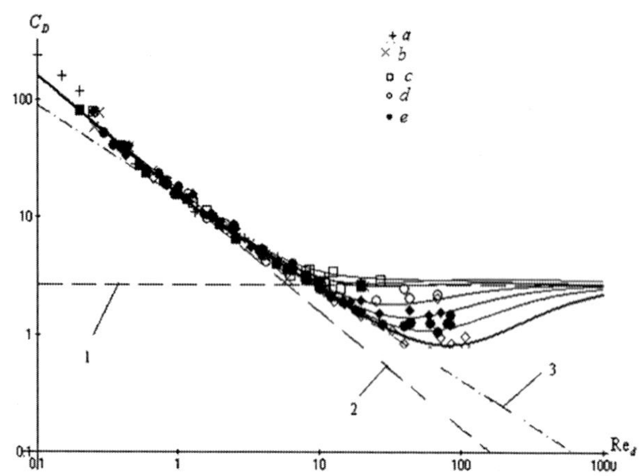

FIGURE 8.16 The change in the bubble resistance coefficient as a function of the number Re_d for different values of the number equal to: (a) 7.0; (b) 1.4; (c) 0.023; (d) 9×10^{-4}; (e) 1.1×10^{-5}; (1) $C_D = 8/3$; (2) $C_D = 24/Re_d$; and (3) $C_D = \frac{24}{Re_d}\left(1+0.1538\,Re_d^{0.687}\right)$.

paper [220] that the drop is not deformed, if $\mathrm{Mo} < 1.2\times10^{-7}\mathrm{Bo}^{8.15}$ for a number $\mathrm{Bo} < 5$ and $\mathrm{Bo} > 5$ for, if $\mathrm{Mo} < 0.2\times10^{-7}\mathrm{Bo}^{2.83}$.

Using the data of Harmati [175], the following formula is

$$\frac{C_{DK}}{C_D} \approx 6\left[1-\exp\left(-0{,}126\mathrm{Bo}\right)\right] \tag{8.147}$$

Here C_D is the drag coefficient of the solid particle. For the values of the number $\mathrm{Bo} < 20$, using the experimental data of Harmati, one can obtain the following formula

$$\frac{C_{DK}}{C_D} \approx 1.275\sqrt{\mathrm{Bo}} \tag{8.148}$$

Starting from (8.147), for large numbers $\mathrm{Re}_d \left(C_D \approx \frac{4}{9}\right)$, and Bo the steady-state value of the drag coefficient for droplets and bubbles can be defined as $C_{D\infty} \approx 6C_D = \frac{8}{3}$. If the coefficients of resistance for a solid spherical particle and a gas bubble are known, the drag coefficient for a liquid drop can be determined in the form [10,21]

$$C_{DK} = \frac{\gamma C_D + C_{Dg}}{1+\gamma} \tag{8.149}$$

where C_{DK}, C_D, and C_{Dg} are the drag coefficients for a drop, a solid particle, and a gas bubble. This formula is accurate for $\mathrm{Re}_d << 1$, as calculations show, is performed with an error of no more than 5% at $\mathrm{Re}_d \le 100$.

Correlations between the drag coefficients of a solid spherical particle, drops, and bubbles for $5 < \mathrm{Re}_d < 1000$ are given below

$$\begin{aligned} C_D\left(\mathrm{Re}_d,\gamma\right) &= \frac{2-\gamma}{2}C_D\left(\mathrm{Re}_d,0\right) + \frac{4\gamma}{6+\gamma}C_D\left(\mathrm{Re}_d,2\right), \quad 0\le\gamma<2 \\ C_D\left(\mathrm{Re}_d,\gamma\right) &= \frac{4}{2+\gamma}C_D\left(\mathrm{Re}_d,2\right) + \frac{\gamma-2}{\gamma+2}C_D\left(\mathrm{Re}_d,\infty\right), \quad 2\le\gamma\le\infty \end{aligned} \tag{8.150}$$

where $C_D\left(\mathrm{Re}_d,0\right)$, $C_D\left(\mathrm{Re}_d,2\right)$, and $C_D\left(\mathrm{Re}_d,\infty\right)$ are the drag coefficients for bubbles, for a drop with $\gamma = 2$ and for a solid particle.

A significant effect on the coefficient of resistance is caused by a change in the shape of droplets and bubbles as a result of their deformation. Studies show that the behavior of drops and bubbles basically obey the same qualitative laws and differ substantially from the behavior of solid particles. The form of droplets and bubbles is not predetermined but is formed during their movement and is determined by the instantaneous balance of the force of pressure acting on the surface of the deformed particle from the side of the surrounding fluid, which tends to compress it in the direction of motion and the surface tension that prevents such compression.

The deformation of droplets and bubbles has a significant effect on the value of the coefficient of resistance. If we assume that the volume of liquid in a single drop does not change as a result of a change in its shape, then the volume of the ellipsoidal drop is equal to $\upsilon = \frac{4}{3}\pi a_0^2 b_0$. Taking $\chi = \frac{a_0}{b_0}$ into account that the volume of the spherical droplet is also equal $\upsilon = \frac{4}{3}\pi a_0^2 b_0$, we obtain $a_0 = (\frac{3}{4\pi}\chi\upsilon)^{1/3} = R\chi^{1/3}$ and $b_0 = \frac{\upsilon}{\frac{4}{3}\pi a_0^2} = R\chi^{-\frac{2}{3}}$. Then for the drag coefficient of a transversely streamlined ellipsoidal drop for small and moderate numbers Re_d, we can write expression

$$C_{DE} \approx C_{DC}\frac{R^2}{a_0 b_0} = \chi^{\frac{1}{3}}C_{DC} \tag{8.151}$$

which connects the drag coefficients of an ellipsoidal $\left(C_{DE}\right)$ and spherical C_{DC} particles. The change in shape and deformation of droplets and bubbles at their constant volume is associated with an increase in their surface

$$S = \frac{8\pi a_0 b_0}{1+\chi} \tag{8.152}$$

Equation (8.152) expresses the dependence of the surface of the deformed drop and bubble on the deformation factor, from which, with increasing deformation (with decreasing deformation factor), the surface of the droplet increases. In the absence of deformation $\chi = 1$ and, $a_0 = b_0 = R$, $S = 4\pi R^2$, that is, the drop has a spherical shape and its surface is equal to the surface of the sphere. The deformation of the drops is mainly due to the capillary number Ca and the viscosity ratio γ. In the paper [182], experimental studies on deformation of droplets for various $\gamma = 1.0$; 3.0; 5.0 and numbers Re_d are proposed, with the ratio of the length of the deformed droplet L to its thickness, that is, $Y = L/a$. Using the experimental data, in this paper we propose the equation of deformation of drops in the form

$$Y = \frac{1}{Y_s + \left(1 - Y_s\right)\exp\left(-\beta_0 a^2\right)} \tag{8.153}$$

where Y_s is the value determined by the steady-state value, $Y_s = \frac{1}{7}\gamma^{\frac{1}{2}}$, $\beta_0 = 1.08\times 10^{-3}\,\mathrm{Re}_d^{\frac{2}{3}}\gamma^{-\frac{1}{4}}$. Below in Figure 8.17 compares the calculated and experimental values of the drag coefficient [182] for different values γ.

The deformation of droplets and bubbles is associated with a change in their shape with further breaking into smaller particles, [243] and the change in size significantly affects the value of the coefficient of resistance. At the same time, the minimum and maximum size of droplets and bubbles depend on the drag coefficient. The effect of various external effects (physical phenomena on the surface) on the coefficient of drag of droplets and bubbles is given in the works [244,245].

At present, a qualitative analysis of the effect of the turbulence of an external hydrodynamic field on the resistance of a solid spherical particle is made when its size is of the order of or larger than the internal scale of turbulence [4,40,219,246,247]. At low turbulence intensity of the external flow and small numbers Re_d, turbulence slightly increases the resistance due to an increase in dissipation in the region of the track.

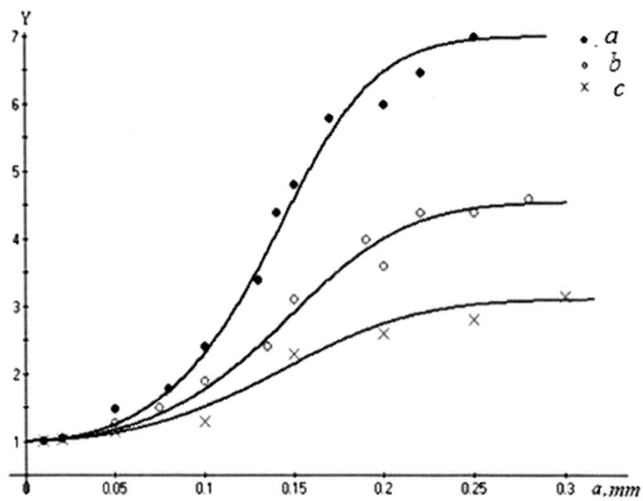

FIGURE 8.17 Change in the droplet deformation factor for different γ, equal: (a) 1.0; (b) 3.0; and (c) 5.0.

As the intensity of turbulence 30% increases, the resistance coefficient of the particle can decrease substantially due to the displacement of the separation point downstream due to the disturbed turbulence of the hydrodynamic track. If the particle flows around the turbulent flow, then the resistance crisis is achieved with smaller numbers Re_d [40]. It was noted in the work [247] that when the intensity of turbulence increases from 0.5% to 2.5%, the critical number decreases from to 1.25×10^5. With turbulence intensity 30% and higher, a resistance crisis can occur much earlier. The effect of turbulence on the coefficient of resistance of solids is estimated as [219]

$$C_D\left(\langle \text{Re}_d \rangle\right)=\frac{\langle F_{TP}\rangle}{\frac{\pi a^2\rho_d}{8}\left(\langle U\rangle\right)^2} \tag{8.154}$$

Taking into account isotropic turbulence, the drag coefficient of solids is defined in the paper [246] as

$$C_{DT}=C_D\left[1+8.76\times10^{-4}\left(\frac{a}{\lambda_0}\right)^3\right] \tag{8.155}$$

where λ_0 is the Kolmogorov scale of turbulence, and C_{DT} is the drag coefficient of particle in turbulent flows.

It should be noted one more difference between the flow of a fixed particle and a particle moving freely in the flow. For a fixed particle, the nature of the flow can depend on the degree of flow turbulence, and turbulence of high intensity can change the structure of the boundary layer and destroy the turbulent trace formed behind the particle, which can significantly change the value of the frictional force and the coefficient of resistance. Freely moving particles in the flow can rotate under the action of the Magnus force, change their orientation, and perform complex non-rectilinear motion. Depending on the size, the particle can be more or less involved in a pulsating turbulent flow, and in this case the Reynolds number is estimated as $\text{Re}_d=\frac{|U-V_d|a}{\nu_c}$. The smaller the particle size, the more it is carried away by turbulent pulsations, that is, the degree of entrainment of particles tends to unity as the particle size decreases. For droplets and bubbles, as the number Re_d increases, the probability of their deformation up to an ellipsoidal shape $\left(\text{We}\sim\text{Re}_d^2\right)$ and the frequency of breaking increase [85,87]. In the paper [10], formulas are proposed for calculating the drag coefficients of ellipsoidal particles for their transverse and longitudinal flow.

The paper [247] presents experimental data on the effect of surface roughness on the magnitude of the resistance crisis. With an increase in the surface roughness of the particles, the laminar boundary layer is more rapidly turbulized, and the resistance crisis sets in at lower values of the number Re_d. Various expressions are proposed in the works [4,48,246] that allow one to estimate the coefficients of resistance and turbulent diffusion of particles, the rate of deposition, and heat transfer taking into account the surface roughness. The force of resistance of the permeable sphere corresponds to the force of resistance of an impenetrable sphere with a smaller diameter a_k equal to $a_k=a(1+\frac{2k}{a^2})^{-1}$ [8], where k is the permeability coefficient. In particular, expressions for calculating the coefficients of resistance of porous cylindrical particles are given in the paper [246].

The mass drag coefficient has a significant effect on the droplet drag coefficient, in particular, evaporation from the droplet surface, at which the drag coefficient decreases $C_{DK}=27\,\text{Re}_d^{-0.84}$, $\text{Re}_d<1$. The following is the formula for calculating the drag coefficient of a spherical droplet with an evaporating surface [122,123]

$$C_{DK}=\frac{24}{\text{Re}_d}\frac{1+0.545\,\text{Re}_d+0.1\,\text{Re}_d^{0.5}\left(1-0.03\,\text{Re}_d\right)}{1+A\,\text{Re}_P^B},\quad 10<\text{Re}_d<200,\ 1<\text{Re}_P<20 \tag{8.156}$$

$$A=0.09+0.077\exp\left(-0.4\,\text{Re}_d\right),\ B=0.4+0.77\exp\left(-0.04\,\text{Re}_d\right),$$

where $\mathrm{Re}_P = \frac{V_P a}{\nu_c}$ is the Reynolds number, determined through the rate of evaporation, and V_P is the rate of evaporation from the surface of the drop. Taking into account the Stefan flow, the drag coefficient of the evaporating spherical droplet for small numbers Re_d is defined as [118,175]

$$C_D = \frac{24}{\mathrm{Re}_d\left(1+B_m\right)}, \qquad B_m = \frac{Y_s - Y_\infty}{1-Y_s} \tag{8.157}$$

where Y_s, Y_∞ are mole fractions of steam on the surface and away from the surface. The calculation of the coefficient of resistance for the case of non-stationary bubble growth in a binary mixture is given in the paper [248]. The paper [17] proposes the following equation for determining the drag coefficient of an evaporating drop

$$\frac{C_D}{C_{D0}} = \frac{1}{\left(1+B_m\right)^{\kappa}}, \quad \kappa = 0.19\mathrm{Sc}^{-0.74}\left(1+B_m\right)^{-0.29} \tag{8.158}$$

Accordingly, the mass flux from the surface of the evaporating drop is determined in the form

$$\mathrm{Sh} = \frac{\mathrm{Sh}_0}{0.3+0.7\left(1+B_m\right)^{0.88}} \tag{8.159}$$

In addition to these factors, the effect of the wall and the interaction with other particles due to the superposition of the hydrodynamic fields, where the wall effect is taken into account by the introduction of the coefficient $f = F\left(\vartheta, \mathrm{Re}_d\right)$, $\vartheta = a/D_k$ (D_k is channel diameter), and affects the magnitude of the frictional force and the drag coefficient [26,91,118]. In particular, for $\mathrm{Re}_d > 100$ the coefficient that takes into account the wall effect [91], the following formula is proposed: $f = (1-\vartheta^2)^{3/2}$. For droplets and bubbles, the presence of adsorbed surfactants on the surface is strongly influenced by the coefficient of resistance. The presence of microimpurities in the liquid adsorbed on the surface of droplets and bubbles strengthens the interface of the phases; thus, the droplets behave like solid particles. A number of formulas are given that take into account the effect of the concentration of surface-active substances adsorbed on the surface on the drag and drop drag coefficient for small numbers Re_d [121].

The expressions for determining the drag coefficient of particles considered above are empirical, and for very small numbers are semi-empirical or theoretical. In addition to the preceding formulas for calculating the coefficient of resistance of solids, droplets, and bubbles, there are many expressions in the literature for various regions of variation of the number Re_d, We, Mo, and Bo [83,91,218,222,233].

Naturally, large problems arise for large values of the number Re_d associated with a change in the nature of the hydrodynamic flow around the particles, which causes the unpredictable character of the change in the derivative of the drag coefficient with respect to the number Re_d. An analysis of the set of expressions and approximations for calculating the drag coefficients of solids, droplets, and bubbles has shown that only a successful choice of empirical functions for determination $\xi\left(\mathrm{Re}_d\right)$, based on the experience and intuition of the researcher, can lead to a successful result. Moreover, the function $\xi\left(\mathrm{Re}_d\right)$ must be continuously differentiated for the whole range of the number Re_d change and in its structure should preferably not be very complicated with a small content of unknown coefficients.

It should be noted that the experimental data of the drag coefficient of solid particles for the region $2\times10^5 \le \mathrm{Re}_d \le 10^7$ are significantly different. An analogous situation is observed for very small values of the number $\mathrm{Re} < 10^{-3}$. From the preceding formulas it follows that, in the hypothetical case $\mathrm{Re}_d \to 0$, the drag coefficient tends to infinity $C_D \to \infty$, that is, we have a singular problem, in particular, the particle size is zero, and their resistance is infinite. In all probability, in order to exclude this singularity, we must put the condition that the drag coefficient at zero $C_D = 0$ is $\mathrm{Re}_d \to 0$. In all probability, the resistance curve must pass through a maximum for some values $\mathrm{Re}_d \to \mathrm{Re}_m$. It is obvious that the singularity boundary determines the lower limit of applicability of the Stokes law.

If the particle sizes are small in comparison with the mean free path of the liquid molecules λ, that is, the Knudsen number $\mathrm{Kn} = \frac{\lambda}{a} << 1$, then there is a molecular slip, leading to a decrease in the drag coefficient $\left(\mathrm{Kn} \sim {}^{M}\!/_{\mathrm{Re}_d}\right)$. In this case, one can use the Kenning am amendment, using which the drag coefficient can be determined from the semi-empirical dependence

$$C_D \approx 7.09\mathrm{Kn}\left[\xi\left(1-\mathrm{Kn}\right)+\mathrm{Kn}\right]\left(\frac{U}{\sqrt{2RT}}\right)^{-1} \tag{8.160}$$

where ξ is the coefficient of slip velocity, $\left(\frac{U}{\sqrt{2RT}}\right)$ is the ratio of the flow rate to the velocity of molecular motion, T is temperature, and R is the gas constant.

It should be noted that great difficulties arise when determining the coefficient of resistance for deformable drops and bubbles for large values of numbers $\mathrm{Re}_d >> 1$, $\mathrm{We} > 1$ and small values of the number $\mathrm{Mo} << 1$. Under these conditions, drops and bubbles can take a form that is not amenable to experimental analysis and theoretical description.

8.7 SEDIMENTATION AND RISING OF SOLIDS, DROPLETS, AND BUBBLES IN GRAVITATIONAL FIELD

In the industrial practice of chemical, petrochemical, and oil-refining technologies, the basis of the processes of purification, separation, phase separation, and classification of polydisperse flows is the deposition of particles or the emergence of droplets and bubbles in the gravitational field, which is the subject of a large number of studies, although listing all the works would constitute a whole bibliography [2,4,8,12,249–252]. The problems of separation, stratification, and classification of disperse systems, which form the basis of deposition and phase transfer, represent a very complex structural problem associated with:

a. The hydrodynamic structure and direction of the flow (ascending, descending, horizontal) and the physical interaction of forces of different nature (Archimedean force, weight and resistance force, dynamic head force, etc.) and non-hydrodynamic forces (electrostatic, thermo, and diffusionphoretic). It is important to note that when moving a finely dispersed component in one direction or another, it is also necessary to take into account the Magnus force [8,12] causing the upward (transverse) migration of particles characteristic of both vertical and horizontal flows and depending on the flow velocity gradient and the particle sizes, with turbulent transport in the flow, diffusion and thermophoretic forces, depending on the gradients of concentration and temperature, respectively, [12,249] etc.;
b. Interaction of particles with each other (coagulation, agglomeration, crushing), with the wall (wall effect) and with the supporting phase. When the turbulence of the flow is high and the particle concentrations are relatively high, the flow of the disperse system is accompanied by collision and interaction of the particles, leading to a change in the number of particles, their deformation and other physical phenomena (structure breakdown, abrasion, etc.), the collision frequency being determined by the flow velocity, specific energy dissipation;
c. Distribution of the sizes of polydisperse particles, their shape and concentration, affecting the nature of their separation and the rate of precipitation as a result of the elution of the polydisperse component from the volume of the stream. At large particle concentrations in the flow volume, deformations of the hydrodynamic fields of each particle arise due to

their mutual interference, which causes their movement to be constrained as a result of the interaction of the particles themselves, leading to their coagulation, crushing, and development of other physical processes;

d. Physicochemical properties of the particles and the carrier medium (viscosity, density, surface tension) and physicochemical transformations (dissolution, evaporation, sublimation, condensation, etc.);
e. Diffusion transfer and sedimentation in turbulent flows (turbulent diffusion coefficient of particles, specific dissipation of energy, scale of turbulent pulsations), very important for deposition in vertical channels; and
f. The stochastic nature of the polydisperse system associated with the variation and dispersion of particle sizes, the fluctuation in the concentration and distribution of particles, and individual parameters (the pulsation of the velocity components, the pulsating nature of the motion of the particles themselves in the turbulent flow, the random change in the shape and size of the deformed particles, etc.). The sedimentation and motion of particles in the gravitational field is stochastic, determined by the fact that the particles settle down, perform erratic descents and ascents simultaneously with pulsation moles.

Therefore, in each specific case, to solve this problem, it is necessary to analyze all the effects accompanying migration of particles, their deposition, and separation. It is obvious that an important role in the migration and deposition of particles belongs to the resistance forces, which depend on the number $\mathrm{Re}_d = {}^{Ua}/_{\nu_c}$, shape and size of the particles, on the physical and chemical properties of particles and the medium. The most complete review on the determination of the coefficient of resistance of solid particles, droplets, and bubbles is described in the paper [24].

When a single particle moves in a turbulized medium, the resistance force is determined according to the complex equation proposed by Bass, Boussinesque, Oseen, and the generalized Chen [5,8,12]. As various studies have shown, the terms entering into this equation, under certain conditions (deviation of the flow from the steady state, the density of the medium of the same order as the particle density, when the particle moves with high acceleration), are very insignificant, and in technical calculations the expression for the force resistance can have a more simplified form. Using this equation, the resistance force acting on a moving spherical particle with velocity $U = U(t)$ can be determined in the form

$$F_s = 2\pi\rho_c R_d^3 \left[\frac{1}{3}\frac{dU}{dt} + \frac{3\nu_c}{R_d^2} U + \frac{3}{R_d}\sqrt{\frac{\nu_c}{\pi}} \int_{-\infty}^{t} \frac{dU}{d\tau}\frac{d\tau}{\sqrt{t-\tau}} \right] \tag{8.161}$$

In particular, this expression for very particles of small particle sizes in a slowly moving medium goes over into the Stokes formula: $F_s = 6\pi\eta_c U R_d$. Certain difficulties in calculating the drag coefficients and the deposition rate arise for deformable droplets and bubbles when their dependence on the numbers Weber $\mathrm{We} = \frac{\rho_c U^2 a}{\sigma}$, Morton $\mathrm{Mo} = \frac{g\eta_c^4}{\rho_c \sigma^3}\frac{\Delta\rho}{\rho_c}$, Bond $\mathrm{Bo} = \frac{ga^2}{\sigma}\Delta\rho$, and Acrivos $\mathrm{Ac} = \frac{\rho_c \nu_c^2}{\sigma a}$, $\mathrm{Ac} = \mathrm{We}\,\mathrm{Re}_d^{-2}$, appears in the region of their deformations. These parameters are very important factors that determine the degree of migration and sedimentation of particles in the turbulent flow. It should be noted that for Stokes spherical solid particles at $\mathrm{Re}_d < 1$, the rate of their deposition from the volume is determined as $V_s = \tau_p g$.

It should be noted that diffusion transport to the surface, which depends on the coefficient of turbulent diffusion of particles, plays an important role in the migration, migration, and deposition of fine particles.

$$D_{Td} = \mu_p D_T \tag{8.162}$$

A set of formulas for calculating the coefficient of turbulent diffusion of particles as a function of the dynamic velocity and deposition rate are proposed in the paper [48]. It should be noted that in industrial practice there are problems of deposition on the internal surface of tubular apparatus (transport tubes, heat exchangers, condensers, tubular furnaces, etc.), accompanied by deposition and formation of a dense layer with a low coefficient of thermal conductivity on the inner surface of the pipes, time, a significant influence on the processes of heat transfer, mass and momentum, and in general, leading to a decrease in the efficiency and productivity of the process. Such sedimentation primarily leads to an increase in the energy costs of the entire process.

8.7.1 Sedimentation of Solid Spherical Particles from the Volume

Sedimentation of solid particles from the flow volume in industrial practice is used for separation, purification, and classification of particles by size or by fraction. The basis of these processes is usually the presence of a gravitational field, although other fields (centrifugal, electric, magnetic, etc.) are sometimes used, or combinations of these fields are used for deeper purification. To calculate such processes, an important parameter is the rate of gravitational sedimentation of particles, determined by the nature of the total effect of forces of different nature on the particle and depending on the number Re_d and physicochemical properties of the medium and particles (the differences in particle and medium densities $\Delta\rho$, viscosity of the medium ν_c, surface tension), shape, and size of particles [6,8,12,249]. The solution of the problem of precipitation of a single particle in a slow flow is sufficiently known for small numbers and complexity only for moderate and large numbers, related to the complexity of the flow around the particle (separation of the boundary layer from the particle surface, resistance crisis, turbulence of the track, etc.) [24,91]. The motion of a single particle in a force field with a slow flow of the medium, taking into account the attached mass, the weight force, corrected for the strength of Archimedes and the drag force, is described by equations [8,12,249]

$$\frac{dV_p}{dt} = \frac{\Delta\rho}{2\rho_d + \rho_c} g - \frac{3}{4} C_D \frac{\rho_c}{2\rho_d + \rho_c} \left|V_p - U\right| \left(V_p - U\right) \tag{8.163}$$

In the steady state, this equation is transformed to a simpler form [5,91,249]

$$V_s = \sqrt{\frac{4}{3} \frac{\Delta\rho\, a\, g}{\rho_c C_D}} \tag{8.164}$$

Here, C_D is the resistance coefficient for various regions of the number change Re_d for solid particles and deformable drops can be found in the works [25,26,91,253]. In the particular case, if $\mathrm{Re}_d < 1$ $\left(C_D = 24/\mathrm{Re}_d\right)$, from equation (8.164), one can obtain the well-known expression for the Stokes sedimentation velocity of fine particles: $V_p = \tau_p g$. In work [254] to solve the problem of the precipitation of solid spherical particles, the drag coefficient is adopted in the form

$$C_D = \left[\left(\frac{A}{\mathrm{Re}}\right)^{1/n} + B^{1/n}\right]^n \tag{8.165}$$

Obviously, this formula within the permissible error is only suitable for $\mathrm{Re}_d \le 10^4$. A similar expression for the drag coefficient is used for drops and gas bubbles in the work [24,253] for the area $\mathrm{Re}_d \le 100$.

Finally, using (8.164) and (8.165), for the deposition rate in [254], expression

$$V_s = \frac{\nu_c}{a}\left[\sqrt{\frac{1}{4}\left(\frac{A}{B}\right)^{2/n} + \left(\frac{4}{3}\frac{a_*^3}{B}\right)^{1/4}} - \frac{1}{2}\left(\frac{A}{B}\right)^{1/n}\right]^n \tag{8.166}$$

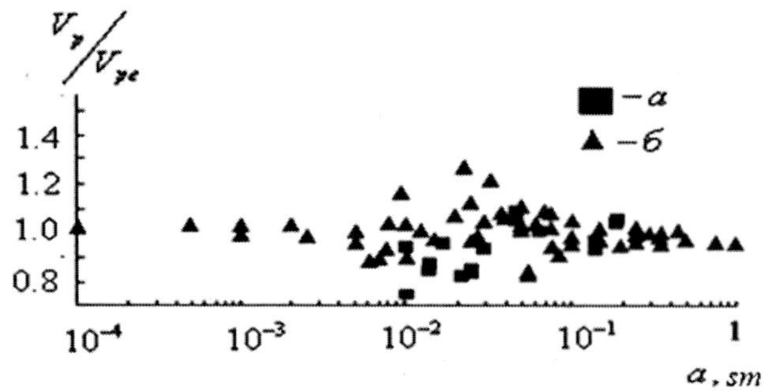

FIGURE 8.18 Comparison of experimental and calculated values of settling velocity of solid spherical particles according to equations (8.165) and (8.166): (a) [263]; and (b) [264] (here V_{pe} are the experimental values of settling velocity).

Here, $a_* = \left(\frac{\Delta\rho g}{\rho_c v_c^2}\right)^{1/3} a$. Comparison with various experimental data on the precipitation of solid spherical particles allowed us to determine the coefficients entering into this equation: $A = 23.2,\ B = 0.4,\ n = 2$ [256]; $A = 30.6,\ B = 0.37,\ n = 2$ [255]; $A = 21.4,\ B = 0.36,\ n = 2$ [257]; $A = 24.0 - 66.0,\ B = 0.4 - 0.7,\ n = 0.8 - 2$ [258].

The next formula for the deposition rate of solid spherical particles is expression [253,259–261]

$$V_s = \frac{\nu_c}{a} a_*^3 \left[\left(\frac{3A}{4}\right)^{2/n} + \left(\frac{3B}{4} a_*^3\right)^{1/n}\right]^{-n/2} \tag{8.167}$$

The values of the coefficients entering into this equation are as follows: $A = 26.2,\ B = 1.91,\ n = 2$ [259]; $A = 24.0,\ B = 4/9,\ n = 2.0$ [260]; $A = 24.0,\ B = 0.44,\ n = 2.22$ [257]; $A = 32,\ B = 0.83,\ n = 2.0$ [262]. The relative error for (8.166) and (8.167) is of the order of 6.36–7.02 (Figure 8.20). It should be noted that, depending on the nature of the expression $C_D(\text{Re}_d)$ for different regions of variation of the number Re_d, equation (8.165) is acceptable for describing the deposition rate of single solid spherical particles, [259,263,264,265] and also with certain changes for cylindrical particles [226,266].

As follows from Figure 8.18, a large scatter of experimental data on the deposition rate of particles is observed with increasing particle size and concentration.

It should be noted that in the literature [173,213,226,267] there are more complex expressions for the drag coefficient, covering a wide range of the number change $\text{Re}_d \le 10^6$, for which the derivation of the expression for the precipitation rate of solid particles in an explicit form is not possible. In this case, the calculation of the sedimentation rate should be carried out numerically by solving the transcendental equation using experimental data.

8.7.2 Sedimentation Particles from a Concentrated Dispersed Stream

In practical applications, often polydisperse systems are characterized by a sufficiently high concentration of particles, hence, are accompanied by the interaction of particles, the constraint of precipitation and the coagulation of particles or the formation of agglomerates. When particles are precipitated from the volume under the action of gravity, the most typical problem is gravitational coagulation, [6,40] that is, the capture of large particles by small particles along the path when they move. Gravitational coagulation is characterized by a large particle capture coefficient and is determined in the form of empirical relationships

$$\vartheta = \frac{\text{Stk}^2}{(\text{Stk}+0.5)^2}, \quad \text{Stk} > 1/12, \quad \text{Re}_d >> 1;$$
$$\vartheta = \left(1 + \frac{3}{4}\frac{\ln(2\text{Stk})}{\text{Stk}-1.214}\right), \quad \text{Stk} > 1.214, \quad \text{Re}_d << 1 \tag{8.168}$$

where $\text{Stk} = \frac{2}{9}\frac{\rho_d R_1^2}{\eta_c R_2} V_s(R_2)$ is the Stokes number, and R_1, R_2 are the radii of small and large particles. It should be noted that the interaction of particles is possible if the distance between particles [12] $l_m \approx 80a\sqrt[3]{\frac{\rho_d}{C_m}}$ (C_m is mass concentration of particles) is less than the particle size. In addition [220,268,269], it is important to note that the precipitation of coarse particles is accompanied by the emergence of ascending flows of the carrier medium as a result of its displacement, leading to the emergence of finely dispersed components of the disperse system.

A detailed analysis is made of the deposition of solid spherical particles describing the rate of their restricted deposition, the essence of which is reduced to the Richardson–Zaki [270]

$$V_p / V_s \approx (1-\varphi)^n \tag{8.169}$$

with various empirical changes associated with the dependence of the exponent n on a number of parameters (particle size, volume fraction, etc.), as well as the Happel and Kuwabara equations [21,271]. Due to the use of various empirical relationships, in the final analysis, for the rate of restricted deposition of a multitude of particles, one can write

$$\frac{V_p}{V_s} = \xi(\varphi, a) \tag{8.170}$$

where $\xi(\varphi, a)$ is the function characterizing the effect of the parameters of the medium and dimensions on the constraint of motion, and the form of which is proposed by empirical dependencies of a different nature [21,40,269,270,272]

$$\xi(\varphi) = \frac{1-\varphi}{1+k\varphi^{1/3}}, \; k = 1.3-1.9; \quad \xi(\varphi) = (1-\varphi)(1-\frac{5}{2}\varphi), \quad \varphi < 0.4$$
$$\xi(\varphi) = (1-\varphi)^{1/2}\Phi(\varphi)\frac{18.67}{1+17.67\left[\Phi(\varphi)\right]^{6/7}}, \; \Phi(\varphi) = (1-\varphi)^{1/2}\frac{\eta_c}{\eta_\varphi}, \; \text{Re}_d > 1000 \tag{8.171}$$
$$\xi(\varphi) = \frac{1 - 3/2\,\varphi^{1/3} + 3/2\,\varphi^{5/3} - \phi^2}{1 + 2/3\,\varphi^{5/3}}, \quad \xi(\varphi) \approx 1 - \frac{9}{5}\varphi^{1/3} + \varphi - \frac{1}{5}\varphi^2$$

The last expressions correspond to the corrections of Happel and Kuwabara [21,271] to the deposition rate. It is not difficult to show that all these dependences have a significant difference, although their values vary within limits $0 < \xi(\varphi) < 1$. A more complete analysis of the function in the preceding equations and possible causes of the scatter of theoretical and experimental data are discussed in the papers [21,270]. Perhaps these reasons are associated with the formation of aggregates and with the violation of the transverse homogeneity of the particle distribution as a result of coagulation, the dependence of this function on the particle sizes and properties of the medium and particles, etc. Thus, in spite of the presence of a set of formulas for determining the function $\xi(\varphi, a)$ that affect the cramped deposition of particles, it is clearly impossible to confirm which of the above formulas best coincides with the experimental data.

Using the Happel equation to calculate mass transfer to constrained particles and the equation for convective transport (5.114) for small numbers Pe<<1, we can write

$$\mathrm{Sh} = \frac{(n+1)^{\frac{n}{n+1}}}{\Gamma\left(\dfrac{1}{n+1}\right)}\left[\frac{\left(\frac{2}{3}\right)\left(3+2\varphi^{5/3}\right)}{2-3\varphi^{1/3}+3\varphi^{5/3}-2\varphi^2}\right]^{1/n+1}\mathrm{Pe}^{\frac{1}{n+1}} \tag{8.172}$$

If we take the linear velocity distribution profile in the boundary layer $n = 1$, then this equation goes into an expression of the type

$$\mathrm{Sh} = \frac{2}{\sqrt{3}\Gamma\left(\frac{1}{2}\right)}\left[\frac{\left(3+2\varphi^{5/3}\right)}{2-3\varphi^{1/3}+3\varphi^{5/3}-2\varphi^2}\right]^{1/2}\mathrm{Pe}^{\frac{1}{2}} \tag{8.173}$$

In particular, taking a parabolic velocity profile along the thickness of the boundary layer $n = 2$, this equation can be represented in the form

$$\mathrm{Sh} = \frac{6^{1/3}}{\Gamma\left(\frac{1}{3}\right)}\left[\frac{\left(3+2\varphi^{5/3}\right)}{2-3\varphi^{1/3}+3\varphi^{5/3}-2\varphi^2}\right]^{1/3}\mathrm{Pe}^{\frac{1}{3}} \tag{8.174}$$

In the literature, one can find a lot of experimental studies on constrained deposition, among which we should mention the work [273], where a lot of experimental data are presented for the rate of precipitation of solid particles of different nature from the flow (Figure 8.19), which are satisfactorily approximated by expression (8.156).

In particular, the lower and upper limits of the distribution of experimental data are satisfactorily described by expressions of the type (8.156): curve 1 corresponds to $n = 3 - 0.9\varphi^{0.92}$ curve 3 is $n \approx 1$. Proceeding from these expressions, the average statistical precipitation curve is also described by

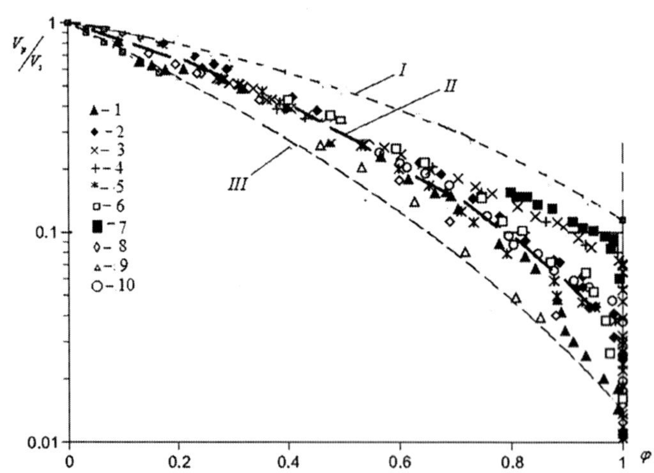

FIGURE 8.19 The velocity of cramped deposition of particles of different types: (1), (2), (3), and (4) glass particles, in sizes, respectively, $a = 0.5\,\mathrm{mm}$, $a = 0.35\,\mathrm{mm}$, $a = 1.85\,\mathrm{mm}$, and $a = 3.0\,\mathrm{mm}$; (5) polymer particles, $a = 2.4\,\mathrm{mm}$; (6) particles anthracite, $a = 1.32$ mm; (7) lead particles, $a = 2.4\,\mathrm{mm}$; (8) and (9) of the particles sand dimensions, respectively, $a = 0.22$ mm and $a = 0.32$ mm; (10) particles gravel, $a = 2.4$ mm; (I) upper limit of experimental data; (II) curve calculated by the formula $n = 1.75 - 1.25\varphi^{8.5}$; and (III) lower limit experimental data.

the expression (8) with $n = 1.75 - 1.25\varphi^{8.5}$ (in Figure 8.19 this curve corresponds to the number 2—fatty dash-dotted curve), although the literature confirms that the exponent is equal $n = 3.5 - 4$ in equation (8.156).

It is possible that this is due to a decrease n in the index with increasing particle sizes. The dependence of the exponent on the particle size using experimental data [273] can be expressed in the form (Figure 8.20)

$$n = \left(0.049 + 0.07a^2\right)^{-1/2} \tag{8.175}$$

with a correlation coefficient equal to $r^2 \approx 0.95$. As follows from Figure 8.16, a large spread of the experimental data is observed in the region $\varphi > 0.8$ where the suspension has high effective viscosity values. Experimental studies show that the greater the density of the particles themselves, the greater the error in the discrepancy between the experimental and calculated values of the sedimentation rate. Obviously, this corresponds to higher values of the index n, depending on the particle density, the effective viscosity of the suspension, the number Ar, etc.

The dependence of the exponent n on the Galilean number in the form [249]

$$n = 2.51\log\left[\frac{\left(1.83\text{Ga}^{0.018} - 1.2\text{Ga}^{-0.016}\right)^{13.3}}{\sqrt{\left(1 + 5.53\times10^{-5}\text{Ga}\right) - 1}}\right], \quad 10 < \text{Ga} < 10^5 \tag{8.175a}$$

where $\text{Ga} = \frac{a^3\rho_c(\rho_d - \rho_c)g}{\eta_c^2}$ is the Galileo number. The dependence of the exponent on the Galilean number is given in Figure 8.20 (curve b). In this paper, we propose the following approximation on the basis of experimental data

$$n = \frac{4.75}{\left(1 + 0.0005\text{Ga}^2\right)^{1/25}} \tag{8.175b}$$

The expression (8.175b) in comparison with (8.175a) is simpler with a small number of unknown coefficients and yields to a better approximation of the experimental data (Figure 8.21).

The set of experimental data for calculation n, depending on the particle size, density, and viscosity of the medium, proposed in the works [269,270,274–276], are too scattered and difficult to give any mathematical description. A sufficiently large number of empirical formulas and

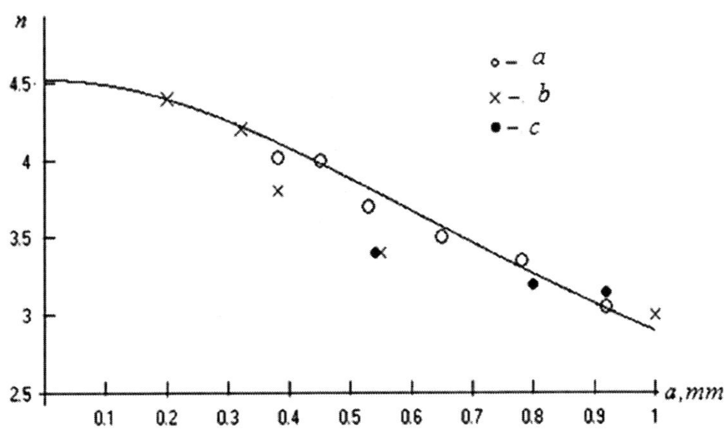

FIGURE 8.20 Dependence of the exponent on the particle size: (a) [43]; (b) [44]; and (c) [42].

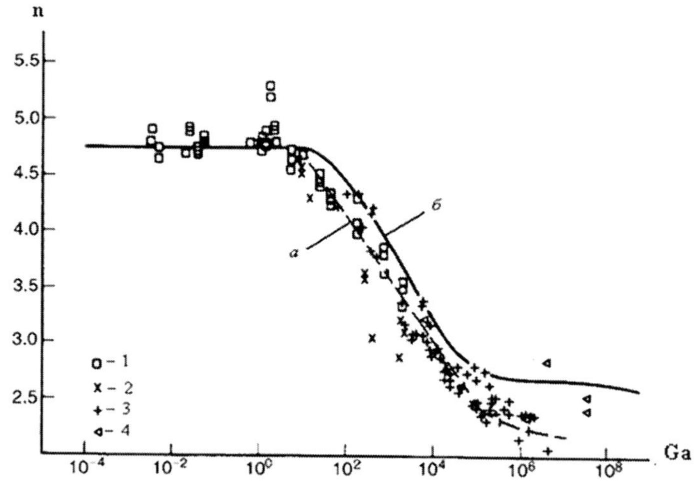

FIGURE 8.21 Comparison of experimental and calculated values for different values of the ratio a/d_T (d_T is diameter of the pipe or stirrer) equal to: (1) <0.001; (2) 0.001–0.01; (3) 0.01–0.1; and (4) >0.1.

experimental data on the determination of the deposition rate of solid spherical particles in a liquid medium are also given in the works [91,256,257,265,276,277] for the range of the number change $2\times10^4 \le \mathrm{Re}_d \le 3\times10^5$, where the drag coefficient is assumed to be equal $C_D \approx 0.44$.

The wide scatter of the experimental data on the deposition of solid particles, given in the literature, does not allow one to unequivocally indicate a specific description of the processes for a large range of the number change, satisfying some regularity (except for Stokes particles), although in practical calculations some modifications of the given formulas are made by introducing empiricism for specific applications.

8.8 SEDIMENTATION OF PARTICLES IN AN ISOTROPIC TURBULENT FLOW

The sedimentation of particles is strongly influenced by the turbulence of the flow, and this effect is perceptible depending on the ratio of particle sizes and the scale of turbulent pulsations. If the particle sizes are more or comparable to the Kolmogorov scale of turbulence, then the behavior of the particles is characterized by turbulent wandering, which increases the probability of collision, coagulation, and deposition rate. Theoretical and experimental problems of gravitational sedimentation of aerosol particles from a volume with a small concentration in an isotropic turbulent flow are considered in the works [250,278–286]. Statistical models of motion, particle deposition, and heat transfer in an isotropic turbulent flow with a constant temperature gradient are considered in the works [286,287]. It should be noted that in an isotropic turbulent flow, if the particle size is smaller than the scale of turbulent pulsations $a < \lambda_0 = (\nu_c^3/\varepsilon_R)^{1/4}$, then such particles, for which $\mu_p \to 1$, follow any pulsations. For the case $a < \lambda_0$ and small concentration $\varphi << 1$, according to the principles of hydrodynamic analogy and similarity, one can write

$$\frac{V_p}{V_s} \sim \frac{U_\lambda}{V_\lambda} \tag{8.176}$$

where U_λ is the Kolmogorov scale of velocity, V_λ is velocity the of turbulent pulsations equal to $V_\lambda = \left(\frac{\varepsilon_R^3}{\nu_c}\right)^{1/4} \tau_\lambda$ the scale λ_0, $\tau_\lambda = \lambda/U_\lambda = \left(\nu_c/\varepsilon_R\right)^{1/2}$, is the time scale of turbulence or the period of turbulent scale fluctuations λ_0, V_s is the rate of deposition of Stokes particles. Then for the particle deposition rate in an isotropic turbulent flow we can write

$$\frac{V_p}{V_s} \sim \frac{U_\lambda}{\tau_\lambda}\left(\frac{\nu_c}{\varepsilon_R^3}\right)^{1/4} = \frac{U_\lambda}{\left(\varepsilon_R \nu_c\right)^{1/4}} \tag{8.177}$$

This equation can also be rewritten in the form

$$\frac{V_p}{U_\lambda} \sim \frac{V_s}{\left(\varepsilon_R \nu_c\right)^{1/4}} = \frac{\tau_p g}{\left(\varepsilon_R \nu_c\right)^{1/4}}\left(1-\frac{\rho_c}{\rho_d}\right) \tag{8.178}$$

For small particle concentrations and particle sizes, a similar equation has been proposed in the paper [278,279] for the rate of gravitational droplet deposition in an isotropic turbulent flow in the form

$$\frac{V_{sk}}{U_\lambda} = \mathrm{Stk}_\lambda\left(\frac{\nu_c}{\varepsilon_R^3}\right)^{1/4} g\left(1-\frac{\rho_c}{\rho_d}\right) \tag{8.179}$$

where $\mathrm{Stk}_\lambda = {}^{\tau_p}\!/_{\tau_\lambda}$ is Stokes number for turbulent flow. Comparison of this expression with the experimental data in the region $24.5 \le \mathrm{Re}_\lambda \le 42.7$, $0.92 \le \mathrm{Stk}_\lambda \le 3.2$, and $0.435 \le \lambda \le 0.606$ gives a fairly satisfactory relative error ~6%–8%. Comparing with the deposition rate in a laminar flow $V_{sk} = \tau_p g\, {}^{\Delta\rho}\!/_{\rho_d}$, it can be noted that the relaxation time for particles in an isotropic turbulent flow can be determined as: $\tau_p = U_\lambda \mathrm{Stk}_\lambda \left({}^{\nu_c}\!/_{\varepsilon_R^3}\right)^{1/4}$. Consequently, for finely dispersed particles that are smaller than the scale of turbulent pulsations, the particle deposition rate will depend on the intensity of turbulent pulsations, the measure of which is the dissipation of turbulent energy ε_R and is characterized by random motion of the particles along with the pulsating moles that carry them. The experimental dependences of the number $\mathrm{Re}_\lambda = {}^{U'\lambda}\!/_{\nu_c}$, scale of turbulence λ, and the velocity U', on the volume fraction of particles $\left(1.5\times10^{-6} < \varphi \le 1.5\times10^{-4}\right)$, are proposed. Using the experimental data [279], the expression for the dependence of the specific dissipation of energy ε_R and the scale of turbulent pulsations on the volume fraction of particles can be expressed by the following empirical correlations (Figure 8.22)

$$\varepsilon_R = 340 - 2.96\times10^6\varphi + 3.75\times10^{10}\varphi^2 - 10^{14}\varphi^3 \tag{8.180}$$

$$\lambda = 0.6 + 1714.5\varphi - 4.94\times10^7\varphi^2 + 2\times10^{11}\varphi^3 \tag{8.181}$$

At low turbulence intensities and small numbers, turbulence slightly increases the resistance to movement of particles and, conversely, at high intensity, turbulence reduces the resistance, due to the narrowing of the trace behind the particle and a sharp decrease in the coefficient of resistance, thereby

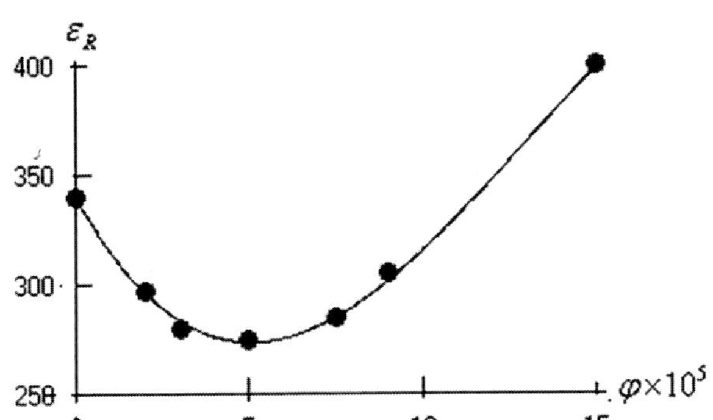

FIGURE 8.22 The dependence of the specific dissipation of energy on the volume fraction of particles.

increasing the deposition rate. Precipitation of large particles with high velocities in the turbulent flow is characterized by the fact that these particles do not remain inside the initial mol of the liquid, but leave it, moving downward under the influence of the force of weight and cross many other moles [6,12].

The increase in the intensity of turbulence contributes to the increase in the probability of collision and coarsening of particles, which is a positive effect for the processes of separation and separation of polydisperse systems.

Sedimentation of particles at their high concentration in a turbulent flow will be determined by their interaction, collision, coagulation, and breaking, as a result of which even a homogeneous dispersed system will always be polydisperse. The phenomena of coagulation and fragmentation of particles in different ways affect the velocity of cramped deposition in horizontal and vertical flows. In vertical streams, the polydispersity of the particles causes their delamination in height, thereby changing the rate of their deposition in height. In an isotropic turbulent flow, the phenomena of coagulation and fragmentation of particles depend on their maximum and minimum size, which determine the threshold of aggregate stability and which depend on the basic characteristics of turbulence and the properties of the particles. In a turbulent flow, the particle sizes exert a significant influence on the relaxation time τ_p and, given that the degree of entrainment of the particles by the pulsating medium $\mu_P \sim \tau_p^{-1/2}$, then with increasing τ_p the entrainment of the particles medium contributes to the growth rate of their deposition.

In conclusion, we note that the calculation of the rate of free deposition in a turbulent flow by numerical methods is given in many papers [288,289]. We note that usually the use of numerical methods of solving is connected with: (1) the nonlinear character of the problem arising from the nonlinearity of the particle resistance coefficient for a wide range of the number change [213,289]; (2) with the need to take into account the parameters of the turbulent flow, the nature of the cross-sectional velocity distribution, the dissipation energy, the turbulent diffusion coefficient, etc. [287]; (3) taking into account the aggregative instability of a polydisperse medium and thermophoretic and other types of particle migration. It is important to note that the numerical solutions obtained, although more accurately reflect the deposition pattern, but they, because of the complexity of the equations themselves, are not always suitable for practical calculations of a wide class of chemical technology problems associated with particle deposition and separation of a disperse system.

8.9 SEDIMENTATION AND RISING OF DROPS AND BUBBLES

The problems of the rising $(\rho_d < \rho_c)$ and deposition $(\rho_d > \rho_c)$ of droplets and bubbles in chemical technology are characteristic for the separation of emulsions, for the absorption of gases in column apparatuses, for liquid extraction, for stratifying a two-phase liquid, etc. The processes of separation and deposition of deformable particles (droplets, bubbles) in principle differ substantially from the settling of solid particles in that:

1. droplets and bubbles under certain conditions undergo deformation in the gravitational and electric fields, thereby changing the shape and nature of the resistance [6,8,21,26,85,290,291];
2. the fluid flowing around the drop and bubble initiates, due to friction, the circulation of the internal fluid from which the drop consists, this effect will depend on the dimensionless parameter representing the ratio of the dynamic viscosity of the internal liquid to the dynamic viscosity of the external liquid and expressing the degree of surface mobility. Naturally, for gas bubbles in a liquid medium, $\gamma \to 0$. For a slow medium flow for $Re_d << 1$, the deposition rate of a spherical droplet in a pure liquid is expressed by the Hadamard–Rybczynski formula [6,8,21,293]

$$V_{sk} = \frac{2}{9}\frac{\Delta\rho g R_d^2}{\eta_c}\frac{1+\gamma}{2+3\gamma} \tag{8.182}$$

From this expression it is not difficult to obtain an equation for the ascent of bubbles of small dimensions $\left(\gamma \to 0, \Delta\rho \to \rho_c\right)$

$$V_{sg} = -\frac{1}{9}\frac{gR_d^2}{\nu_c} \tag{8.183}$$

If the number $\mathrm{Bo} < 13$, then the bubble resistance coefficient can be expressed as [24,25]: $C_D \approx 0.568\mathrm{Bo}^{1/2}$, using which, we get the bubble ascent rate $\left(\rho_d << \rho_c, \Delta\rho \approx \rho_c\right)$

$$V_{sg} \approx 1.53\left(g\mathrm{K}\right)^{1/2} \tag{8.184}$$

where $K = \left(\frac{\sigma}{\rho_c g}\right)^{1/2}$ is the capillary number. For a number $\mathrm{Bo} > 40$, the drag coefficient stabilizes at a level $C_D \approx 8/3$. Using expression (8.184) for the rate of ascent of bubbles in the form of "spherical caps," we obtain

$$V_{sg} = \left(\frac{a_e g}{2}\right)^{1/2} \tag{8.185}$$

where a_e is the equivalent diameter of the bubble, corresponding to the radius of curvature of the spherical part of the cap.

3. the deposition of droplets and the emergence of bubbles $\left(\gamma \to 0\right)$ in concentrated streams are accompanied by their collision, coalescence (fusion) and fragmentation [6,65,250,251,292], which substantially changes the size spectrum and the nature of the particle motion. With an increasing coalescence rate, the average droplet size increases, which leads to an increase in the rate of deposition or ascent. In an isotropic turbulent flow, the phenomena of coagulation and fragmentation of liquid particles depend on their maximum and minimum size, which determine the threshold of aggregate stability and which depend on the main characteristics of turbulence and the properties of the particles.
4. droplet precipitation is accompanied by various side effects, such as the Marangoni effect [294–296], creating additional thermocapillary and convective currents on the drop surface due to the difference in temperature and surface tension at different points of the droplet surface. The Marangoni effect significantly changes the circulation flow of liquid inside the drop, creating a certain chaos in the flow, as a result of the formation of many local convective flows on the surface penetrating deep into the volume of the drop, and exerts a retarding effect on the deposition rate. Ultimately, the rate of deposition of deformable particles (droplets and bubbles) is determined by the nature of the hydrodynamic flow around the particles, the properties of the medium, and the deformation of their shape (Figure 8.22). Moreover, for small numbers $\mathrm{Re}_d < 1$, the droplets and bubbles have a high spherical shape, and the deposition rate obeys the Hadamard–Rybchinsky law (8.182), the shape of the particles changes Re_d to an ellipsoidal shape with increasing, and then acquires forms that cannot be described. Such a change in the shape of the deformable particles is determined by turbulence and separation of the boundary layer and by many other factors (Figure 8.23).

The paper [8,25,297] proposes empirical formulas for calculating the bubble surfacing velocity in a liquid medium in the form of piecewise approximation

$$V_{sg} = 0.13\left(g\frac{\Delta\rho}{\rho_c}\right)^{0.72}\frac{a^{1.18}}{\nu_c^{0.45}},\quad 1.4 < \mathrm{Re}_d < 500\ ;\ V_{sg} = 1.74\left(g\frac{\Delta\rho}{\rho_c}a\right)^{1/2},\quad \mathrm{Re}_d > 500 \tag{8.186}$$

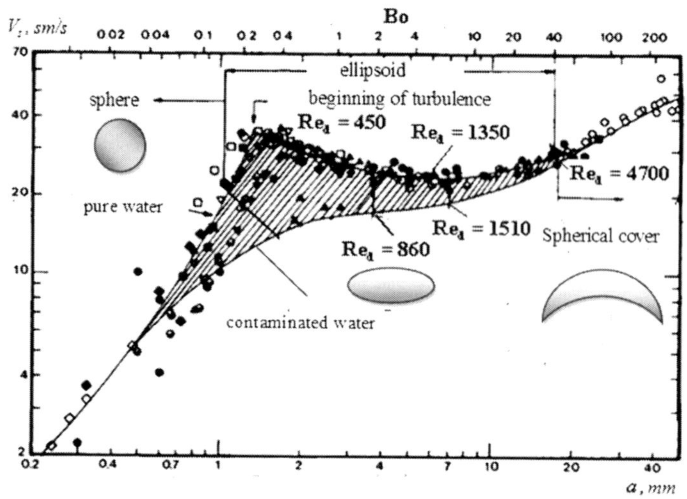

FIGURE 8.23 Dependence of the ascent rate of air bubbles in water (points are experiments [91]).

In addition to the above formulas for calculating the rate of deposition or rising (8.182 through 8.186), we propose a piecewise approximation of the bubble rise rate for a large range of variation of numbers Re_d, Mo [18]

$$V_{sg} = \chi\left(\mathrm{Ac}, \mathrm{Mo}, \mathrm{R}_d\right)\left(gR_d \frac{\Delta\rho}{\rho_c}\right)^{1/2}, \tag{8.187}$$

Assuming that the number $\mathrm{Ar}=\mathrm{Ac}^{-3}$ Mo, the equation for calculating the function in (8.187), taking into account the influence of the droplet and medium viscosity ratio $\gamma = {}^{\eta_d}\!/_{\eta_c}$, as well as the effect of numbers Mo, Ac, We, or droplet deformation, are given in Table 8.8, where $\mathrm{K} = \left(\frac{\sigma\Delta\rho}{\rho_c^2 g}\right)^{1/2}$. In particular, the rate of ascent of bubbles in various liquids using (8.187), the expressions given in Table 8.1, and the experimental data is proposed in the form of a semi-empirical expression

$$V_p = -\frac{\left(9ag\,{}^{\Delta\rho}\!/_{2\rho_c}\right)^{1/2} a^{3/2}\mathrm{Mo}^{-1/4}}{1+\mathrm{k}_0 a^{3/4}+\mathrm{k}_1 a^{7/3}}\,\frac{1+\gamma}{1+{}^{3}\!/_{2}\gamma},\quad 9\times10^{-7}\le \mathrm{Mo}\le 78,\ \mathrm{Re}_d<3\times10^{3}$$
$$\mathrm{k}_0 = 0.2\left(0.1+\mathrm{Mo}^{-1/5}\right),\quad \mathrm{k}_1 = \left(0.02+5\mathrm{Mo}^{2/5}\right)^{-1} \tag{8.188}$$

TABLE 8.8
Formulas for Calculation χ

Expressions to Define the Value χ	**Conditions**
$\chi = 2\mathrm{Ar}^{1/2}\frac{1+\gamma}{6+9\gamma} = 2\mathrm{Ac}^{-3/2}\mathrm{Mo}^{1/2}\frac{1+\gamma}{6+9\gamma}$,	$\mathrm{Re}_d<1$, $\mathrm{Ar}<4.5$
$\chi = \frac{1}{3}\mathrm{Ar}^{1/4} = \frac{1}{3}\mathrm{Ac}^{-3/4}\mathrm{Mo}^{1/4}$,	$4.5<\mathrm{Ar}<100\left(\frac{\mathrm{R}_d}{\mathrm{K}}\right)^{1/2}$
$\chi \approx 1.4\left(\frac{R_d}{K}\right)^{-5/6}$,	$\mathrm{Ar}>2000\left(\frac{\mathrm{R}_d}{\mathrm{K}}\right)^{3.8}$, $\frac{R_d}{K}<1.5$
$\chi \approx 1.3\left(\frac{R_d}{K}\right)^{-1/2}$,	$\mathrm{Ar}>2000\left(\frac{\mathrm{R}_d}{\mathrm{K}}\right)^{-3.8}$, $1.5<\frac{R_d}{K}<2$

Figure 8.24 shows the calculated values of the bubble rising velocity [149] in various liquids (for different numbers Mo) calculated from formula (8.188) and experimental values. As follows from the experimental data on the rising of particles (Figure 8.24) and from (8.188), the influence of the number $\gamma \approx 0.04$ on the rising rate of bubbles is significant only at its small values Mo << 1. As follows from the experimental data and equation (8.188) for large values of the number $1.4 < \mathrm{Mo} < 78$, the bubble emerges according to the Stokes law and is proportional $\sim a^2$, with a decrease in the number $\mathrm{Mo} < 9\times10^{-4}$, the ascent rate is proportional $\sim a^{5/4}$, and for the values of the number $\mathrm{Mo} < 10^{-7}$ is proportional $\sim a^{-1/3}$.

For a large range of air bubble sizes, their rising in an aqueous medium using experimental data [8,297] can be described by equation

$$V_p = \frac{\left(9ag\,\Delta\rho/2\rho_c\right)^{1/2} a^{3/2}\mathrm{Mo}^{-1/4}}{1+1.2a^{3/4}+4550a^{10/3}} + 98\left[1-\exp\left(-0.18a^2\right)\right] \tag{8.189}$$

Where you can take $\left(\frac{9}{2}g\frac{\Delta\rho}{\rho_c}\right) \approx 22$ for an air-to-water system. Figure 8.25 shows a comparison of the calculated (8.189) and experimental values of the rising rate [8,17].

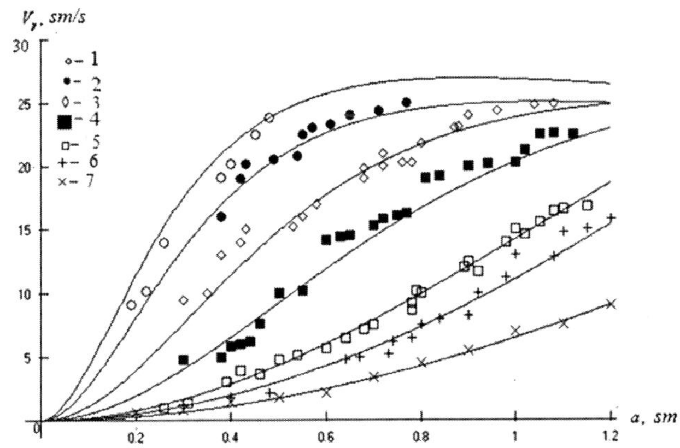

FIGURE 8.24 The rate of rising of bubbles in various media, depending on their size and number Mo (points are experiment [149]), equal to: (1) 9×10^{-7}; (2) 10^{-4}; (3) 9×10^{-4}; (4) 0.023; (5) 1.4; (6) 7.0; and (7) 78.0.

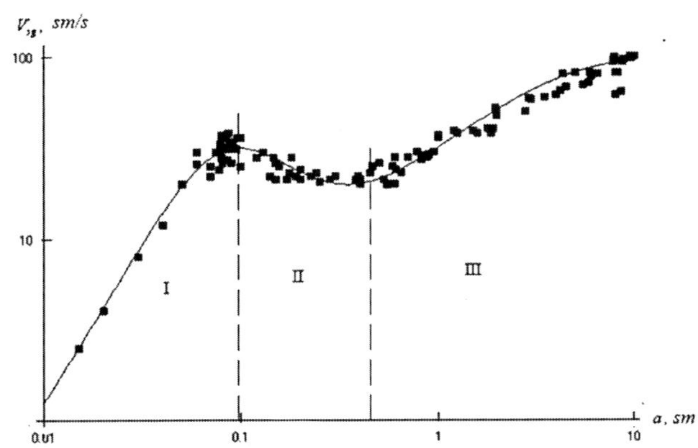

FIGURE 8.25 Comparison of the estimated ascent rate of air bubble (8.189) with experimental data. (From Soo, S.L., *Fluid Dynamics of Multiphase Systems*, Blaisdell Publishing Company, Waltham, MA, 1970; Koisi Asano Mass Transfer. From Fundamentals to modern Industrial Application, WILEY –VSH Verlag GmbH and Co, Weinheim, 2006.)

As follows from formula (8.189) for very small values of the number Mo (in particular, for water $Mo \approx 3\times10^{-11}$), the rising rate of large bubbles is proportional $\sim a^{-4/3}$. Thus, we can assume that the dependence of the exponent of the particle size in the equation for the rising rate of bubbles on the number Mo is the following $\sim Mo^{-1/8}$. This index will not depend on the number Mo for those liquids whose value is of the order of.

As follows from Figure 8.25, with an increase in the value of the number, the area of deformation of the bubbles shifts toward their larger dimensions. Figure 8.25 shows: (I) region of laminar flow around a particle, where the shape of the bubble is spherical; (II) intermediate region, where the shape of the bubble is close to an oval or ellipsoid; and (III) turbulence of the boundary layer, where the bubbles deform to spherical caps (Figure 8.23).

When flowing in vertical channels at high values $\lambda_s = {}^{a}\!/_{d_m}$, the wall of the pipe influences the velocity of the bubble ascent. When $\lambda_s \approx 1$ the motion of the bubble is characterized by a "piston" mode of rising, and when $\lambda_s << 1$ the effect of the wall becomes negligible. In this paper [83,91,298], the effect of the wall upon droplet deposition is taken into account as follows

$$\frac{V_{sk}}{V_{s\infty}} = (1-\lambda_s^2)^{3/2} \tag{8.190}$$

where $V_{s\infty}$ is the settling velocity without the influence of the wall, $\lambda \le 0.06,\ \text{Re} \le 0.1$; $\lambda_s \le 0.12$, $\text{Re} \ge 10$; $\lambda_s \le 0.08 + 0.02\log \text{Re},\ 0.1 < \text{Re} < 100$.

The rate of precipitation of water droplets in air on the basis of experimental data under normal conditions $\left(T = 25°\text{C},\ P = 1.0\ at,\ \eta_d = 0.01\,{}^{\text{gr}}\!/_{\text{sm s}},\ \eta_A = 1.8\times10^{-4}\,{}^{\text{gr}}\!/_{\text{sm s}}\right)$ [40], using the Stokes precipitation equation, can be represented in the form

$$V_{sk} = \frac{2}{9}\frac{\Delta\rho g R_d^2}{\eta_c}\varsigma(R_d),$$
$$\varsigma(R_d) = \left[1 + 701.55R_d^{3/2} - 388.7R_d^2\right]^{-1} \tag{8.191}$$

Figure 8.26 shows the experimental data on the deposition of droplets in air and the calculated curves calculated from equation (8.191).

As follows from expressions (8.189 through 8.191) for a wide range of changes in the number or large range of particle size changes, it becomes necessary to use an empirical approach that allows satisfactory description of the whole region within the allowed accuracy. This is due to the

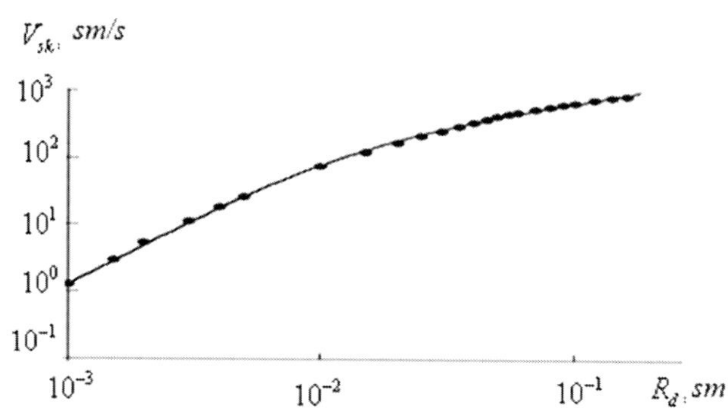

FIGURE 8.26 Dependence of the settling velocity of water droplets in air from their size (points are experiment data [40]).

fact that the presence of different droplet flow situations with increasing number Re_d and size of particles determines the complex nature of the resistance curve, which also affects the rate of rising and deposition. In this regard, to cover a large area of the number change, the most effective is the description of the resistance curve by different rows [24,25]. In contrast to the piecewise approximation of the ascent rate (8.187) and (8.188), the formulas using certain approximation series can unite the expression describing a large range of the number change.

It should be noted that, unlike bubbles, large droplet sizes are unstable, due to which they are subjected to breaking to the limiting size a_{min} [85,250,272], depending on the specific dissipation of energy and the properties of droplets and media.

8.10 SEDIMENTATION OF PARTICLES IN VERTICAL AND HORIZONTAL PIPES FROM OF AN ISOTROPIC TURBULENT FLOW

The nature of the deposition of particles from the polydisperse turbulent flow differs significantly from the free precipitation of them from the volume. As noted in the paper [12], when depositing aerosol particles on the walls of pipes and channels, the following mechanisms and models are distinguished: (1) free-inertial, based on the principle of free inertial ejection of particles to the wall; (2) lifting-migration, connecting the deposition of particles with their upward migration (the Magnus effect); (3) convective-inertial, which relate the rate of deposition of particles with inertial effects; (4) effective diffusion; and (5) turbulent migration, where the turbulent migration of particles to the wall is considered as the driving force of sedimentation, etc. The paper [12] contains a large number of experimental studies on the deposition of aerosol particles, carried out by the author himself, and borrowed from literature sources [48,250,272]. Here it should be noted the work, where the models of particle deposition are constructed on the basis of the migration-gravitational mechanism, which takes into account the diffusion and gravitational component of particle migration. The pattern of deposition of particles with the formation of a layer on the inner surface of horizontal and vertical pipes is different (Figure 8.27).

A symmetrical deposition layer is formed in the section of vertical pipes (Figure 8.27a), which is explained by the diffusion mechanism of the transverse migration of particles to the surface. In the cross section of horizontal pipes, an asymmetric deposition layer is formed (Figure 8.27b), which is explained by the migration-gravity mechanism, with the gravitational component of the sedimentation rate dominating the bottom of the pipe.

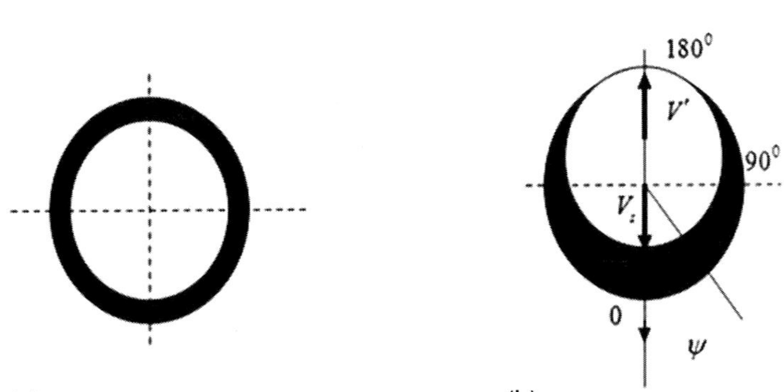

(a) (b)

FIGURE 8.27 The nature of the formation of a layer of particles in the cross section of (a) vertical and (b) horizontal pipes.

8.10.1 Sedimentation of Particles in Vertical Pipes and Channels from a Turbulent Flow

The basis for the deposition of particles from the turbulent flow in vertical pipes and channels with insignificant gravitational component is the turbulent transfer of particles to the surface of the wall from the flow, complicated by the forces of transverse migration. In this connection, in solving this problem, an important role is played by the distribution of the transverse pulsation flow velocity V' and the coefficient of turbulent diffusion. In this paper [48], using the experimental measurements [299–301], the expressions for the distribution of the transverse pulsation velocity and the distribution of the turbulent diffusion coefficient over the channel cross section are given in the form of expressions (4.44) and (4.45).

In the literature, there are many empirical formulas describing the rate of deposition of aerosol particles in vertical pipes, which reduce to an expression of the form

$$V_{s+} = A\tau_+^2 \tag{8.192}$$

where $V_{s+} = V_s/U_*$, $\tau_+ = \tau_p U_*^2/\nu_c$, is the dimensionless relaxation time, A is some coefficient equal to $A = 5.3\times10^{-4}$ [301], $\tau_+ < 5.5$; $A = 4.67\times10^{-4}$ [302], $\tau_+ < 8$; $A = 2.8\times10^{-4}$ [303]; $A = 6.0\times10^{-4}$ [304], $\tau_+ < 10$; $A = 5.3\times10^{-4}$ [301], $A = 3.25\times10^{-4}$ [305], $\tau_+ < 12$. A model of sedimentation in the form of [12]

$$V_{s+} = 7.25\times10^{-4}\left(\frac{\tau_+}{1+\omega\tau_+}\right)^2 \tag{8.193}$$

Taking into account the foregoing arguments permitting the influence of the transverse velocity of turbulent transport and turbulent diffusion, the following equation for the mass transfer of particles in a vertical pipe [250,285]

$$V_r\frac{\partial c}{\partial r} = \frac{1}{r}\frac{\partial}{\partial r}\left(rD_{Td}\frac{\partial c}{\partial r}\right)$$
$$r = 0,\ c = c_0,\ \frac{\partial c}{\partial r} = 0 \tag{8.194}$$

To simplify the analytical solution, the diffusion coefficients of particles for an isotropic turbulent flow in (8.194) are taken as (8.2a), that is, taken constant over the cross section of the pipe. For small dimensions of aerosol particles, the value of the transverse velocity component is defined in the form

$$V_{\mathrm{r}} \approx \frac{\Delta r}{\tau} = \frac{\theta_L V_{\mathrm{m}}}{\tau} \tag{8.195}$$

where $\theta_L = \int_0^\infty R_L(\theta)d\theta$ is the Lagrangian scale of turbulence, R_L is Lagrangian time correlation coefficient, and $\Delta r = \theta_L V_m$. Then, by introducing dimensionless variables

$$X = \frac{c - c_0}{c_0},\ \rho = \frac{\mathrm{r}}{\mathrm{R}}, \tag{8.196}$$

and taking into account the expressions (8.195) and (8.196), equation (8.194) is transformed to the form $(\alpha_1 \approx 1)$

$$\frac{R^{2/3}\theta_L}{\mu_{\mathrm{p}}^2\varepsilon^{1/3}}\frac{\partial X}{\tau\,\partial\tau} = \rho^{4/3}\frac{\partial^2 X}{\partial\rho^2} + \frac{7}{3}\rho^{1/3}\frac{\partial X}{\partial\rho} \tag{8.197}$$

$$\rho = 0,\ X = 0,\ \frac{\partial X}{\partial \rho} = 0 \tag{8.198}$$

The solution of equation (8.197) with the boundary conditions (8.198) is possible by the method of separation of variables, by introducing an expression $X(\tau,\rho) = \psi(\tau)\phi(\rho)$. By substituting this expression, we obtain two equations

$$\frac{\partial \psi}{\partial \tau} = -v^2 \tau \frac{\mu_p^2 \varepsilon^{1/3}}{R^{2/3}\theta_L}\psi,\ \psi = B_1 \exp\left[-\frac{v^2}{2}\frac{\varepsilon^{1/3}}{R^{2/3}\theta_L}\mu_p^2\tau^2\right] \tag{8.199}$$

$$\rho^{4/3}\frac{\partial^2 \varphi}{\partial \rho^2} + \frac{7}{3}\rho^{1/3}\frac{\partial \varphi}{\partial \rho} + v^2\varphi = 0 \tag{8.200}$$

Here, v^2 are the eigenvalues of the boundary value problem. To solve the second equation, we introduce a new variable $Z = \frac{3}{2}\rho^{2/3}$, as a result of which we have

$$Z^2\frac{\partial^2 \phi}{\partial Z^2} + \frac{7}{2}Z\frac{\partial \phi}{\partial Z} + \frac{3}{2}v^2 Z\phi = 0 \tag{8.201}$$

The solution of equation (8.201) is represented as

$$\varphi = Z^{-5/4}J_{5/2}\left(\sqrt{6}vZ^{1/2}\right) \tag{8.202}$$

where $J_{5/2}(z)$ the Bessel function is of fractional order. The general solution (8.196) is represented in the form

$$X(r,t) = \sum_{n=0}^{\infty}(-1)^n B_n\left(\frac{R}{r}\right)^{5/6} J_{5/2}\left[3.674v_n\left(\frac{r}{R}\right)^{1/3}\right]\exp\left[-\frac{v_n^2}{2}\frac{\varepsilon^{1/3}}{R^{2/3}\theta_L}\mu_p^2\tau^2\right] \tag{8.203}$$

Here, v_n are the positive roots of the equation $J_{5/2}(3.674v_n) = 0$ or $v_0 = 0$, $v_1 = 1.498$, $v_2 = 2.588$, etc.

From the solution of this equation, the mass flow of particles to the inner surface of the pipe according to the first Fick's law can be determined in the form

$$J = D_{Td}\frac{\partial c}{\partial r}\Big|_{r=R} = c_0(\varepsilon_R R)^{1/3}\sum_{n=0}^{\infty}(-1)^n k_n \exp\left[-\frac{v_n^2}{2}\frac{\varepsilon_R^{1/3}}{R^{2/3}\theta_L}\mu_p^2\tau_p^2\right] \tag{8.204}$$

and the settling velocity of particles on the surface in the form

$$V_p = \frac{J}{c_0} = (\varepsilon_R R)^{1/3}\sum_{n=0}^{\infty}(-1)^n k_n \exp\left[-\frac{v_n^2}{2}\frac{\varepsilon_R^{1/3}}{R^{2/3}\theta_L}\mu_p^2\tau_p^2\right] \tag{8.205}$$

Since the series converges rapidly, it suffices to restrict ourselves to three terms $V_{s+} = {}^{V_p}\!/_{U_*}$ and $\tau_+ = {}^{\tau_p U_*^2}\!/_{v_c}$ introducing the dimensionless quantities u, expression (8.205) simplifies to form

$$V_{s+} = A_0 + A_1\exp(-m_1\mu_p^2\tau_+^2) - A_2\exp(-m_2\mu_p^2\tau_+^2) \tag{8.206}$$

where $A_i = k_i\frac{(\varepsilon_R R)^{1/3}}{U_*}$, $m_i = \frac{\xi_i^2}{\theta_L}\left(\frac{\varepsilon_R}{R^2}\right)^{1/3}\left(\frac{v_c}{U_*^2}\right)^2$, $i = 0,1,2,\ldots$ are the coefficients of the series. Using experimental studies on the deposition of aerosol particles from a turbulent gas stream [301,304,306,307],

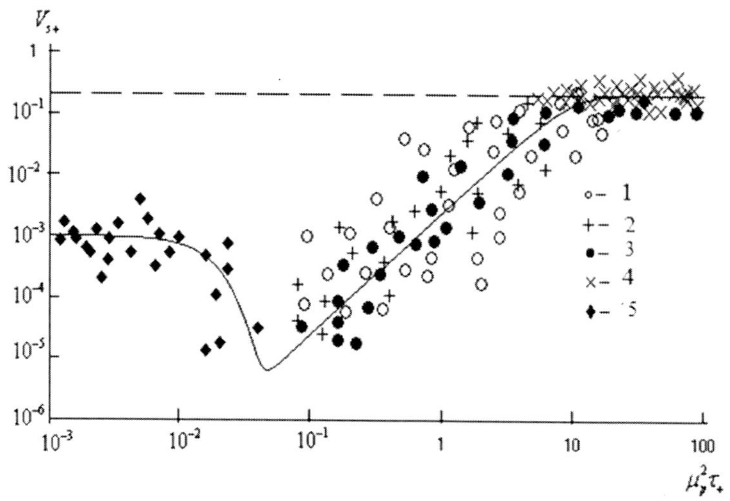

FIGURE 8.28 Comparison of the calculated values of the dimensionless particle deposition rate (8.199) with the experimental data: (1) [275]; (2) [278]; (3) [281]; (4) [12,281]; and (5) [306].

the coefficients in expression (8.206) are estimated as: $A_0 \approx 0.20$, $A_1 \approx 0.001$, $A_2 \approx 0.202$, $m_1 \approx 2\times10^3$, $m_1 \approx 0.002$ The second term in equation (8.206) is significant for the region $10^{-3} < \tau_+ < 10^{-1}$ and characterizes the deposition of submicron particles, for which $\mu_p \to 1$, $\mu_g \to 0$. In Figure 8.28, a comparison of the experimental data [12,307] is borrowed from the studies of various authors for $\tau_+ > 10^{-3}$ the rate of precipitation of aerosol particles $a = 0.1-2.04\,mkm$ with its calculated values (8.206).

As shown by the calculated and experimental studies in Figure 8.27, the relative particle settling velocity is set at and is defined as $\mu_p^2\tau_+ \geq 16.6$

$$V_p \approx k_2\frac{(\varepsilon_R R)^{1/3}}{U_*} \approx 0.2 \tag{8.207}$$

For small values $10^{-1} < \mu_p^2\tau_+^2 << 1$ of expression (8.206) simplifies to the form

$$V_{s+} \approx 2.4\times10^{-3}\mu_p^2\tau_+^2 \tag{8.208}$$

coincidental with the expression (8.195). When sedimentation fine particles from a turbulent liquid flow [87,250], the following precipitation rate formula

$$V_p = k_1\left(\frac{\varepsilon_R R^2}{\nu_c}\right)^{1/2}\left(b_0 - \exp(-m_1\mu_p^2\tau_p^2)\right) \tag{8.209}$$

As follows from equation (8.209), the rate of precipitation from the turbulent flow of a liquid medium on a vertical surface, in addition to other parameters, is inversely proportional to the viscosity of the medium $\nu_c^{1/2}$.

In addition to the studies listed above, papers dealing with the theoretical and experimental investigation of aerosol particle deposition on the inner surface of vertical tubes are also proposed [299,308,309], where a lot of empirical formulas are also proposed for the deposition rate, and stochastic modeling of sedimentation in a turbulent flow is proposed [310]. Most of the experimental data [301,304,307–309] available in the literature are characterized by a significant spread, not subject to deterministic description.

8.10.2 Sedimentation of Particles in Horizontal Channels

Unlike vertical channels, the rate of deposition in horizontal channels, along with the diffusion transfer, is significantly dependent on the gravitational component, which increases the deposition rate on the bottom of the pipe and reduces it to the upper ceiling part. This creates an asymmetric picture of the cross section of the deposited layer of particles (Figure 8.27b).

In this paper [311], the flux of particles on the inner surface of a horizontal pipe is represented by an empirical expression

$$J(\psi) = k_0\left[1+10\exp\left[2(\cos\psi - 1)\right]\right] \tag{8.210}$$

and in work [312] in the form

$$J(\psi) = k_0(V_d\psi)\exp\left[B(V_d)(\cos\psi - 1)\right] \tag{8.211}$$

where k_0, B are the coefficients are determined on the basis of experimental data. Experimental studies of the processes of particle deposition on the inner surface of horizontal tubes are devoted to work that experimentally confirms the asymmetry of the cross section of the deposited layer of particles [313,314]. In this paper [250,285], the equation for the rate of particle deposition in horizontal tubes for the case of neglecting diffusion transport is proposed in the form

$$\frac{\nu_c}{R^2}\frac{\partial^2 V_p}{\partial\psi^2} + F\left(V' - V_p\right) - FV_s\cos\psi = 0 \tag{8.212}$$

where $F = \tau_p^{-1}$ is the time constant, ψ – the angle calculated from the bottom of the pipe in radiance. As a result of solving equation (8.212), in the first approximation, the following expression is obtained

$$V_p = \alpha_0\left(1 - \alpha_1\sin^2\frac{\psi}{2}\right) \tag{8.213}$$

where $\alpha_0 = \tau_p g + U'$, $\alpha_1 = \frac{2U'}{\tau_p g + U'}$. Using the experimental data [250,285], these coefficients are expressed in the form of empirical relationships

$$\alpha_0 = \tau_p g + 4.55\times10^{10}\frac{\tau_+ U_*^{3/2}}{\mathrm{Re}^{2.45}}$$
$$\alpha_1 = \left(1 + 3.763\times10^{-10}\tau_+ U_*^{4.3}\right)^{-1} \tag{8.214}$$

For the deposition thickness along the pipe cross section [250,285], an expression of the form

$$\delta(t,\psi) = \frac{\alpha_0\nu_c}{U_*^2}\left(1 - \alpha_1\sin^2\frac{\psi}{2}\right)\left[1 - \exp(-mt)\right] \tag{8.215}$$

where m is the coefficient, the value of which is determined on the basis of experimental data. The flux of particles on the inner surface of a horizontal pipe is defined as

$$J \approx \mu_p^2\left(\varepsilon_R R\right)^{1/3}\rho_0 + \left(V' + V_s\cos\psi\right)\rho_0 \tag{8.216}$$

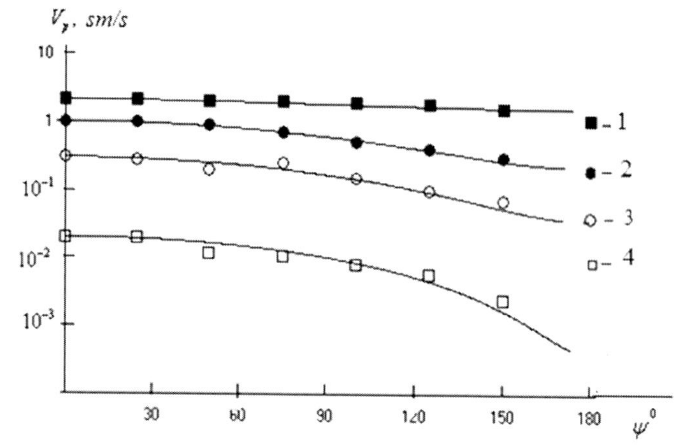

FIGURE 8.29 A comparison of the distribution of the particle deposition rate in a horizontal channel over a cross section at a dynamic flow velocity: (1) $U_* = 108\ ^{sm}/_{s}$; (2) $87.9\ ^{sm}/_{s}$; (3) $68\ ^{sm}/_{s}$; and (4) $28.7\ ^{sm}/_{s}$.

Figure 8.29 shows a comparison of the distribution of the deposition rate (8.213) in a horizontal tube of diameter $d_m = 10-20\ sm$ with experimental data on the sedimentation of zinc solid particles $\rho_d = 7140\ ^{kq}/_{m^3}$, $a = 1.69\,\text{mkm}$ and tungsten particles ($\rho_d = 1730\ ^{kq}/_{m^3}$, $a = 1.46\,\text{mkm}$ and $a = 2.04\,\text{mkm}$) over the pipe cross section for different values $0 < \psi < 180°$ of the dynamic flow velocity.

As follows from Figure 8.29 at the bottom of the pipe, the particle deposition rate for all values of the dynamic flow velocity is higher than for. At high values of the flow velocity (air), the rate of turbulent deposition is almost constant (Figure 8.29 [1] and [2]) and only slightly decreases at $\psi > 120°$. Perhaps this is due to the increase in the velocity of convective particle entrainment by the turbulent flow. At low values of the flow velocity, the decrease in the rate of turbulent deposition begins at $\psi < 30°$. It should be noted that for coarsely dispersed particles, the rate of gravitational deposition is greater than the rate of turbulent transfer. Finally, we note that the rate of deposition of particles from the turbulent flow is significantly affected by the surface quality of the pipes (roughness). For large values of the surface roughness, the viscous layer disappears, which affects the distribution of the transverse flow velocity and the diffusion transport of particles $\left(D_{Td} \sim \varepsilon^{0.094},\ \tau_+ \le 20,\ D_{Td} \sim \varepsilon^{-0.094},\ \tau_+ > 20\right)$, where ε is the height of the roughness protrusions.

8.10.3 Precipitation and Formation of a Dense Layer of Particles on the Inner Surface of Pipes and Its Effect on Heat and Mass Transfer

Many processes in chemical technology that take place in tubular apparatus (heat exchangers, condensers, transport pipes, tubular furnaces, homogeneous tubular reactors, etc.) [315–322] are accompanied by the precipitation and adherence of various impurity particles to the wall and the formation of a dense heterogeneous layer on the inner surface pipes with a certain roughness of the surface of the layer. Examples of such formations are paraffinic deposits in pipes during oil transportation, formation and deposition of ice in pipes during the transportation of natural gas, coke deposition in tubular homogeneous pyrolysis and heating furnaces, deposition of crystals of mineral salts on the surface of refrigerant condenser tubes, deposition of various particles in pipes heat exchangers, etc. It should be noted that the presence on the surface of a dense layer with a thickness $\delta(t)$ [315–322] that grows with the time of the process and is characterized by low thermal conductivity in comparison with the pipe wall greatly influences and alters the parameters of mass and heat transfer and flow hydrodynamics and, on the whole, creates non-stationary conditions for the entire process. Such deposition for technological processes is forced, which leads to an increase in the energy and material costs of the process.

The formation of a dense layer on the surface of pipes is carried out in the following ways: (1) direct deposition of various particles from the volume of the flow due to the migration-gravity mechanism; and (2) by transferring substances to the boundary layer and subsequent flow of various chemical and physical processes directly on the surface with further adherence of reaction products to the wall surface. This refers to the chemical formation of coke on the inner surface of homogeneous tubular furnaces and reactors, the formation of paraffin deposits in pipes for oil transport, the crystallization of various components (mineral salts) from the liquid medium on the surface of tubular capacitors, etc. The complex structure of the deposited layer will depend on the nature and type of particles, on their properties and dimensions, and the roughness of the surface of the layer will be determined by the size and shape of the particles.

The formation of a dense layer of particles on heat and mass exchange surfaces is determined by a variety of factors, among which it is important to note the hydrodynamic and thermodynamic conditions, the rheological properties of the disperse medium, the adhesion compatibility of particles with a streamlined surface, the physical and chemical transformations in the boundary layer, the particle size and concentration, the orientation of the streamlined surfaces, etc. [317–319]. Since the deposition processes are continuous, this causes a continuous increase in the thickness of the layer until the pipe is completely clogged. Analysis of such processes has shown that deposition on the inner surface of pipes with the formation of a dense layer of particles with a low thermal conductivity coefficient worsens heat exchange with the external environment, which results in a decrease in heat transfer and heat transfer coefficients. Introducing the dimensionless thickness of the deposits in the form $\beta = 1 - \delta/R_0$ (δ is the thickness of the deposited layer, R_0 is the radius of the clean pipe), we express the main transport parameters in the pipes β as: heat transfer coefficient for turbulent flow $\alpha/\alpha_0 \approx \beta^{-1.8}$ and for laminar flow $\alpha/\alpha_0 \approx \beta^{-3/2}$; flow velocity $U/U_0 = \beta^{-2}$; number $\mathrm{Re}/\mathrm{Re}_0 = \beta^{-1}$; velocity $V_\lambda/V_{\lambda 0} \approx \beta^{-15/8}$ and frequency of turbulent pulsations $\omega/\omega_0 \approx \beta^{-15/4}$; intensity of turbulence; $I/I_0 = \beta^{1/4}$ scale of turbulent pulsations $\lambda/\lambda_0 \approx \beta^{1.88}$; the degree of entrainment of particles by a pulsating flow $\mu_p/\mu_{p0} \approx \beta^{15/4}$; coefficient of resistance in pipes for laminar flow $\xi_L/\xi_{L0} = \beta$; and turbulent flow $\xi_T/\xi_{T0} = \beta^{1/4}$ (in these expressions, the subscript "0" refers to a clean pipe). As follows from these formulas, if in the case of laminar flow the growth of the thickness of the deposits in the pipes leads to a hydrodynamic instability, then for the turbulent flow the same factor increases the energy dissipation, the turbulence scale decreases, and, correspondingly, the turbulence intensity decays. As noted in the works [104,105], the increase in the thickness of the deposits in the pipes reduces over time the particle deposition rate to a certain extent, since the rate of detachment and entrainment of particles increases with increasing velocity of the main stream.

Assuming that the specific dissipation of energy in an isotropic turbulent flow is $\varepsilon_R \sim U^3/d_m$, it can be written down $\varepsilon_R = \varepsilon_{R0}\beta^{-5}$. It is noted in the paper [71] that in an isotropic turbulent flow the mass-transfer coefficient or number is defined as

$$\mathrm{Sh} = C_{13}\mathrm{Sc}^{1/3}\left(\frac{\varepsilon_R R^4}{\nu_c^3}\right)^{1/3} \tag{8.217}$$

Then, taking into account the above formulas, we can write $\mathrm{Sh} = \mathrm{Sh}_0\beta^{1/3}$, that is, with increasing thickness of sediments, the number or mass-transfer coefficient.

8.11 THE INFLUENCE OF THE DEPOSITION RATE ON THE EVOLUTION OF THE FUNCTION DISTRIBUTION

In practice, as a result of the deposition of particles, the particle size distribution function significantly changes as a result of the washout of the spectrum of large particles. The evolution of the distribution function is primarily affected by the loss of aggregate stability of the disperse system, expressed in continuous collision, coalescence, and crushing of droplets and bubbles, and in the variation of the dispersed composition. Particle sedimentation or sedimentation instability of the system has an inverse effect on the evolution of the distribution function, with time shifting the distribution curve toward small particles by washing away large particles from the spectrum. The evolution of the particle size distribution function and time can be determined on the basis of the solution of the stochastic Focker–Planck equation [71,86,203,323]. Considering the precipitation from the volume of the flow, and neglecting the diffusion of the particles, their interaction, the evolution of the particle distribution can be expressed by a simple equation

$$\frac{\partial P(a,t)}{\partial t}+V_p\frac{\partial P(a,t)}{\partial x}=0, \qquad P(a,t)\big|_{t=0}=P_0(a) \tag{8.218}$$

The solution of this hyperbolic equation can be represented in the form

$$P(a,t)=P_0(a)f(x-V_pt) \tag{8.219}$$

Let us determine the change in the mean particle size with time in the form

$$a_s=\int_0^\infty aP(a,t)da=a_{s0}f(x-V_pt) \tag{8.220}$$

where $a_{s0}=\int_0^\infty aP_0(a)da$ is the average particle size at the initial time. As follows from this formula, the current particle size decreases with time, if $f(x-V_pt)$ the decreasing function. In the work [40], more complicated cases of the influence of diffusion and deposition rate on the evolution of the distribution function are considered. For free Stokes particles in a stationary medium, in the absence of external forces, the solution of the Fokker–Planck equation has the form of a symmetric normal distribution. Obviously, the presence of interaction of particles leading to their coagulation and fragmentation, the influence of external forces, and many other factors presuppose the solution of the stochastic equation in the form of an asymmetric distribution curve with several extreme values [85,250,251].

Analysis of the processes of deposition of particles allows us to distinguish the following features:

1. Despite the existence of a large number of formulas for calculating the rate of deposition and the emergence of particles, it is not possible to isolate and propose a single formula uniquely, with the exception of Stokes particles $(\mathrm{Re}_d << 1)$. This is primarily due to the flow conditions of the particles (the separation of liquid from the surface of the particle, the turbulence of the boundary layer, etc.) for different numbers Re_d, the inconstancy of the size and shape of the particles, and also the possible physical phenomena, the state and properties of particles and the medium, leading to a complex dependence of the coefficient of resistance on the number. At the same time, to increase the interfacial surface in order to improve the phenomena of mass and heat transfer, it is necessary to have physical phenomena of crushing and grinding the particles. As follows from this study, for the rate of precipitation of solid particles from the volume, the following formulas can be proposed: (8.165, 8.166) for $\mathrm{Re}_d \le 10^4$; (8.167), with corrections for the tightness of precipitation and the particle size for $2 \times 10^4 \le \mathrm{Re}_d \le 3 \times 10^5$; (8.189) for the deposition and ascent of deformable drops and gas bubbles at $9 \times 10^7 \le \mathrm{Mo} \le 78$, $\mathrm{Re}_d < 3 \times 10^3$; (8.206 through 8.208) for the deposition of particles in vertical and horizontal pipes at $10^{-3} < \tau_+ < 10^2$. Sedimentation of the particles is complicated by: (a) the Magnus effect, which acts in the rotational motion of particles; (b) the Marangoni effect, characteristic of droplets and bubbles and leading to the appearance of convective currents on the surface; (c) the presence

of various physical evaporation processes (the Fasi effect, the Stefan current from the surface), condensation growth of particles, crystallization associated with the complexity of phase transitions; (d) interaction of hydrodynamic fields of particles of different sizes and gravitational coagulation; and (e) dissipative phenomena arising in the process of turbulent flow and many other factors. The calculation of the deposition rate of solid particles and deformable droplets and bubbles should correspond to the hydrodynamic conditions under which the process takes place.

2. As a result of fluctuations in the particle concentration and distribution in an inhomogeneous turbulent field, fluctuations in the transverse velocity, energy dissipation, and many other factors, the deposition processes are stochastic, which results in a large scatter of the experimental data, and this imposes a special empirical approach to solving this problem especially for large areas of number change $\left(\text{Re}_d, \text{Mo}, \text{and We}\right)$. Under the action of turbulent pulsations and mixing, the presence of a shift in the hydrodynamic fields of the particles and the carrier of the turbulent medium, the particles acquire an additional relative velocity that affects the probability of their collision, coagulation, and deposition rate. The need to use empirical modeling to describe sedimentation processes is primarily due to the non-constant nature of the hydrodynamic flow around the particles by the external flow, which creates certain nonlinearities that cannot be determined deterministically. It is important to note that all processes of chemical technology are characterized by randomness of behavior with inherent statistical characteristics. However, in any random process there is a deterministic component that characterizes, at least on an average, the features and essence of the phenomena taking place. Only the reasonable use of stochastic estimation algorithms (including stochastic filtering and approximation algorithms) under conditions of uncertainty of statistical parameters (correlation coefficient or covariance, standard deviation, error distribution, etc.) makes it possible to obtain certain results qualitatively different from the deterministic description. In this study, many models, (8.164 through 8.167) (8.206 through 8.209), are given for the average statistical description of the processes of particle deposition, suitable for practical calculations within the permissible accuracy.
3. Investigating the deposition of particles in an isotropic turbulent flow for different scales of turbulence made it possible to express the sedimentation rate through the main turbulence parameters-specific energy dissipation ε_R, turbulence scale λ, and viscosity of the medium ν_c. Consequently, for an isotropic turbulent flow, an important parameter is the specific energy dissipation, for the calculation of which various expressions are proposed [71]. It should be noted that in mass-exchange processes involving dispersed particles, in order to intensify and increase the efficiency of mass transfer, it is necessary that the rate of deposition or ascent is minimal, that is, particles were more in suspension. This is achieved by crushing droplets and bubbles to a certain dispersion, which increases the interfacial contact surface.
4. The sedimentation of polydisperse particles is characterized by the inconstancy of the dimensions or the distribution functions of their dimensions related to the constraint of precipitation, the collision and interaction of particles with each other, accompanied by such physical phenomena as coagulation, deformation of their shape, destruction and fragmentation, leading to the evolution of the particle spectrum. The influence of the constraint of particles in concentrated flows on the sedimentation rate is taken into account by introducing the appropriate corrections, (8.166) (8.167), which depend on the volume fraction of particles in the flow, particle sizes, etc.

9 Stochastic Analogue of Mass Transfer Equations and Practical Applications

The physicochemical mechanics of the basic processes of chemical technology studies the general laws of the transfer of a substance (mass, heat, and momentum) based on the traditional concepts of continuum mechanics. For heterophase polydisperse media, where there are random (stochastic) fluctuations, in particular a change in the particle size, this description principle turns out to be somewhat inadequate. The change in the size and shape of particles in the processes of chemical technology caused by phase (dissolution, evaporation, melting, condensation, crystallization) and chemical transformations, mechanical phenomena (agglomeration, destruction, wear), as well as physical phenomena of coagulation (coalescence) and crushing significantly deform the function density distribution of particles in size and time, thereby exerting a significant influence on the phenomena of mass, heat, and momentum transfer in polydisperse systems. In this chapter, we will consider stochastic equations that are an analog of the mass-transfer equations and determine the nature of the distribution of particles in a dispersed medium, taking into account the course of various physical processes.

9.1 DIMENSIONS AND DISTRIBUTION FUNCTIONS OF POLYDISPERSE PARTICLES

In practical applications of disperse systems in chemical technology, solid particles, droplets, and bubbles are characterized by polydispersity of the state, that is, the particle sizes can vary from minimum to maximum values, although mass and heat transfer calculations always use the average size. The shape of the particles that form part of the disperse systems is generally nonspherical, although the spherical shape is a special case or an idealization of an irregular shape. Strictly spherical in the absence of deformation of the form are drops and bubbles that take a spherical shape under the action of surface tension in the absence of external fields (gravitational, electrical, etc.). Drops and small bubbles also retain a spherical shape. The granules of fertilizers and drugs, the catalyst particles and the adsorbent, which are obtained as a result of the technological process, approach the spherical form. Obviously, even with the technological production of particles, there is always a deviation of the shape from the spherical (Figure 9.1). Along with this, in practice there are particles of cylindrical, planar and irregular shape.

Here, a_s is the average or equivalent diameter of a particle, determined by various methods: (1) the median diameter, that is, the sum of the diameters of all particles divided by their total number $a_s = \frac{\sum n_i a_i}{\sum n_i}$; (2) by volume: $a_s = \left[\frac{\int a^3 P(a)da}{\int P(a)da}\right]^{1/3}$; (3) on the surface: $a_s = \left[\frac{\int a^2 P(a)da}{\int P(a)da}\right]^{1/2}$; by volume to surface ratio: $a_{32} = \left[\frac{\int a^3 P(a)da}{\int a^2 P(a)da}\right]$; by weight: $a_{43} = \left[\frac{\int a^4 P(a)da}{\int a^3 P(a)da}\right] = \frac{\int aP(a)da}{\int P(a)da}$.

For particles of irregular shape, the average size is defined as: $a_s = \left[\frac{6\upsilon}{\pi}\right]^{1/3}$, where υ is the volume of the particle, and $P(a)$ is the probability function of the particle size distribution, characterizing polydisperse systems. Depending on the nature of the problem being solved, different average diameters and spectra in size and mass should be used. It is important to note that in analyzing and solving problems of heat and mass transfer and chemical reaction on the surface, it is desirable to use the average diameter

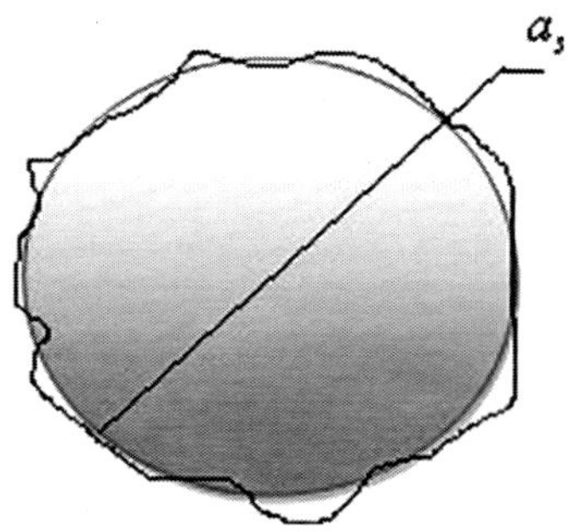

FIGURE 9.1 Deviation of the shape of the particle from the spherical.

over the surface. The state of a polydisperse system determines the particle size distribution function or the evolution of the distribution function in size and time. Usually, in processes accompanied by physical phenomena (droplet evaporation, condensation, agglomeration coagulation, crushing, etc.), the particle sizes are not constant. The most effective representation of information on the state of a polydisperse system is the characteristic change and evolution of the particle distribution function in size and residence time. In this case, the character and form of the distribution function changes with time, beginning with the initial distribution and ending with the limiting value. In the steady state, which corresponds to the constant particle size, the known equations for the continuous particle distribution function are used: the normal and lognormal distributions, the Rosen–Rammler distribution, the gamma distribution, etc., each of which is characterized by its own parameters. In particular, the density of the normal and lognormal particle size distributions is widely used in applied problems for various fields, including in the problems of mass and heat transfer, and the Rosen–Rammler distribution is used to construct the distribution function for finely dispersed particles and nanoparticles. It is important to note that these types of distribution functions are characteristic for the steady state or the constant size of a polydisperse system. Most chemical technology problems associated with the presence of dispersed inclusions are characterized by variable particle size values, that is, evolution of the distribution function in size and time. Table 9.1 shows some of these distributions for the steady state.

TABLE 9.1
Probability Distribution Functions

No.	Distribution Functions $P(a)$	Parameters
1.	The density of the normal distribution $P(a)=\frac{1}{\sigma_a\sqrt{2\pi}}\exp\left[-(a-\mu_a)^2/2\sigma_a^2\right]$ 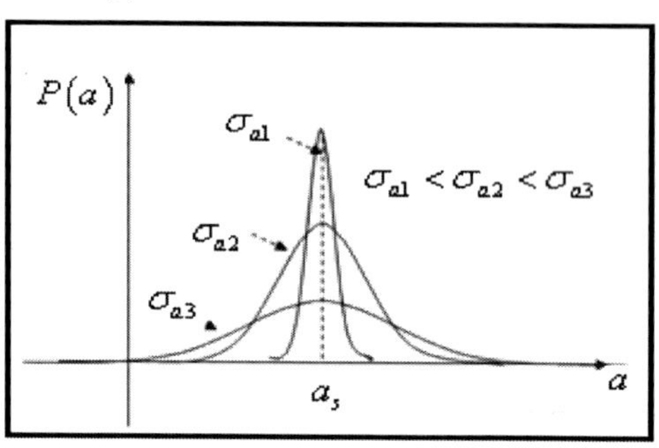	σ_a is standard deviation, $\mu_a=a_s$ is mean value of the variable. The maximum of the function is: $P(a)=\left(\sigma_a\sqrt{2\pi}\right)^{-1}$ Distribution function is $F(a)=\frac{1}{2}\left[1+\operatorname{erf}\left(\frac{a-\mu_a}{\sqrt{2}\sigma_a}\right)\right]$

(Continued)

TABLE 9.1 (*Continued*)
Probability Distribution Functions

No.	Distribution Functions $P(a)$	Parameters
2.	Lognormal distribution $P(a)=\frac{1}{a\sqrt{2\pi}\ln\sigma_a}\exp\left[-\frac{(\ln a-\ln a_s)^2}{2\ln^2\sigma_a}\right]$ 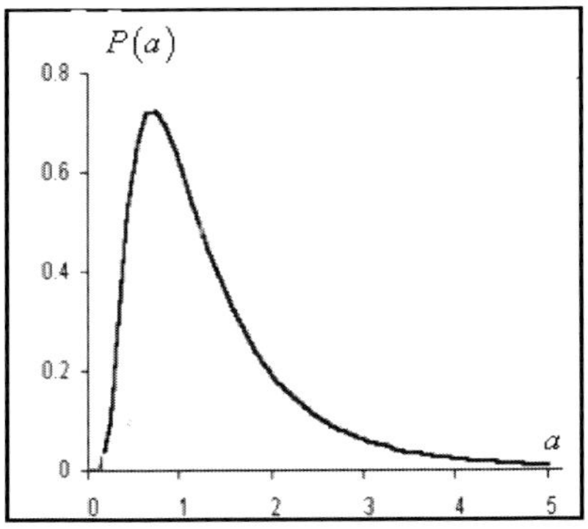	Median is: $\exp(\mu_a)$, Mode is: $\exp\left(\mu_a+\sigma_a^2/2\right)$ Distribution function is $F(a)=\frac{1}{2}\left[1+\operatorname{erf}\left(\frac{\ln a-\ln\mu_a}{\sqrt{2}\sigma}\right)\right]$
3.	Exponential distribution $P(a)=\lambda\exp(-\lambda a)$ 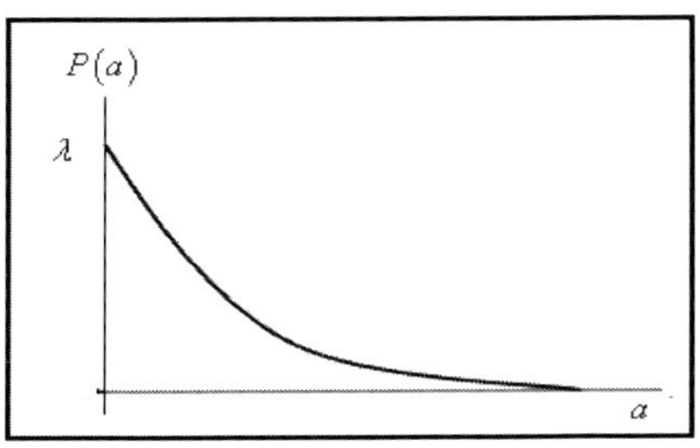	λ is constant positive quantity; Median is: $(\ln 2)/\lambda$, variance is: λ^{-2}. Distribution function is: $F(a)=1-\exp(-\lambda a)$
4.	Gamma distribution $P(a)=\frac{1}{\Gamma(\alpha+1)\beta^{\alpha+1}}a^{\alpha}\exp\left(-\frac{a}{\beta}\right)$ 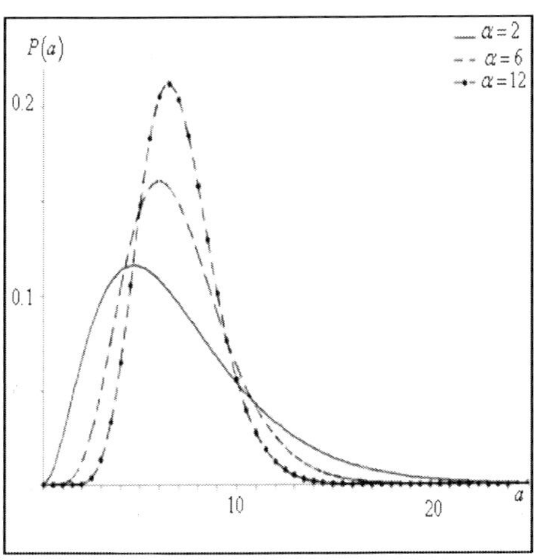	a is particle size, β is scale parameter of the gamma distribution, α is form parameter, The average size is $a_s=\beta(\alpha+1)$ The average volume of a particle is $\upsilon_s=\frac{4\pi}{3}\beta^3(\alpha+1)(\alpha+2)(\alpha+3)$ Average surface area of the particle is $S_s=4\pi\beta^2(\alpha+1)(\alpha+2)$

As a result of the evolution of the particle distribution function, distribution curves can be obtained for each time point, depending on the initial distribution assignment, which does not obey the standard distribution.

In more complex cases, the distribution function is characterized by a set of extrema, that is, is multimodal.

9.2 STOCHASTIC DIFFERENTIAL EQUATIONS

The use of methods of statistical physics for the reconstruction and description of the evolution of the particle size distribution function is based on the stochastic differential Fokker–Planck equation and the integral and differential kinetic equation. It is known that the generalized Fokker–Planck equation constructed for Markov processes can serve as a basis for a theoretical study of the majority of processes of chemical technology involving polydisperse systems. At the same time, the presence of the phenomena of interaction and collision between particles places significant restrictions on the use of the Fokker–Planck equation, adapted to describe the evolution of the distribution function and characterized by a smooth variation of the generalized coordinates (spatial coordinates, mass and particle size, etc.). In this respect, it is of particular interest to use the kinetic equation to describe processes with a sudden change in the generalized coordinates. The Fokker–Planck equation for polydisperse systems can be represented in the form of a vector expression [8,40].

$$\frac{\partial P}{\partial t}=-\frac{\partial}{\partial x}\Big[\Big(f\big(a(t),t\big)P\big(a(t)\big)\Big)\Big]+\frac{1}{2}\frac{\partial}{\partial a}\left(\frac{\partial}{\partial a}\right)^{T}\Big[G\big(a(t),t\big)BG\big(a(t),t\big)P\big(a(t)\big)\Big] \tag{9.1}$$

Here, B is the vector of stochastic diffusion coefficients. In the derivation of equation (9.1), it is assumed that the random process under consideration obeys the nonlinear equation

$$\frac{da(t)}{dt}=f\big(a(t),t\big)+G\zeta(t) \tag{9.2}$$

with normal white noise, zero mathematical expectation $M\big[\zeta(t)\big]=0$, and a given covariance matrix $\mathrm{Cov}\big[\zeta(t)\zeta(\tau)\big]=B\delta(t-\tau)$ ($\delta(t-\tau)$ is delta function). It should be noted that the function $f\big(a(t),t\big)$ expresses the rate of change in particle size, depending on the ongoing processes of mass transfer and heat transfer. If $f\big(a(t),t\big)>0$, then it characterizes the growth of particle sizes in the processes of crystallization, condensation, swelling, coagulation, etc. If $f\big(a(t),t\big)<0$, then it characterizes the reduction in particle sizes in such processes as crushing and grinding, particle drying, evaporation of droplets, etc. If the function $f\big(a(t),t\big)$ is strongly nonlinear, then great difficulties arise in the analytic solution of equation (9.1). An important condition for using the distribution function and the reliability of the solution of equation (9.1) is the normalization condition. The solution of stochastic equations differs significantly from the solution of ordinary differential equations $\int_0^\infty P(a)da=1$. The solution of ordinary differential equations reduces to the determination of unknown functions at an arbitrary instant of time from the given initial conditions. The solution of the stochastic differential equation is connected with the definition of the distribution of the values of the unknown functions at an arbitrary instant of time. The analytic solution (9.1) is allowed only for particular cases and in general represents great difficulties related to the structure of the function $f(a,t)$ and the character of the initial distribution. However, in some cases it is possible to approximate (9.1) by reducing the latter to a system of ordinary differential equations. In particular, if we linearize a function $f\big(a(t),t\big)$ in a neighborhood of some mean

$$f\big(a(t),t\big)\approx f\big(\mu_a(t),t\big)+\frac{\partial f\big(\mu_a(t),t\big)}{\partial\mu_a(t)}\big(a(t)-\mu_a(t)\big) \tag{9.3}$$

and assume that the nature and form of the distributional density function remains constant throughout the entire evolution period, then the system of differential equations, in particular for the normal distribution, allowing one to determine the change in the variance $\sigma^2(a)$ and mean elements $\mu_a(a)$, will be presented in the form

$$\frac{d\mu_a(t)}{dt} = f\left(\mu_a(t),t\right) \tag{9.4}$$

$$\frac{d\sigma_a^2}{dt} = \frac{\partial f\left(\mu_a(t),t\right)}{\partial \mu_a}\sigma_a^2 + \sigma_a^2\frac{\partial f^T\left(\mu_a(t),t\right)}{\partial \mu_a} + G\left(\mu_a(t),t\right)BG^T\left(\mu_a(t),t\right) \tag{9.5}$$

with a given initial value $\mu_a(0) = \mu_{a0}, \sigma_a^2(0) = \sigma_{a0}^2$. Solutions (9.4) and (9.5) ensure that the distribution function is constant over the entire evolution period, with the exception of the maximum and mean.

Example 9.1

Let us assume that the distribution function for the entire evolutionary period does not change the character and obeys the normal law

$$P(a,t) = \frac{1}{\sigma_a\sqrt{2\pi}}\exp\left[-\frac{(a-\mu_a)^2}{2\sigma_a^2}\right] \tag{9.6}$$

and the coordinate change is described by a linear equation

$$\frac{da}{dt} = -Aa + \zeta(t) \tag{9.7}$$

Such an equation describes the change in droplet sizes during evaporation from the surface, in the sublimation of solid particles, in the fragmentation of droplets and bubbles, in the grinding of solid particles, etc., where the particle size decreases monotonically to a minimum value with time.

It is required to determine the laws of variation of the variance $\sigma_a^2(t)$ and $\mu_a(t)$ the mean and to construct the evolution of the distribution function. The Fokker–Planck equation can be represented in the form

$$\frac{\partial P(a)}{\partial t} = -\frac{\partial}{\partial a}\left(\frac{da}{dt}P(a)\right) + \frac{B}{2}\frac{\partial^2 P(a)}{\partial a^2} \tag{9.8}$$

Having determined the derivatives entering into this equation from (9.3) after separating the variables for variance and mean, we obtain $(a_s = \mu_a)$

$$\begin{aligned} \frac{d\sigma_a^2}{dt} &= -2A\sigma_a^2 + B, \qquad \sigma_a^2(t)\big|_{t=0} = \sigma_{a0}^2 \\ \frac{d\mu_a}{dt} &= -A\mu_a, \qquad \mu_a(t)\big|_{t=0} = \mu_{a0} \end{aligned} \tag{9.9}$$

The solution of this system of equations with initial values gives

$$\begin{aligned} \sigma_a^2 &= \sigma_{a0}^2 + \frac{B}{2A}\left[1 - \exp(-2At)\right] \\ \mu_a &= \mu_{a0}\exp(-At) \end{aligned} \tag{9.10}$$

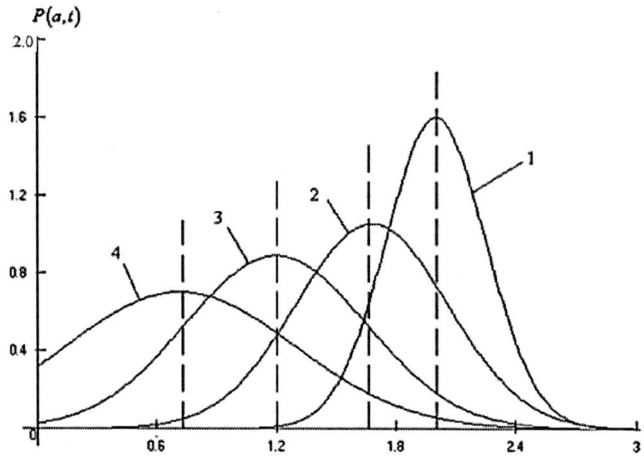

FIGURE 9.2 Evolution of the particle distribution function in size and time, equal to: (1) 0; (2) 6.0; (3) 10.0; and (4) 20.0.

Thus, substituting these values for the dispersion and mean in (9.6), we determine the value of the distribution function at any time. Figure 9.2 shows the distribution curves of finely dispersed particles, calculated from formulas (9.8) and (9.10), for values of $\mu_{a0} = 2$, $\sigma_{a0}^2 = 0.25$, $B/2A = 0.5$.

This graph clearly shows the evolution of the distribution function in time and the change in the average particle size and dispersion over time, with the character of the distribution curve being constant. We note that the nature of the distribution function over the entire evolutionary period does not always remain constant and obeys the normal distribution.

However, it is not always possible to use the linearization method to effectively solve the stochastic equation in practical applications, since the distribution function is deformed, as will be discussed in the next section.

In contrast to the Fokker–Planck equation, the kinetic integro-differential equation forms the basis for describing the evolution of the particle distribution function for a discrete and instantaneous change in their sizes, taking into account the processes of coagulation (coalescence) and crushing. One of the main characteristics of a polydisperse system is the distribution of particles at a time t. Then the quantity

$$N(t) = \int_0^\infty P(m,t)\,dm \tag{9.11}$$

is the total number of particles that are in a unit mass at a time t. Considering only pair collisions, the change $P(m,t+\Delta t) - P(m,t)$ in the time interval Δt of the number of particles by mass m can be represented in the form $\Delta t\left(N^+ - N^-\right)$. Since pair collisions are considered, the number of particles of mass m formed per unit time as a result of coagulation of small particles is equal to

$$N^+ = \frac{1}{2}\int_0^m \omega(m',m-m')P(m',t)P(m-m',t)\,dm' \tag{9.12}$$

and the number of particles of mass m that collide per unit time with other particles is equal to

$$N^- = P(m,t)\int_0^\infty \omega(m,m')P(m',t)\,dm' \tag{9.13}$$

TABLE 9.2
Frequency Functions of Collisions for Fine Particles

	Conditions	Formulas	Notes
1	Brownian coagulation	$\omega(a_i,a_j)=\dfrac{2k_BT}{3\eta_c}\left(\dfrac{1}{a_i}+\dfrac{1}{a_j}\right)(a_i+a_j)$	
2	For very small particles transported by gas	$\omega(a_i,a_j)=\left(\dfrac{1}{8}\right)^{1/6}\left(\dfrac{6k_BT}{\rho_d}\right)^{1/2}\left(\dfrac{1}{a_i^3}+\dfrac{1}{a_j^3}\right)^{1/2}\times(a_i+a_j)^2$	k_B is Boltzmann's constant s
3	The interpolation formula	$\omega(a_i,a_j)=\dfrac{R_{ij}}{R_{ij}+\left(l_i^2+l_j^2\right)^{1/2}}+\dfrac{4D_{ij}(1+AKn_i)(1+AKn_j)}{\left(\bar{C}_i^2+\bar{C}_j^2\right)^{1/2}R_{ij}}$	$R_{ij}=R_i+R_j$, $A=1.257+0{,}4e^{-1.1/Kn_i}$, $\bar{C}_i=\left(\dfrac{8k_BT}{\pi m_i}\right)$
4	Laminar shear flow	$\omega(a_i,a_j)=\dfrac{4}{3}\left(\dfrac{a_i}{2}+\dfrac{a_j}{2}\right)\dfrac{\partial U}{\partial y}$	
5	In turbulent flow of a viscous fluid	$\omega(a_i.a_j)=1.294(a_i+a_j)^3\left(\varepsilon_R/\nu_C\right)^{1/2}$	ε_R is specific energy dissipation
6	Coalescence of small bubbles	$\omega(a_i,a_j)=\zeta\bar{U}\dfrac{\pi}{4}(a_i+a_j)^2$	ζ is efficiency of coagulation; $\bar{U}$ is relative velocity

where $\omega(m,m')=\omega(m',m)$ the symmetric function characterizing the frequency of collisions of particles with masses m and m' or the function of the coagulation nucleus, which depends on the hydrodynamic conditions and the physicochemical properties of the dispersed and carrier phases (Table 9.2).

In the literature, you can find a large number of formulas expressing the core of coagulation, which will be described in the next section. Using formulas (9.12) and (9.13), we determine the change in the number of particles per unit mass when they coagulate in the form

$$\frac{\partial P(m,t)}{\partial t}=\frac{1}{2}\int_0^m \omega(m',m-m')P(m-m',t)P(m',t)dm'-P(m,t)\int_0^\infty \omega(m,m')P(m',t)dm' \tag{9.14}$$

Expression (9.14) is the kinetic equation of coagulation and crushing and forms the basis of the population mass balance for a multitude of particles. This equation is quite complicated mathematically, and therefore the issues related to the existence and uniqueness of the solution, restrictions on the form of the function $\omega(m,m')$ have not yet been solved in the required scope. When the coagulation nucleus is constant $\omega(m',m)=k$, the simplified equation (9.14) for the coagulation of particles in an inhomogeneous unbounded medium is represented as

$$\frac{dN}{dt}=-\frac{k}{2}N^2(t) \tag{9.15}$$

analytic solution, which will be presented in the form

$$N(t)=\frac{N_0}{\left(1+{N_0kt}/{2}\right)} \tag{9.16}$$

where N_0 is the initial number of particles. Experimental data, both for monodisperse particles and for polydisperse ones, follow this dependence, at least in the initial period. The rate of average growth of the volume of particles due to coagulation is determined by the relation

$$\upsilon = \upsilon_0\left(1+\frac{N_0 k}{2}t\right) \tag{9.17}$$

An important feature of these solutions is their independence from the nature and properties of the initial distribution. A similar confirmation of the independence of the solution from the initial distribution is characteristic for various solutions related to the deformation and evolution of the distribution function.

9.3 EVOLUTION OF THE DISTRIBUTION FUNCTION OF DROPLETS AND BUBBLES IN ISOTROPIC TURBULENT FLOW

Evolution of the distribution function of droplets and bubbles in a limited volume of flow, which changes as a result of coalescence and breaking in an isotropic turbulent flow, is an important indicator of the disperse system and gives its qualitative and quantitative characteristics. Two stochastic equations form the basis for describing the evolution of the distribution function: (1) the integro-differential equation of coagulation and fragmentation, constructed on the basis of the population balance [40,85,168,169,188,191,324]; and (2) the stochastic differential Fokker–Planck equation [40,85–87,205,325].

9.3.1 Kinetic Differential Equation of Coagulation and Breaking

Equation (9.8), describing the phenomena of coalescence and breaking of droplets and bubbles, constructed on the basis of population balance, is a nonlinear integro-differential equation that has a shift in the argument and depends on the frequency of collisions and fragmentation, on the variable sizes (volumes) of particles, and can be represented in a general form

$$\frac{\partial P(\upsilon,\upsilon')}{\partial t} + \mathrm{U_c}\nabla P(\upsilon,\upsilon') = \mathrm{C}^+ + \mathrm{C}^- + \mathrm{B}^+ + \mathrm{B}^- \tag{9.18}$$

where $\mathrm{C}^+ = 1/2\int_0^{\upsilon}\bar{\beta}(\upsilon',\upsilon-\upsilon')P(\upsilon')P(\upsilon-\upsilon')d\upsilon'$ is the expression taking into account the creation of new particles in a certain time; $\mathrm{C}^- = -P(\upsilon)\int_0^{\infty}\bar{\beta}(\upsilon',\upsilon)P(\upsilon')d\upsilon'$ is the expression describing the escape of particles from a given range of sizes; $\mathrm{B}^+ = \int_0^{\infty}\varpi(\upsilon,\upsilon')P(\upsilon')d\upsilon'$, $B^- = -P(\upsilon)/\upsilon\int_0^{\upsilon}\varpi(\upsilon',\upsilon)\upsilon' d\upsilon'$ are two integral components describing the arrival and departure of particles from a given range during their crushing, $\bar{\beta}(\upsilon',\upsilon)$ is the frequency of a two-particle collision of particles by volume υ_1 and υ_2 at their unit concentration over a period of time Δt; and $\varpi(\upsilon,\upsilon')$ is a function characterizing the density of the distribution of volume particles formed per unit time as a result of the destruction of a particle by volume υ. The function $\bar{\beta}(\upsilon',\upsilon)$ is symmetric with respect to the arguments $\bar{\beta} = \bar{\beta}(\upsilon_1,\upsilon_2)$, $\upsilon_1 \geq \upsilon_2$ and $\bar{\beta} = \bar{\beta}(\upsilon_2,\upsilon_1)$, $\upsilon_2 \geq \upsilon_1$. In contrast $\bar{\beta}(\upsilon',\upsilon)$, the function $\varpi(\upsilon,\upsilon')$ is an asymmetric function, and, based on physical considerations, we can assume that for $\varpi(\upsilon,\upsilon') \geq 0$, $\upsilon < \upsilon'$ and $\varpi(\upsilon,\upsilon') = 0$ with $\upsilon > \upsilon'$ [40]. In particular, the collision frequency of droplets in a viscous fluid is determined by a symmetric function of the form [326]

$$\bar{\beta} \approx 1.294 N_0 (a_i + a_j)^3 \left(\frac{\varepsilon_R}{\nu_c}\right)^{1/2} \tag{9.19}$$

In the work [139] for the calculation $\bar{\beta}(\upsilon,\upsilon')$, a symmetric function is used that is analogous to the normal distribution

$$\bar{\beta}(\upsilon,\upsilon') = \frac{2.4}{\upsilon'}\exp\left[-4.5\frac{(2\upsilon-\upsilon')^2}{\upsilon'^2}\right] \tag{9.20}$$

A more complicated formula for calculating the collision frequency of particles (bubbles) with dimensions a_i and a_j proposed in work [139,140,326–328]

$$\bar{\beta}(a_i,a_j) = C_{ij}\varepsilon_R^{1/3}(a_i+a_j)^2\left(a_i^{2/3}+a_j^{2/3}\right)^{3/2}\lambda(a_i,a_j) \tag{9.21}$$

Here, $C_{ij} \approx 0.28-1.11$, $\lambda(a_i,a_j)$ is the coalescence efficiency, which is defined as [139]

$$\lambda(a_i,a_j) = \exp\left[-C_{11}\frac{\rho_c\eta_c\varepsilon_R}{\sigma^2}\left(\frac{a_ia_j}{a_i+a_j}\right)^4\right] \tag{9.22}$$

and refined to a simpler form in the work [8]

$$\lambda(a_i,a_j) = \exp\left[-2.3\sqrt{\frac{\rho_c\varepsilon_R^{2/3}a_{ij}^{5/3}}{\sigma}}\right] \tag{9.23}$$

where $a_{ij} = 2\left(\frac{1}{a_i}+\frac{1}{a_j}\right)^{-1}$. In the work [168] for secondary crushing, that is, for "daughter" drops the following formula is proposed

$$\varpi(\upsilon,\upsilon') = \frac{\left(A\upsilon^{2/9}-B\right)\left[A(\upsilon'-\upsilon)^{2/9}-B\right]}{\int\limits_{\upsilon_{\min}}^{\upsilon'-\upsilon_{\min}}\left(A\upsilon^{2/9}-B\right)\left[A(\upsilon'-\upsilon)^{2/9}-B\right]d\upsilon} \tag{9.24}$$

where $A = \frac{8.2}{2}\left(\frac{6}{\pi}\right)^{2/9}\rho_c\varepsilon_R^{2/3}$, $B = \frac{6\sigma}{(6/\pi)^{1/3}\upsilon'^{1/3}}$. A review of the statistical analysis of the processes of crushing droplets and bubbles is given in the papers [157,168,169], where different mechanisms of breaking droplets and bubbles and various expressions for the functions $\bar{\beta}(\upsilon_1,\upsilon_2)$ and $\varpi(\upsilon,\upsilon')$ are also analyzed [168,169,329,331–342].

The solution of (9.23), in view of its nonlinearity, presents great difficulties both theoretically and practically. Some particular analytical solutions of the equation of fragmentation

$$\frac{\partial P(\upsilon,\upsilon')}{\partial t} = \int\limits_0^\infty \varpi(\upsilon,\upsilon')P(\upsilon')d\upsilon' - \frac{P(\upsilon)}{\upsilon}\int\limits_0^\upsilon \varpi(\upsilon',\upsilon)\upsilon'd\upsilon' \tag{9.25}$$

$$t=0,\ P(\upsilon,0)=P_0(\upsilon);\ \ \upsilon\to\infty,\ P(\upsilon,t)\to 0;\ \upsilon\to 0,\ P(\upsilon,t)\to 0$$

(only the last two terms of the equation [9.23] are taken into account, $C^+ = C^- = 0$) for a constant value and a given initial distribution in the form of a gamma distribution, and also using the method of moments are given in the paper [40], using the method of separation of variables in the paper [85]. In particular, the solution of this equation can be sought by the method of separation of variables for constant values $\varpi(\upsilon,\upsilon') = \text{const}$ in the form $P(\upsilon,t) = \psi(t)\varphi(\upsilon)$, we obtain two equations

$$\frac{\partial \psi}{\partial t} = -\mu^2 \psi, \qquad \psi(t) = B_1 \exp(-\mu^2 t)$$

$$\int_0^{\infty} \varpi(\upsilon,\upsilon')\varphi(\upsilon)d\upsilon - \frac{\varphi}{\upsilon}\int_0^{\upsilon} \upsilon'\varpi(\upsilon,\upsilon')d\upsilon' = -\mu^2\varphi \tag{9.26}$$

Using the boundary conditions [85], the general solution (9.25) can be represented in the form

$$P(\upsilon,t) = \sum_{n=0}^{\infty} B_n \exp(-\mu_n^2 t)\exp\left[\int \frac{\varpi(\upsilon,\upsilon')\upsilon^2 + \alpha(\upsilon) - \left(\partial\alpha/\partial\upsilon\right)\upsilon}{\alpha(\upsilon)\upsilon - \mu_n^2\upsilon^2} d\upsilon\right] \tag{9.27}$$

where $\alpha = \int_0^{\upsilon} \upsilon'\varpi(\upsilon,\upsilon')d\upsilon'$. The solution (9.27) is very complicated, although for a certain choice of the form of the function of the frequency of fragmentation and the initial distribution $P_0(\upsilon)$, it is simplified to a specific form. In particular, if the function of the fragmentation frequency is given in the form $\varpi(\upsilon) = K_0\upsilon^{-m}$, then from this equation we have the following form of the distribution function

$$P(\upsilon,t) = \sum_{n=0}^{\infty} B_n \exp\left[-\mu_n^2 K_0 t + \int \frac{d\upsilon}{\upsilon - \mu_n^2(2-m)\upsilon^m}\right] \tag{9.28}$$

If we take as the initial distribution the gamma function of the form $P_0(0) = 2\frac{N_0}{\upsilon_0}\upsilon\exp(-2\upsilon)$, then, from (9.28), we have

$$P(\upsilon,t) = 2\frac{N_0}{\upsilon_0}\exp\left(-2\frac{\upsilon}{\upsilon_0} - 2\mu_0^2 K_0 t\right) \tag{9.29}$$

Taking into account the normalization conditions $\int_0^{\infty} P(\upsilon,t)d\upsilon = 1$ and $\upsilon \sim a^3$, we obtain the cumulative distribution function in the form

$$P_K(a,t) = \int_0^a \frac{P(\upsilon,t)}{N_0} da = 1 - \exp(-k_0 a^3) \tag{9.30}$$

The solution is very satisfactory with the experimental data for the cumulative distribution function [168,169], if $k_0 = 0.00105(x/a)^2$ (where x is the length of the droplet moving when the liquid is sprayed under pressure) (Figure 9.3).

The solution of the coalescence equation obtained from (9.18) and having a shift in the argument, in the form

$$\frac{\partial P(\upsilon,\upsilon')}{\partial t} = \frac{1}{2}\int_0^{\upsilon} \bar{\beta}(\upsilon',\upsilon-\upsilon')P(\upsilon')P(\upsilon-\upsilon')d\upsilon' - P(\upsilon)\int_0^{\infty} \bar{\beta}(\upsilon',\upsilon)P(\upsilon')d\upsilon' \tag{9.31}$$

The solutions (9.31) essentially depend on the specification of the initial distribution and the function of the coalescence nucleus $\bar{\beta}(\upsilon)$.

A number of particular analytic solutions for certain initial distributions and functions $\bar{\beta}(\upsilon)$ are given in Table 9.3. Numerical solutions of the coalescence and breaking equations are given in the works, and a general analysis of the results of the numerical and analytical solutions is given in the papers [85,157,160,162,205,325].

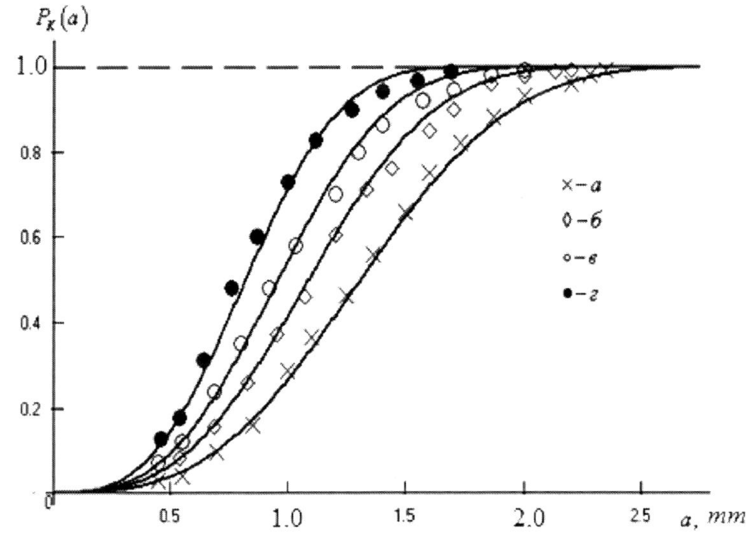

FIGURE 9.3 Change in the cumulative droplet distribution function when the liquid is sprayed at distances equal: (1) 17.2; (2) 22.65; (3) 27.39; and (4) 34.07.

TABLE 9.3
Partial Analytical Solutions of the Coagulation Equation (9.31) for a Given Nucleus $\bar{\beta}: \left(\psi(x,\tau) = \frac{\upsilon_0}{N_0} P(x,t),\ x = \frac{\upsilon}{\upsilon_0},\ \tau = \frac{N_0 - N(t)}{N_0} = f(\bar{\beta}, t, N_0)\right)$

The Nucleus of Coagulation	Initial Distribution	Particular Solutions	References
1. $\bar{\beta} = \beta_0$ $\tau = \frac{\lambda t}{1+\lambda t}$, $\lambda = \frac{N_0 \beta_0}{2}$	a. Gamma distribution $\psi(x,0) = \frac{(1+v)^{(1+v)}}{\Gamma(1+v)} \times x^{v} \exp[-(1+v)x]$	$\psi(x,t) = \frac{(1-\tau)^2}{\tau x} \exp[-(1+v)x] \times \sum_{k=1}^{\infty} \frac{(1+v)^{k(1+v)}}{\Gamma[(1+v)k]} \tau^k x^{(1+v)k}$	[40,328]
		a. $v = 0$ $\psi(x,t) = (1-\tau)^2 \exp[-x(1-\tau)]$	
		b. $v = 1$ $\psi(x,t) = \frac{2(1-\tau)^2}{\tau^{1/2}} \exp(-2x) \mathrm{sh}(2x\tau^{1/2})$	
	b. Superposition two delta functions $\psi(x,0) = a_1 \delta(x - x_0) + a_2 \delta(x - m x_0)$	$\psi(x,t) = (1-\tau)^2 \sum_{k=1}^{\infty} \tau^{k-1} \times \sum_{j=-}^{k} \frac{k!}{(k-j)!\,j!} a_1^{k-j} a_2^{j} \delta[x - (mj + k - j)x_0]$	

(*Continued*)

TABLE 9.3 (*Continued*)
Partial Analytical Solutions of the Coagulation Equation (9.31) for a Given Nucleus
$$\bar{\beta}:\left(\psi(x,\tau)=\frac{\upsilon_0}{N_0}P(x,t),\ x=\frac{\upsilon}{\upsilon_0},\ \tau=\frac{N_0-N(t)}{N_0}=f\left(\bar{\beta},t,N_0\right)\right)$$

The Nucleus of Coagulation	Initial Distribution	Particular Solutions	References
2. $\bar{\beta}=\beta_1(x_1+x_2)$ $\tau=1-\exp(-2\lambda\upsilon_0 t)$	a. The sum of two delta functions $\psi(x,0)=a_1\delta(x-x_0)$ $+a_2\delta(x-mx_0)$	$\psi(x,t)=(1-\tau)^2\exp(-\tau x)\sum_{k=1}^{\infty}\frac{(\tau x)^k}{(k+1)!}$ $\times\sum_{j=-}^{k}\frac{k!}{(k-j)!\,j!}a_1^j a_2^{k-j}\delta\left[x-(mj+k-j)x_0\right]$	[40,330]
	b. Gamma distribution	$\psi(x,t)=(1-\tau)\exp\left[-(\tau+\nu+1)x\right]$ $\times\sum_{k=0}^{\infty}\frac{(1+\nu)^{(1+\nu)(k+1)}\tau^k}{(k+1)!\Gamma\left[(k+1)(1+\nu)\right]}x^{\nu+k(\nu+2)}$ At $\nu=0$ $\psi(x,t)=(1-\tau)\exp\left[-(\tau+1)x\right]$ $\times\frac{\mathrm{I}_1\left(2x\sqrt{\tau}\right)}{x\sqrt{\tau}}$ $\mathrm{I}_1(x)$ – Bessel function of the first order of the imaginary argument.	
3. $\bar{\beta}=\beta_0\upsilon_1\upsilon$	a. The sum of two delta functions $\psi(x,0)=a_1\delta(x-x_0)$ $+a_2\delta(x-mx_0)$	$\psi(x,\tau)=\frac{1}{x}\exp(-2\tau x)$ $\times\sum_{k=0}^{\infty}\frac{(2\tau x)^k}{k!}\left(1+\frac{\tau x}{k+1}\right)x_0^{k+1}$ $\times\sum_{j=0}^{k+1}\frac{(k+1)!}{j!(k+1-j)!}a_1^j a_2^{k+1-j}$ $\times\delta\left[x-\left(j+(k+1-j)m\right)x_0\right]$	[40]
	b. Gamma distribution	$\psi(x,\tau)=\exp\left[-(2\tau+\nu+1)x\right]$ $\times\sum_{k=0}^{\infty}\frac{(2\tau x)^k}{k!\Gamma\left[(\nu+2)(k+1)\right]}$ $\times\left(1+\frac{\tau x}{k+1}\right)(1+\nu)^{(k+1)(\nu+2)}x^{(k+1)(\nu+2)-2}$ $\Gamma(n)$ – Gamma function. At $\nu=0$ $\psi(x,\tau)=\exp\left[-(2\tau+1)x\right]$ $\times\sum_{k=0}^{\infty}\frac{(2\tau x)^k}{k!\Gamma\left[2(k+1)\right]}\left(1+\frac{\tau x}{k+1}\right)x^{2k}$	

9.3.2 The Simultaneous Occurrence of the Phenomena of Coalescence and Breaking of Droplets in an Isotropic Turbulent Flow

Simultaneous processes of coalescence and breaking of droplets and bubbles are observed in transport pipes, including in the trunk of an oil well, taking into account gravitational coagulation, in mixing devices (liquid-phase extraction), in gas–liquid reactors, etc. In practice, the phenomena of coalescence and breaking can interact and influence each other. Such interaction in the framework of stochastic models is described by an integral and differential kinetic equation, which can be approximated as

$$\frac{\partial P(\upsilon,t)}{\partial t}=-\frac{1}{2}\int_0^{\upsilon}\omega_k(\upsilon-\upsilon_i)P(\upsilon-\upsilon_i,t)P(\upsilon_i,t)d\upsilon_i+\frac{P(\upsilon,t)}{\upsilon}\int_0^{\upsilon}\omega_D(\upsilon_i,\upsilon)d\upsilon \tag{9.32}$$

where $P(\upsilon,t)$ is the probability density of the particle volume distribution. Assuming that the number of particles is

$$N(t)=\int_0^{\upsilon}P(\upsilon,t)d\upsilon \tag{9.33}$$

and reintegrating (9.32), we have

$$\frac{\partial N}{\partial t}=\int_0^{\upsilon}\frac{\partial P}{\partial t}d\upsilon=-\frac{1}{2}\int_0^{\infty}\int_0^{\upsilon}\omega_k(\upsilon-\upsilon_i)P(\upsilon-\upsilon_i,t)P(\upsilon_i,t)d\upsilon_i+\int_0^{\infty}\frac{P(\upsilon,t)}{\upsilon}\int_0^{\upsilon}\omega_D(\upsilon_i,\upsilon)d\upsilon \tag{9.34}$$

At constant frequencies of coalescence and fragmentation of droplets, we can write expression (9.34) for the number of particles in the form

$$\begin{gathered}\frac{dN}{dt}=-\frac{1}{2}\omega_K\upsilon N^2+\omega_D N\\ N(t)\big|_{t=0}=N_0\end{gathered} \tag{9.35}$$

where N is the number of particles per unit volume. The general solution of (9.35) is represented in the form

$$N(t)=\frac{N_0\exp(\omega_D t)}{1+\dfrac{\omega_K}{2\omega_D}\upsilon N_0(\exp(\omega_D t)-t} \tag{9.36}$$

This solution reflects the change in the number of particles in the volume of the dispersed stream as a result $\beta=\omega_D/(\omega_K\upsilon N_0)$ of their coalescence and breaking. Depending on the ratio, coalescence or crushing processes may predominate (Figure 9.4).

The coalescence of droplets in the flow, with a large difference in the phase densities, leads to a stratification of the flow, although this can be prevented by the processes of fragmentation of the droplets. The phenomenon of breaking of droplets is associated with an increase in the interfacial surface, which improves the rate of mass-exchange processes in the flow. This phenomenon is most conducive to the intensity of turbulence, which increases the collision frequency of droplets. In the processes of breaking, the presence of turbulence promotes the deformation of the droplet shape, and the increase in surface stresses leading to their fragmentation.

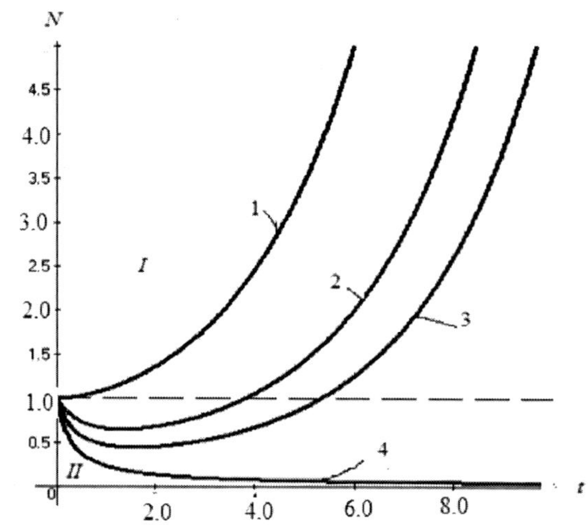

FIGURE 9.4 The change in the number of drops during their coalescence and breaking: (1) $\beta >> 1$; (2) $\beta = 1/3$; (3) $\beta = 1/5$; and (4) $\beta << 1$. I is area pure crushing of drops; II is the region of pure coalescence of droplets.

It follows from equation (9.36) that $\omega_K >> \omega_D$, under the condition, coalescence processes prevail and the solution $\exp(\omega_D t) \approx 1 + \omega_D t$, provided that, is represented in the form

$$N(t) = \frac{N_0}{1 + \frac{1}{2}\omega_K \upsilon N_0 t} \tag{9.37}$$

If $\omega_D >> \omega_K$, then the processes of fragmentation predominate and the solution is presented in the form

$$N(t) = N_0 \exp(\omega_D t) \tag{9.38}$$

In Figure 9.5, a comparison of the experimental data for the number of particles in the coalescence of spherical no deformable droplets for small values of the Weber number We < 1 with equation (9.37) is given.

Here, $t^* = (\omega_K \upsilon N_0)^{-1}$. In the event of a collision and coalescence of two drops in a turbulent flow with a change in the total mass, the time scale of the turbulence is defined in the form

$$t_R = \int_0^t R(\theta) d\theta \tag{9.39}$$

where $R(\theta)$ is the coefficient of time correlation, which is determined for the coalescence of droplets in the form $R(\theta) = (\omega_K \theta)^{-1/2}$. Then we have

$$t_R = \int_0^t (\omega_K \theta)^{-1/2} d\theta = 2\left(\frac{t}{\omega_K}\right)^{1/2} \tag{9.40}$$

The nature of the change in the correlation coefficient is shown in Figure 9.6.

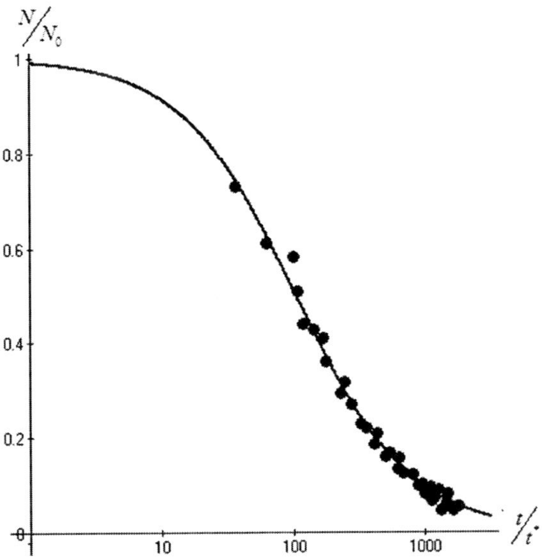

FIGURE 9.5 The change in the number of droplets upon their coalescence in the volume of the stream for small values of the number We.

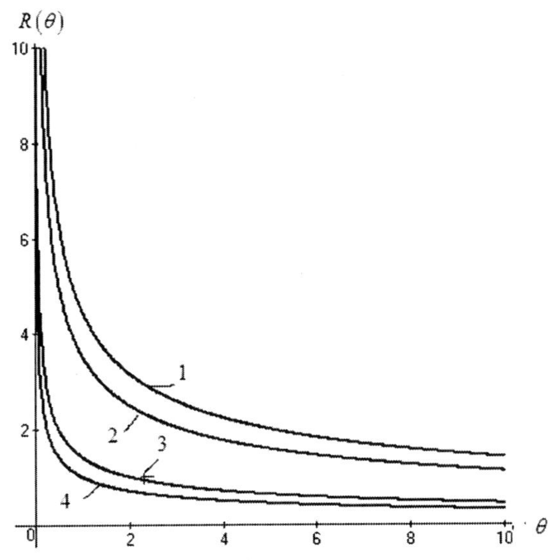

FIGURE 9.6 The coefficient of time correlation at frequencies ω_K equal to: (1) 0.05; (2) 0.08; (3) 0.5; and (4) 1.0.

The change in the average mass of droplets, taking into account their coalescence, will be determined as

$$\frac{dM}{dt} = k_0 \frac{M}{t_R} = \frac{k_0}{2}\left(\frac{\omega_K}{t}\right)^{1/2} M \tag{9.41}$$

where k_0 is the coefficient of coalescence rate. Assuming droplets are spherical with masses $M = \pi/6a^3\rho_d$, the expression (9.41) is transformed to the form for changing the average droplet size

$$\frac{d\langle a\rangle}{dt} = \frac{k_0}{6}\left(\frac{\omega_K}{t}\right)^{1/2}\langle a\rangle, \langle a\rangle(t)\Big|_{t=0} = a_0 \tag{9.42}$$

The solution of this equation is represented in the form

$$\langle a\rangle = a_0 \exp\left(\frac{k_0}{3}\omega_K^{1/2}t^{1/2}\right) = a_0 \exp\left(\frac{k_0}{3}\left(t/t^*\right)^{1/2}\right) \tag{9.43}$$

Here, $t^* = \omega_K^{-1}$. In Figure 9.7 we compare the experimental data on the dimensions of the coalescence of spherical non-deformable droplets for small values of the Weber We < 1 number with the equation (9.43).

The fragmentation of the droplets increases the number of particles and the probability of their collision, that is, in a dispersive system, the conditions of aggregative instability are always ensured.

9.3.3 The Stochastic Fokker–Planck Equation

The Fokker–Planck equation (9.1) describes disperse systems with a continuous change in the size of dispersed particles and properties of the medium. Although the processes of coalescence and fragmentation are characterized by an abrupt change in the properties (sizes) of particles, in principle, for a sufficiently large time interval, the change in the mean properties can be assumed to be quasi-continuous with an infinitesimal jump. In particular, it can be assumed that the size of droplets and bubbles varies continuously in time and obeys the equation of variation of the average particle mass in time

$$\frac{dm}{dt} = \pm\omega(a)m \tag{9.44}$$

The sign corresponds to the process of coalescence, and the sign corresponds to the fragmentation of drops and bubbles. A lot of experimental studies on breaking and coalescence of droplets and bubbles in a turbulent flow show that the average droplet size is set at the minimum (breaking)

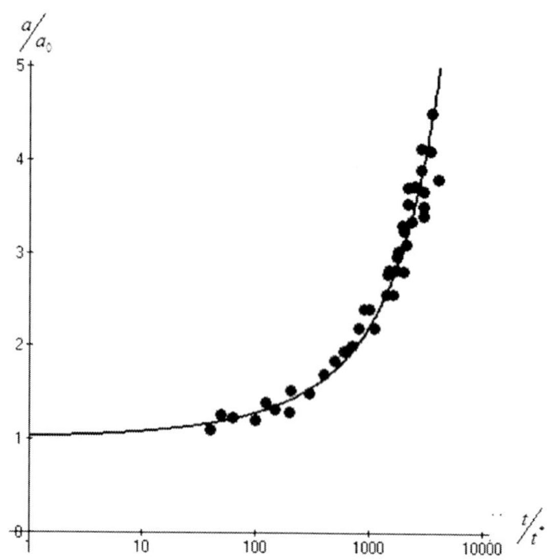

FIGURE 9.7 The change in the average droplet size upon their coalescence in volume of turbulent flow.

or maximum (coalescence) level, which corresponds to the aggregative stability of the dispersed medium. In view of the foregoing, in equation (9.44) it should be considered as a reduced mass with respect to the extreme values of drops and bubbles, $m = \pi/6\,\rho_d\left(a - a_{\min}\right)^3$ that is, for breaking and $m = \pi/6\,\rho_d\left(a_{\max} - a\right)^3$ for coalescence. In particular, starting from equation (9.38), we obtain an expression for changing the droplet sizes when they are fragmented in the form

$$\frac{da}{dt} = -k(\omega, a)\left(a - a_{\min}\right) = f(a)$$
$$t = 0, \quad a = a_0 \tag{9.45}$$

which is a continuous process with respect to time (where $k(\omega, a) = \omega(a)/3$). Experimental measurements of medium sizes during the fragmentation of droplets in mixing devices in a turbulent flow are given in the works [175,176,240]. In particular, using the experimental data [176], the following equations for determining the initial and minimum sizes of oil droplets in an aqueous medium are presented in the works [87,175] for the velocity of the mixer $n = 1000-2500\,\text{min}^{-1}$: $a_0 = 1.75\times10^2\left(n\,d_M\right)^{-1.5}$ and $a_{\min} = 1.695\times10^4\left(n\,d_M\right)^{-1.75}$.

A comparison of the calculated and experimental values of the change in the average droplet size is shown in Figure 9.8.

Thus, considering the change in particle sizes in the form of a continuous function, the Fokker–Planck equation with boundary conditions in the simplest case can be written in the form

$$\frac{\partial P(a,t)}{\partial t} = -\frac{\partial}{\partial a}\left[P(a,t)\frac{da}{dt}\right] + \frac{B}{2}\frac{\partial^2 P(a,t)}{\partial a^2}$$
$$t = 0, \quad P(a,0) = P_0(a); \quad a \to 0, \quad P(a,t) \to 0 \tag{9.46}$$

The solution of this equation presents great difficulties associated with specifying the form of the function $f(a)$, although some particular analytic solutions of equation (9.46), depending on the function definition $f(a)$, can be found in the works [85,87,175,205,325] (Table 9.4).

Stationary solutions of the Fokker–Planck equation are proposed in the paper. Numerous experimental studies on the coalescence and fragmentation of droplets and bubbles [40,65,168,169,324]

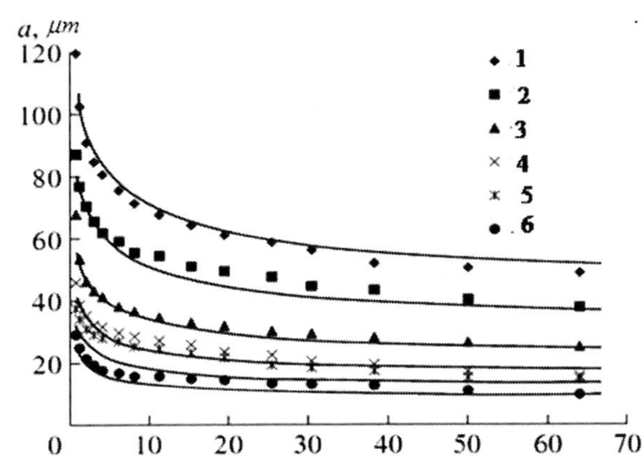

FIGURE 9.8 Change in the average droplet size at different rotational velocities: (1) 1000 min^{-1}; (2) 1200 min^{-1}; (3) 1500 min^{-1}; (4) 1800 min^{-1}; (5) 2100 min; and (6) 2500 min^{-1}.

TABLE 9.4
Analytical Solutions of the Fokker–Planck Equation (9.46) for Various Function Expressions *f*(*a*)

	Expression for $f(a)$ in Equation (9.46)	Analytical Solutions	References
1	$f(a)=K$	$P(a,t)=\frac{\sqrt{\pi}}{2}P_0(a)\exp\left(\frac{-K^2t}{2B}\right)\mathrm{erf}\left(\frac{a}{\sqrt{2Bt}}\right)$	
2	$f(a)=m_0-m_1a$	$P(a,t)=\sum_{n=0}^{\infty}C_n\exp\left(-\frac{m_0^2t}{B}\right)\exp(-m_1nt)\mathrm{H}_{n+1}\left(a\sqrt{\frac{m_1}{B}}\right)$ $C_n=\frac{1}{\sqrt{\pi}\,2^n n!}\sqrt{\frac{m_1}{B}}\int_{-\infty}^{\infty}P_0(a)\mathrm{H}_n\left(a\sqrt{\frac{m_1}{B}}\right)da$ $\mathrm{H}(x)$ is Hermite functions.	
3	$f(r)=-Kr+\frac{m}{r}$, $m_0={}^{m}\!/\!_{K}$, $r={}^{a}\!/\!_{2}$	$P(r,t)=r^{\theta}\exp\left(-\frac{Kr^2}{2B}\right)\sum_{n=0}^{\infty}C_n\mathrm{L}_{\mathrm{n}}^{\left(\frac{\theta-1}{2}\right)}\left(\frac{Kr^2}{2B}\right)\exp(-2Knt)$ $C_n=\frac{\theta^{\frac{\theta+1}{2}}\int_0^{\infty}P_0(r)\mathrm{L}_{\mathrm{n}}^{\left(\frac{\theta-1}{2}\right)}\left(\frac{Kr^2}{2B}\right)dr}{m_0^{\frac{\theta+1}{2}}2^{\frac{\theta-1}{2}}\Gamma\left(n+\frac{\theta+1}{2}\right)n!},\qquad \theta=\frac{m}{B}$ $\mathrm{L}_{\mathrm{n}}(r)$ is Laguerre functions.	[86,87,205,325]
4	$f(r)=b_0r-\frac{b_1}{r}$, $\tilde{r}={}^{r}\!/\!_{R}$, $\tilde{\tau}=b_0t$	When $t\to\infty$, we have $P_{\infty}(r)=C_0r^{\theta}\exp\left(-\frac{Kr^2}{2B}\right),\quad C_0=2\left(\frac{\theta}{2m_0}\right)^{\frac{\theta+1}{2}}$ (9.46a) $\alpha_1=R\sqrt{\frac{b_0}{B}},\quad \alpha_2=\frac{b_1}{(b_0R^2)}$, $A_i=\frac{1}{k}\int_0^1P_0(\tilde{r})\tilde{r}^{2\alpha_1^2\alpha_2}\exp(-\alpha_1^2\tilde{r})\tilde{r}\times F\left[\left(1+\frac{\lambda_i}{2}\right);\left(\frac{3}{2}+\alpha_1^2\alpha_2^2\right);\alpha_1^2\tilde{r}^2\right]d\tilde{r}$ $F(a,c,z)$ is hypergeometric function. $P(\tilde{r},\tilde{\tau})=\sum_iA_i\tilde{r}F\left[\left(1+\frac{\lambda_i}{2}\right);\left(\frac{3}{2}+\alpha_1^2\alpha_2^2\right);\alpha_1^2\tilde{r}^2\right]\exp(\lambda_i\tilde{\tau})$	[205,325]

show that the distribution function is multimodal, which is explained by the presence of secondary, tertiary, etc. in the flow phenomena of coalescence and breaking. And the number of extreme values in the "tail" of the distribution curve characterizes additional acts of coalescence of particles, and the number of extreme values at the beginning of the distribution curve is the number of acts of repeated fragmentation.

Obviously, the deterministic description of these phenomena without regard to their stochastic nature is incomplete and can lead to significant deviations from the true nature of these processes. The use of the kinetic equation of coalescence and crushing (9.32) and the Fokker–Planck equation (9.46) make it possible in a broad sense to interpret and analyze these phenomena at different instants of time. It should be noted that some analytical solutions of these equations (Tables 9.3 and 9.4) make it possible to obtain important theoretical results for the study and analysis of these phenomena. In particular, from the solution of the Fokker–Planck equation for the asymptotic case (9.46a) (Table 9.3) it follows that for an infinite time $t \rightarrow \infty$ of flow the finite particle distribution does not depend on the value and nature of the initial distribution.

And, for $\theta \rightarrow 0$, this distribution coincides with the distribution of Rosen–Rammler used in the work [167], with $\theta = 1$ is the Rayleigh distribution, and with $\theta = 2$ is the Maxwell distribution. Thus, this conclusion allows us to judge the stationarity of the distribution function for a long time of the coalescence processes and the fragmentation of the character of the initial distribution that are invariant to the task. At the same time, numerous experimental studies indicate multimodality of the distribution function, which is explained by the presence of repeated acts of coalescence and breaking. Moreover, the set of maxima of the distribution function on the left correspond to the repeated fractions of the drops, and in the vibrating "tail" part, the repeated coalescence. To describe such distribution functions, the most appropriate expression is (9.46), which is a superposition of the set of lognormal functions [175] (Figure 9.6). It should be noted that the spectra of large and small droplets are practically bound to shift with respect to each other in time. However, after a while, when the resources of the large-droplet spectrum are exhausted, it is possible that the spectrum begins to behave like a single-shot distribution. It is important to note that the nature of the evolution of the distribution function and the coefficient of turbulent diffusion is significantly influenced by the dropout of particles from the volume considered as a result of their rising or deposition [48]. In this case, the distribution spectrum changes significantly with the rate of deposition or ascent. Obviously, the presence of several maximum values of the distribution function is explained by secondary and tertiary phenomena of crushing and coalescence of droplets and bubbles. It follows from Figure 9.9 that the displacement of the maximum values of the distribution function is observed at different times. The evolution of the distribution function makes it possible to interpret the general picture at any time flow of the process, that is, characterizes the kinetics of coalescence and breaking processes.

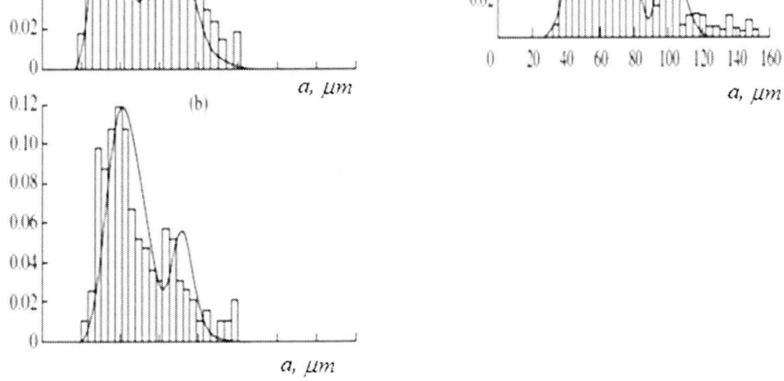

FIGURE 9.9 Evolution of the distribution function of oil droplets in an aqueous medium in a mixer at different times: (a) 7 min; (b) 60 min; and (c) 100 min.

9.3.4 An Analogy of Stochastic Equations with Mass-Transfer Phenomena

With the Fokker–Planck equation and its stochastic properties, we first learned (9.1) and (9.46) and noted the diffusion character of this equation, which most characterizes the mass-transfer processes.

We represent the parabolic differential equation (9.1) for the one-dimensional case in the form

$$\frac{\partial P(a,t)}{\partial t} = -\frac{\partial}{\partial a}\left[P(a,t)\frac{da}{dt}\right] + \frac{1}{2}\frac{\partial^2}{\partial a^2}\left(B(a)P(a,t)\right)$$
$$t=0,\ P(a,0)=P_0(a);\ a\to 0,\ P(a,t)\to 0 \tag{9.47}$$

Introducing the notion of probability flow

$$J_P = f(a)P(a,t) - \frac{1}{2}\frac{\partial}{\partial a}\left(B(a)P(a,t)\right) \tag{9.48}$$

we rewrite equation (9.47) in the form

$$\frac{\partial P(a,t)}{\partial t} + \frac{\partial}{\partial a}\left(J_P(a,t)\right) = 0 \tag{9.49}$$

Equation (9.48) is analogous to the diffusion flux of mass through a given surface. Equation (9.49) has the form of a local conservation equation similar to the equation of conservation of mass. Introducing the integral density of the number of particles in the form

$$N = \int_0^\infty P(a,t)\,da \tag{9.50}$$

We integrate (9.47) in the range from 0 to the constant of the stochastic diffusion coefficient $B(a)=B$, as a result of which we obtain

$$\frac{\partial N}{\partial t} = \frac{\partial}{\partial t}\int_0^\infty P(a,t)\,da = -\frac{\partial}{\partial a}\left[\int_0^\infty P(a,t)\,da\frac{da}{dt}\right] + \frac{B}{2}\frac{\partial^2}{\partial a^2}\left(\int_0^\alpha P(a,t)\,da\right) \tag{9.51}$$

Equation (9.51) can be rewritten in the form

$$\frac{\partial N}{\partial t} = -\frac{\partial}{\partial a}\left(N f(a)\right) + \frac{B}{2}\frac{\partial^2 N}{\partial a^2} \tag{9.52}$$

Expression (9.52) represents the equation of mass transfer of particles with variable dimensions and is characterized by boundary conditions similar to the mass-transfer equations. Thus, the Fokker–Planck equation is an analog of the mass-transfer equation, which represents a more extended interpretation and takes into account the distribution of the random measurement error.

9.4 APPLIED PROBLEMS OF MASS TRANSFER IN POLYDISPERSE MEDIA

In this section, we will consider the problems of mass transfer associated with the presence of polydisperse particles and droplets, accompanied by coalescence, crushing, and agglomeration. In contrast to traditional mass-exchange processes, in the practice of chemical technology, there are such processes in which mass and heat-transfer phenomena are not basic, but concomitant, although the occurrence of such processes significantly affects transport phenomena, changing their basic parameters. Such problems in the practice of chemical and oil-refining technologies are encountered in various processes related to processing and preparation of feedstock, separation of disperse systems, agglomeration of fine particles, grinding of solid particles, separation and separation of emulsions, etc.

In the oil-emulsion separation processes, the coalescence of the droplets plays an important role in their coarsening and deposition, which corresponds to the interfacial film rupture [85,87,175,343]. The coalescence of droplets in the flow depends on the frequency of their collision, the concentration, the specific energy of the turbulent flow, and the properties of the droplet and medium. The presence of an external field (electric, magnetic) and turbulence have a significant effect on the collision frequency of droplets. In this case, the diffusion flux density in isotropic turbulent flow depends on the turbulent diffusion coefficient, which is a complex function of droplet size, flow pattern, and deposition rate.

In Section 8.4 it is shown that under the conditions of isotropic turbulence, if the turbulence scale $\lambda > \lambda_0$ $\lambda_0 = \left(v_C^3/\varepsilon_R\right)^{1/2}$ is the Kolmogorov scale of turbulence under the condition that droplets are entrained by the pulsating medium, the collision frequency of the droplets is equal to $\omega \sim \varphi_0 \mu_P \left(\varepsilon_R/a^2\right)^{1/3}$, and if $\lambda < \lambda_0$ is $\omega \sim \varphi_0 \left(\varepsilon_R/v_C\right)^{1/2}$ where ε_R is the specific dissipation energy, μ_P is the droplet entrainment by the pulsating medium, and φ_0 is the volume fraction of droplets in the flow. With $\lambda > \lambda_0$ relatively large-scale pulsations, the emulsion is stirred vigorously, thereby ensuring a uniform distribution of water droplets in the oil volume. In turbulent flow, the collision frequency increases substantially in comparison with the laminar flow.

Dispersed emulsion, the size of the droplets $a = a_{\min}$, which, due to excess energy, associated with a large interfacial surface, tends to coalesce, that is, to a decrease in this surface. The increase in droplet sizes due to their collision and coalescence is possible only up to $a = a_{\max}$, further growth of droplets leading to their precipitation from the concentrated stream or to their deformation and breaking. Consequently, there is a minimum droplet size $a_{\min}$, characterizing the breaking threshold and the maximum droplet size $a_{\max}$, characterizing the coalescence threshold. Section 8.2 shows that the maximum droplet size in an isotropic turbulent flow is defined as: $a_{\max} \sim \left(\sigma/\rho_c^{1/3}\rho_d^{2/3}\right)^{0.6} \varepsilon_R^{-0.4}$, a minimum size as: $a_{\min} \sim \left(\rho_C \rho_d^2\right)^{1/3} v_C^2/\sigma$.

9.4.1 Evolution of the Droplet Distribution Function in Oil Emulsion

Coalescence and breaking of droplets significantly change the dispersity of oil emulsions, which is characterized by the evolution of the probability distribution function over time and in dimensions described by the kinetic equation and the stochastic Fokker–Planck equation. For the phenomena of coalescence and fragmentation of drops characterized by the presence of a set of maxima in the distribution function, the evolution of the distribution function can be described by the sum of log-normal distributions [175].

$$P(a,t) = \sum_{m=0}^{\infty} \sum_{n=0}^{\infty} E_{mn}(t).a^{n/2} \exp\left[-\alpha_n(t)\left(\ln a - a_s(t)\right)^2\right] \tag{9.53}$$

where $P(a,t)$ is the probability distribution function, $a_s = \ln a_{\max}$, $a_{\max}$ is the droplet size corresponding to different maxima of the probability distribution function, and E_{mn} is the coefficients of the series (9.53). It is important to note that an equation of the form (9.53) describes the polygonal curves of the distribution function for a set of maxima and minima.

Experimental studies have shown that the distribution of droplets during their crushing and coalescence in a turbulent flow is multimodal (two-humped or multihumped), and a specific interaction of the two humps of the distribution is observed (associated with a change in the value of the maxima and coordinates).

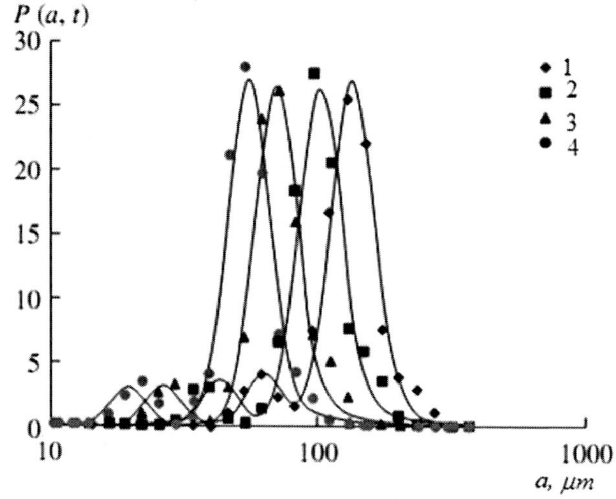

FIGURE 9.10 Evolution of the distribution function of oil droplets in an aqueous medium in a mixer at different times: (1) 0.5 min; (2) 2 min; (3) 15 min; and (4) 60 min.

The spectrum of large and small drops is practically bound to shift relative to each other (Figure 9.10).

However, after a while, when the resources of the large-drop or small droplet spectrum are exhausted, the spectrum begins to behave like a single-shot spectrum. In practice, the behavior of the multihump distributions in the model representation is confirmed when the distribution is represented by the sum of two or more distribution functions. The nature of the evolution of the distribution function and the change in the coefficient of turbulent diffusion can also be significantly influenced by the precipitation of particles from the turbulent flow. The particle size distribution spectrum varies significantly with the rate of droplet deposition.

The preceding analysis allows us to consider important properties of coalescence and droplet fracturing processes—aggregate instability, in which spatial inhomogeneity, deformation and deposition of droplets, the nucleation of new particles, and many other factors play a major role.

Obviously, under real conditions, there may exist a quasi-equilibrium state between the processes of coalescence and crushing, which leads to stationary distribution functions. The intensification of oil-emulsion separation processes is primarily related to the turbulence of the flow in the intermediate layer, since turbulent velocity pulsations contribute to weakening of intermolecular bonds between the adsorbed components of the armor shells, to decreasing and destroying them, to increasing the collision frequency of droplets, etc.

9.4.2 Modeling the Process of Granule Formation of Powdery Materials

Processes of granulation of powdered materials are widely used in the chemical, pharmaceutical, and food industries. The need for granulation of powdered materials and the requirements for their quality made it possible to develop various types of devices and apparatus: agitators with high external stresses, rotary drum apparatuses, fluidized bed apparatuses, and other structures [344]. Numerous experimental studies of granulation processes in granulator agitators [344–349] and in drum machines [349–353] have shown that the final granule size is determined by a number of parameters, among which it is important to note the size of the nucleus–nucleus formed, the particle size of the powder and the droplets, agglomeration conditions, powder and liquid properties, and from the granulation method. The influence of the particle size of droplets of binder on the formation and further growth of granules and on morphology of the structure is considered in the works [346,347,353,354–356]. It should be noted that the prediction of an appropriate amount of liquid (droplet size) to obtain the desired granule size is very difficult, due to the fact that besides the above factors, the size of the formed granules depends on the adhesive properties of the powder and on the physical properties of the liquid (viscosity, surface tension). The thickness of the layering and the conditions for the completion of the structure of the granule are determined by the moisture capacity or wettability of the surface. In this connection, the influence of droplet size on the rate of growth and the formation of embryos in granulator agitators was investigated [355,356].

The most important problem in the industrial processes of granulation of powdered materials is the detection of the distribution function of polydisperse granules in size, which makes it possible to determine in practical calculations the change in their average size along the length of the granulator. Experimental and theoretical study of the formation of granules of polydispersed composition and, related to this, the distribution of granules in size and measurement of their dimensions and porosity, are devoted to the work [345,357,358]. The experimental distributions of granules in size in granulator mixers showed their two-humped nature of the distribution curve, the maxima of which are determined in the region of nucleation and in the structure of the granule [345,351,354,357–364]. In drum machines, the most effective description is the evolution of the probability density function of the granule size distribution using the stochastic differential Fokker–Planck equation on the basis of experimental data characterizing the continuous layering and growth of the granule [359–363].

The processes of granulation are accompanied by compaction, deformation, and deterioration of the granules [344,345,347,365,366], leading to a change in their size, degree of polydispersity, and physical properties—density, strength, and porosity [351,353,354].

A lot of empirical formulas for calculating the physical properties of granules are given in the works [345,352–354]. It should be noted that the density of the granules is defined as $\rho = \rho_d\left(1-\varepsilon\left(t\right)\right)$, where ρ_d is the density of the material, and $\varepsilon\left(t\right)$ is the porosity of the granule, depending on the time and on the rheology of lamination, compaction, and deformation, as will be discussed below. Thus, as follows from this formula, the change in density is associated with a change in the porosity of the granule in time.

In general, the process of granulation, which at first sight seems to be simple, is a very complex phenomenon, including the study and description of such phenomena as nucleation, the structural formation of the granule itself, the rheology of compaction, deformation, etc. Figure 9.11 shows the scheme of granulation of powdery materials in the drum apparatus by the rolling method.

The mechanism of granulation of powdery materials by the rolling method is determined by the following steps: (1) mixing the powder with droplets of binder and forming the core of the granule. The nucleation in the granulation process is determined by the character of the capillary interaction in the layer of powder particles with liquid, the size of the droplets of the binder, the number of contacts per unit volume of the material (coordination number), and the relaxation time τ_p.

In practical cases, the source of nucleation can: (a) also be the coarse particles present in the original powder; (b) growth and formation of granules as a result of their rolling over the surface of the powder. In this stage, the particle size of the powder (the thickness of the layering) and the particle size of the droplets of binder, the velocity of rolling play an important role. The final size of the granules is determined by the degree of liquid distribution due to capillary forces in the pores and the liquid content in the volume of the granule (moisture capacity); (c) compaction of granules under the action of deforming external stresses and own weight. As a result of the consolidation of the granule, the liquid contained in the pores is extruded to the surface, which increases the layering velocity of the powder; (d) stabilization and consolidation of the granule structure as a result of strengthening of internal bonds between individual particles in the volume of the granule and stabilization of the final shape.

In the final analysis, the geometry of the granule determines the geometry of two factors: the geometry of the dynamics of the movement of the granule in the process of its pelletization and the geometry of the anisotropy of strength and other properties, more precisely its resistance to abrasion

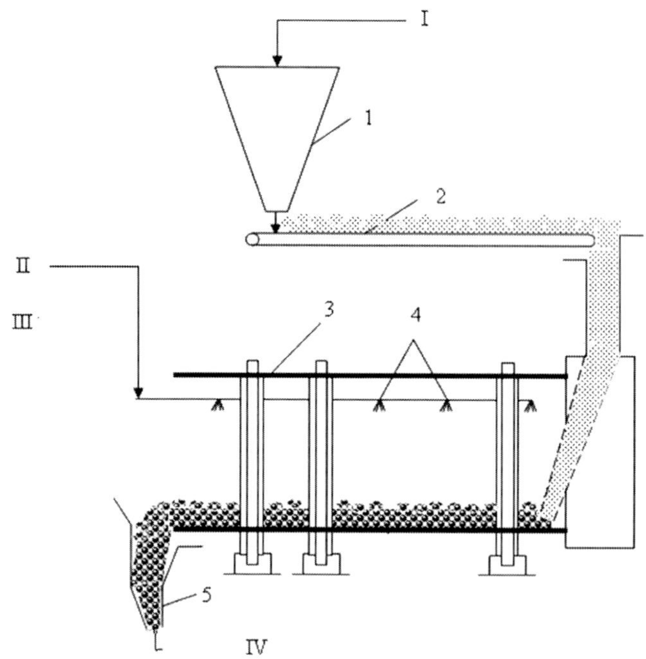

FIGURE 9.11 Scheme of installation for granulation of powdery materials in drum machines: (1) bunker; (2) conveyor; (3) reel apparatus; (4) nozzles for water injection; (5) hopper for granules; (I) supply of powdered material; (II, III) water supply line with additives; and (IV) the output of the granular material.

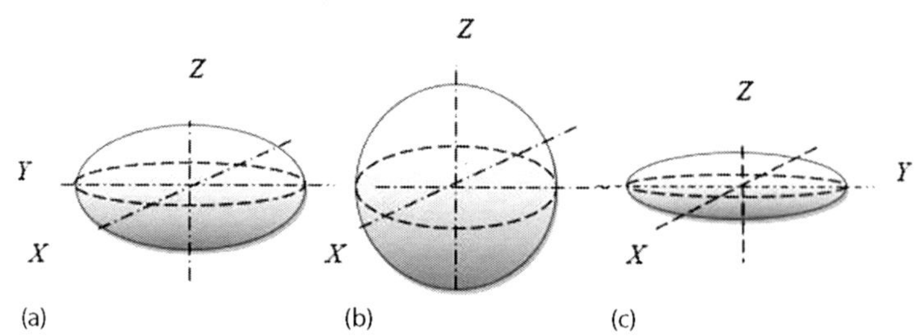

FIGURE 9.12 Characteristic shapes of granules (*X, Y, Z* are anisotropy axis): (a) ellipsoid; (b) sphere; and (c) elongated ellipsoid.

and deformation. The dynamics of the process consists of the rotational movement of the granule along the surface of the powder and the apparatus, with rotation being carried out in all directions, as a result of which the geometry of the rolling dynamics has the symmetry of the rotating sphere or sphere, that is, consists of an infinite number of symmetry axes of infinite order. Because of this, with this geometry, the dynamics of the rolling process itself gives the granule a rounded shape (Figure 9.12): (a) oval, (b) sphere, (c) ellipse. On the other hand, the geometry of the anisotropy of the granule structure is determined by abrasion, wear, and deformation in all directions, which creates conditions for distortion and displacement of the symmetry axes.

However, in all cases of the combined action of the geometry of the dynamics of rotational motion and the anisotropy of the structure, the shape of the granule approaches a round shape, with the exception of strongly deformed ones. At the same time, as a result of the action of various external and internal forces, the granule loses stability and collapses, while losing a certain symmetry and shape.

In spherical coordinates (r,θ,ψ), the surface area of the granule is defined in the form [362,364]

$$S=\iint\sqrt{1+\frac{1}{r^2}\left(\frac{\partial r}{\partial\theta}\right)^2+\frac{1}{r^2\sin^2\theta}\left(\frac{\partial r}{\partial\psi}\right)^2}\,r^2\sin\theta d\theta d\psi \tag{9.54}$$

where θ,ψ are the polar angles. If the granule has a radius R, then as a result of layering, the radius of the variable surface increases by the amount of layering thickness λ or $R+\lambda(\theta,\psi)$. If we assume that $\lambda << R$, then the variable surface area can be represented in the form

$$S\approx\iint\left[1+\frac{1}{2R^2}\left(\frac{\partial\lambda}{\partial\theta}\right)^2+\frac{1}{R^2\sin^2\theta}\left(\frac{\partial\lambda}{\partial\psi}\right)^2\right](R+\lambda)^2\sin\theta.d\theta d\psi \tag{9.55}$$

If the thickness of the layering changes $\Delta\lambda$ by an amount, the surface area of the granule will change by an amount

$$\Delta S=\iint\left[2(R+\lambda)\Delta\lambda+\frac{\partial\lambda}{\partial\theta}\frac{\partial\Delta\lambda}{\partial\theta}+\frac{1}{2R^2\sin^2\theta}\left(\frac{\partial\lambda}{\partial\psi}\right)\frac{\partial\Delta\lambda}{\partial\psi}\right]\sin\theta.d\theta d\psi \tag{9.56}$$

Expression (9.56) can be regarded as a variation in the surface area of the granule as the thickness of the layering changes. Integrating the last two terms by parts from 0 to π, we have

$$\int\frac{\partial\lambda}{\partial\theta}\frac{\partial\Delta\lambda}{\partial\theta}\sin\theta.d\theta=-\int\Delta\lambda\frac{\partial}{\partial\theta}\left(\sin\theta\frac{\partial\lambda}{\partial\theta}\right)d\theta \tag{9.57}$$

With these expressions in mind, and assuming that the length of the surface area changes as a result of pure layering is equal $\iint\lambda\Delta l d\theta d\psi$, equation (9.56) is represented in the form

$$\Delta S=\iint\left[\left(2(R+\lambda)-\frac{1}{\sin\theta}\frac{\partial}{\partial\theta}\left(\sin\theta\frac{\partial\lambda}{\partial\theta}\right)-\frac{1}{\sin^2\theta}\frac{\partial^2\lambda}{\partial\psi^2}\right)\Delta\lambda\sin\theta+\lambda\Delta l\right]d\theta d\psi \tag{9.58}$$

The change in the volume of the granule is defined as

$$\Delta\upsilon=\iint\Delta\lambda(R+\lambda)^2\sin\theta.d\theta d\psi \tag{9.59}$$

Dividing the integral (9.58) by (9.59), taking $\lambda << R$ into account the integral, we obtain the expression corresponding to the change in the surface area of the granule with respect to the change in volume. Then, passing to the limit, we obtain the following expression

$$\frac{dS}{d\upsilon}=\frac{2}{R}-\frac{1}{R^2\sin\theta}\frac{\partial}{\partial\theta}\left(\sin\theta\frac{\partial\lambda}{\partial\theta}\right)-\frac{1}{R^2\sin^2\psi}\frac{\partial^2\lambda}{\partial\psi^2}+\lambda\frac{\partial l}{\partial\upsilon} \tag{9.60}$$

Multiplying and dividing both sides by Δt and denoting $d\upsilon/dt=\alpha$ and $V=\partial l/\partial t$, equation (9.60) can be rewritten in the form

$$\frac{dS}{dt}=\alpha\left[\frac{2}{R}-\frac{1}{R^2\sin\theta}\frac{\partial}{\partial\theta}\left(\sin\theta\frac{\partial\lambda}{\partial\theta}\right)-\frac{1}{R^2\sin\theta}\frac{\partial^2\lambda}{\partial\psi^2}\right]+\lambda V \tag{9.61}$$

where V is the velocity of movement of the granule, which determines the geometry of dynamics. Thus, in equation (9.61), the first term reflects the geometry of the anisotropy of the shape, and the second term reflects the dynamics of the movement of the granule. In the general case, equation (9.61) reflects the dynamics of the asymmetric growth of the granule as a result of its rolling. For a symmetrical spherical granule, putting $S = \pi a^2$ (a is diameter of the granule), we obtain a simpler equation for the change in the average granule size as a result of its rolling

$$\frac{da}{dt} = \frac{2\alpha}{\pi a^2} + \frac{\lambda V}{2\pi a} \tag{9.62}$$

It should be noted that the spherical shape of the granule corresponds to an isotropic and uniform structure and is distinguished by the invariance of the strength and other properties of the granule in different directions. The general solution of equation (9.62) is a transcendental expression in connection with which, we consider special cases of its solution.

1. If $a << 4\alpha/\lambda V$, then equations (9.62) can be represented in the form

$$a^2 \frac{da}{dt} = \frac{2\alpha}{\pi}, \; a(t)\big|_{t=0} = a_0, \tag{9.63}$$

with the solution

$$a(t) = \left(a_0^3 + \frac{6\alpha}{\pi} t \right)^{1/3} \tag{9.64}$$

2. If $a > 4\alpha/\lambda V$, then equation (9.62) is represented in the form

$$\frac{da}{dt} = \frac{\lambda V}{2\pi a} \tag{9.65}$$

with the solution:

$$a(t) = \left(a_0^2 + \frac{\lambda V}{\pi} t \right)^{1/2} \tag{9.66}$$

Expression (9.66) coincides with the rolling equation in the drum granulator given in the paper [349], if $V = \omega R_B$ (where ω is the angular velocity of rotation and the radius of the drum). Proceeding from the solutions (9.64) and (9.66), the process of granule formation can be divided into two regions: $0 \le t \le \tau_P$ where the nucleus of the granule is structured and $t > \tau_P$ where the granule is grown by layering the powder onto the surface.

3. Consider the stationary case of equation (9.61) for $t > \tau_P$

$$\frac{\alpha}{R^2}\left[\frac{1}{\sin\theta} \frac{\partial}{\partial \theta}\left(\sin\theta \frac{\partial \lambda}{\partial \theta} \right) + \frac{1}{\sin^2\theta} \frac{\partial^2 \lambda}{\partial \psi^2} \right] = \lambda V \tag{9.67}$$

If we assume that the thickness of the layer to be layered does not change with respect to the angle ψ, then equation (9.67) can be rewritten in the form

$$\frac{\partial}{\partial \theta}\left(\sin\theta \frac{\partial \lambda}{\partial \theta} \right) = \frac{\lambda V R^2}{\alpha} \sin\theta \tag{9.68}$$

In view of the insignificance of the second derivative $\partial^2\lambda/\partial\theta^2$, this equation can be written in the form

$$\cos\theta\frac{\partial\lambda}{\partial\theta}=\frac{\lambda VR^2}{\alpha}\sin\theta \tag{9.69}$$

Dividing both sides by and integrating both sides, we have

$$Ci\theta=\beta.Si\theta \tag{9.70}$$

where $Ci\theta, Si\theta$ are the integral cosine and the integral sine, $\beta=\frac{R^2V}{\alpha\partial\ln\lambda/\partial\theta}$, and θ is the angle in radians. Expression (9.70) represents the equation of stationary layering in the process of granulation of powdery materials. Figure 9.13 shows the numerical trajectories of layering the powder on the surface of the granule for different values. The curves presented give a graphic interpretation of layering on the surface of a spherical granule.

Rheology of compaction of granules: The resulting granule as a result of rolling under the action of external deforming forces due to the interaction of the granules between themselves, and the wall of the apparatus is subjected to compaction due to the rearrangement of the powder particles in the granule. Each particle of the powder in the granule contacts a variety of others, and in the immediate vicinity of contact between the spherical particles a circular capillary gap is formed where the binder is retained. If we assume that the particle size of the powder is approximately the same, as a result of their agglomeration, ordered structures with different coordination numbers N_k can form into a single system. In the presence of a large number of finely dispersed fractions, a denser and more stable granule structure results from compaction. Porosity of the granule with loose packing $(N_k=6)$, is $\varepsilon=0.476$, but with a sufficiently dense structure $(N_k=12)$ is $\varepsilon=0.259$ [366]. As a result of the compaction of the granule, the solid phase (powder particles) flows into the pores, compressing the liquid in the pores and displacing it through a continuous closed pore channel system. If isolated closed pores are included in the impermeable or impermeable medium of the granule volume, then the volume flow leads to an increase in the

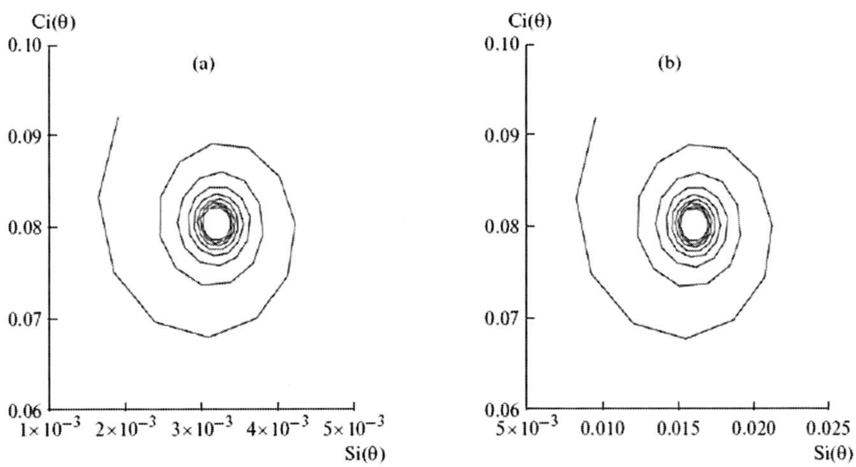

FIGURE 9.13 The nature of the change in the dimensions of the granules when layering, depending on β, equal to: (a) 0.5 and (b) 1.0.

liquid pressure ΔP_f, thereby distorting the shape, generating an anisotropy of the internal structure. The nature of the external stresses acting on the granule or on the particles in the granule is determined by the presence of centrifugal forces appearing during the rotation of the drum and the mass forces due to the weight of the overlying layers, as well as a number of forces that arise when the pellets collide with each other and with the wall of the apparatus. By analogy with the theory of elasticity [367], the strain tensor is analogous to the tensor of viscous strain rates, and the effective coefficients of bulk and shear viscosity are analogous to the moduli of all-round compression and shear. Thus, the rheology of deformation and compaction of porous granules is characterized by effective macro-rheological characteristics: the shear viscosity coefficient and the relaxation volume viscosity coefficient associated with the volumetric flow of the porous medium. Suppose that in each local element of the porous granule a macroscopically homogeneous stress state is realized. According to the theory of elasticity, elastic volumetric deformations of the skeleton of the porous granule are connected with the external deforming stress by the relation

$$\frac{\Delta V_S}{V_S} = \frac{\sigma_D}{\eta_S}\Delta t \tag{9.71}$$

where $V_S = 1/\rho_P$ is the specific volume of the porous medium, ΔV_S is the change in the volume of the local element of the medium, $\rho_P = \rho_d(1-\varepsilon)$ is the bulk density of the granule, ρ_d is the density of the dense phase material, $\sigma_D = \Gamma_D - \Delta P_f + g(\rho_d - \rho)z$ is the deforming stress composed of external stress, the pressure drop inside the granule, and the weight of the overlying layers is thick z. Differentiating (9.71) with respect to t, we obtain

$$\frac{1}{1-\varepsilon}\frac{d\varepsilon}{dt} = -\frac{\sigma_D}{\eta_S} \tag{9.72}$$

Passing from the substantial derivative to the local derivative in equation (9.72), we obtain

$$\frac{\partial \varepsilon}{\partial t} + div\left(\varepsilon \bar{U}_S\right) = -(1-\varepsilon)\eta_S^{-1}\sigma_D \tag{9.73}$$

where $\bar{U}_S$ is the velocity vector of particles in the volume of the granule as a result of its compaction.

For one-dimensional radial compaction of a granule, equation (9.73) is simplified to the form

$$\frac{\partial \varepsilon}{\partial t} + U_{Sr}\frac{\partial \varepsilon}{\partial r} = -(1-\varepsilon)\eta_S^{-1}\sigma_D \tag{9.74}$$

Expressing the bulk viscosity η_S through the shear viscosity ξ_S in the form [8]

$$\eta_S = \frac{4}{3}\xi_S\frac{1-\varepsilon}{\varepsilon} \tag{9.75}$$

we get

$$\frac{\partial \varepsilon}{\partial t} + U_{Sr}\frac{\partial \varepsilon}{\partial r} = -\frac{3}{4}\varepsilon\xi_S^{-1}\sigma_D \tag{9.76}$$

The solution of equation (9.76) can be represented in the form

$$\varepsilon(t) = \varepsilon_0 \exp\left[-\frac{3}{4}\xi_S^{-1}\sigma_D\left(t + \frac{r}{U_{Sr}}\right)\right] \tag{9.77}$$

For the average porosity by volume of the granule, one can write [365]

$$\varepsilon(t) = \varepsilon_0 \exp\left(-\frac{3}{4}\xi_S^{-1}\sigma_D \bar{t}\right) \tag{9.78}$$

Equation (9.78) allows us to estimate the porosity of the granule at various times of residence, taking into account the compaction, and taking into account their values when calculating physical characteristics. However, if the initial and current porosity values of the granule are known, then from equation (9.78) we can estimate the shear viscosity value as the solution of the inverse problem

$$\xi_S = \frac{3}{4}\frac{\sigma_D \bar{t}}{\ln \varepsilon_0/\varepsilon} \tag{9.79}$$

For the granulation process, we estimate the shear viscosity at the following values: for a loose granule structure $\varepsilon_0 = 0.476$, for a dense structure is $\varepsilon = 0.259$, granulation time $t = 480\,s$ and deforming stress $\sigma_D \approx 100\,\text{N/m}^2$. With these data, the shear viscosity is $\xi_S = 6.10^4\,\text{N s/m}^2$.

Assuming that the rate of change of the granule surface as a result of compaction is equal $S\,{}^{d\varepsilon}\!/_{dt}$, the layering equation can be written in the form

$$\frac{da}{dt} = \frac{2\alpha}{\pi a^2} + \frac{\lambda V}{2\pi a} + \frac{a}{2}\frac{d\varepsilon}{dt} \tag{9.80}$$

Expression (9.80) describes the change in the size of the granules in the process of the granule formation of powdery materials in the drum granulators. This equation includes three stages of granule formation: nucleation $(t < \tau_P)$, formation of granules by lamination $(t > \tau_P)$ and their compaction, determined by the last term.

The analysis of the processes of granulation of powdery materials described in this study made it possible to describe the main stages of granule formation: nucleation, layering as a result of the rotation of the drum and growth of the granule, compaction and the creation of a strong granule structure. Moreover, the models (9.74) and (9.80) take into account the dynamics of motion and the anisotropy of the shape and strength properties of the granule. In particular, setting in equation (9.80) the condition

$$a >> \left(\frac{4\lambda V}{3\pi\varepsilon}\frac{\xi_S}{\sigma_D}\right)^{1/2} \tag{9.81}$$

Taking into account (9.78), we obtain an equation for the change in the size of the granules in the region of pure compaction and stabilization of the structure

$$\frac{da}{dt} = \frac{\lambda V}{2\pi a} - \frac{3}{8}\frac{\sigma_D\varepsilon_0}{\xi_S}\exp\left(-\frac{3}{4}\frac{\sigma_D}{\xi_S}t\right) \tag{9.82}$$

In view of the insignificance $\sigma_D/\xi_S << 1$, linearizing this equation, we determine the approximate solution (9.82) in the form

$$a \approx \sqrt{\beta_0 - \left(\beta_0 - a_0^2\right)\exp\left(-\beta_1 l + \beta_2 l^2\right)} \tag{9.83}$$

where $\beta_0 = \frac{4}{3\pi}\frac{\lambda V \xi_S}{\sigma_D \varepsilon_0}$, $\beta_1 = \frac{3}{4}\frac{\sigma_D\varepsilon_0}{\xi_S V}$, $\beta_2 = \frac{9}{32}\left(\frac{\sigma_D}{\xi_S V}\right)^2 \varepsilon_0$, a_0, is the average size of the embryo, and l is the length of the apparatus. Using the industrial data, we estimated the coefficients: $\beta_0 \approx 1332{,}5$; $\beta_1 \approx 2.10^{-3}$ and $\beta_2 \approx 3{,}3.10^{-4}$.

Figure 9.14 shows the numerical solutions of equations (9.83) and the experimental values of the average granule size along the length of the drum apparatus. Moreover, the expression $V = \omega R_B$, $(\omega = 0.59\,\text{min}^{-1},\ R_B = 0.9\text{ m})$ is used as the linear velocity. As follows from Figure 9.14,

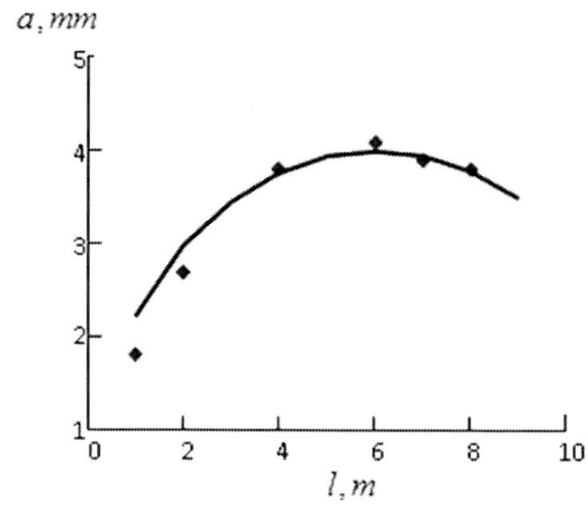

FIGURE 9.14 Changing the size of granules along the length of a cylindrical apparatus (points are experimental data, solid curve-calculation by equation [9.83]).

when the condition is satisfied $l > 6\mathcal{M}$ the size of the granules decreases as a result of its compaction and strengthening of the structure. Strengthening, stabilization, and fixing of the granule structure is carried out in two ways: (1) using surface coatings (encapsulation) [368,369]; and (2) Strengthening the internal bonds between the powder particles in the granule, by using various additives to the powder. Consequently, the main factor in stabilizing the structure of the granule is its strength and resistance to fracture and deformation as a result of the action of external loads. In principle, the strength value can be estimated from the following semi-empirical equation

$$\frac{d\ln\Delta}{d\tau_D} = f(k,\Delta) \tag{9.84}$$

where Δ is the strength or parameter characterizing the strength of the granule, τ_D is the deformation time or the parameter on which the strength depends $f(k,\Delta)$ is a function chosen on the basis of experimental studies, although somewhat analogous to (9.84), is given in [365]. It should be noted that these phenomena significantly affect the strength of the granule, the value of which, in addition to other parameters, depends on the porosity of the granule. This dependence is expressed by various formulas given in the works [370–372]

$$\frac{d\ln(\Delta-\Delta_0)}{dC_d} = k_0\frac{\Delta}{C_d^2} \tag{9.85}$$

In addition, various empirical equations for determining the strength of granules [373]

$$\Delta = \Delta_0\exp(-b\varepsilon), \tag{9.86}$$

$$\Delta = \frac{1-\varepsilon}{\varepsilon}\frac{F}{a^2}, \tag{9.87}$$

$$\Delta = 3.7(1-\varepsilon)^4\frac{F}{a^{3/2}m^{1/2}}, \tag{9.88}$$

Here, F is the average strength per unit cross-sectional area, the characteristic size. The above dependences (9.86) through (9.88) show that the strength of granules increases with decreasing porosity.

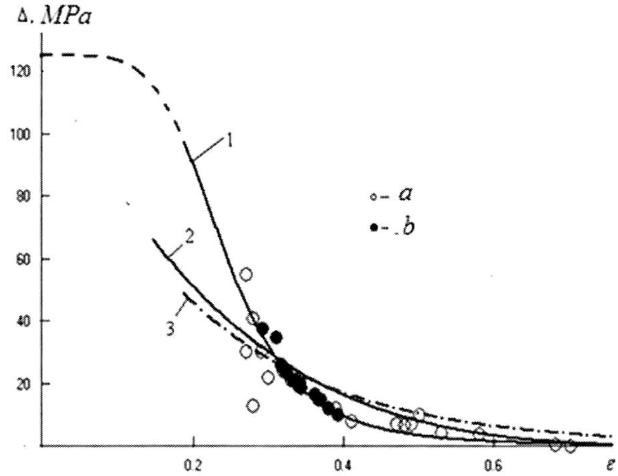

FIGURE 9.15 Dependence graph of granules strength on porosity: (1) calculation by formula (9.89); (2) calculation by the formula (9.86); and (3) calculation by the formula (9.87). a = experimental data of the work and b = experimental studies on the strength of superphosphate granules.

Using the experimental data [373], we determine the dependence of the strength of the granule on the porosity in the form

$$\Delta = (0.008 + 9.6\varepsilon^5)^{-1} \tag{9.89}$$

with a correlation coefficient equal to $r^2 \approx 0.9421$ (Figure 9.15), that is, with increasing porosity of the granules, the strength is significantly reduced.

Evolution of the granule size distribution function: The stochastic description of the process of granulation of powdery materials is based on the stochastic differential Fokker–Planck equation written in the form

$$\frac{\partial P(r,t)}{\partial t} = -\frac{\partial}{\partial r}\left[f(r)P(r,t)\right] + B\frac{\partial^2 P(r,t)}{\partial r^2} \tag{9.90}$$

where $P(r,t)$ is the density of the distribution function of granules in size and time, $f(r) = dr/dt$ is the rate of granule formation, B is the stochastic diffusion coefficient, and $r = a/a_0$ is the dimensionless diameter of the granules. Starting from equation (9.82) and setting $\sigma_D/\xi_S << 1$, the velocity of granule formation in a dimensionless form, we define in the form

$$\frac{dr}{dt} = \frac{m_R}{r} - kr \tag{9.91}$$

where $m_R = \frac{V\lambda}{2\pi a_0^2}$, $k = \frac{3}{8}\frac{\sigma_D}{\xi_S}\varepsilon_0$. Taking into account expression (9.91), the Fokker–Planck equation can be written in the form

$$\frac{\partial P(r,t)}{\partial t} - k\frac{\partial}{\partial r}\left[\left(\frac{m}{r} - r\right)P(r,t)\right] + \frac{B}{a_0^2}\frac{\partial^2 P(r,t)}{\partial r^2} \tag{9.92}$$

where $m = m_R/k$ and the solution of equation (9.92) by the method of separation of variables is represented in the form

$$P(r,t) = r^{\theta}\exp\left(\frac{ka_0^2 r^2}{2B}\right)\sum_{n=0}^{\infty} C_n L_n^{(\alpha)}\left(\frac{ka_0^2 r^2}{2B}\right)\exp(-2knt) \tag{9.93}$$

where $\theta = m_R a_0^2 / B$, $\alpha = \frac{m_R a_0^2 - B}{2B}$, $L_n^{(\alpha)}$ are the Laguerre functions,

$$C_n = \frac{\theta^{\frac{\theta+1}{2}} \int_0^{\infty} P_0(r) L_n^{(\alpha)}\left(\frac{k a_0^2 r^2}{2B}\right) dr}{2^{\frac{\theta-1}{2}} \Gamma\left(n + \frac{\theta+1}{2}\right) m^{\frac{\theta+1}{\theta}} n!} \tag{9.94}$$

Solutions (9.93) and (9.94) characterize the evolution of the distribution function of the probability density of granules in size and in time. The asymptotic value of the distribution is obtained from the solution of (9.94), taking $t \to \infty$ into account the properties of the Laguerre function, in the form

$$P(r) = C_R r^{\theta} \exp\left(-\frac{k a_0^2 r^2}{2B}\right) = C_{PR} a^{\theta} \exp\left(-b a^2\right) \tag{9.95}$$

$$C_{PR} = 2 a_0^{-\theta} \left(\frac{\theta}{2m}\right)^{\frac{\theta+1}{2}}, \quad b = k/2B \tag{9.96}$$

Thus, as follows from equation (9.95), for a long residence time of granules, the limit distribution is established in the apparatus, which does not depend on the initial distribution. For large values t of the compaction rate of granules, such a state is achieved at lower values t. Experimental studies on the distribution of granules by size and length of the drum granulator using the sieve analysis were carried out using the apparatus shown in Figure 9.11. Using the experimental data of granule formation, we determine the following values of the coefficients entering into equation (9.95)

$$\begin{aligned} C_{PR} &= 0.234 \exp(0.798l - 0.3788l^2 + 0.0312l^3) \\ \theta &= 0.1206 \exp(-1.964l - 0.422l^2 + 0.0284l^3) \\ b &= 1.0267 \exp(-0.084l - 0.1288l^2 + 0.014l^3) \end{aligned} \tag{9.97}$$

where l is the current length of the drum unit.

Figure 9.16 shows the calculated values of the evolution of the probability density distribution function and its comparison with experimental data on the distribution of granules along the length of the apparatus [362–364].

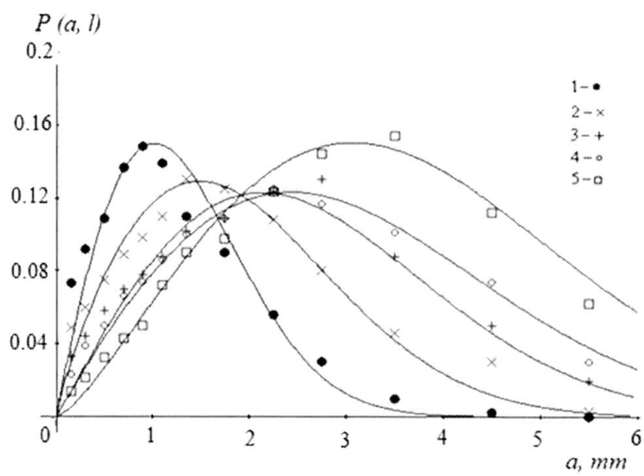

FIGURE 9.16 Experimental and theoretical curves for the distribution of superphosphate granules in size and length of the apparatus, equal to: (1) 0; (2) 10 cm; (3) 20 cm; (4) 40 cm; and (5) 80 cm.

As follows from this figure, the distribution functions are of single-shot nature, since the probability of formation of two-humped distributions for cylindrical devices is observed in the region of nucleation, that is, at $t < \tau_P$.

Thus, the granulation process of powdered materials is stochastic, since the granulometric composition of the granules obtained is polydisperse, which is determined by the uneven granular completeness, depending on the size of the droplet droplets (spray pattern), powder particles, and such phenomena as coagulation and fracture, wear, and deformation. At the same time, as a result of the rotation of the drum, the nuclei formed can also coagulate, which affects the final size of the granules. Obviously, with the increase in the size of liquid droplets and the size of the nucleus, the probability of formation of large-sized granules and large lumps increases.

References

1. Coulson J.M., Richardson J.F. *Chemical Engineering. V.1. Fluid Flow, Heat Transfer and Mass Transfer.* 6th ed., Oxford, UK: Reed Educational and Professional Publishing, 1999.
2. Ronald F. Probstein P. *Physicochemical Hydrodynamics.* New York: John Wiley & Sons, 1994.
3. Kutepov A.M., Polyanin A.D., Zapryanov Z.D., Vyazmin A.V., Kazenin A.D. *Chemical Hydrodynamics.* Moscow, Russia: Quantum, 1996.
4. Kelbaliyev G.I., Rasulov S.R. *Hydrodynamics and Mass Transfer in Disperse Medium.* Sankt Petersburg, Russia: Chemizdat, 2014.
5. Loitsyansky L.G. *Mechanics of Fluid and Gas.* Moscow, Russia: Nauka, 1987.
6. Levich V.G. *Physicochemical Hydrodynamics.* Englewood Clifts, NJ: Prentice Hall, 1962.
7. Kelbaliyev G.I., Rasulov S.R., Rzayev A.G. *Oil Hydrodynamics.* Moscow, Russia: Maska, 2015.
8. Soo S.L. *Fluid Dynamics of Multiphase Systems.* Waltham, MA: Blaisdell Publishing Company, 1970.
9. Nigmatulin R.I. *Fundamentals of Mechanics of Heterogeneous Media.* Moscow, Russia: Nauka, 1977.
10. Polyanin A.D., Kutepov A.M., Kazenin D.A. *Hydrodynamics, Mass and Heat Transfer in Chemical Engineering.* London, UK: CRC Press, 2002.
11. Grank J. *The Mathematics of Diffusion.* Oxford, UK: Clarendon Press, 1975.
12. Mednikov E.P. *Turbulent Transport and Deposition of Aerosols.* Moscow, Russia: Nauka, 1980.
13. Hinze J.O. *Turbulence.* New York: McGraw-Hill. 1975.
14. Luikov V.V. *Heat and Mass Transfer: Handbook.* Moscow, Russia: Energia, 1986.
15. Kutateladze S.S. *Heat Transfer and Hydrodynamic Drag. Handbook.* Moscow, Russia: Nauka, 1990.
16. *Perry's Chemical Engineers Handbook.* Section 5: Heat and Mass transfer, 8th ed., New York: McGraw-Hill Companies, 2008.
17. Koisi Asano Mass Transfer. From Fundamentals to modern Industrial Application. 2006. WILEY–VSH Verlag GmbH and Co, Weinheim.
18. Bennet C.O., Myers J.E. *Momentum, Heat and Mass Transfer.* New York: McGraw-Hill Book CO, 1962.
19. Matvienko V.N., Kirsanov E.A. Viscosity and structure of disperse systems. *Proceedings of Moscow University*, 2011, 52, 4, 243–276.
20. Mewis J., Wagner J. *Colloidal Suspension Rheology.* Cambridge, UK: Cambridge University Press, 2012.
21. Brounshtein B.I., Schegolev V.V. *Hydrodynamics, Mass and Heat Transfer in the Column Apparatuses.* Sankt-Petersburg, Russia: Chemistry, 1988.
22. Colak Y., Pehlivan D., Sarimeseli A., Kelbaliyev G. Rheological properties of asphalt-plastic blends. *J. Pet. Sci. Technol.*, Taylor & Francis Group, USA: 2003, 21, 9–10, 1427–1438.
23. Verberg R., de Schepper I.M., Cohen E.G.D. Viscosity of Colloidal Suspensions. 1996, http://citeseerx.ist.psu.edu/viewdoc/download?doi=10.1.1.285.6617&rep=rep1&type=pdf.
24. Chistos Vassilicos J. Dissipation in turbulent flow. *Annu. Rev. Fluid Mech.*, 2015, 47.
25. Kelbaliyev G.I. Coefficients of resistance of solid particles, droplets and bubbles of various shapes. *Theor. Found. Chem. Eng.*, 2011, 45, 2, 1–20.
26. Kelbaliyev G., Ceylan K. Development of new empirical equations for estimation of drag coefficient, shape deformation and rising velocity of gas bubbles or liquid drops. *Chem. Eng. Comm.*, 2007, 194, 194–206.
27. Ceylan K., Altunbas A., Kelbaliyev G. A new model for estimation of drag force in the flow Newtonian fluids around rigid of deformable particles. *Powder Technol.*, 2001, 119, 250–261.
28. Scheidegger A.E. *The Physics of Trough Porous Media.* Revised edition. New York: Macmillan, 1960.
29. Böckh P., Wetzel T. *Heat Transfer: Basics and Practice.* New York: Springer, 2012. 289 p.
30. Cao E. *Heat Transfer in Process Engineering.* New York: The McGraw-Hill Companies, 2010.
31. Tikhonov A.N., Samarsky A.A. *Equations of Mathematical Physics.* M.: Nauka, 1965.
32. Li W., Webb R.I. Fouling characteristics of internal helical-rib roughness using low velocity cooling tower. *Int. J., Heat and Mass Transfer*, 2002, 43, 1685–1691.
33. Bergles A.E. Heat transfer enhancement–the encouragement and accommodation of high fluxes. *Trans. ASME J. Heat Transfer*, 1995, 119, 8–19.

34. Eckert E.R., Drake R.M. *Analysis of Heat and Mass Transfer.* New York: McGraw-Hill, 1972, 378–412.
35. Ceylan K., Kelbaliyev G. The roughness effects on friction and heat transfer in the fully developed turbulent flow in pipes. *Appl. Therm. Eng.*, 2003, 23, 557–570.
36. Frank-Kamenetsky D.A. *Basics of Macrokinetics. Diffusion and Heat Transfer in Chemical Kinetics.* Moscow, Russia: Intellect Publishing House, 2008.
37. Cawood W. The movement of dust or smoke particles in a temperature. *Trans. Faraday Soc.*, 1936, 36, 1068–1073.
38. Waldmann L. Über die krafteines in homogenen gases auf kleinen suspensierte kugeln. *Z. Naturforsch.*, 1959, 7, 589–598.
39. Deryagin B.V., Bakanov S.P. Theory of large solid thermophoresis aerosol particles. *Rep. Acad. Sci. USSR*, 1962, 1, 139–142.
40. Voloshchuk V.M., Sedunov Y.S. Processes of coagulation in the disperse systems. Sankt-Petersburg, Russia: Hydrometizdat, 1975.
41. Dilman V.V., Polyanin A.D. *Methods of Model Equations and Analogies.* Moscow, Russia: Chemistry, 1988.
42. Lin S.P. Turbulence in viscous sublayers. *Ind. Eng. Chem. Fundam.*, 1975, 14, 3, 246–247.
43. Owen P. Dust deposition from turbulent airstreams. In: *Aerodynamic Capture of Particles.* Oxford: Pergamon Press, 1960, 6–25.
44. Mizushina T., Ogino F. Eddy viscosity and universal velocity profile in turbulent flow in a straight pipe. *J. Chem. Eng. Jap.*, 1970, 3, 2, 166–170.
45. Laufer J., Investigation of turbulent flow in a two-dimensional channel. NASA TN N 2123, Washington, DC: National Advisory Committee for Aeronautics, 1950.
46. Reichardt H.,Vollstandigedarstellung der turbulenten geachwindigkeit–averteilung in platen leitungen. *Z. angew. Math.und Mech.*, 1951, 7,8, 208–210.
47. Nunner W. Warmeubergang und druskabfall in rauchenrohren. *VDI–Forschungsh.*, 1956, 455, 1–39.
48. Altunbas A., Kelbaliyev G., Ceylan K. Eddy diffusivity of particles in turbulent flow in rough channels. *J. Aerosol Sci.*, 2002, 33, 1075–1086.
49. Dasvies C.N. Deposition from moving aerosols. In *Aerosol Science*, London, UK: Academic Press, 1966, 468.
50. Soo S.L., Ihrig H.K. Experimental determination of statistical properties of two–phase turbulent motion. *Trans. ASME*, 1960, 82, 3, 609–621.
51. Aerov M.E., Todes O.M. *Hydraulic and Thermal Bases of Work Devices with a Stationary and Boiling Granular Layer.* Sankt-Petersburg, Russia: Chemistry, 1968.
52. Wakao N., Kaguei S. *Heat and Mass Transfer in Packed Beds.* New York: Cordon and Breach Science Publishers, 1982.
53. Wakao N. Particle to fluid transfer coefficient and diffusivities at low flow rate in packed beds. *Chem. Eng. Sci.*, 1976, 31, 1115–1122.
54. Wakao N., Kaguei S., Funazkri T. Effect of fluid dispersion coefficients on particle to fluid rate in packed beds: Correlation of Nusselt numbers. *Chem. Eng. Sci.*, 1979, 34, 325–336.
55. Welty J.R., Wiks C.E., Wilson R.E., Rorrer G.I. *Fundamentals of Momentum, Heat and Mass Transfer,* New Jersey: Wiley, 2008.
56. Sterling S.V., Scriven L.E. Interfacial turbulence: hydrodynamic instability and the Marangoni effect. *AIChE J.*, 1959, 5, 4, 514–523.
57. Marra J., Huethorst J.A. Physical principles of Marangoni drying. *Langmuir*, 1991, 7, 2748–2755.
58. Semkov K., Kolev N. On the evaluation of the interfacial turbulence (the Marangoni effect) in gas (vapour)—Liquid mass transfer. Part 1. A method for estimating interfacial turbulence effect. *Chem. Eng. Progress*, 1991, 29, 77–82.
59. Lu H., Yang H.V., Maa J.R. On the induction of the Marangoni convection of the gas/liquids interface. *Ind. Eng. Chem. Res.*, 1997, 43, 7, 1909–1913.
60. Brain P.L., Ross J.R. The effect of Gibbs adsorption on Marangoni instability in penetration mass transfer. *AIChE J.*, 1972, 18, 3, 582–591.
61. Van Kloster H.W., Drinkeburg A.A. The influence of gradients in surface tension on the mass transfer in a packed column. *Ind. Chem. Eng. Symp. Series*, 56, 1979, 2 (56), 21–37.
62. Crumzin Y.N., Kvashnin S.Y., Lotkov V.A., Malyusov V.A. A method for taking into account the effect of surface tension film columns. *Theor. Found. Chem. Eng.*, 1982, 16, 5, 579–584.
63. Nicolis G., Prigogine I. *Exploring Complexity.* New York: W.H. Freeman & Company, 1996.
64. Golovin A.A. Mass transfer under interfacial turbulence: Kinetic regularities. *Chem. Eng. Sci.*, 1992, 47, 8, 2069–2080.

65. Tayaraman K., Gupta D.K. Drying of fruits and vegetable. In: *Handbook of Industrial Drying*. New York: Merket Dekker, 2005, 643–690.
66. Hayaloglu A., Karabulut I., Kelbaliyev G. Mathematical modeling of drying characteristics of strained yoghurt a convective type tray-dryer. *J. Food Eng.*, 2007, 78, 109–117.
67. Lykov A.V. *The Theory of Drying*. Moscow, Russia: Energy, 1968.
68. Wang N., Brennan J.G. Changes in structure density and porosity of potato during dehydration. *Food Eng.*, 1995, 24, 61–76.
69. Kelbaliev G.I., Manafov M.R. Mass transfer in the processes of drying porous materials. *J. Eng. Phys. Thermophys.*, 2009, 82, 3, 1–8.
70. Memmedov A., Kelbaliyev G.I., Alisoy G.T. Solution of an inverse problem for mass transfer in a drying process in a magnetic field. *Inverse Probl. Sci. Eng.*, 2010, 8, 5, 723–736.
71. Kelbaliyev G.I. Mass transfer between a drop or a gas bubble in isotropic turbulent flow. *Theor. Found. Chem. Eng.*, 2012, 46, 5, 554–562.
72. Calderbank P.H., Moo-Young M.B. The continuous and mass transfer properties of dispersions. *Chem. Eng. Sci.*, 1961, 16, 37.
73. Scott D.S., Hayduk W. Gas absorption in horizontal co current bubble flow. *Can. J. Chem. Eng.*, 1966, 44, 130–142.
74. Jepsen J.C. Mass transfer in two-phase glow in horizontal pipelines. *AIChE J.*, 1970, 16, 705–714.
75. Walter J.F., Blanch H.W. Bubble break-up in gas- liquid bioreactors: Break-up in turbulent flows. *Chem. Eng. J.*, 1986, 32, 7–16.
76. Lamant J.C., Scott D.C. An eddy cell model mass-transfer into surface of a turbulent liquid. *AIChE J.*, 1970, 16, 513–521.
77. Prasher B.D., Wills G.B. Mass transfer in an agitated vessel. *Ind. Eng. Chem. Process Des. Dev.*, 1973, 12, 351–362.
78. Kawase Y., Halard B., Moo-Yong M. Theoretical prediction of volumetric mass transfer coefficient in bubble columns for Newtonian and non-Newtonian fluids. *Chem. Eng. Sci.*, 1987, 42, 1600–1614.
79. Sebastiao S.A., Jorge M.T. Vasconkelas S.P. Orvalho Mass transfer to clean bubbles at low turbulent energy dissipation. *Chem. Eng. Sci.*, 2006, 61, 1324–1332.
80. Linek V., Kordac M., Fujasova M., Moucha T. Mass transfer coefficient in stirred tanks interpreted through idealized eddy structure of turbulence in the bubble vicinity. *Chem. Eng. Proc.*, 2004, 43, 1511–1521.
81. Kawase Y., Moo-Young M. Mathematical models for design of bioreactors applications of Kolmogorov's theory of isotropic turbulence. *Chem. Eng. J.*, 1990, 43, 5, 19–27.
82. Kawase Y., Halard B., Moo-Young M. Liquid-phase mass transfer coefficients in bioreactors. *Biotech. Bioeng.*, 1992, 39, 1130–1141.
83. Michaelides E.E. *Particles. Bubbles and Drops: Their Motion, Heat and Mass Transfer.* New Jersey: World Scientific, 2006.
84. Taneda S. Studies on wake vortices. III: Experimental investigation of the wake behind a sphere at low Reynolds numbers. *Rep. Res. Inst. Appl. Mech. Kyushu Univ.*, 1956, 4, 99–108.
85. Sarimeseli A., Kelbaliyev G. Modeling of the break-up particles in developed turbulent flow. *Chem. Eng. Sci.*, 2004, 59, 1233–1243.
86. Ceylan S., Kelbaliyev G. Estimation of the maximum stable drop sizes, coalescence frequencies and the size distributions in isotropic turbulent dispersions. *Colloid Surf. A Physicochem. Eng. Asp.*, 2003, 212, 285–294.
87. Kelbaliev G.I., Ibragimov Z.I. Coalescence and fragmentation of droplets in isotropic turbulent flow. *Teor. Found. Chem. Eng.*, 2009, 43, 3, 16–26.
88. Protodyakonov I.O., Lublinskaya I.E., Ryzhkov A.E. *Hydrodynamics and Mass Transfer in Dispersed Liquid-solid Systems.* Sankt-Petersburg, Russia: Chemistry, 1987.
89. Clusser E.L. *Diffusion, Mass Transfer in Fluid Systems* (3rd ed.). Cambridge, UK: Cambridge University Press, 1996.
90. Kress T.S. Mass transfer between small bubbles and liquids in co-current turbulent pipeline flow. Dissertation of University of Tennessee (UK) for degree doctor of Philosophy. 1972. p. 167.
91. Clift R., Grace J.R., Weber M.E. *Bubbles, Drops and Particles.* New York: Academic Press, 1978.
92. Schlichting H. *Boundary–Layer Theory.* New York: McGraw-Hill, 1968.
93. Raymond F., Rosant J.M. A numerical and experimental study of terminal velocity and shape of bubbles in viscous liquids. *Chem. Eng. Sci.*, 2002., 55, 943–954.
94. Vascocelos J.M.T., Orvalho S.P., Alves S.S. Gas-liquids mass transfer to single bubbles: Effect of surface contamination. *AIChE J.*, 2002, 48.

95. Resnick W., Gal-Or B. *Gas-liquid Dispersions in Chemical Engineering.* New York: Academic Press, 1968, 7, 295–304.
96. Basmadjian D. *Mass Transfer: Principles and Applications.* New York: CRC Press, 2005.
97. Kozloborodov A.N. The flow of a nonlinear viscoelastic fluid along the lateral the surface of a circular cylinder. *Proceedings of the Tomsk Polytechnic University,* 2007, 310, 1, 182–185.
98. Van Rossum I.N. Study of wave inception of falling liquid film. *J. Appl. Sci. Res.,* 1958, 47, 121, 125–132.
99. Luikov A.V., Schulman Z.P., Puris B.I. External convective mass transfer in non-Newtonian fluid. *Int. J. Heat Mass Transfer,* 1969, 12, 377–391.
100. James D.F., Acosta A.J. The laminar flow of dilute polymer solution around circular cylinders. *J. Fluid Mech.,* 1980, 42, 269–288.
101. Rao B.K. Heat transfer to non—Newtonian flows over a cylinder in cross flow. *Int. Heat Fluid Flow,* 2000, 21, 693–700.
102. Mizushima T., Usui H., Veno K., Kato T. Experiments of pseudoplastic fluid cross around a circular cylinder. *Heat Transfer Jpn.* 1978, 7, 3, 92–101.
103. Ghosh V.K., Gupta S.N., Kumar S., Upadhay S.N. Mass transfer in gross flow of non—Newtonian fluid a circular cylinder. *Int. J. Heat Mass Transfer,* 1986, 29, 955–960.
104. Chabra R.P., Ricardcon J.F. *Non–Newtonian Flow in the Process Industries. Fundamentals and Engineering Application.* Oxford, UK: Butter–Heinemann Lunarce House, Jordan Hill, 1999.
105. Kawase Y. Particle—fluid heat/mass transfer: Newtonian and non-Newtonian fluids. *Warme und Stoffubertragung,* 1992, 27, 73–76.
106. Chamka A.R. Heat and mass transfer for non-Newtonian fluid along a surface Embedden porous medium with uniform wall heat and mass. *Int. J. Energy,* 2007, 1, 3, 97–104.
107. Ram Prakash Bharti, R.P. Chhabra, V.Eswaran. Steady forced convection heat transfer from a heated circular cylinder to power-law fluids. *Int. J. Heat Mass Transf.,* 2007, 50, 977–990.
108. Bian W., Vasseur P., Bilgin J. Natural convection of non–Newtonian fluids in annular porous layer. *Chem. Eng. Comm.,* 1994, 129, 79–97.
109. Khan W.A., Culham J.R., Yovanovich M.M. Fluid flow and heat transfer in power–law fluids across circular cylinders: analytical study. *Trans. ASME,* 2006, 128, 870–878.
110. Tiqian J., Decheng H., Guren Q., Yingnong X. A theoretical study of mass and heat in film flow of non-Newtonian power law fluids. *J. Chem. Ind. Eng.,* 1984, 3, 52–69.
111. Shenoy A. *Heat Transfer to Non-Newtonian Fluids: Fundamentals and Analytical Expressions,* New Jersey: Wiley-VCH, 2018.
112. Olajuwon B.I. Flow and natural convection heat transfer in a power law fluid past a vertical plate with heat generation. *Int. J. Nonlin. Sci.,* 2009, 7, 1, 50–56.
113. Vinogradov I., Khezzar L., Siginer D. Heat transfer of non-Newtonian dilatant power law fluids in square and rectangular cavities. *J. Appl. Fluid Mech.,* 2011, 4, 2, 37–42.
114. Turana O., Chakrabortya N., Poolea RJ. Laminar natural convection of Bingham fluids in a square enclosure with differentially heated side walls. *Non-Newton. Fluid Mech.,* 2010, 165, 901–913.
115. Hady F.M., Ibrahim F.S., Abdel-Gaied S.M., Eid M.R. Influence of chemical reaction on mixed convection of non-Newtonian fluids along non-isothermal horizontal surface in porous media. *Proceedings of the World Congress on Engineering,* London, UK, 2008, Vol. 3.
116. Berkovsky B.M., Polevikov V.K. Numerical study of problems on high-intensive free convection. In: D.B. Spalding, H. Afgan (eds.). *Heat Transfer and Turbulent Buoyant Convection.* Washington, DC: Hemisphere, 1977, 443–455.
117. Chabra R.P. *Bubbles, Drops, and Particles in Non-Newtonian Fluids.* Boca Raton, FL: CRC Press, 2007.
118. Haberman W.L., Morton R.K. An experimental investigation of the drag and shape of air bubbles rising in various liquids. D.W. Taylor Model Basin Report, V.802, Department of the Navy. Washington, DC, 1953.
119. Hovenkamp B. Investigation of porous media flow with regard to the emulsion science. Dissertation for Doctor of Technical Sciences. Swiss Federal Institute of Technology, Zurich, 2002.
120. Yih C.-S. *Advances in Applied Mechanics.* New York: Academic Press, 1972.
121. Crowe C.T. *Multiphase Flow Handbook.* New York: CRC Press, 2006.
122. Clift K.A., Lever D.A. Isothermal flow past a blowing sphere. *Int. J. Numer. Methods Fluids,* 1985, 5, 709–811.
123. Karamanev D.G., Equation for calculation of the terminal velocity and drag coefficient of solid spheres and gas bubbles. *Chem. Eng. Comm.,* 1996, 147, 73–79.
124. Dewsbury K.H., Karamanev D.G., Margaritis A. Rising solid hydrodynamics at high Reynolds numbers in non-Newtonian fluids. *Chem. Eng. Sci.,* 2002, 87, 120–133.

125. Ceylan K., Herdem S., Abbasov T., Theoretical model for estimation of drag force in the flow non-Newtonian fluids around spherical solid flow particles. *Powder Technol.*, 1999, 103, 286–295.
126. Dewsbury K., Karamanev D.G., Margaritis S.A. Hydrodynamic characteristics of free rise of light solid particles and gas bubbles in non-Newtonian liquids. *Chem. Eng. Sci.*, 199, 54, 4825–4834.
127. Grace J.R. Hydrodynamics of liquid drops in immiscible liquids. In: Cheremisinoff N.P., Gupta E., *Handbook of Fluids in Motion*. London, UK: Ann Arbor Science, 1983, 273.
128. Yih C.-S. *Advances in Applied Mechanics*. New York: Academic Press, 1972.
129. Karamanev D.G. Equation for calculation of the terminal velocity and drag coefficient of solid spheres and gas bubbles. *Chem. Eng. Comm.*, 1996, 147, 73–81.
130. Dewsbury K.H., Karamanev D.G., Margaritis A. Rising solid hydrodynamics at high Reynolds numbers in non-Newtonian fluids. *Chem. Eng. Sci.*, 2002, 87, 120–133.
131. Ockedon J.R., Evens G.A. The drag on the spheres at low Reynolds numbers flow. *J. Aerosol. Sci.*, 1972, 3, 4, 237–245.
132. Dewsbury K., Karamanev D.G., Margaritis S.A. Hydrodynamic characteristics of free rise of light solid particles and gas bubbles in non-Newtonian liquids. *Chem. Eng. Sci.*, 1999, 54, 4825–4836.
133. Loth E. *Particles, Drops and Bubbles: Fluid Dynamics and Numerical Methods*. Cambridge, UK: Cambridge University Press, 2000.
134. Dewsbury K.H., Karamanev D.G., Margaritis A. Rising solid hydrodynamics at high Reynolds numbers in non-Newtonian fluids. *Chem. Eng. Sci.*, 2002, 87, 120–133.
135. Oil Emulsions. http://petrowiki.org/Oil emulsions. Petroleum Engineering Handbook, Kokal, Sandi, Aramco, 533–570.
136. Klaus Kroy, Isabelle Capron, Madeleine Djabourov, Physique thermique, ESPCI. 10, rue Vauquelin, Paris. France. 1999.
137. Krassimir D. On the viscosity of dilute emulsions. *J. Colloid Interface Sci.*, 2001, 235, 144–149.
138. Colaloglou C.A., Tavlarides L.L. Description of interaction process in agitated liquid-liquid dispersion. *Chem. Eng. Sci.*, 1977, 32, 1289–1296.
139. Prince M.J., Blanch H.W. Bubble coalescence and break-up in air-spared columns. *AIChE J.*, 1990, 36, 1485–1496.
140. Hesketh R.P., Ethells A.W., Russell T.W.F. Bubble breakage in pipeline flow. *Chem. Eng. Sci.*, 1991, 46, 1–12.
141. Hesketh R.P., Etchells A.W., Russell T.W.F. Experimental observations of bubble breakage in turbulent flow. *Ind. Eng. Chem. Res.*, 1991, 30, 835–846.
142. Tsouris C., Tavlarides L. Breakage and coalescence models for drops in turbulent dispersions. *AIChE J.*, 1994, 40, 3, 395–408.
143. Luo H., Svendsen H. F. Theoretical model for drop and bubble breakup in turbulent dispersions. *AIChE J.*, 1996, 42, 5, 1225–1236.
144. Liu S., Li D. Drop coalescence in turbulent dispersions. *Chem. Eng. Sci.*, 1999, 54, 23, 5667–5673.
145. Braginsky L.N., Belevitskaya M.A. On the fragmentation of drops under mechanical stirring in the absence of coalescence. *Theor. Found. Chem. Eng.*, 1990, 24, 4, 509–517.
146. Belevitskaya M.A., Barabash V.M. Obtaining stable emulsions in apparatus with stirrers. *Theor. Found. Chem. Eng.*, 1994, 28, 4, 342–351.
147. Barabash V.M., Belevitskaya M.A. Mass transfer from bubbles and drops in machines with agitators. *Theor. Found. Chem. Eng.*, 1995, 29, 4, 3623–3632.
148. Raymond F., Rozant J.M. A numerical and experimental study of the terminal velocity and shape of bubbles in viscous fluids. *Chem. Eng. Sci.*, 2000, 57, 943–953.
149. Galinat S., Masbernat O., Guiraud P., Daimazzonne C., Noik C. Drop break-up in turbulent pipe flow downstream of a restriction. *Chem. Eng. Sci.*, 2005, 60, 23, 6511–6520.
150. Blanchette F., Bigioni, T.P. Dynamics of drop coalescence at fluid interfaces. *J. Fluid Mech.*, 2009, 620, 333–342.
151. Narhe R., Beysens D., Nikolayev V.S. Dynamics of drop coalescence on a surface: The role of initial conditions and surface properties. *Int. J. Thermophys.*, 2005, 26, 6, 1743–1752.
152. Balmforth N.J., Llewellyn Smith S.G., Young W.R. Dynamics of interfaces and layers in a stratified turbulent fluid. *J. Fluid Mech.*, 1998, 355, 329–337.
153. Tarasov V.V., Shilin S.A. Models of heterogeneous hydrodynamic coagulation in the presence of a second solvent. *Theor. Found. Chem. Eng.*, 2007, 41, 2, 191–202.
154. Rosenzweig A.K., Strashinsky C.S. Coalescence of low-boiling dispersed phase in the turbulent flow of the cooling emulsion. *J. Appl. Chem.*, 2008, 81, 9, 1567–1574.
155. Kelbaliev G.I. Coagulation of dispersed particles in an inhomogeneous field. *Theor. Found. Chem. Eng.*, 1992, 26, 3, 390–342.

156. Liao Y., Lucas D. A literature review of theoretical models for drop and bubble breakup in turbulent dispersions. *Chem. Eng. Sci.*, 2009, 64, 15, 3389–3395.
157. Walter J.F., Blanch H.W. Bubble break-up in gas–liquid bioreactors: Break-up in turbulent flows. *Chem. Eng. J.*, 1986, 32, 7–16.
158. Narsimhan G. Model for drop coalescence in a locally isotropic turbulent flow field. *J Colloid Interface Sci.*, 2004, 272, 1, 197–207.
159. Wong D.C.Y., Simmons M.J.H., Decent S.P., Parau E.I., King A.C. Break-up dynamics and drop size distributions created from spiraling liquid Jets. *Int. J. Multiph. Flow*, 2004, 30, 5, 499–510.
160. Kraume M., Gabler A., Schulze K. Influence of physical properties on drop size distribution of stirred liquid-liquid dispersions. *Chem. Eng. Tech.*, 2004, 27, 3, 330–346.
161. Revankar S.T. Coalescence and breakup of fluid particles in multi-phase flow. *ICMF–4th International Conference on Multiphase Flow*, New Orleans, LA, 2001.
162. Vanni M. Approximate population balance equations for aggregation-breakage processes. *J. Colloid Interface Sci.*, 2000, 221, 2, 143–152.
163. Attarakih M.M., Bart H.J., Faqir N.M. Solution of the droplet breakage equation for interacting liquid–liquid dispersions: A conservative discretization approach. *Chem. Eng. Sci.*, 2004, 59, 12, 2547–2556.
164. Tobin T., Muralidhar R., Wright H., Ramkrishna D. Determination of coalescence frequencies in liquid–liquid dispersions: Effect of drop size dependence. *Chem. Eng. Sci.*, 1990, 45, 12, 3491–3502.
165. Simmons M.J.H., Azzopardi B.J. Drop size distribution in dispersed liquid–liquid pipe flow. *Int. J. Multiph. Flow*, 2001, 27, 843–854.
166. Angeli P., Hewitt O.F. Drop size distribution in horizontal oil–water dispersed flow. *Chem. Eng. Sci.*, 2000, 55, 3133–3145.
167. Martinez–Bazan C., Montanes J.I., Lasheras J.C. On the break-up of air bubble injected into fully developed turbulent flow. Part 1: Break–up frequency. *J. Fluid Mech.*, 1999, 401, 157–176.
168. Martinez–Bazan C., Montanes J.I., Lasheras J.C. On the break-up of air bubble injected into fully developed turbulent flow. Part 2: Size population density. *J Fluid Mech.*, 1999, 401, 183–204.
169. Kostoglou M., Karabelas A.J. A contribution towards predicting the evolution of droplet size distribution in flowing dilute liquid/liquid dispersions. *Chem. Eng. Sci.*, 2001, 56, 14, 4283–4292.
170. Hinze J.O. Fundamentals of the hydrodynamic mechanism of splitting in dispersion processes. *AIChE J.*, 1955, 1, 289–297.
171. Shinnar R. On the behavior of liquid dispersions in mixing vessels. *J. Fluid Mech.*, 1961, 10, 259–271.
172. Baldyga J., Bourne J.R. Interpretation of turbulent mixing using fractals and multifractals. *Chem. Eng. Sci.*, 1995, 50, 381–393.
173. Evans G.M., Jameson G.J., Atkinson B.W. Prediction of bubble size generated by a plunging liquids jet bubble column. *Chem. Eng. Sci.*, 1992, 47, 3265–3276.
174. Sis H., Kelbaliyev G., Chander S. Kinetics of drop breakage in stirred vessels under turbulent conditions. *J. Dispersion Sci. Techn.*, 2005, 26, 565–576.
175. Sis H., Chander S. Kinetics of emulsification of dodecane in the absence and presence of nonionic surfactants. *Colloids Surf. A Physicochem. Asp.*, 2004, 235, 113–121.
176. Pilch M., Erdman C.A. Use of breakup them data and velocity history data to predict the maximum size of stable fragments for acceleration-induced breakup of a liquid drop. *Int. J. Multiph. Flow*, 1987, 13, 6, 741–752.
177. Arai K., Konno M., Matunaga Y., Saito S. Effect of dispersed-phase viscosity on the maximum stable drop size for breakup in turbulent flow. *J. Chem. Eng. Japan*, 1977, 10, 4, 232–241.
178. Vankova N., Tcholakova S, Denkov N.D, Ivanov I.B., Vulchev V.D., Danner T. Emulsification in turbulent flow. Part 1: Mean and maximum drop diameters in inertial and viscous regimes. *J. Colloid Interface Sci.*, 2007, 312, 2, 363–372.
179. Sleicher C.A. Maximum stable drop size in turbulent flow. *AIChE J.*, 2004, 8, 4, 471–485.
180. Kelbaliyev G., Ceylan K. Estimation of the minimum stable drop sizes, break-up frequencies, and size distributions in turbulent dispersions. *J. Disper. Sci. Techn.*, 2005, 26, 4, 487–496.
181. Wellek R.M., Angrawal A.K. Skelland A.H. Shape of liquid drop mowing in liquid media. *AIChE J.*, 1966, 12, 854–866.
182. Helenbrook B.T., Edwards C.F. Quasi–steady deformation and drag of uncontaminated liquid drops. *Int. J. Multiph. Flow*, 2002, 28, 1631–1657.
183. Taylor T.D., Acrivos A. On the deformation and drag of falling viscous drop at low Reynolds number. *J. Fluid Mech.*, 1964, 18, 466–476.
184. Narsimhan G., Gupta J.P., Ramkrishna D. A model for transitional breakage probability of droplets in agitated lean liquid-liquid dispersions. *Chem. Eng. Sci.*, 1979, 34, 257–266.

185. Lee C.H., Erickson L.E., Glasgow L.A. Bubble breakup and coalescence in turbulent gas-liquid dispersions. *Chem. Eng. Comm.*, 1987, 59, 65–78.
186. Hagesaether L., Jakobsen H.A., Svendsen H.F. A model for turbulent binary breakup of dispersed fluid particles. *Chem. Eng. Sci.*, 2002, 57, 3251–3265.
187. Wang T., Wang J., Jin Y. A novel theoretical breakup kernel function for bubbles/droplets in a turbulent flow. *Chem. Eng. Sci.*, 2003, 58, 4629–4636.
188. Chatzi, E. Analysis of interactions for liquid–liquid dispersions in agitated vessels. *Ind. Eng. Chem. Res.*, 1987, 26, 2263–2272.
189. Chatzi E., Kiparissides C. Dynamic simulation of bimodal drop size distributions in low-coalescence batch dispersion systems. *Chem. Eng. Sci.*, 1992, 47, 445–456.
190. Alopaeus V. Simulation of the population balances for liquid–liquid systems in a nonideal stirred tank. Part 2: Parameter fitting and the use of the multiblock model for dense dispersions. *Chem. Eng. Sci.*, 2002, 57, 1815–1826.
191. Lehr F., Milles M., Mewes D. Bubble-size distributions and flow fields in bubble columns. *AIChE J.*, 2002, 48, 2426–2436.
192. Konno M., Aoki M., Saito S. Scale effect on breakup process in liquid–liquid agitated tanks. *J. Chem. Eng. Japan*, 1983, 16, 312–321.
193. Kelbaliyev G., Sarimeseli A. Modeling of drop coalescence in isotropic flow. *J. Disper. Sci. Techn.*, 2005, 26, 443–452.
194. Fanton X., Cazabat A.M., Quyru D. Thickness and shape of films driven by a Marangoni flow. *Langmuir*, 1996, 12, 5875–5886.
195. Leo L.Y., Matar O.K., Perez de Ortir E.S., Hewitt G.F. A description of phase inversion behavior in agitated liquid-liquid dispersions under the Marangoni effect. *Chem. Eng. Sci.*, 2002, 57, 3505–3516.
196. Scheludko A. Thin liquid film. *Adv. Colloid Interface Sci.*, 1967, 11, 391–423.
197. Kelbaliev G.I., Safarov F.F. Thinning of the interphase film in processes of oil emulsion separation. *Chem. Technol. Fuels Oils*, 2011, 4 (566), 18–25.
198. Chen J.D., Slattery J.C. Effects of London-van der Waals forces on the thinning of a dimpled liquid films as a small drop or bubble approaches a horizontal solid phase. *AIChE J.*, 1996, 28, 6, 955–965.
199. Sherman P. *Emulsion Science*. London, UK: Academic Press, 1968.
200. Petrov A.A., Blatov S.A. Study of the stability of hydrocarbon layers on the border with aqueous solutions of demulsifiers. *Chem. Technol. Fuels Oils*, 1969, 5, 25–33.
201. Kelbaliyev G., Safarov F. An analysis of the coalescence in oil emulsion separation. Energy Sources, Part A: *Recov. Utiliz. and Envir. Effects*, 2012, V.34, 2203–2012.
202. Yeo L.Y., Matar O.K., Perez de Ortiz E.S., Hewitt G.F. A description of phase inversion behavior in agitated liquid–liquid dispersions under the influence of the Marangoni effects. *Chem. Eng. Sci.*, 2002, 57, 3505–3516.
203. Burrill K.A., Woods D.R. Film shapes for deformable drops at liquid-liquid interfaces. II. The mechanisms of film drainage. *J. Colloid Interface Sci.*, 1973, 42, 1, 15–26.
204. Gardiner C.W. Handbook of stochastic methods for Physics. In: Haken (ed.). *Chemistry and the Natural Sciences*. New York: Springer, 1985, 13.
205. Poston T., Stuart I. *Catastrophe Theory and Its Applications*. Moscow, Russia: Fizmatlit, 1980.
206. Sevik M., Park S.H. The splitting of drops and bubbles by turbulent fluid flow. *J. Fluids Eng.*, 1973, 95, 53–67.
207. Risso F., Farbe J. Oscillation and breakup of bubble immersed in a turbulent field. *J. Fluids Eng.*, 1998, 372, 323–332.
208. Kelbaliyev G.I., Suleimanov G.Z., Phariborz A.Z., Gasanov A.A., Rustamova A.I. Extraction separation and cleaning of sewage waters by organic solvents with recirculation. *Russ. J. Appl. Chem.*, 2011, 84, 6, 1114–1125.
209. Proudman I., Pearson J.R., Expansion at small Reynolds number for the flow past a sphere and circular cylinder. *J. Fluid Mech.*, 1957, 2, 3, 237–248.
210. Chester W., Breach D.R. On the past a sphere at low Reynolds numbers. *J. Fluid Mech.*, 1968, 37, 4, 751–763.
211. Taneda S. Studies on wake vortices. III: Experimental investigation of the wake behind a sphere at low Reynolds numbers. *Rep. Res. Inst. Appl. Mech. Kyushu Univ.*, 1956, 4, 99–106.
212. Kelbaliyev G.I., Rasulov S.R., Tagiyev D.B., Mustafaeva G.R. *Mechanics and Rheology of Oil Disperse Systems*. Moscow, Russia: Maska. 2017.
213. Zapryanov Z., Tabakova S. *Dynamics of Bubbles, Drops and Rigid Particles*. Dordrecht, the Netherlands: Kluwer Academic Publishers, 1999.

214. Ван Дейк М. Методы возмущений в механике жидкости. М.: Мир, 1967.
215. Pruppacher H.R., Steinberger E.H. An experimental determination of the drag on a sphere at low Reynolds numbers. *J. Appl. Phys*., 1968, 38, 49, 4129–4138.
216. Hay K.J., Liu Z.C., Hanratty T.J. Relation of deposition to drop size when the rate law is nonlinear. *Int. J. Multiphase Flow*, 1986, 22, 829–837.
217. Majumder A.K., Barnwal J.P. A computational method to predict particle free terminal settling velocity. *The Indian Mining and Eng. J*., 2004, 85, 17–28.
218. Almedeij J. Drag coefficient of around a sphere: Matching asymptotically the wide trend. *Powder Technol*., 2008, 186, 2, 218–226.
219. Bagchi P., Balachandar B. Effect of turbulence on the drag and lift of a particle. *Phys. Fluids*, 2003, 15, 11, 3496–3508.
220. Concha F., Barrientos A. Settling velocities of particulate systems. 3. Power-series expansions for the drag coefficient of a sphere and production of the settling velocity. *Int. J. Mineral Processing*, 1982, 9, 2, 167–178.
221. Cheng N.-S. Comparison of formulas for drag coefficient and settling velocity of spherical particles. *Powder Technol*., 2009, 3 (189), 395–398.
222. Brown P.P., Lawler D.F. Sphere drag and settling velocity revisited. *J. Environ. Eng*., 2003, 129, 3, 222–231.
223. Peria E., Anta J., Puertas J., Teijero T. Estimation of drag coefficient and setting velocity of the Cockle Cenastodermaedule using particle image velocimetry. *J. Coastal Res*., 2008, 24, 150–163.
224. Flemmer R.L.C., Banks C.L. On the drag coefficient a sphere. *Powder Techn*., 1986, 48, 3, 217–226.
225. Turton R., Levenspiel O.A., A short note on the drag correlation for spheres. *Powder Technol*., 1986, 47, 1, 83–92.
226. Gabito J., Tsouris C. Drag coefficient and settling velocity for particles of cylindrical shape. *Powder Technol*., 2008, 183, 2, 314–324.
227. Kurose R., Makino H. Effect of out low from the surface of a sphere on drag, shear lift and scalar diffusion. *Phys. Fluids*, 2003, 15, 3, 2338–2346.
228. Balduga J., Henczka M., Shekunov B.Y. Fluid dynamics, mass transfer and particle formation. In: York P., Kampella U.B., Shekunov B.Y. (eds.). *Supercritical Fluid Technology for Drug Product Development*. Inform a Health Care, 2004, 91.
229. Kondratiev A.S., Naumova E.A. Determining the velocity of free sedimentation of solid particles in a Newtonian fluid. *Theor. Found. Chem. Eng*., 2003, 36, 6, 606–615.
230. Chien S.F. Settling velocity of irregularly shaped particles. *SPE 69th Annual Technical Conf. and Exhibition*. New Orleans, LA, 1994, 18, 10.
231. Haider A., Levenspiel O. Drag coefficient and terminal velocity of spherical and nonspherical particles. *Powder Technol*., 1989, 58, 63–72.
232. Taylor T., Acrivos A. On the deformation and drag of a falling drop at low Reynolds number. *J. Fluid Mech*. 1964, 18, 466.
233. Ockedon J.R., Evens G.A. The drag on the spheres at low Reynolds numbers flow. *J. Aerosol Sci*., 1972, 3, 4, 237–248.
234. LeClair B.P., Hamielek A.E., Pruppacher H.R. A numerical study of the drag on a sphere at low Reynolds numbers. *J. Atmos. Sci*., 1970, 27, 308–316.
235. Bhaga D., Weber M.E. Bubbles in viscous liquids: Shape, wakes and velocities. *J. Fluid Mech*., 1981, 105, 61–74.
236. Bozzano G., Dente M. Shape and velocity of single bubbles motion: A novel approach. *Comput. Chem. Eng*., 2001, 25, 571–584.
237. Kelbaliev G.I., Tagiev D.B., Rasulov S.R., Mustafaeva G.R., Kerimli V.I. Rheology of structured oil dispersed system. *Teor. Found. Chem. Eng*. 2017, 51, 5, 582–588.
238. Helenbrook B.T., Edwards C.F. Quasi-steady deformation and drag of contaminated liquid drops. *Int. J. Multiphase Flow*, 2002, 28, 1631–1642.
239. Maxworty T., Grann C., Kurten M., Durst F. Experiments on the rise of air bubbles in clean viscous liquids. *J. Fluid. Mech*., 1996, 321, 421–432.
240. Sajjadi S., Zerfa M., Brooks B.M., Dynamic behaviors of drops in oil/water dispersion. *Chem. Eng. Sci*., 2002, 57, 663–674.
241. Rodi W., Fueyo N. Engineering turbulence modeling and experiments. *Proceedings of the 5th International Symposium on Engineering Measurements*, Mallorca, Spain, 2002.
242. Fenn Z.G., Michaelides E.E. Heat and mass transfer coefficient of viscous spheres. *Int. J. Heat and Mass Transfer*, 2001, 44, 23, 4445–4456.

243. Ceylan K., Kelbaliyev G. A theoretical model for the particle distribution in polydispersed solid mixture under hydrodynamic and gravitational effects. *Powder Technol.*, 2001, 115, 84–89.
244. Haberman W.L., Morton R.K. An experimental investigation of the drag and shape of air bubbles rising in various liquids. D.W. Taylor Model Basin Report, V.802. Department of the Navy. Washington, DC, 1953.
245. Hovenkamp B. Investigation of porous media flow with regard to the emulsion science. Dissertation for Doctor of Technical Sciences. Swiss Federal Institute of Technology. Zurich, 2002.
246. Bricato A., Ciofalo M., Grisafi F., Micale G. Numerical prediction of flow fields in baffled stirred vessels. *Chem. Eng. Sci.*, 1998, 53, 21, 3653–3663.
247. Tropea C., Yarin A.L., Foss J.F. *Springer Handbook of Experimental Fluids Mechanics*. New York: Springer, 2007.
248. Ackovic R. Drag of a growing bubbles an rectilinear accelerated ascension in purer liquids and binary solutions. *Theor. Appl. Mech.*, 2003, 3, 177–186.
249. Coulson J.M., Richardson J.F., Hakker J.H., Backhurst J.R. *Chemical Engineering. V.2. Particle and Separation Process*. New York: Butterworth Heinemann, 2002.
250. Sarimeseli A., Kelbaliyev G. Deposition of dispersed particles in isotropic turbulent flow. *J. Disper. Sci. Techn.*, 2008, 29, 307–314.
251. Sarimeseli A., Kelbaliyev G. Sedimentation of solid particles in turbulent flow in horizontal channels. *Powder Technol.*, 2004, 140 (1), 79–89.
252. Varaskin A.Y., Polezhaev Y.V., Polyakov A.F. Effect of particle concentration on fluctuation velocity the disperse phase for turbulent pipe flow. *Heat Fluid Flow*, 2000, 21, 562–572.
253. Davies C.N. Deposition from moving aerosols. In C.N. Davies (ed.). *Aerosol Science*, London, UK: Academic Press, 1966, 392–446.
254. Zhiyao S., Tingting W., Fumin X., Ruijie L. A simple formula for predicting settling velocity of sediment particles. *Water Sci. Eng.*, 2008, 1, 1, 37–43.
255. Concha F., Almendra E.R. Settling velocities of particle systems: 1. Settling velocities of individual spherical particles. *Int. J. Miner. Process.*, 1979, 75, 4, 349–358.
256. Zigrang D.J., Sylvester N.D. An explicit equation for particle settling velocities in solid-liquid systems. *AIChE J.*, 1981, 27, 6, 1043–1052.
257. Brown P.P., Lawler D.F. Sphere drag and settling velocity revisited. *J. Environ. Eng.*, 2003, 129, 3, 222–232.
258. Camenen B. Simple and general formula for the settling velocity of particles. *J. Hydraul. Eng.*, 2007, 133, 2, 229–238.
259. Swanson V. F. The development of a formula for direct determination of free settling velocity of any size particle. *Transactions, SME/AIME*, 1967, 238, 160–172.
260. Guo J. Logarithmic matching and its applications in computational hydraulics and sediment transport. *J. Hydraul. Res.*, 2002, 40, 5, 555–564.
261. Khan A.R, Richardson J.F. The resistance to motion of a solid sphere in a fluid. *Chem. Eng. Commun.*, 1987, 62, 1–6, 135–145.
262. Jimenez J.A., Madsen O.S. A simple formula to estimate settling velocity of natural sediments. *J Waterw. Port Coast. Ocean Eng.*, 2003, 129, 2, 70–83.
263. Engelund F., Hansen E. A monograph on sediment transport in alluvial streams (3rd ed.). Copenhagen, Denmark: Technical Press, 1972.
264. Cheng N.S. Simplified settling velocity formula for sediment particle. *J. Hydraul. Eng.*, 1997, 123, 2, 149–158.
265. Turton R., Clark N.N. An explicit relationship to predict spherical-particle terminal velocity. *Powder Technol.*, 1987, 53, 2, 127–136 .
266. Yin C., Rosedae L., Kaer S.K., Sorenson H. Modeling the motion of cylindrical particles in nonuniform flow. *Chem. Eng. Sci.*, 2003, 58, 3489–3497.
267. Kurose R., Makino H. Effect of out low from the surface of a sphere on drag, shear lift and scalar diffusion. *Phys. Fluids*, 2003, 15, 3, 2338.
268. Concha F., Almendra E.R. Settling velocities of particulate systems. 2. Settling velocities of suspensions of spherical particles. *Int. J. Mineral Process*, 1979, 6, 31–42.
269. Concha F., Christiansen A. Settling velocities of particulate systems. 5. Settling velocities of suspensions of particles of arbitrary shape. *Int. J. Mineral Process*, 1986, 18, 309–317.
270. Richardson J.F., Zaki W.N. Sedimentation and fluidization: Part I. *Trans. Inst. Chem. Eng.*, 1954, 32, 35–47.
271. Happel J., Brenner H. *Low Reynolds Number Hydrodynamics: With Special Applications to Particulate Media*. Prentice-Hall, 1965.

272. Kelbaliyev G.I., Rzayev A.G., Kasumov A. Precipitation of particles from concentrated dispersed stream. *Ing. Phys. Zh.*, 1991, 61, 3, 365–373.
273. Baldock T.E., Tomkins M.R., Nielsen P., Hughes M.G. Setting velocity of sedimentation at high concentrations. *Coast. Eng.*, 2004, 51, 91–99.
274. Cleasby J.L., Woods C.F. Intermixing of dual media and multi–media granular Filters. *J. Am. Water Works Assoc.*, 1975, 67, 197–208.
275. Cleasby J.L., Fan K. Predicting fluidization and expansion of filter media. *J. Environ. Eng.*, 1981, 107, 455–468.
276. Concha F. Settling velocities of particulate systems. *Powder Part. J.*, 2009, 27, 18–27.
277. Turian R.M., Yuan T.F., Mauri G. Pressure drop correlation for pipeline flow of solid-liquid suspensions. *AIChE J.*, 1971, 17, 4, 809–817.
278. Wilhelm R.H., Kwauk M. Fluidization of solid particles. *Chem. Eng. Prog.*, 1948, 44, 201–212.
279. Bosse T., Kleiser L. Small particles in homogenous turbulence: Settling velocity enhancement by two–way coupling. *Phys. Fluid*, 2006, 18, 1–11.
280. Bosse T., Hartel E., Kleiser L. Numerical simulation of finite Reynolds number suspension drop settling under gravity. *Phys. Fluid*, 2005, 17, 14–21.
281. Maxey M.R. The gravitational settling of aerosol particles in homogeneous turbulence and random flow fields. *J. Fluid Mech.*, 1987, 174, 441–452.
282. Yang C.Y., Lei U. The role of the turbulent scales in the settling velocity of heavy particles in homogeneous isotropic turbulence. *J. Fluids Mech.*, 1998, 371, 179–190.
283. Mei R. Effect of turbulence on the particle settling the nonlinear drag range. *Int. J. Multiphase Flow*, 1994, 20, 273–285.
284. Yang T.S., Shy S.S. The settling velocity of heavy particles an aqueous near–isotropic turbulence. *Phys. Fluids*, 2003, 15, 868–879.
285. Kelbaliyev G.I., Ibragimov Z.I., Kasimova R.K. Precipitation of aerosol particles in vertical channels from an isotropic turbulent flow. *J. Eng. Phys. Thermophys.*, 2010, 83, 5, 1–9.
286. Derevich I.V., Zaichik L.I. Precipitation of particles from the turbulent flow. *Mech. Fluid Gas*, 1988, 5, 96–109.
287. Zaichik L.I., Alipchenkov V.M. *Statistical Models of Particle Motion in Turbulent Fluid.* Moscow, Russia: Fizmatlit, 2007.
288. Majumer A.K. Prediction of fine coal particles terminal settling velocities in aqueous media. *Indian Min. Eng. J.*, 2003, 42, 7, 29–37.
289. Zhang Y., Seigneur C., Seinfeld J.H., Jacobson M.Z., Binkowski F.S. Simulation of aerosol dynamics: A comparative review of algorithms used in air quality models. *Aerosol Sci. Technol.*, 1999, 31, 487–497.
290. Xu X., Homsy M.M. The settling velocity and shape distortion of drops in a uniform electric field. *J. Fluid Mech.*, 2006, 564, 395–407.
291. Kelbaliyev G., Sarimeseli A. Modeling of drop coalescence in isotropic turbulent flow. *J. Disper. Sci. Tech.*, 2006, 27, 443–452.
292. Kelbaliyev G., Ceylan K. Estimation of the minimum stable drop sizes. Break-up frequencies and distribution in turbulent dispersions. *J. Disper. Sci. Tech.*, 2005, 26, 487–496.
293. Shankar R., Balasubramaniam R. *The Motion of Bubbles and Drops in Reduced Gravity.* Cambridge, UK: Cambridge University Press, 2001.
294. Wegener M., Fevre M., Wang Z., Paschedag A., Kraune M. Marangoni convection in single drop flow—Experimental investigation and 3D-simulation. *6th International Conference of Multiphase Flow, ICMF*, Leipzig, Germany, 2007, 9.
295. Fanton X., Cazabat A.M., Quyru D. Thickness and shape of films driven by a Marangoni flow. *Langmuir*, 1996, 12, 5875–5886.
296. Leo L.Y., Matar O.K., Perez de Ortir E.S., Hewitt G.F. A description of phase inversion behavior in agitated liquid-liquid dispersions under the of Marangoni effect. *Chem. Eng. Sci.*, 2002, 57, 3505–3515.
297. Revill B.K. Jet Mixing. In: N. Harnby, M.F. Edwarda, A.W. Nienov (eds.). *Mixing in the Industries.* Boston, MA: Butteworth–Heinmann, 1992.
298. Wegener M., Kraume M., Pashedag A.R. Terminal and transient drop rise velocity of single toluene droplets in water. *AIChE J.*, 2010, 56, 1, 2–12.
299. Reichard H. Vollstandige darstellung der turbulenten geschwindigsverteilung in platen leitungen. *Zeitschrift fur Angewante Mathematik und Mechanik*, 1951, 31, 7, 208–214.
300. Laufer J. The structure of turbulence in fully developed pipe flow. 1954, NACA Report 1174.
301. Friedlander S.K., Johnstone H.F. Principles of gas–solids separation in dry streams. Reaction kinetics and unit operations. *Chem. Eng. Progr. Symp. Ser.*, 1959, 55, 25, 135–146.

302. Kneen T., Strauss W. Deposition of dust from turbulent gas stream. *Atmos. Environ.*, 1969, 3, 1, 55–65.
303. Owen P.R. Pneumatic transport. *J. Fluid Mech.*, 1969, 39, 2, 407–416.
304. Liu B.Y.H., Agarwal J.K. Experimental observation of aerosol deposition in turbulent flow. *J. Aerosol Sci.*, 1974, 5, 12, 145–158.
305. McCoy D.D., Hanratty T.J. Rate of deposition of droplets in annular two–phase flow. *Inter. J. Multiphase Flow*, 1977, 3, 4, 319–329.
306. Liu B.Y., Ilori T.A. Aerosol deposition in turbulent pipe flow. *Environm. Sci. Technol.*, 1974, 5, 3, 257–268.
307. Sippola M.R. Deposition in ventilation duct. Ph. Doctor Dissertation, University of California, Berkeley, CA, 2002.
308. Weiss C. The liquid deposition fraction impinging vertical walls and following films. *Int. J. Multiphase Flow*, 2005, 31, 1, 115–126.
309. Okawa T., Kotani A., Kataoka I. Experiments for liquid phase mass transfer rate in annular regime for a small vertical tube. *Int. J. Heat Mass Transfer*, 2005, 48, 3–4, 858–896.
310. Mito Y., Hanraty T.J. A stochastic description of wall sources in turbulent field. *Int. J. Multiphase Flow*, 2003, 29, 1373–1384.
311. Anderson R.J., Russel T.W.F. Circumferential variation of interchange in horizontal annular two–phase flow. *Int. Eng. Chem. Fundam.*, 1970, 9, 340–351.
312. Mols B., Oliemans R.V.A. Turbulent diffusion model for particle dispersion in horizontal tube flow. *Int. J. Multiphase Flow*, 1988, 24, 1, 55–67.
313. Sehmel B.A. Particle deposition from turbulent air flow. *J. Geophys. Res.*, 1970, 75, 9, 1766–1778.
314. Yoshioka N., Karaoka C., Emi H. On the deposition of aerosol particles to the horizontal pipe wall from turbulent stream. *Kagaku Kogaku*, 1972, 36, 9, 1010–1021.
315. Kelbaliyev G.I. Modeling of non-stationary processes in heat-exchange machines. *Theor. Found. Chem. Eng.*, 1982, 16, 1, 38–47.
316. Kelbaliyev G.I. Modeling of nonstationary processes in homogeneous tubular furnaces. *Theor. Found. Chem. Eng.*, 1983, 17, 3, 330–341.
317. Kelbaliyev G.I. Heat transfer during the flow of multiphase systems with deposition of the solid phase. *Theor. Found. Chem. Eng.*, 1985, 19, 5, 616–624.
318. Kelbaliyev, G.I. The flow of disperse systems in the boundary layer with precipitation of the solid phase. *Theor. Found. Chem. Eng.*, 1988, 22, 5, 706–715.
319. Shakhtakhtinsky T.N., Kelbaliyev G.I., Nosenko LV. Heat exchange with the wall of the tube under conditions of precipitation of the solid phase. *Theor. Found. Chem. Eng.*, 1997, 31, 1, 11–21.
320. Shakhtakhtinsky T.N., Kelbaliev G.I. The phenomenon of self-inhibition in processes of deposition taking into account phase transformations. *Rep. Acad. Sci. USSR*, 1987, 21, 132–138.
321. Shakhtakhtinsky T.N., Kelbaliyev G.I. The effect of self-inhibition in processes of deposition of particles from a dispersed stream. *Rep. Acad. Sci. USSR*, 1986, 288, 948–954.
322. Kelbaliyev G.I. Thermal and hydro-abrasive action of two-phase flow onto the deposition layer. *Theor. Found. Chem. Eng.*, 1988, 22, 2, 212–217.
323. Ekmekyapar A., Künkül A., Yüceer M., Kelbaliyev G. Modeling of the size distribution resulting from dissolution of spherical solid particles in turbulent flow. *J. Disper. Sci. Technol.*, 2012, 33, 521–532.
324. Lasheras J.C., Eastwood C., Martin–Bazan C., Montanes J.I. A review of statistical models for the break–up of an immiscible fluid immersed into a fully developed turbulent flow. *Inter. J. Multiphase Flow*, 2002, 28, 247–262.
325. Protodyakonov I.O., Bogdanov S.R. *Statistical Theory of Phenomena Transfer in the Processes of Chemical Technology*. Sankt-Petersburg, Russia: Chemistry, 1983.
326. Melzak Z.A. A scalar transport equation. I. *Trans. Amer. Math. Soc.*, 1957, 85, 547–556.
327. Higashitani K., Yamanchi K., Matsuno Y.J. Coalescence of drop in viscous fluid. *Chem. Eng. Japan*, 1983, 18, 299–310.
328. Melzak Z.A. The effect of coalescence in certain collision processes. *Quart. Appl. Soc.*, 1953, 11, 2, 231–241.
329. Lehr F., Milles M., Mewes D. Bubble-size distributions and flow fields in bubble columns. *AIChE J.*, 2002, 48, 2426–2436.
330. Golovin A.M. On the issue of solving the equations of coagulation of rain drops with allowance for condensation. *Report of Academy Sci. USSR*, 1963, 148, 6, 1290–1304.
331. Lissant L. *Emulsion and Emulsion Technology*. New York: Marsel Dekker, 1976.
332. Sjoblom J., Urdahl O., Hoiland H., Christy A.A., Johansen E.J. Water in crude oil emulsions formation, characterization and destabilization. *Progress Colloid Polym. Sci.*, 1990, 82, 131–142.

333. Pozdnieshev G.N. *Stabilization and Destruction of Oil Emulsions*. Moscow, Russia: Nedra, 1982.
334. Tronov V.P. *Destruction of Emulsions During Oil Production*. Moscow, Russia: Nedra, 1974.
335. Tirmizi N.P., Raghurankan B., Wiencek J. Demulsification of water oil solid emulsions by hollow-fiber membranes. *AIChE J.*, 2004, 42, 5, 1263–1271.
336. Danae D., Lee C.H., Fane A.G., Fell C.J.D. A fundamental study of the ultrafiltration of oil-water emulsions. *Membrane Sci.*, 1987, 36, 161–172.
337. Hirschberg A., DeJong N.L., Schipper B.A., Meijer J.G. Influence of temperature and pressure on asphaltene flocculation. *SPE*, 1984, 24, 3, 283–293.
338. Laux H., Rahiman I., Browarzik D. Flocculation of asphaltenes at high pressure. I. Experimental determination of the onset of flocculation. *Petroleum Sci. Techn.*, 2001, 19, 9/10, 1155–1163.
339. Mulhins O.C., Shey E.Y. *Structures and Dynamics of Asphaltenes*. New York: Plenum Press, 1998.
340. Mc Lean J., Kilpatrick P.K. Effects of asphaltene solvency on stability of water crude oil emulsions. *J. Colloid Interface Sci.*, 1997, 189, 2, 242–253.
341. Ermakov S.A., Mordvinov A.A. Effect of asphaltenes on stability water-oil emulsions. *Oil Field Eng.*, 2007, 10, 1–9.
342. Omae T., Furukava J. Kogua Jassy, 1953, 56, 10, 727.
343. Kelbaliyev G.I., Rzaev A.G. Use of the Fokker-Planck equation for describing coagulation and crushing processes in a turbulent flow. *J. Eng. Phys. Thermophysic.*, 1993, 64, 2, 150–153.
344. Salman A., Hounslow M., Seville J.P.K. Granulation, 11. In: *Handbook of Powdered Technology*. London, UK: Elsevier, 2006, p. 1402.
345. Knight P.C. Structuring agglomerated products for improved performance. *Powder Technol.*, 2001, 122, 212–221.
346. Bouwman A.M. Form, formation: The influence of material properties and process conditions of the shape of granules by high shear granulation. Dissertations University of Groningen, 2005.
347. Badawy S.I.F., Hussain M.A. Effect of starting material particle size on its agglomeration behavior in high shear wet granulation. *AAPS Pharm. Sci. Techn.*, 2004, 53, 1–7.
348. Kristensen H.G. Agglomeration of powdered. *Acta Pharm. Suec.*, 1988, 25, 1–7.
349. Kibbe A.H., Lactose. In: *Handbook of Pharmaceutical Excipients*. Washington, DC: American Pharmaceutical Association, 2000, p. 278.
350. Klassen P.P., Grishaev I.T. *Fundamentals of the Technique of Granulation*. Moscow, Russia: Khimiya, 1988.
351. Keirens D. Granulation. Analysis of size distribution and porosity during consolidation in a batch drum granulator. The University of Queensland. Individual Inquiry, 2000.
352. Heim A., Obraniak A., Gluba T. Changes of feed bulk density during drum granulation of bentonite. *Physicochem. Probl. Mi. Process.*, 2005, 2, 877–886.
353. Gluba T. The effect of wetting liquid droplet size on the growth of Agglomerates during wet drum granulation. The 7th Symposium on agglomeration (France), 2001, 2, 877–876.
354. Gluba T. The effect of wetting droplet size on the growth of agglomerates during wet drum granulation. *Powder Technol.*, 2003, 130, 219–224.
355. Ivensen S.M., Listen J.D. Liquid-bound granule impact deformation and coefficient of restitution. *Powder Technol.*, 1988, 99, 234–242.
356. Abberger T., Seo A., Shaefer T. The effect of droplet size and powdered particle size on the mechanisms of nucleation and growth in liquid bed melt agglomeration. *Int. J. Pharm.*, 2002, 249, 185–197.
357. Hapgood K.P., Lister J., Smith R. Nucleation regime map for liquid bound granules. *AIChE J.*, 2003, 49, 2, 350–361.
358. Ceylan K., Kelbaliyev G. Stochastical modeling of the granule size distribution in the agglomeration processes of powdered materials. *Powder Technol.*, 2001, 119, 173–180.
359. Kelbaliyev G.I., Mamedov M.I., Guseinov A.S. Deterministic-stochastic modeling of the granulation process. *Theor. Found. Chem. Eng.*, 1986, 20, 4, 514–520.
360. Kelbalyiev G.I., Samedli V.M., Samedov M.M. Kasimova R.K. Evolution functions of the distribution of granules in the drum apparatus. *J. Appl. Chem.*, 2010, 83, 10, 1831–1836.
361. Kelbaliyev G.I., Samedli V.M., Samedov M.M. Process modeling granulation of powdery materials by the rolling method. *Theor. Found. Chem. Eng.*, 2011, 45, 5, 571–578.
362. Kelbaliyev G.I., Kasimova R.K., Samedov M.M., Samedli V.M. Analysis of dispersity and temporary evolution of the distribution function of granules in drum apparatus. *J. Disper. Sci. Technol.*, 2011, 32, 799–808.
363. Kelbaliyev G.I., Samedli V.M., Samedov M.M. Modeling of granule formation process of powdered materials by the method of rolling. *Powder Technol.*, 2009, 194, 1–2, 87–94.

364. Kelbaliev G.I, Guseinov A.S. Rheological model of compaction and wear of granules by pelletizing by pelletizing. *Theor. Found. Chem. Eng.*, 1986, 9, 41–47.
365. Kelbaliyev G.I. The dislocation mechanism of compaction of bound particles in the process of granulation of powdery materials. *Theor. Found. Chem. Eng.*, 1992, 26, 6, 749–754.
366. Reiner M. *Rheology of Suspensions*. Moscow, Russia: Nauka, 1965, 224.
367. Muraveva E.L., Yankin G.B. Increasing the strength of refractory granules by applying protective silicate coating. *Class and Ceramics*, 2002, 59, 9–10.
368. Gluba T. The effect of wetting conditions on the strength of granules. *Physicochem. Probl. Mi. Process.*, 2002, 36, 238–242.
369. Birudaraj R., Goskonda S., Pande P.G. Granulation characterization. In: D.M. Parikh (ed.). *Handbook of Pharmaceutical Granulation Technology*, Taylor and Francis Group, USA, 2010, 513–534.
370. Gómez B. Gordo E., Ruiz-Navas E.M., Torralba J.M. *J. Ach. Mater. Manufacturing Eng.*, Gliwice, Poland: OCSCO World Press, 2006, 17, 1–2, 57–62.
371. Ryshkewitch E. *J. Am. Ceram. Soc.*, 1953, 36, 2, 65–68.
372. Rumpf H. The strength of granules and agglomerates. In: W.A. Knepper (ed.). *From Agglomeration*, New Jersey: Wiley-Interscience, 1962, 379–418.
373. Kendall K. *Teratology in Particulate Technology*. Bristol, UK: IOP Publishing, 1987.

Index

Note: Page numbers in italic and bold refer to figures and tables, respectively.

Z